Fish Cytogenetics

Fish Cytogenetics

Editors

E. Pisano
Dipartimento di Biologia, Università di Genova
Genova, Italy

C. Ozouf-Costaz
Museum national d' histoire naturelle
CNRS et Département de Systématique et Evolution Paris, France

F. Foresti
Departamento de Morfologia, Instituto de Biociências
Universidade Estadual Paulista, Botucatu, Brazil

B.G. Kapoor
Formerly Professor of Zoology, The University of Jodhpur
Jodhpur, India

Science Publishers

Enfield (NH)   Jersey   Plymouth

SCIENCE PUBLISHERS
An imprint of Edenbridge Ltd., British Isles.
Post Office Box 699
Enfield, New Hampshire 03748
United States of America

Website: *http://www.scipub.net*

sales@scipub.net (marketing department)
editor@scipub.net (editorial department)
info@scipub.net (for all other enquiries)

Library of Congress Cataloging-in-Publication Data

Fish cytogenetics / editors, E. Pisano ... [et al.].
   p. cm.
  Includes bibliographical references and index.
  ISBN 978-1-57808-330-5
1. Fishes--Cytogenetics. I. Pisano, E. (Eva), 1951-

QL638.99F568 2006
572.8'17--dc22

2006045070

ISBN 978-1-57808-330-5

© 2007, Copyright reserved

Cover illustration: Cytogenetic mapping of ribosomal genes onto a metaphase from a triploid female of the *Squalius alburnoides* complex; drawing by Filipa Filipe—reproduced with her permission.

Published by Science Publishers, Enfield, NH, USA
An imprint of Edenbridge Ltd.
Printed in India

Preface

In the past 20 years, fish cytogenetics has become an essential tool in fields as diverse as systematics and evolution, conservation, aquaculture and more recently, genomics. Recent technical advances, especially in molecular procedures, such as chromosomal gene mapping by Fluorescence In Situ Hybridization (FISH), have contributed to enhancing the potential of fish cytogenetic research, and thereby resulting in a growing number of scientific papers.

At present, there is a recognized need to put together data and opinions of the various groups of specialists who are spread over several countries and laboratories and often operating within different research contexts, all sharing cytogenetics as a common tool to approach questions in both fundamental and applied ichthyology.

The book has been also prepared to provide a comprehensive summary of the current state of research in this field by overcoming the limits imposed by the constraints of current scientific journals.

Fish Cytogenetics is organized in four sections (systematics and evolution; biodiversity conservation; stock assessment and aquaculture; genomics) covering the major fields of present fish cytogenetic research. Although we are very aware that much remains to be done before obtaining a complete view of the fish chromosome evolution, the eighteen contributions from thirteen countries which make up this book, provide a comprehensive picture of the ongoing research around the world.

The different chapters of this volume demonstrate, without any ambiguity, the wide ranging role that cytogenetics plays in fish biological science. In particular, some chapters of the last section indicate the complementarity between cytogenetics and genomics, and reflect the growing interest toward a "cytogenomic" approach, as a new strategy to integrate sequencing data coming from the genome projects that are completed or in progress for some fish species, into a broader structural context.

Due to the diversified arrays of themes approached, including evolution, biodiversity and genomics, the book is addressed not only to specialists in cytogenetics but to all scientists interested in fish biology.

We would like to express our gratitude to all the scientists that have contributed to the publication of **Fish Cytogenetics**. First off all to the Authors, whose expertise in the various topics is the major value of present book. The majority of them also provided valuable help as referees.

The following scientists have aided in the revision of manuscripts at different stages of the editing work: Roland Billard, Gauthier Dobigny, Laura Ghigliotti, Percy Alexander Hulley, Federico Mazzei, Indrajit Nanda, Radu Suciu.

E. Pisano, C. Ozouf-Costaz, F. Foresti and **B.G. Kapoor**

Contents

Section 2
Fish Cytogenetics and Biodiversity Conservation

Section 3
Fish Cytogenetics in Stock Assessment and Aquaculture

Section 4
Fish Cytogenetics and Genomics

List of Contributors

Almeida-Toledo Lurdes Foresti de
  Departamento de Genética e Biologia Evolutiva, Instituto de Biociências, Universidade de São Paulo, São Paulo, SP, Brazil.
  E-mail: lftoledo@ib.usp.br

Bertollo Luiz Antonio Carlos
  Cytogenetics Laboratory, Department of Genetics and Evolution, Federal University of São Carlos, 13565-905 São Carlos, SP, Brazil.
  E-mail: bertollo@power.ufscar.br

Bohlen Jörg
  Laboratory of Fish Genetics, Section of Evolutionary Biology and Genetics, Institute of Animal Physiology and Genetics (IAPG), AS CR, 27721 Liběchov, Czech Republic.
  E-mail: bohlen@iapg.cas.cz

Bouneau Laurence
  Genoscope/Centre National de Séquençage and CNRS-UMR 8030, 2 rue Gaston Crémieux, CP5706, F-91057 Evry Cedex 06, France.
  E-mail: Itarrago@toulouse.inra.fr

Bouza Carmen
  Universidad de Santiago de Compostela, Spain. Departamento de Genética, Universidad de Santiago de Compostela, 27002 Lugo, Spain.
  E-mail: genelugo@lugo.usc.es

Collares-Pereira Maria João
  Universidade de Lisboa, Faculdade de Ciências, Departamento de Biologia Animal/Centro de Biologia Ambiental, Campo Grande, 1749-016 Lisboa, Portugal.
  E-mail: mjpereira@fc.ul.pt

Congiu Leonardo

Dipartimento di Biologia, Università di Ferrara, Via L. Borsari 46, I-44100 Ferrara, Italy.
E-mail: leo@dns.unife.it

Coutanceau Jean-Pierre

CNRS UMR 7138 "Systématique, Adaptation, Evolution" MNHN, Département Systématique et Evolution, C.P. 26, 57 rue Cuvier, 75231 Paris Cedex 05, France.
E-mail: coutance@mnhn.fr

Daniel-Silva Maria de Fatima Zambelli

Departamento de Genética e Biologia Evolutiva, Instituto de Biociências, Universidade de São Paulo, São Paulo, SP, Brazil.
E-mail: fdaniel@directnet.com.br

Dettai Agnès

Biofuture Research Group, Physiologische Chemie I, Biozentrum, University of Würzburg, Am Hubland, D-97074 Würzburg, Germany. Muséum National d'Histoire Naturelle UMR 7138, Département de Systématique et Evolution 43, rue Cuvier, F-75231 Paris Cedex 05, France.
E-mail: adettai@mnhn.fr

Fischer Cécile

Genoscope/Centre National de Séquençage and CNRS-UMR 8030, 2 rue Gaston Crémieux, CP5706, F-91057 Evry Cedex 06, France.
E-mail: fischer@genoscope.cns.fr

Flajšhans Martin

Department of Genetics and Breeding, Research Institute of Fishery and Hydrobiology (RIFH), University of South Bohemia, Vodňany, Czech Republic. Joint Laboratory of Genetics, Physiology and Reproduction of Fishes of IAPG AS CR Liběchov and RIFH USB Vodňany, Czech Republic.
E-mail: flajshans@vurh.jcu.cz

Fontana Francesco

Dipartimento di Biologia, Università di Ferrara, Via L. Borsari 46, I-44100 Ferrara, Italy.
E-mail: fon@unife.it

Foresti Fausto

Departamento de Morfologia, Instituto de Biociências, Universidade Estadual Paulista, 18618-000, Botucatu, SP, Brazil.
E-mail: fforesti@ibb.unesp.br

Fortes Gloria G.

Departamento de Genética, Universidad de Santiago de Compostela, 27002 Lugo, Spain.
E-mail: ggfortes@hotmail.com

Gornung Ekaterina

Department of Animal and Human Biology, University of Rome 'La Sapienza' Via A. Borelli 50, 00161, Rome, Italy.
E-mail: ekaterina.gornung@uniroma1.it

Gromicho Marta

Universidade de Lisboa, Faculdade de Ciências, Departamento de Biologia Animal/Centro de Biologia Ambiental, Campo Grande, 1749-016 Lisboa, Portugal.
E-mail: mlsilva@fc.ul.pt

Jankun Malgorzata

Department of Ichthyology, University of Warmia and Mazury in Olsztyn, ul. Oczapowskiego 5, 10-957 Olsztyn, Poland.
E-mail: mjpw@uwm.edu.pl

Kalous Lukáš

Laboratory of Fish Genetics, Section of Evolutionary Biology and Genetics, Institute of Animal Physiology and Genetics (IAPG), AS CR, 277 21, Libéchov, Czech Republic. Department of Zoology and Fisheries, Faculty of Agrobiology, Food and Natural Resources, Czech University of Agriculture, Praha 6 - Suchdol, Czech Republic.
E-mail: kalous@af.czu.cz

Lajus Dmitry L.

St. Petersburg State University, Faculty of Biology and Soil Sciences, Department of Ichthyology and Hydrobiology, 16 Linia V.O., 29. 199178. St. Petersburg, Russia.
E-mail: dlajus@yahoo.com

Luczynski Miraslau

Department of Environmental Biotechnology, University of Warmia and Mazury in Olsztyn, ul. Oczapowskiego 5, 10-957 Olsztyn, Poland.
E-mail: mirehl@uwm.edu.pl

Mannarelli Maria Elena

Department of Animal and Human Biology, University of Rome 'La Sapienza' via A. Borelli 50, 00161, Rome, Italy.

Martínez Paulino

Departamento de Genética, Universidad de Santiago de Compostela, 27002 Lugo, Spain.
E-mail: paumarti@lugo.usc.es

Martins Cesar

UNESP - Universidade Estadual Paulista, Instituto de Biociências, Departamento de Morfologia, CEP 18618-000, Botucatu, SP, Brazil.
E-mail: cmartins@ibb.unesp.br

Moysés Cinthia Bachir

Departamento de Genética e Biologia Evolutiva, Instituto de Biociências, Universidade de São Paulo, São Paulo, SP, Brazil.
E-mail: cintilan@ib.usp.br

Molina Wagner Franco

Departamento de Biologia Celular e Genética, Universidade Federal do Rio Grande do Norte, Campus Universitário, 59078-970 Natal, RN, Brazil.
E-mail: molinawf@yahoo.com.br

Ocalewicz Konrad

Department of Ichthyology, University of Warmia and Mazury in Olsztyn, ul. Oczapowskiego 5, 10-957 Olsztyn, Poland.
E-mail: con@uwm.edu.pl

Oliveira Claudio

Laboratório de Biologia e Genética de Peixes, Departamento de Morfologia, Instituto de Biociências de Botucatu, Universidade Estadual Paulista 18618-000, Botucotu, SP, Brazil.
E-mail: claudio@ibb.unesp.br

Ozouf-Costaz Catherine

CNRS UMR 7138 "Systématique, Adaptation, Evolution" MNHN, Département Systématique et Evolution, C.P. 26, 57 rue Cuvier, 75231 Paris Cedex 05, France.
E-mail: ozouf@mnhn.fr

Pepe Anastasia

Dipartimento di Biologia, Università di Padova, Via Ugo Bassi 58/B, 35121, Padova, Italy.

Phillips Ruth B.

Department of Biological Sciences, Washington State University of Vancouver, Vancouver, Washington 98686, USA.
E-mail: phllipsr@vancouver.wsu.edu

Porto-Foresti Fábio

Laboratório de Genética de Peixes, Departamento de Ciências Biológicas, Faculdade de Ciências de Bauru, Universidade Estadual Paulista, Brazil.
E-mail: fpforesti@fc.unesp.br

Ráb Petr

Laboratory of Fish Genetics, Section of Evolutionary Biology and Genetics, Institute of Animal Physiology and Genetics (IAPG), AS CR, 277 21, Liběchov, Czech Republic.
E-mail: rab@iapg.cas.cz

Rábová Marie

Laboratory of Fish Genetics, Section of Evolutionary Biology and Genetics, Institute of Animal Physiology and Genetics (IAPG), AS CR, 277 21, Liběchov, Czech Republic.
E-mail: rabova@iapg.cas.cz

Rigolino Marcos Guilherme

Estação Experimental de Salmonicultura de Campos do Jordão, Secretaria de Agricultura e Abastecimento do Estado de São Paulo, Instituto de Pesca, São Paulo.
E-mail: marcos@aquicultura.br

Rocco Lucia

Dipartimento di Scienze della Vita, Seconda Università di Napoli, Via Vivaldi 43 – 81100 Caserta, Italy.
E-mail: lucia.rocco@unina2.it

Rossi Anna Rita

Department of Animal and Human Biology, University of Rome 1 Via A. Borelli 50, 00161, Rome, Italy.
E-mail: annarita.rossi@unirome1.it

Sánchez Laura

Departamento de Genética, Facultad de Veterinaria, Universidad de Santiago de Compostela,27002 Lugo, Spain.
E-mail: lasanche@lugo.usc.es

Schmidt Cornelia

Biofuture Research Group, Physiologische Chemie I, Biozentrum, University of Würzburg, Am Hubland, D-97074 Würzburg, Germany.
E-mail: conny@biozentrum.uni-wuerzburg.de

Schultheis Christina

Biofuture Research Group, Physiologische Chemie I, Biozentrum, University of Würzburg, Am Hubland, D-97074 Würzburg, Germany.
E-mail: C.Schultheis@biozentrum.uni-wuerzburg.de

Sola Luciana

Department of Animal and Human Biology, University of Rome 1, Via A. Borelli 50, 00161 Rome, Italy.
E-mail: luciana.sola@uniroma1.it

Tabata Yara Aiko

Estação Experimental de Salmonicultura de Campos do Jordão, Secretaria de Agricultura e Abastecimento do Estado de São Paulo, Instituto de Pesca, São Paulo, Brazil.
E-mail: yara@aquicultura.br

Tóth Balázs

Duna-Ipoly National Park Directorate, Budapest, 1021, Hungary.
E-mail: bt109o@nih.gov

Ueda Takayoshi

Department of Biology, Faculty of Education, Utsunomiya University, 350 Mine, Utsunomiya 321-8505, Japan.
E-mail: ueda@cc.utsunomiya-u.ac.jp

Várkonyi Eszter Patakiné

Institute for Small Animal Research, Gödöllő, P.O. Box 417, Hungary.
E-mail: eszter@katki.hu

Viñas Ana

Departamento de Genética, Universidad de Santiago de Compostela, 27002 Lugo, Spain.
E-mail: avinas@lugo.usc.es

Volff Jean-Nicolas

Equipe "Génomique Evolutive des Vertébrés", Institut de Génomique Fonctionnelle de Lyon, Ecole Normale Supérieure de Lyon, 46 allée d'Italie, 69364 Lyon Cedex 07, France.
E-mail: Jean-Nicolas.Volff@ens-lyon.fr

Zane Lorenzo

Dipartimento di Biologia, Università di Padova, Via Ugo Bassi 58/B, 35121, Padova, Italy.
E-mail: lorenzo.zane@unipd.it

Zhou Qingchun

Biofuture Research Group, Physiologische Chemie I, Biozentrum, University of Würzburg, Am Hubland, D-97074 Würzburg, Germany.
E-mail: qczhou@ufl.edu

SECTION

1

Fish Cytogenetics in Systematics and Evolution

Chromosomal Differentiation in Bitterlings (Pisces, Cyprinidae)

Takayoshi Ueda

INTRODUCTION

Bitterlings (Pisces, Cyprinidae, Acheilognathinae) include more than 40 species and subspecies widely distributed in Eurasia, particularly East Asia (Nakamura, 1969; Choi *et al.*, 1990; Holcik and Jedlicka, 1994; Lin, 1998). A characteristic feature of the reproductive behavior exhibited by bitterlings is the development of an ovipositor in the females of most species at the onset of the spawning condition. At the time of spawning, the ovipositor is used to lay eggs in the mantle cavity of freshwater bivalves where the larvae develop inside the gills (Nakamura, 1969). Although the high number of synapomorphies and characteristic reproductive behavior support the monophyly nature of bitterlings, the phylogenetic relationships among bitterlings are not clear and members of the Acheilognathinae have been variously placed in from one to six genera (Okazaki *et al.*, 2001). I followed the classification of Arai and Akai (1988)

Address for Correspondence: Department of Biology, Faculty of Education, Utsunomiya University, 350 Mine, Utsunomiya 321-8505, Japan. E-mail: ueda@cc.utsunomiya-u.ac.jp

and used three genera of Acheilognathus, Rhodeus and Tanakia. Karyotypes of bitterlings are atypical among cyprinid fishes. Although the diploid chromosomal number of several cyprinid fish is generally 100, 50 or 48, it varies in the Acheilognathinae from 42 to 48: $2n=48$ in Tanakia, $2n=44$ or 42 in Acheilognathus and $2n=48$ or 46 in Rhodeus (Arai and Akai, 1988).

Several conventional Giemsa-stained karyotype studies have been undertaken in bitterlings in Japan (Ojima et al., 1972, 1973), China (Yu et al., 1987; Arai et al., 1988, 1992) and South Korea (Lee et al., 1982, 1983; Lee, 1983; Ueno and Ojima, 1984). Reports of banding karyotypes have also been presented and C-banding karyotypes have been shown in A. gracilis (Hong and Zhou, 1985b), A. rhombeus (Takai and Ojima, 1988) and also in six South Korean and Chinese bitterlings (Ueda et al., 2001a). Intraspecific variability of silver-stained nucleolar organizer regions (Ag-NORs) has been found in six Japanese bitterlings (Takai and Ojima, 1986; Inafuku et al., 2000; Kikuma et al., 2000). DNA-replication bands in A. macropterus (Hong and Zhou, 1985a) and R. ocellatus (Kikuma et al., 2000), and the B-bands in R. ocellatus (Ueda and Naoi, 1999) have also been studied. Furthermore, the 5S ribosomal genes of R. ocellatus and A. tabira were mapped (Inafuku et al., 2000; Kikuma et al., 2000). Moreover, R. ocellatus kurumeus and T. limbata were cytogenetically studied by Ag- and chromomycin A_3 (CMA_3)-staining, C-banding and by mapping of 18S ribosomal genes and telomeric sequences through fluorescence in situ hybridization (FISH) (Sola et al., 2003).

It has been suggested that the quantitative and positional changes observed in constitutive heterochromatin are among the most important factors for speciation (John and Miklos, 1979). Moreover, evidence supporting the hypothesis that the interstitial sites of the (TTAGGG)n telomeric sequence enable greater flexibility for karyotype changes has been reported (Meyne et al., 1990; Ashley and Ward, 1993; Slijepcevic et al., 1996). In some fishes that exhibit quite different karyotypes from closely related species such as Salmo salar (Ueda and Kobayashi, 1990) and Chrysiptera hemicyanea (Takai and Ojima, 1999), terminal and intermediate C-banding heterochromatin have been observed in many chromosomes. Telomeric sequences were found close to heterochromatin in Salmo salar (Abuin et al., 1996) and scattered along the NOR in Anguilla anguilla, A. rostrata (Salvadori et al., 1995) and Oncorhynchus mykiss (Abuin et al., 1996).

In this chapter, the chromosomal differentiation in bitterlings is discussed on the basis of data from cytogenetic studies by silver (Ag) and chromomycin A_3 (CMA_3) staining and C-banding. The results of molecular cytogenetic approaches such as the chromosomal mapping of 18S ribosomal genes and telomeric sequences have also been considered.

Centromeric classification in metacentric (M), submetacentric (SM), subtelocentric (ST), or acrocentric (A)chromosomes followed that of Levan *et al.* (1964).

AG-NOR PATTERN

Takai and Ojima (1986) reported two Ag-NOR bearing chromosomes in six bitterling species (*viz.*, *T. lanceolata*, *T. limbata*, *A. rhombeus*, *A. cyanostigma*, *A. tabira tabira* and *R. ocellatus ocellatus*) with diversity in the location of Ag-NORs among species-subspecies. In another study on *T. limbata*, Ag-NORs were seen on the short arms of two and three ST-A chromosomes in 11 individuals and 1 female, respectively (Ueda and Kobayashi, 1991). Furthermore, in bitterlings, not only a variety in number and position of the Ag-NORs have been found, but also heteromorphic Ag-NORs were often observed (Takai and Ojima, 1986; Ueda *et al.*, 1997, 2001a). An Ag-banded karyotype of *R. atremius suigensis* is shown in Fig. 1.1.1. Since telomeric sequences were found to be scattered along the NORs in *T. limbata* (Sola *et al.*, 2003) it can be hypothesized that the scattering of interstitial telomeric sequences might permit NOR variation in bitterlings.

Only actively transcribed NORs can be positively stained with silver (Miller *et al.*, 1976). The embryo cells contained more Ag-NORs than the kidney cells in the four mature Chinese bitterlings, *R. atremius*, *R. lighti*, *T. himantegus* and *T. signifer* (Ueda *et al.*, 1996, 1997, 2001a). Suzuki *et al.* (1999) observed variation in the number of Ag-NORs during development in *R. ocellatus ocellatus* (Table 1.1.1). It has also been suggested that rDNA is more actively transcribed during the gastrula stage than when compared to the adult stage (Ueda *et al.*, 1996, 1997, 2001a; Ueda, 1997; Suzuki *et al.*, 1999).

In *R. ocellatus ocellatus*, *R. atremius suigensis* (Suzuki *et al.*, 1999) and *R. ocellatus kurumeus* and *T. limbata* (Sola *et al.*, 2003), major ribosomal genes were recognized on Ag-NORs, after FISH with a 18S rDNA probe.

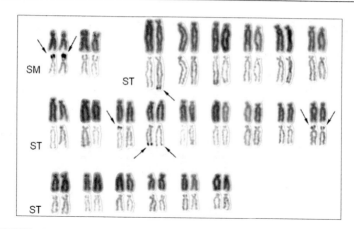

Fig. 1.1.1 Karyotype of *Rhodeus atremius suigensis*. 2n=46; 4 SM and 42 ST. Routine Giemsa stain (upper row) and Ag-stain (lower row). Arrows indicate Ag-NORs.

Table 1.1.1 The variation in the number of Ag-NORs during the development in *Rhodeus ocellatus ocellatus*.

Specimens	Stages	Number of Ag-NORs							Total number of analyzed cells
		0	1	2	3	4	5	7	
A	Early Blastula	4	2	5	1	1	0	0	13
B	Early Blastula	3	3	11	3	2	0	0	22
C	Early Blastula	0	4	22	14	1	0	0	41
D	Blastula	0	2	3	25	3	0	0	33
E	Blastula	0	1	5	21	3	2	0	32
F	Blastula	0	2	5	11	2	4	1	25
G	Blastula	1	2	5	11	2	4	0	25
H	Blastula	0	1	1	15	7	1	0	25
I	Blastula	0	1	3	41	11	1	0	57
J	Gastrula	0	1	4	7	2	1	0	15
K	Neurula	0	1	55	8	4	0	0	68
L	Neurula	0	3	67	20	0	0	0	90
M	Neurula	0	2	44	8	0	0	0	54
N	Hatching	0	0	9	1	0	0	0	10
O	Hatching	0	0	13	2	0	0	0	15
P	Adult	0	9	9	0	0	0	0	18
Q	Adult	0	1	2	0	0	0	0	3

ROBERTSONIAN-TYPE TRANSLOCATIONS

In A. *macropterus* (2n=44), C-bands at the centromeric region on 3 or 4 M chromosome pairs were larger than those on other M chromosome pairs (Ueda *et al.*, 2001a). Assuming that the direction of karyotype evolution in bitterlings is from 2n=48 to 2n=44 (Arai, 1978; Yu *et al.*, 1987), it is reasonable to suggest that centric fusions are related to the karyotypic changes observed in A. *macropterus*, resulting in the large C-bands in some M chromosome pairs as well as possible inactivation of the centromere. Conversely, the inter–individual polymorphism reported for A. *gracilis* where the karyotype change from 2n=44 to 2n=42 was attributed to centric fusion (Hong and Zhou, 1985b).

In A. *rhombeus* specimens from five locations in South Korea, inter–individual variations in diploid numbers (44, 45, 46, 47) were observed (Ueda *et al.*, 2002). Given the fact that all karyotypes had the same fundamental arm number (NF), 74, the observed chromosomal variations can be considered to have occurred in response to Robertsonian-type translocation through chromosomal differentiation. It was presumed that two M-SM chromosome pairs were involved in these chromosomal changes, and that the gametes with 22 chromosomes were formed from fish with 2n=44, those with 22 and 23 chromosomes from fish with 2n=45, those with 22, 23 and 24 chromosomes from fish with 2n=46 and those with 23 and 24 chromosomes from fish with 2n=47. Assuming that 2n=44 is the basic chromosomal number in *Acheilognathus* (Arai and Akai, 1988), these chromosomal polymorphisms would be due to centric fission from a M-SM chromosome into two ST chromosomes, because there were no A chromosomes in any of the karyotypes. While it might be possible to speculate that the activation of the centromere must have accompanied these changes, molecular analyses of DNA base composition in the centromeric region would be required to demonstrate this theory with certainty.

CHROMOSOMAL INVERSIONS

Geographic chromosomal polymorphisms were observed in R. *ocellatus* collected from more than 10 locations in China, Taiwan, South Korea and Japan (Ueda *et al.*, 2001b). All of the individuals had 48 chromosomes (2n), consisting of 8 M, 20 SM and 20 ST. C-banding heterochromatin was observed at the centromeric region of all chromosomes and also at the terminal region of many chromosomes (Fig. 1.1.2). In some individuals, an additional intense C-band in the middle region of the long arm in ST chromosomes was observed (Fig. 1.1.3). The number of pairs with this C-band varied from 0 to 3, depending on the geographic location from where they had been collected. These C-bands probably resulted from paracentric inversions, as illustrated in Fig. 1.1.4, a, b. It is thought that the telomeric sequence played a role in this alteration, although direct evidence for this is yet to be obtained (Ueda *et al.*, 2001b).

The karyotypes with 2n=48 consist mainly of M and SM chromosomes, which is a common feature in cyprinids. However, the chromosomal constitution of 2n=46 (4S M + 42 ST), as shown in Fig. 1.1.5, is quite different from the 2n=48 condition (8 M + 20 SM + 20 ST, or, 8 M + 20 SM + 14-16 ST + 4-6 A). The karyotype with 2n=46

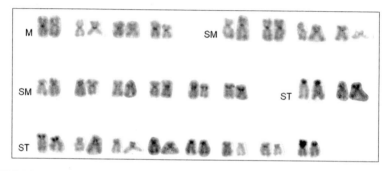

Fig. 1.1.2   C-banded karyotype of *Rhodeus ocellatus ocellatus*. 2n=48; 8 M, 20 SM and 20 ST.

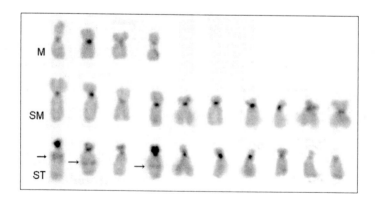

Fig. 1.1.3   C-banded karyotype of *Rhodeus ocellatus ocellatus*. n=24; 4 M, 10 SM and 10 ST. Arrows indicate interstitial C-bands.

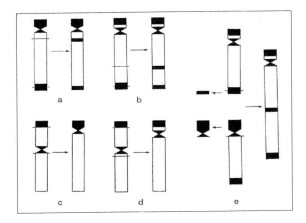

Fig. 1.1.4 Schematic representation of five major types of chromosomal changes. a,b: paracentric inversion; c,d: pericentric inversion; e: tandem fusion.

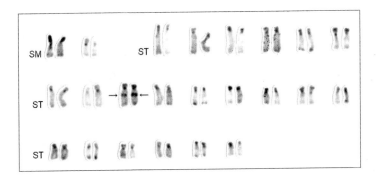

Fig. 1.1.5 C-banded karyotype of *Rhodeus atremius fangi*. 2n=46; 4 SM and 42 ST. Arrows indicate interstitial C-bands.

in *R. atremius atremius* and *R. a. suigensis* from Japan (Ojima *et al.*, 1973), *R. atremius* and *R. atremius fangi* from China (Ueda *et al.*, 1996, 2001a), showed only one chromosomal constitution (4 SM + 42 ST), and the intermediate chromosomal formula between 2n=48 (NF=76) and 2n=46 (NF=50) has not yet been observed in bitterlings. These findings suggest that the direction of karyotypic change is from 2n=48 to 2n=46, which corroborates the findings of Arai (1978), Yu *et al.* (1987) and Okazaki *et al.* (2001). Furthermore, the increase in the number of ST chromosomes during these changes could have been achieved by a series of pericentric inversions in 24 M-SM (Ueda *et al.*, 2001a). More detail of the two types of inversion that can occur from M-SM to ST chromosomes involving C-banding heterochromatin, have been depicted in Fig. 1.1.4, c, d. The fact that the numbers of chromosomes with the intense interstitial C-band are similar in specimens with 2n=46 and 2n=48, seems to indicate that the observed pericentric inversions were mainly due to the type d change illustrated in Fig. 1.1.4.

In *R. ocellatus kurumeus*, telomeric sequences were found in the natural telomeres and the pericentromeric regions of many chromosomes, but in *T. limbata* FISH with the same telomeric probe did not reveal any interstitial sites in addition to those at the terminal ends of the chromosome arms. Of the bony fish assayed to date, *R. ocellatus kurumeus* has the highest number of interstitial sites for telomeric sequences. If one considers the complete absence of interstitial telomeric sequences in a species such as *T. limbata*, then the hypothesis that the interstitial sites might provide the clue for the reduction observed in a karyotype with 2n=48 and NF=76, to a karyotype with 2n=46 and NF=50 in related species of *Rhodeus* is particularly appealing (Sola *et al.*, 2003).

TANDEM FUSION AND OTHER CHANGES

The karyotype with 2n=46 had a large intense C-band in the middle region of the long arm in the 9th ST chromosome pair (Ueda *et al.*, 2001a). These ST chromosomes also had intense C-bands in the centromeric and terminal regions of both the short and the long arms (Fig. 1.1.5). In *Salvelinus malma* (Ueda and Ojima, 1983; Ueda *et al.*, 1991) and *Salmo salar* (Ueda and Kobayashi, 1990), it was suggested that the intense interstitial C-band in the long arm was the result of a tandem fusion. Accordingly, the 9th ST chromosome pair in the karyotype with 2n=46 could have resulted from 4 ST chromosomes in the 2n=48 karyotype by tandem fusion, as illustrated in Fig. 1.1.4 e, with a consequent decrease in the diploid number to 2n=46. As a result, the intense C-band in the middle region of the long arm on the 9th ST chromosome pair in 2n=46 may have originated from the C-band at the terminal region of the long arm on the corresponding ST chromosomes in the karyotypes with 2n=48.

Further studies assessing the occurrence of other possible non-telomeric sites of (TTAGGG)n sequences in bitterling chromosomes will be important to corroborate the present tandem fusion hypothesis.

Four-six unquestionable A chromosomes were observed in *T. signifer* (2n=48) (Fig. 1.1.6). The number of ST-A with an intense C-band

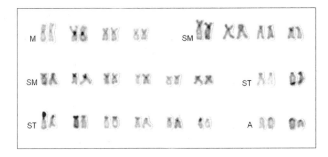

Fig. 1.1.6 C-banded karyotype of *Tanakia signifer*. 2n=48; 8 M, 20 SM, 16 ST and 4 A.

occupying the entire short arm was similar in all 2n=48 specimens analyzed (Ueda *et al.*, 2001a). This suggests that the karyotypic increase or decrease in the number of A chromosomes occurs in response to a proportionate increase or decrease in C-banding heterochromatin.

FACTORS INDUCING CHROMOSOMAL CHANGES

Searching for the factors that induce chromosomal changes could be a way to study karyotype differentiation. I am of the opinion that the appearance of embryos with chromosomal aberrations can be used to support the notion of karyotype differentiation.

Until recently, chromosomal aberrations in fish were reported in response to hybridization (Ueda *et al.*, 1984, 1994; Yamazaki *et al.*, 1989; Ueda and Takahashi, 1995), by alterations in temperature, or hydrostatic pressure shock during early development (Ueda and Kobayashi, 1988; Ueda and Aoki, 1994), as well as aberrations induced by the conditions under which the eggs are stored (Yamazaki *et al.*, 1989; Ueda, 1996; Várkonyi *et al.*, 1998; Aegerter and Jalabert, 2004).

The chromosomes of hybrids between female R. *ocellatus ocellatus* and male R. *atremius* were analyzed at approximately the gastrula stage (Ueda *et al.*, 1994). All of the hybrids in eight groups had intermediate karyotypes between the parents. In one group, 1 hybrid contained intermediate karyotype, 4 were tetraploid and two were diploid-tetraploid mosaics.

The chromosomes of the hybrids between female R. *ocellatus ocellatus* and male A. *typus* were analyzed at the blastula stage (Ueda and Takahashi, 1995). In this crossing, all embryos died before invagination. Embryos with structural chromosomal aberrations and karyotypes that were markedly different from the intermediate-type were recognized.

These findings indicate either that hybridization promotes the occurrence of chromosomal aberrations, or that chromosomal reconstitution occurs during early development.

Evidence of chromosomal aberrations caused by cold-shock during early development has been previously reported in *R. ocellatus ocellatus* embryos (Ueda and Aoki, 1994). Subsequent to cold-shock treatment—administered 11 times during the 1-celled stage of early development — many embryos with chromosomal aberrations such as aneuploidy, polyploidy and structural modifications were observed. Finally, the relationship between the age of the eggs and chromosomal aberrations in bitterlings was demonstrated through an analysis of the chromosomes from embryos using eggs that had been stored either in fresh water or a preservative prior to fertilization. The percentage of embryos that reached gastrulation decreased, while chromosomal aberrations increased, in response to the duration of storage of unfertilized eggs (Ueda, 1996).

Given the fact that the karyotypes among interspecies generally exhibit a tendency towards discontinuity, it seems likely that karyotypic changes are capable of bringing about marked changes in a short period. I imagine that karyotype differentiation is an expression of chromosomal aberrations, and that many inductive factors of chromosomal aberrations are the factors of karyotype differentiation. Moreover, I consider the possibility that karyotype differentiation can be promoted during growth and differentiation of cells in response to chromosome changes during the early development and, consequently, that considerable chromosomal changes are induced in a short period. Furthermore, I regard the chromosomal aberrations caused by hybridization, physical shock during early development, and the aging of eggs as triggers (Ueda, 1993).

BITTERLINGS AND ENVIRONMENTAL EDUCATION

During the previous decade, many countries have made considerable improvements to their sewage disposal facilities in order to protect the environment from pollution. In addition, environmental management has become a high priority among most governments. Nonetheless, there is still much that must be done in terms of increasing public awareness of environmental problems. In response to this need I have searched for ways to apply most of the results acquired from the fundamental research on bitterlings in the field to environmental education. By studying the life history of bitterlings, we have had to collate information from many disciplines, including genetics, morphology, embryology, physiology and ecology. Moreover, the propagation of bitterlings is closely related to agricultural activities, and they have been adapting to nature modifications induced by men. The drastic change in our life style in the latter half of the twentieth century has disturbed the ecological balance leading to a marked increase in the number of bitterling species facing extinction. Bitterlings are useful indicators of both ecosystem health and of the way we should treat the environment. They also provide us with a good model for future learning programs dealing with environmental education.

Acknowledgements

This study was supported in part by Grants-in-Aid for Scientific Research (B(2)10041156 and B(1)12575009) from the Ministry of Education, Science, Sports and Culture, Japan, and for Eminent Research at Utsunomiya University.

References

Abuin, M., P. Martinez and L. Sanchez. 1996. Localization of the repetitive telomeric sequence (TTAGGG)n in four salmonid species. *Genome* 39: 1035–1038.

Aegerter, S. and B. Jalabert. 2004. Effects of post-ovulatory oocyte ageing and temperature on egg quality and on the occurrence of triploid fry in rainbow trout, *Oncorhynchus mykiss*. *Aquaculture* 231: 59–71.

Arai, R. 1978. Classification and karyotypes in fishes. *Idem* 32: 39–46. (In Japanese).

Arai, R. and Y. Akai. 1988. *Acheilognathus melanogaster*, a senior synonym of *A. moriokae*, with a revision of the genera of the subfamily Acheilognathinae (Cypriniformes, Cyprinidae). *Bulletin of the National Science Museum, Tokyo*, (A) 14: 199–213.

Arai, R., A. Suzuki and Y. Akai. 1988. A karyotype of a Chinese bitterling, *Paracheilognathus himantegus* (Cypriniformes, Cyprinidae). *Bulletin of the National Science Museum, Tokyo*, (A) 14: 43–46.

Arai, R., Y. Akai and N. Suzuki. 1992. The karyotype of a Chinese bitterling, *Acheilognathus tonkinensis* (Pisces, Cyprinidae). *Bulletin of the National Science Museum, Tokyo*, (A) 18: 73–77.

Ashley, T. and D.C. Ward. 1993. A "hot spot" of recombination coincides with an interstitial telomeric sequence in the American hamster. *Cytogenetics and Cell Genetics* 62: 169–171.

Choi, K.C., S.R. Jeon, I.S. Kim and Y.M. Son. 1990. Coloured illustrations of the freshwater fishes of Korea. *Hyangmun Sa, Seoul, Korea*. (In Korean with English abstract).

Holcik, J. and L. Jedlicka. 1994. Geographical variation of some taxonomically important characters in fishes: the case of the bitterling *Rhodeus sericeus*. *Environmental Biology of Fishes* 41: 147–170.

Hong, Y. and T. Zhou. 1985a. Chromosome banding in fishes I. An improved BrdU-Hoechst-Giemsa method for revealing DNA-replication bands in fish chromosomes. *Acta Genetica Sinica* 12: 67–71. (In Chinese with English abstract).

Hong, Y. and T. Zhou. 1985b. Studies on the karyotype and C-banding patterns in *Acheilognathus gracilis* with a discussion on the evolution of acheilognathid fishes. *Acta Genetica Sinica* 12: 143–148. (In Chinese with English abstract).

Inafuku, J., M. Nabeyama, Y. Kikuma, J. Saitoh, S. Kubota and S. Kohno. 2000. Chromosomal location and nucleotid sequences of 5S ribosomal DNA of two cyprinid species (Osteichthyes, Pisces). *Chromosome Research* 8: 193–199.

John, B. and G.L.G. Miklos. 1979. Functional aspects of satellite DNA and heterochromatin. *International Review of Cytology* 58: 1–114.

Kikuma, Y., J. Inafuku, S. Kubota and S. Kohno. 2000. Banding karyotype and 5S ribosomal DNA loci in the Japanese bitterling, *Rhodeus ocellatus* (Cyprinidae). *Chromosome Science* 3: 101–103.

Lee, G. Y. 1983. Karyotypes of the acheilognathine fishes (Cyprinidae) in Korea (II). *Annual Report of Biological Research, Jeonbug National University* 4: 1–9. (In Korean with English abstract).

Lee, G.Y., J.N. So and S.J. Kim. 1982. Karyotypes of the acheilognathine fishes (Cyprinidae) in Korea (I). *Annual Report of Biological Research, Jeonbug National University* 3: 19–24. (In Korean with English abstract).

Lee, H.Y., C.H. Yu, S.K. Jeon and H.S. Lee. 1983. The karyotype analysis on 29 species of fresh water fish in Korea. *Bulletin of the Institute of Basic Science, Inha University* 4: 79–86. (In Korean with English abstract).

Levan, A., K. Fredga and A.A. Sandberg. 1964. Nomenclature for centromeric position on chromosomes. *Hereditas* 52: 201–220.

Lin, R.D. 1998. Acheilognathinae. In: *Fauna Sinica, Osteichthyes, Cypriniformes II.* Y. Chen (ed.). *Science Press, Beijing, China*, pp. 413–454 and 506–506. (In Chinese with English key).

Meyne, J., R.J. Baker, H.H. Hobart, T.C. Hsu, O.A. Ryder, O.G. Ward, J.E. Wiley, D.H. Wurster-Hill, T.L. Yates and R.K. Moyzis. 1990. Distribution of non-telomeric sites of the (TTAGGG)n telomeric sequences in vertebrate chromosomes. *Chromosoma* 99: 3–10.

Miller, D.A., V.G. Dev, R. Tantravahi and O.J. Miller. 1976. Suppression of human nucleolus organizer activity in mouse-human somatic hybrid cells. *Experimental Cell Research* 101: 235–243.

Nakamura, M. 1969. Cyprinid fishes of Japan. Studies on the life history of cyprinid fishes of Japan. *Special Publication of the Research Institute of National Resources* 4: 1–455. (In Japanese with English summary).

Ojima, Y., M. Hayashi and K. Ueno. 1972. Cytogenetic studies in lower vertebrates. X. Karyotype and DNA studies in 15 species of Japanese Cyprinidae. *Japanese Journal of Genetics* 47: 431–440.

Ojima, Y., K. Ueno and M. Hayashi. 1973. Karyotypes of the acheilognathine fishes (Cyprinidae) of Japan with a discussion of phylogenetic problems. *Zoological Magazine (Tokyo)* 82: 171–177. (In Japanese with English abstract).

Okazaki, M., K. Naruse, A. Shima and R. Arai. 2001. Phylogenetic relationships of bitterlings based on mitochondrial 12S ribosomal DNA sequences. *Journal of Fish Biology* 58: 89–106.

Salvadori, S., A.M. Deiana, E. Coluccia, G. Floridia, E. Rossi and O. Zuffardi. 1995. Colocalization of (TTAGGG)n telomeric sequences and ribosomal genes in Atlantic eel. *Chromosome Research* 3: 54–58.

Slijepcevic, P., Y. Xiao, I. Dominguez and A. T. Natarajan. 1996. Spontaneous and radiation-induced chromosomal breakage at interstitial telomeric sites. *Chromosoma* 104: 596–604.

Sola, L., E. Gornung, H. Naoi, R. Gunji, C. Sato, K. Kawamura, R. Arai and T. Ueda. 2003. FISH-mapping of 18S ribosomal RNA genes and telomeric sequences in the Japanese bitterlings *Rhodeus ocellatus kurumeus* and *Tanakia limbata* (Pisces, Cyprinidae) reveals significant cytogenetic differences in morphologically similar karyotypes. *Genetica* 119: 99–106.

Suzuki, T., H. Yasue, R. Arai and T. Ueda. 1999. FISH-mapping of 18S ribosomal RNA genes in two aceilognathid fishes. Meeting of the Japanese Society of Fisheries Science, Abstracts: 116. (In Japanese).

Takai, A. and Y. Ojima. 1986. Some features on nucleolus organizer regions in fish chromosomes. In: *Indo-Pacific Fish Biology: Proceedings of the Second International Conference of Indo-Pacific Fishes,* T. Uyeno, R. Arai, T. Taniuchi and K. Matsuura (eds.). Ichthyological Society of Japan, Tokyo, pp. 899–909.

Takai, A. and Y. Ojima. 1988. Chromosomal distribution of C-banded heterochromatin in cyprinid fishes. *Proceedings of the Japan Academy* 64B: 49–52.

Takai, A. and Y. Ojima. 1999. Constitutive heterochromachin distribution in the chromosomes of pomacentrid fishes (Perciformes). *Cytologia* 64: 87–91.

Ueda, T. 1993. How do fish karyotypes change? *Biological Science (Tokyo)* 45: 182–186. (In Japanese).

Ueda, T. 1996. Chromosomal aberrations induced by retention of cyprinid fish unfertilized eggs in freshwater. *Cytologia* 61: 423–430.

Ueda, T. 1997. Nucleolar organizer regions on fish chromosomes in Salmonidae and Acheilognathinae. *Fish Genetics and Breeding Science* 25: 1–10. (In Japanese with English abstract).

Ueda, T. and K. Aoki. 1994. Chromosomal aberrations induced by cold shock in fertilized eggs of rose bitterling. In: *Proceedings of Fourth Indo-Pacific Fish Conference,* Faculty of Fisheries, Kasetsart University, Thailand (ed.). pp. 473–483.

Ueda, T. and J. Kobayashi. 1988. Heteroploidy induced by production of triploids in the kokanee, *Oncorhynchus nerka. Chromosome Information Service* 44: 7–8.

Ueda, T. and J. Kobayashi. 1990. Karyotype differentiation of Atlantic salmon, *Salmo salar*; especially the sequential karyotype change. *La Kromosomo* 58: 1967–1972.

Ueda, T. and J. Kobayashi. 1991. Chromosomal polymorphisms in oily bitterling *Acheilognathus limbatus. Chromosome Information Service* 50: 13–14.

Ueda, T. and H. Naoi. 1999. BrdU-4Na-EDTA-Giemsa band karyotypes of 3 small freshwater fish, *Danio rerio, Oryzias latipes,* and *Rhodeus ocellatus. Genome* 42: 531–535.

Ueda, T. and Y. Ojima. 1983. Karyotypes with C-banding patterns of two species in the genus *Salvelinus* of the family Salmonidae. *Proceedings of the Japan Academy* 59B: 343–346.

Ueda, T. and Y. Takahashi. 1995. Chromosomal studies of three kinds of hybrids in acheilognathin fishes. *Bulletin of the Faculty of Education Utsunomiya University* 45II: 59–65. (In Japanese with English abstract).

Ueda, T., R. Sato and Y. Fukuda. 1984. Triploid hybrids between female rainbow trout and male brook trout. *Bulletin of the Japanese Society of Scientific Fisheries* 50: 1331–1336.

Ueda, T., H. Fukuda and J. Kobayashi. 1991. Karyological study of the dolly varden *Salvelinus malma* from Alaska. *Japanese Journal of Ichthyology* 37: 354–357.

Ueda, T., H. Onozaki and F. Hayashi. 1994. Triploid hybrids between female rose bitterling and male Kyushu rose bitterling. *Bulletin of the Faculty of Education Utsunomiya University* 44II: 65–72. (In Japanese with English abstract).

Ueda, T., R. Arai, Y. Akai, T. Ishinabe and H. Wu. 1996. Karyological study of a bitterling *Rhodeus atremius* (Teleostei: Cyprinidae) collected in Zhejiang, China. *Cytobios* 86: 265–268.

Ueda, T., N. Mashiko, H. Takizawa, Y. Akai, T. Ishinabe, R. Arai and H. Wu. 1997. Ag-NOR variation in chromosomes of Chinese bitterings, *Rhodeus lighti* and *Tanakia himantegus* (Cypriniformes, Cyprinidae). *Ichthyological Research* 44: 302–305.

Ueda, T., H. Naoi and R. Arai. 2001a. Flexibility on the karyotype evolution in bitterlings (Pisces, Cyprinidae). *Genetica* 111: 423–432.

Ueda, T., H. Naoi, R. Gunji, S. Ohtake, R. Arai, S. Jeon, Gornung E. and L. Sola. 2001b. Geographic chromosomal polymorphisms of *Rhodeus ocellatus ocellatus* (cyprinid fishes) in East Asia. 73th Annual Meeting of the Genetics Society of Japan, *Genes & Genetic Systems* 76: 457.

Ueda, T., H. Naoi, M. Aoyama, R. Arai and S. Jeon. 2002. Chromosomal polymorphisms of *Acheilognathus rhombeus* (Pisces, Cyprinidae) in South Korea. In Abstracts of the 53rd Annual Meeting of the Society of Chromosome Research. *Chromosome Research* 6: 123.

Ueno, K. and Y. Ojima. 1984. A chromosome study of nine species of Korean cyprinid fish. *Japanese Journal of Ichthyology* 31: 338–344.

Várkonyi, E., M. Bercsényi, C. Ozouf-Costaz and R. Billard. 1998. Chromosomal and morphological abnormalities caused by oocyte ageing in *Silurus glanis*. *Journal of Fish Biology* 52: 899–906.

Yamazaki, F., J. Goodier and K. Yamano. 1989. Chromosomal aberrations caused by aging and hybridization in charr, masu salmon and related salmonids. *Physiology and Ecology Japan, Special Volume* 1: 529–542.

Yu, X., T. Zhou, K. Li, Y. Li and M. Zhou. 1987. On the karyosystematics of cyprinid fishes and a summary of fish chromosome studies in China. *Genetica* 72: 225–236.

Chromosome Variability in Gymnotiformes (Teleostei: Ostariophysi)

Lurdes Foresti de Almeida-Toledo[1]*, Maria de Fatima Zambelli Daniel-Silva[1], Cinthia Bachir Moysés[1] and Fausto Foresti[2]

INTRODUCTION

The fish order Gymnotiformes is a monophyletic group of electrogenic freshwater fish with a wide Neotropical distribution (Nelson, 1994). These fish are endemic to the freshwaters of South and Central America (Lundberg *et al.*, 1987; Mago-Leccia, 1994), have nocturnal habits, and are characterized by extreme territoriality (Knudsen, 1974), low vagility, and small populations. Gymnotiform fishes continually emit weak electric discharges, which they use in object location and communication (Albert, 2001).

Address for Correspondence: [1]*Departamento de Genética e Biologia Evolutiva, Instituto de Biociências, Universidade de São Paulo, São Paulo, SP, Brazil. E-mail: lftoledo@ib.usp.br
[2]Departamento de Morfologia, Instituto de Biociências, Universidade Estadual Paulista, Botucatu, SP, Brazil.

These American knifefishes or Neotropical electric fishes are important components of the nocturnal ichthyofauna of Middle and South American freshwaters. They range in latitude from the Salado River in the Pampas of Argentina to the San Nicolas River in southwestern Chiapas, Mexico (Albert, 2001). Gymnotiform species are very useful for the establishment of biogeographic hypotheses, since an expressive number of species are found in the great South American rivers, indicating a uniform ancestral geographical distribution (Mago-Leccia, 1994).

According to Mago-Leccia (1994), the order Gymnotiformes encompasses 6 fish families: Sternopygidae, Apteronotidae, Rhamphichthyidae, Hypopomidae, Gymnotidae and Electrophoridae. Albert (2001) proposed that *Electrophorus*, the only genus of Electrophoridae, and *Gymnotus* could form a natural taxon, the Gymnotidae, reducing the number of families to 5.

A growing number of gymnotiform species have been described over the last years: in 1994, there were 94 valid species (Mago-Leccia, 1994) but in 2001, Albert considered that the number of valid species of Gymnotiformes in the literature was 108, apart from about 34 undescribed species recognized in museum collections. This author called the attention to the fact that with the intensification of the survey in deep river channels and flood-plain floating meadows in lowland Amazon, this number might increase. Fernandes *et al.* (2004) made a comprehensive survey of the main channels of the Amazon River and of its major tributaries and reported the occurrence of 43 electric fish species.

A survey of the gymnotiform species published in Reis *et al.* (2003) presents 117 as the total number: 19 species of Gymnotidae (including *Electrophorus*) (Campos da Paz, 2003); 27 species of Sternopygidae; 13 species of Rhamphicthyidae (Ferraris Jr., 2003); 14 species of Hypopomidae (Albert and Crampton, 2003) and 44 species of Apteronotidae (Albert, 2003a). More recently a new species of *Gymnotus* was described by Fernandes *et al.* (2005), raising the total number of species to 118.

Species diversity of Gymnotiformes has sometimes been overlooked, because the fishes are often difficult to identify, and a number of cryptic species and species complexes may exist that are not evident by their morphological characteristics. In such cases, cytogenetic analysis may provide very efficient tools for the characterization of species and populations.

Cytogenetic studies have been carried out in these electric fishes for the last 30 years, and species-specific chromosome numbers and formulae have been found for about all the species analyzed, allowing a clear identification of karyotypical differences among species and populations. The chromosome numbers in Gymnotiformes range from 2n=24, found in *Apteronotus albifrons*, to 2n=54, in *Gymnotus carapo*.

In this chapter we shall present a survey of the chromosome studies carried out in the five families of Gymnotiformes and discuss the chromosome variability in this group. A synthesis of the available cytogenetic data is presented in Table 1.2.1.

Gymnotidae

The family Gymnotidae, considered monotypic for a long time, formed only by genus *Gymnotus*, includes now, after an extensive re-evaluation of the gymnotiform higher-level systematics and taxonomy, the genus *Electrophorus* (Campos da Paz, 2003). While populations of *Electrophorus* are found throughout northern South America, especially in the Amazon and Orinoco River basins, the genus *Gymnotus* stands as the most widespread genus in the order Gymnotiformes, being reported in waters from Argentina in South America, to Mexico in North America. Genus *Gymnotus* currently includes nineteen valid species, thus representing one of the most diverse gymnotiform genera (Campos da Paz, 2003; Fernandes *et al.*, 2005).

Cytogenetic analysis in this family has already been performed in 6 out of the 20 species of this group: *Electrophorus electricus*, collected in the Amazon River, at Almeirim (PA), in the Araguaia River, Nova Crixás (GO), and five species of *Gymnotus* sampled from the Paraná River basin. *E. electricus* has 2n=52 with 42 metacentric/submetacentric (M/SM) and 10 acrocentric chromosomes (A) (Fonteles *et al.*, unpublished). For the genus *Gymnotus*, cytogenetic data are available for *G. carapo*, *G. pantherinus*, *G. inaequilabiatus* and *G. sylvius* (Fernandes-Matioli *et al.*, 1998) and for a recently described species from the Pantanal Matogrossense region, *G. pantanal* (Fernandes *et al.*, 2005). *G. carapo* specimens from 15 populations of the Upper Paraná River system were analyzed, and the chromosome number and formula found were 2n=54 (52 MSM + 2 STA). In specimens collected in the Amazon River, at Humaitá and Belém, they were 2n=48 (34 MSM + 14 STA) and 2n=42

Table 1.2.1 Available cytogenetic data on Gymnotiformes species in South American rivers (modified from Oliveira et al., present volume).

Taxon	Locality	n	2n	Karyotype	Sex Chrom.	Ref.
Family Apteronotidae						
Apteronotus						
A. albifrons	(not available)	11		(not available)		29
A. albifrons	(not available)		24	14m+2sm+2st+6a		30
A. albifrons	Ilha de Marajó, Pará		24	12m+4sm+2st+6a		17
A. anas	Manaus, Amazonas		52	30m+12sm+10a		33
A. hasemani	Manaus, Amazonas		52	26m+16sm+10a		33
Apteronotus sp.	R Mogi-Guaçu, São Paulo		52	46m,sm+6st,a		10
Family Gymnotidae						
Electrophorus						
E. electricus	Amazonas, Brazil		52	(not described)		32
E. electricus	Almeirim, Amazonas		52	42m,sm+10a		26
E. electricus	Araguaia, Nova Crixás - GO		52	42m,sm+10a		26
Gymnotus						
G. carapo	Botucatu, São Paulo		54	54m,sm		27
G. carapo	Miracatu, São Paulo		54	54m,sm		28
G. carapo	Iguape, São Paulo		54	52m,sm+2st		24
G. carapo	Macallé, Argentina		54	(not available)		19
G. carapo	Torres, Rio Grande do Sul		54	46m,sm+8st,a		27
G. carapo	Jundiaí, São Paulo		54	52m,sm+2st,a		25
G. carapo	Pirassununga, São Paulo		54	52m,sm+2st,a		25
G. carapo	Salto Grande, São Paulo		54	52m,sm+2st,a		25
G. carapo	Primeiro de Maio, Paraná		54	52m,sm+2st,a		25

(Table 1.2.1 contd.)

(Table 1.2.1 contd.)

Species	Locality	2n	Karyotype		No.
G. carapo	Americana, São Paulo	54	52m,sm+2st,a		25
G. carapo	Rio Claro, São Paulo	54	52m,sm+2st,a		25
G. carapo	Brotas, São Paulo	52	50m,sm+2st,a		28
G. carapo	Humaitá, Amazonas	48	34m,sm+14st,a		28
G. carapo	Belém, Pará	42	32m,sm+10st,a		28
G. carapo	Porto Rico, Paraná	40	36m,sm+4st,a		20
G. carapo	Porto Rico, Paraná	54	(not described)		20
G. carapo	R Paranapanema, São Paulo	54	(not described)		23
G. carapo	R Mogi-Guaçu, São Paulo	3n=81	78m,sm+3st,a		24
G. carapo	R Capivara, São Paulo	40	(not described)		39
G. carapo	R Capivara, São Paulo	50	(not described)		39
G. carapo	R Capivara, São Paulo	54	(not described)		39
G. carapo	R Paraná, Guairá, Paraná	54	52m,sm+2st,a		36
G. carapo	R Paraná, Guairá, Paraná	54	52m,sm+2st,a	X_1X_2Y	37
G. carapo	Santa Fé, L. Rodeo, Argentina	40/54	(not described)		35
G. aff. carapo	R Araguaia, Mato Grosso	42	22m+6m+14st,a		40
G. inaequilabiatus	R Tietê, São Paulo	52	50m,sm+2st,a		23
G. pantherinus	East basin	52	38m+8sm+6st,a		31
G. pantherinus	Paraná basin	52	38m+8sm+6st,a		31
G. sylvius	R Paraná, Guairá, Paraná	40	30m,sm+10st,a		38
G. sylvius	R Ribeira do Iguape, São Paulo	40	30m,sm+10st,a		23
G. sylvius	R Mogi-Guaçu, São Paulo	40	38m,sm+2st,a		23
Gymnotus sp.	Miracatu, São Paulo	52	50m,sm+2st,a		28
Gymnotus sp. A	Porto Rico, Paraná	54	(not described)		20
Gymnotus sp. B	Itanhaém, São Paulo	52	52m,sm		27

(Table 1.2.1 contd.)

(Table 1.2.1 contd.)

Gymnotus sp. C	Paranapiacaba, São Paulo	52	52m,sm		27
Gymnotus sp. 1	Americana, São Paulo	40	30m,sm+10st,a		24
Gymnotus sp. 2	Rio Claro, São Paulo	52	50m,sm+2st,a		24
Gymnotus sp. 3	Porto Primavera, São Paulo	52	50m,sm+2st,a		24
Gymnotus sp.	Miracatu, São Paulo	40	30m,sm+10st,a		24
Gymnotus sp. C	R Paraná, Guairá, Paraná	40	14m,sm+26st,a		36
Gymnotus pantanal	Pantanal, Mato Grosso do Sul	40	14m,sm+26st,a		41
Gymnotus sp.	R Paraná, Guairá, Paraná	39/40	M 14m,sm+25st,a / F 14m,sm+26st,a	X_1X_2Y	38
Family Hypopomidae					
Brachyhypopomus					
B. brevirostris	Humaitá, Amazonas	36	4m+2sm+8st+22a		3
B. pinnicaudatus	R Tietê, São Paulo	41/42	M 1m+40a / F42a	X_1X_2Y	13
Brachyhypopomus sp. A	R Miranda, Mato Grosso do Sul	42	42a		21
Brachyhypopomus sp. B	R Miranda, Mato Grosso do Sul	40	4st+36a		21
Hypopomus					
H. artedi	R Negro, Amazonas	38	32m,sm+6st,a		41
Hypopomus sp.	Sta Maria da Serra, São Paulo	41/42	M 1m+40a / F42a	X_1X_2Y	11
Hypopygus					
H. lepturus	Belém, Pará	50	16m+20sm+10st+4a		3
Family Rhamphichthyidae					
Rhamphichthys					
R. cf. marmoratus	Humaitá, Amazonas	52	38m+10sm+4st		3

(Table 1.2.1 contd.)

(Table 1.2.1 contd.)

Family Sternopygidae						
Eigenmannia						
E. humboldtii	R Jari, Paraná		40	8m,sm+32st,a		27
E. virescens	(not available)		32	(not available)		22
E. virescens	Rio Claro, São Paulo		38	22m,sm+16st,a	XY	27
E. virescens	Mogi-Guaçu, São Paulo		38	22m,sm+16st,a		14
E. virescens	Icém, São Paulo		38	22m,sm+16st,a		42
E. virescens	Corumbataí, São Paulo		38	22m,sm+16st,a	XY	27
E. virescens	Ipeúna, São Paulo		38	22m,sm+16st,a	XY	27
E. virescens	Pirassununga, São Paulo		38	22m,sm+16st,a	XY	27
E. virescens	Botucatu, São Paulo		38	22m,sm+16st,a	XY	27
E. virescens	Salto Grande, São Paulo		38	22m,sm+16st,a	XY	27
E. virescens	Ilha de Marajó, Pará		38	M 16m,sm+22st,a / F 17m, sm+21st,a	ZW	27
E. virescens	R Mogi-Guaçu, São Paulo		38	16m,sm+22st,a		15
E. virescens	R Tietê, São Paulo		38	16m,sm+22st,a	XY	15
E. virescens	R São Francisco, Minas Gerais		38	M 22m,sm+16st,a / F 23m, sm+15st,a	ZW	2
E. virescens	R Amazonas, Almeirim, Pará		38	M 16m,sm+22st,a / F 17m, sm+21st,a	ZW	2
Eigenmannia sp.	(not available)	17		(not available)		29
Eigenmannia sp.	R Jari, Pará		31/32	M 13m,sm+18st,a / F 12m, sm+20st,a	X_1X_2Y	5
Eigenmannia sp.	R Jari, Pará		3n/46	20m,sm+26st,a		5
Eigenmannia sp.	Ilha de Marajó, Pará		29/30	M 7m,sm+22st,a/F 8m, sm+22st,a		27

(Table 1.2.1 contd.)

(Table 1.2.1 contd.)

Eigenmannia sp.	Ilha de Marajó, Pará	31/32	M 11m,sm+20st,a/F 14m, sm+18st,a		27
Eigenmannia sp.	Ilha de Marajó, Pará	30	12m,sm+18st,a		27
Eigenmannia sp.	Ilha de Marajó, Pará	28	14m,sm+14st,a		27
Eigenmannia sp.	Torres, Rio Grande do Sul	30	6m,sm+24st,a		27
Eigenmannia sp. 1	R Mogi-Guaçu, São Paulo	28	14m,sm+14st,a		6
Eigenmannia sp. 1	R Mogi-Guaçu, São Paulo	28	14m,sm+14st,a		12
Eigenmannia sp. 1	Pirassununga, São Paulo	28	14m,sm+14st,a		7
Eigenmannia sp. 1	Araras, São Paulo	28	14m,sm+14st,a		7
Eigenmannia sp. 1	Araras, São Paulo	28	14m,sm+14st,a		12
Eigenmannia sp. 1	Luís Antônio, São Paulo	28	14m,sm+14st,a		42
Eigenmannia sp. 1	Guaíra, São Paulo	28	14m,sm+14st,a		42
Eigenmannia sp. 1	Icém, São Paulo	28	14m,sm+14st,a		42
Eigenmannia sp. 2	Botucatu, São Paulo	31/32	M 9m,sm+22a / F 8m,sm+24a	X_1X_2Y	4
Eigenmannia sp. 2	Icém, São Paulo	31/32	M 9m,sm+22a / F 8m,sm+24a	X_1X_2Y	42
Eigenmannia sp. 2	Sta.Albertina, São Paulo	31/32	M 9m,sm+22a / F 8m,sm+24a	X_1X_2Y	42
Eigenmannia sp. 2	Piquerobi, São Paulo	31/32	M 9m,sm+22a / F 8m,sm+24a	X_1X_2Y	42
Eigenmannia sp. 2	Auriflama, São Paulo	31/32	M 9m,sm+22a / F 8m,sm+24a	X_1X_2Y	42
Eigenmannia sp. 2	Alto Paraná, São Paulo	31/32	M 9m,sm+22a / F 8m,sm+24a	X_1X_2Y	14
Eigenmannia sp. 2	R Tietê, São Paulo	31/32	M 9m,sm+22a / F 8m,sm+24a	X_1X_2Y	6
Eigenmannia sp. 2	Corumbataí, São Paulo	31/32	M 9m,sm+22a / F 8m,sm+24a	X_1X_2Y	27
Eigenmannia sp. 2	R Capivara, Botucatu, São Paulo	31/32	M 9m,sm+22a / F 8m,sm+24a	X_1X_2Y	8
Eigenmannia sp. 2	Promissão, São Paulo	31/32	M 9m,sm+22a / F 8m,sm+24a	X_1X_2Y	8
Eigenmannia sp.	Promissão, São Paulo	36	10m,sm+26st,a		42
Eigenmannia sp.	Sta. Albertina, São Paulo	36	10m,sm+26st,a		42
Eigenmannia sp.	R Capivara, Botucatu, São Paulo	36	M 7m,sm+29st,a / F 8m, sm+28st,a	ZW	27

(Table 1.2.1 contd.)

(Table 1.2.1 contd.)

Eigenmannia sp.	Piquerobi, São Paulo		36	14m,sm+22st,a	42
Eigenmannia sp.	R Itabapoana, Rio de Janeiro		38	(not available)	34
Eigenmannia sp.	R São Francisco, Minas Gerais		36	16m,sm+20st,a	16
Eigenmannia sp.	R São Francisco, Minas Gerais		34	24m,sm+10st,a	1
Sternopygus					
S. macrurus	(not available)	24			
S. macrurus	(not available)		46	(not available)	29
S. macrurus	Manaus, Amazonas		46	(not described)	18
S. macrurus	Manaus, Amazonas		46	30m+16sm	9
S. macrurus	Rio Claro, São Paulo		46	28m+18sm	9
S. macrurus	Três Marias, Minas Gerais		46	32m+14sm	9

Table References:

(1) Almeida-Toledo et al., 2001b; (2) Almeida-Toledo et al., 2002; (3) Almeida-Toledo, 1978; (4) Almeida-Toledo et al., 1984; (5) Almeida-Toledo et al., 1985; (6) Almeida-Toledo et al., 1988; (7) Almeida-Toledo et al., 1990; (8) Almeida-Toledo et al., 1991; (9) Almeida-Toledo et al., 1993; (10) Almeida-Toledo et al., 1994; (11) Almeida-Toledo et al., 1995; (12) Almeida-Toledo et al., 1996; (13) Almeida-Toledo et al., 1998; (14) Almeida-Toledo et al., 1999; (15) Almeida-Toledo et al., 2001a; (16) Almeida-Toledo et al., 2004; (17) Almeida-Toledo et al., 1981; (18) Bertollo et al., 1980; (19) Bordenave et al., 1992; (20) Borin and Júlio Jr., 1994; (21) Cereali et al., 2002; (22) Denton, 1973; (23) Fernandes-Matioli et al., 2000; (24) Fernandes-Matioli, 1999; (25) Fernandes-Matioli et al., 1997; (26) Fonteles, et al., unpublished (27) Foresti, 1987; (28) Foresti et al., 1984; (29) Hinegardner and Rosen, 1972; (30) Howell, 1972; (31) Marchetto et al., 1998; (32) Medaglia et al., 1995; (33) Nakayama et al., 2000; (34) Oliveira et al., present volume; (35) Rolón, 2002; (36) Silva and Margarido, 2001; (37) Silva et al., 2002a; (38) Silva et al., 2002b; (39) Souza et al., 1997; (40) Vénere, 1998; (41) Fernandes, et al., 2005; (42) present volume.

(32 MSM + 10 STA), respectively. The chromosome number variation found for this species in the Paraná River system (Oliveira *et al.*, present volume) may be attributed to species misidentification. G. *inaequilabiatus* presents 2n=52 (50 MSM + 2 STA) and was found in the Upper Paraná River basin; the two samples already analyzed are from the Tietê river, G. *pantherinus* is found in rivers of the southeastern Brazilian coast. This species presents 2n=52 chromosomes (46 MSM + 6 STA) and a very peculiar distribution of constitutive heterochromatin on the long arms of four chromosome pairs. The chromosome number of G. *sylvius* is 2n=40 (38 MSM + 2 STA) (Albert *et al.*, 1999), the lowest among the *Gymnotus* species analyzed so far.

The species *Gymnotus pantanal* from the Pantanal Matogrossense recently described, presents 2n=40 chromosomes, (28 MSM + 12 STA) (Fernandes *et al.*, 2005). Specimens of this species collected in the Paraná River, at Foz do Iguaçu, PR, presented a morphologically differentiated sex chromosome system (Silva and Margarido, 2002a). *Gymnotus pantanal* differs from G. *sylvius* that presents the same chromosome number, but a different chromosome formula.

The data presented above show that in the family Gymnotidae, the chromosome numbers of the species vary from 2n=40 to 2n=54, the latter being the highest chromosome number found in Gymnotiformes so far. The occurrence of 2n=52 in three species, *Electrophorus electricus*, *Gymnotus pantherinus* and G. *inaequilabiatus*, reinforces the suggestion presented by Fernandes-Matioli and Almeida-Toledo (2001) that the diploid number 2n=52 may be a basal, albeit not plesiomorphic trait in the family Gymnotidae.

The constitutive heterochromatin pattern was analyzed for G. *carapo*, G. *pantherinus*, G. *inaequilabiatus*, G. *sylvius* and G. *pantanal*. G. *carapo* and G. *sylvius* present very conspicuous C-positive blocks in all the chromosomes at the pericentromeric regions. G. *pantherinus* and G. *inaequilabiatus* have, in addition, large heterochromatic blocks along the chromosome arms of other chromosome pairs; G. *pantanal* presents small pericentromeric blocks in all the chromosome pairs. *Electrophorus electricus* has medium sized heterochromatic blocks at the pericentromeric region of all the chromosomes. One pair of NORs was present in all the species analyzed with the exception of G. *pantanal* that had three chromosomes bearing NORs (Fernandes *et al.*, 2005).

Rhamphichthyidae

This is a small family of Gymnotiformes composed of 13 species divided into three genera (Ferraris Jr., 2003). Only one species, *Rhamphichthys* cf. *marmoratus* from the Amazon basin, at Humaitá (AM), was analyzed cytogenetically. The chromosome number of this species is 2n=52, and its karyotype is composed only of metacentric and submetacentric chromosomes.

Hypopomidae

The known diversity of this family is 14 species (Albert and Crampton, 2003), four of which were analyzed cytogenetically. The chromosome numbers vary from 2n=36 to 2n=50. In genus *Brachyhypopomus*, the species *B. brevirostris* (from the Amazon river) presents 2n=36 chromosomes (4 MSM + 32 STA). In *B. pinnicaudatus* (from the Paraná River), the females have 2n=42 acrocentric chromosomes, and the males 41 acrocentrics and one metacentric, characterizing a morphologically differentiated sex chromosome system with male heterogamety (Almeida-Toledo *et al.*, 2000a). *Hypopomus artedi*, from the Negro River (AM), was found to have 2n=38 chromosomes (32 MSM + 6 STA) and *Hypopygus lepturus*, from the Amazon River (PA), had 2n=50, (36 MSM + 14 STA) (Almeida-Toledo, 1978). The cytogenetic data show the occurrence of morphologically differentiated sex chromosomes, of low chromosome numbers and a prevalence of acrocentrics in the genus *Brachyhypopomus*, and higher chromosome numbers in genus *Hypopygus*.

The constitutive heterochromatin pattern was analyzed for *Brachyhypopomus pinnicaudatus* and the presence of small C-banded positive blocks was reported. Only one pair of NORs was detected in this species (Almeida-Toledo *et al.*, 2000a).

Apteronotidae

Apteronotidae is the most speciose family of gymnotiform fishes, with 44 species described thus far (Albert, 2003a). Three species of genus *Apteronotus* have been analyzed cytogenetically, and the chromosome numbers varied from 2n=24 to 2n=52 (Oliveira *et al.*, present volume). *Apteronotus albifrons* presents 2n=24, the lowest chromosome number among the Gymnotiformes. The three other species analyzed so far,

Apteronotus anas and *A. hasemani* from the Amazon basin and *Apteronotus* sp. from the Paraná River basin, presented 2n=52 chromosomes.

Constitutive heterochromatin blocks were present at the pericentromeric region of all the chromosomes of *Apteronotus albifrons* and a large C-band positive block was present along the long arm of the NOR-bearing chromosome pair. Only one pair of NORs was detected in this species (Almeida-Toledo *et al.*, 1981).

Sternopygidae

This family comprises 27 species, and cytogenetic analysis is available for two genera: *Sternopygus* and *Eigenmannia*. From the cytogenetic point of view, this is the best studied group of Gymnotiforms.

Genus *Sternopygus*

Among the Sternopygidae, *Sternopygus* is the most widely distributed genus, in terms of both geography and habitats, extending the full range of the family (Albert, 2003b). Chromosome data are available for a single species, *Sternopygus macrurus*. Samples of this species from the Amazon, São Francisco and Paraná River basins were analyzed, and remarkable chromosome conservation was detected. The three populations presented 2n=46 bi-armed chromosomes and a similar constitutive heterochromatin pattern. The chromosome formulae were found to be different in the three populations due to slight centromere shifts, or maybe due to the difficulty in establishing the exact arm ratio for the small chromosomes in the karyotype. A morphological difference among populations was found in the NORs-bearing chromosome pair, due to a gradual loss of the heterochromatic region associated to NORs (Almeida-Toledo *et al.*, 1993).

Genus *Eigenmannia*

Eigenmannia is a ubiquitous genus in South America and its northern range reaches Panamá. Its taxonomy is confused due to its large variation within species and within populations in a geographical cline (Mago-Leccia, 1994). Eight nominal species of this genus are recognized (Albert, 2003b) and three of them were assigned to the Paraguai, Paraná, São Francisco river basins and to coastal rivers from the eastern Brazilian coast (Campos da Paz, 1997). The same author also emphasized the morphological complexity and the uncertainty about the species taxonomy of this group (Campos da Paz, 1997). These characteristics are associated with a high degree of chromosome variability (Almeida-Toledo, 1978; Foresti, 1987).

In contrast to other gymnotiform genera, the representatives of genus *Eigenmannia* have been the object of very intense cytogenetic studies. Specimens were sampled from locations in the Amazon, Paraná, São Francisco and eastern Brazilian coastal rivers basins (Fig. 1.2.1).

In the Amazon River basin, different cytotypes were found: $2n=30$ (6 MSM + 24 STA) in Almeirim (PA) (unpublished results); $2n=31/32$ X_1X_2Y (male: 9 MSM+22 STA/female: 8 MSM + 24 STA) (Foresti, 1987); $2n=38$ ZW (male: 14 MSM + 24 STA/female: 15 MSM + 23 STA) on the Marajó Island and $2n=38$ ZW (female: 17 MSM + 21 STA) in Almeirim (PA), the two latter presenting morphologically differentiated sex chromosomes with female heterogamety (Almeida-Toledo *et al.*, 2002); and $2n=40$ (4 SM + 36 STA) in Belém (PA) (Almeida-Toledo,1978). The chromosome number $2n=40$ was found only in this Amazonian population and is the highest chromosome number found in the genus *Eigenmannia* so far.

In the Três Marias reservoir ot the São Francisco River basin (Fig. 1.2.1a), three different cytotypes were found in sympatry and syntopy (Fig. 1.2.2 a-d): (a) $2n=34$ (24 MSM + 10 STA); (b) $2n=36$ (16 MSM + 20 STA); (c) $2n=38$ ZZ (male: 22 MSM + 16 STA) and (d) $2n=38$ ZW (female: 23 MSM + 15 STA).

In the Upper Paraná River basin (Fig. 1.2.1b), *Eigenmannia* samples were analyzed from 32 localities on the Tietê, Paranapanema, Mogi-Guaçu, Paraná, Pardo and Grande, Rivers, and 7 different cytotypes were detected (Fig. 1.2.3a-j): (a) $2n=28$ (14 MSM + 14 STA) in the Mogi-Guaçu (Almeida-Toledo *et al.*, 1996) and Grande Rivers; (b) $2n=32$ X_1X_2 (female: 8 MSM + 24 A); (c) $2n=31$ X_1X_2Y (male: 9 MSM + 22 A) in the Tietê River (Almeida-Toledo *et al.*, 2000b); (d) $2n=36$ (14 MSM + 22 STA) in Piquerobi, on the Paraná river; (e) $2n=36$ ZZ (male: 6 MSM + 30 STA); (f) $2n=36$ ZW (female: 7 MSM + 29 STA) in the Tietê River; (g) $2n=36$ (8 MSM + 28 STA) in Santa Albertina on the Grande River; (h) *Eigenmannia virescens* with $2n=38$ (16 MSM + 22 STA) in the Mogi-Guaçu River; (i) *E. virescens* $2n=38$ XX (female: 16 MSM + 22 STA) and (j) $2n=38$ XY (male: 16 MSM + 22 STA) in the Tietê, Grande and Paranapanema Rivers (Almeida-Toledo *et al.*, 2001a).

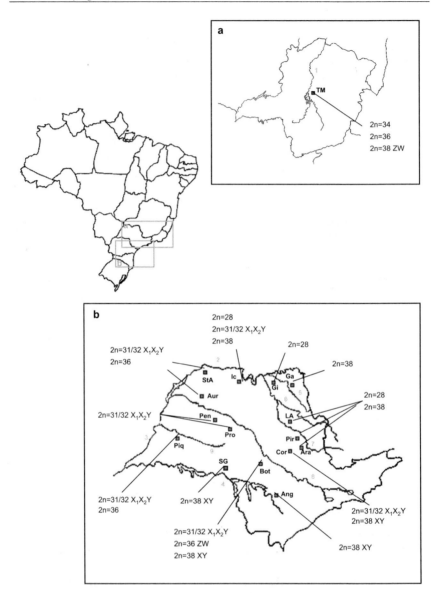

Fig. 1.2.1 Geographic distribution of the cytotypes of *Eigenmannia* in two Brazilian river basins: (**a**) São Francisco River basin (State of Minas Gerais) and (**b**) Upper Paraná river system (State of São Paulo).

Localities sampled:
Ang: Angatuba; **Ara:** Araras; **Aur:** Auriflama; **Bot:** Botucatu; **Cor:** Corumbataí; **Ga:** Guará; **Gi:** Guaíra; **Ic:** Icém; **LA:** Luís Antônio; **Pen:** Penápolis; **Pir:** Pirassununga; **Piq:** Piquerobi; **Pro:** Promissão; **SG:** Salto Grande; **StA:** Santa Albertina; **TM:** Três Marias; **1.** São Francisco River; **2.** Grande River; **3.** Paraná River; **4.** Paranapanema River; **5.** Sapucaí River; **6.** Pardo River; **7.** Mogi-Guaçu River; **8.** Tietê River; **9.** Peixe River.

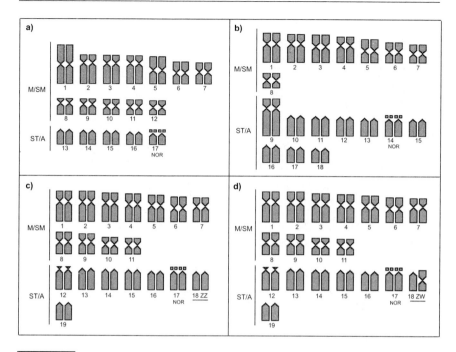

Fig. 1.2.2 Karyograms of *Eigenmannia* cytotypes of São Francisco River basin: **(a)** 2n=34; **(b)** 2n=36; **(c)** 2n=38 ZZ (male); **(d)** 2n=38 ZW (female).

The constitutive heterochromatin is usually present in the form of small dots at the pericentromeric regions of all the chromosomes in the *Eigenmannia* cytotypes. In *E. virescens* the short arms of one submetacentric pair is also heterochromatic. The X-chromosome in this species, when present, has a large heterochromatic portion, which is not found in the Y chromosome. The cytotypes with 2n=31/32 present G-C rich heterochromatic blocks, that stain brightly with Chromomycin A_3, at the pericentromeric regions of some acrocentrics (Almeida-Toledo *et al.*, 2000b). All the cytotypes are characterized by the presence of only one pair of NORs.

According to Campos da Paz (1997), three valid species have been recognized in the region encompassing the São Francisco, Paraná, Paraguai and Brazilian East coastal rivers: *Eigenmannia microstoma*, *E. trilineata* and *E. virescens*.

Our data on cytogenetic constitution reveal ten different cytotypes in this region (Figs. 1.2.2 and 1.2.3), indicating the probable occurrence of three different species complexes: *E. virescens* complex, formed by species

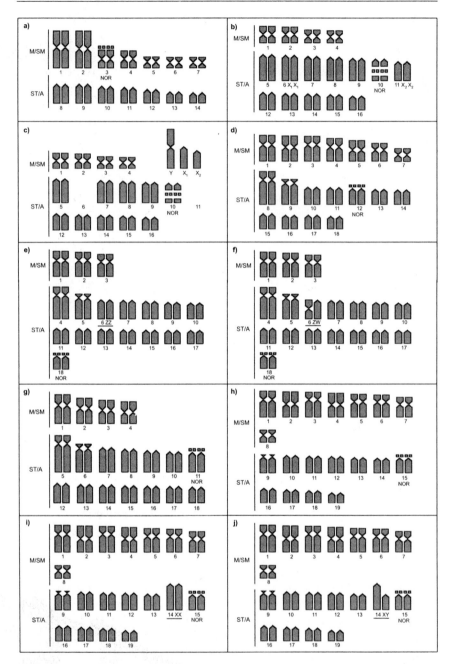

Fig. 1.2.3 Karyograms of *Eigenmannia* cytotypes of Upper Paraná River system: **(a)** 2n=28; **(b)** 2n=32 $X_1X_1X_2X_2$ (female); **(c)** 2n=31 X_1X_2Y (male); **(d)** 2n=36 (Piquerobi); **(e)** 2n=36 ZZ (male); **(f)** 2n=36 ZW (female); **(g)** 2n=36 (Santa Albertina); **(h)** 2n=38; **(i)** 2n=38 XX (female); **(j)** 2n=38 XY (male).

with a chromosome number of 2n=38, including species from the Paraná, São Francisco and Amazon River basins, whose main characteristics would be the chromosome number (2n=38) and the presence of a morphologically differentiated sex chromosome system, as already proposed by Almeida-Toledo *et al.* (2002). A second species complex would be the *E. microstoma* group, composed of species from the Upper Paraná system bearing chromosome numbers of 2n=31/32 and 2n=28, and from the São Francisco basin, with 2n=34 chromosomes. Chromosome homologies have already been found by comparing the R-band patterns of the cytotypes 2n=31/32 and 2n=28 (Almeida-Toledo *et al.*, 1988), which corroborated their similarity. A third complex would be the *E. trilineata* group, formed by the cytotypes with 2n=36 chromosomes, one from the São Francisco River basin and three others from different locations in the Upper Paraná basin.

Cytotypes with chromosome numbers 2n=31/32 were found in specimens from the Amazon basin and also from rivers of the coastal region of South Brazil, in Torres (RS). These fish may be part of *E. microstoma* species complex. In addition, cytotypes with 2n=38 chromosomes and morphologically differentiated sex chromosomes are probably part of a *virescens* group, as previously proposed (Almeida-Toledo *et al.*, 2002).

The relationships among *Eigenmannia* species, with emphasis on the Upper Paraná and São Francisco River basins, have also been studied by employing molecular markers and mitochondrial DNA (mtDNA) sequences. The profiles of inter simple sequence repeats (ISSR) obtained by single primer amplification reaction (SPAR) (Gupta *et al.*, 1994), a technique based on the amplification by PCR (polymerase chain reaction) of DNA fragments flanked by microsatellites, were found to be specific to *Eigenmannia* cytotypes described for the Upper Paraná and São Francisco river basins. When applied to *Gymnotus* species, this technique provided species-specific molecular markers (Fernandes-Matioli *et al.*, 2000). In addition, a preliminary phylogenetic analysis of *Eigenmannia* cytotypes employing mtDNA sequences (Moysés, unpublished data) indicates the existence of two groups of species, involving the populations of the Paraná and the São Francisco rivers. One group includes species with 2n=28, 32 and 34 chromosomes; a second cluster includes the cytotypes with 2n=36 and 2n=38, both from the Paraná and São Francisco rivers.

Final Considerations

Here, the cytogenetic information presented about Gymnotiformes discloses a high karyotypic variability of chromosome number and/or chromosome formula for all the species analyzed. All the species have specific karyotypes. The chromosome variability is particularly evident in genus *Eigenmannia,* for which 7 different cytotypes were described in the Upper Paraná river system populations. This is probably due to the characteristics of the populations of this group, that are usually small, enabling the fixation of chromosome rearrangements.

On the other hand, for other genera such as *Sternopygus,* karyotypic conservation was detected in specimens from the Paraná, São Francisco and Amazon river basins. Genus *Gymnotus,* for which a considerable number of populations have already been analyzed, also presents very conservative cytogenetic species. In the Upper Paraná basin tributaries, the genus *Gymnotus* presented interspecific karyotypic variability but intraspecific conservation of karyotypes, for all the populations analyzed so far, in a situation that strongly contrasts to those of *Eigenmannia* for which wide variability was reported.

The constitutive heterochromatin patterns and NORs distribution in gymnotiforms are very conservative, with the exception of some *Gymnotus* species that present larger blocks of heterochromatin in some chromosomes. C-bands in *Eigenmannia* are usually restricted to small blocks at the pericentromeric regions of chromosomes; larger blocks are usually associated with sex chromosomes.

Only one pair of NORs is present in all the gymnotiform species with the single exception of *Gymnotus pantanal,* that had exceptional chromosomes bearing three NORs (Fernandes *et al.,* 2005).

A very important occurrence in the Gymnotiformes is the presence of morphologically differentiated sex chromosomes in some species. In species of Gymnotidae such as *Gymnotus pantanal,* in an expressive number of populations of *Eigenmannia* and in species of *Brachyhypopomus,* morphologically differentiated sex chromosomes were found, including XX:XY, ZZ:ZW, $X_1X_1X_2X_2$:X_1X_2Y. The scattered occurrence of these sex chromosome systems among species and the finding of both male and female heterogamety and of simple and multiple systems in closely related species, indicate that these sex chromosome systems are in their early stages of differentiation.

Considering the relevant information obtained in the small number of species of the order Gymnotiformes already studied such as the evidence of cryptic species, the chromosome variability among populations and the occurrence of different cytotypes in sympatry and syntopy, cytogenetic analysis can be considered an important tool that may contribute to a better understanding of this group. Moreover, the fact that gymnotiform species are found in almost every river of the Neotropical region, and in small water streams, shows that some populations may be endangered and in risk of extinction.

The integration of taxonomic, cytogenetic and molecular studies may provide a comprehensive view of this fish group, thus contributing to conservation programs of this important component of the Neotropical fish fauna.

References

Albert, J.S. 2001. Species diversity and phylogenetic systematics of American knifefishes (Gymntiformes, Teleostei). *Miscellaneous Publications Museum of Zoology*. University of Michigan, No. 190. Ann Arbor, Michigan, USA.

Albert, J.S. 2003a. Family Apteronotidae. In: *Check List of the Freshwater Fishes of South America*, R.E. Reis, S.O. Kullander and C.J. Ferraris (eds.). EdiPUCs, Porto Alegre, pp. 497–502.

Albert, J.S. 2003b. Family Sternopygidae. In: *Check List of the Freshwater Fishes of South America*, R.E. Reis, S.O. Kullander and C.J. Ferraris (eds.). EdiPUCs, Porto Alegre, pp. 487–491.

Albert, J.S. and W.G.R. Crampton. 2003. Family Hypopomidae. In: *Check List of the Freshwater Fishes of South America*, R.E. Reis, S.O. Kullander and C.J. Ferraris (eds.). EdiPUCs, Porto Alegre, pp. 494–496.

Albert, J.S., F.M.C. Fernandes-Matioli and L.F. Almeida-Toledo. 1999. New species of *Gymnotus* (Gymnotiformes, Teleostei) from Southeastern Brazil: Toward the deconstruction of *Gymnotus carapo*. *Copeia* 1999: 410–421.

Almeida-Toledo, L.F. 1978. Contribuição a citogenética dos Gymnotoidei (Pisces, Ostariophysi). Ph.D. Thesis. University of São Paulo, SP Brazil.

Almeida-Toledo, L.F., F. Foresti and S.A. Toledo-Filho. 1981. Constitutive heterochromatin and nucleolus organizer region in the knifefish *Apteronotus albifrons* (Pisces, Apteronotidae). *Experientia* 37: 953–954.

Almeida-Toledo, L.F., F. Foresti and S.A. Toledo-Filho. 1984. Complex sex chromosome system in *Eigenmannia* sp. (Pisces, Gymnotiformes). *Genetica* 64: 165–169.

Almeida-Toledo, L.F., F. Foresti and S.A. Toledo-Filho. 1985. Spontaneus triploidy and NOR activity in *Eigenmannia* sp. (Pisces, Sternopygidae) from the Amazon basin. *Genetica* 66: 85–88.

Almeida-Toledo, L.F., E. Viegas-Péquignot, F. Foresti, S.A. Toledo-Filho and B. Dutrillaux. 1988. BrdU replication patterns demonstrating chromosome homeologies in two fish species, genus *Eigenmannia*. *Cytogenetics and Cell Genetics* 48: 117–120.

Almeida-Toledo, L.F., M.F.Z. Daniel, F. Foresti and S.A. Toledo-Filho. 1990. Diferença na localização das regiões organizadoras de nucléolo em duas populações próximas de *Eigenmannia*. *Proceedings of the III Simpósio de Citogenética Evolutiva e Aplicada de Peixes Neotropicais*, p. 34.

Almeida-Toledo, L.F., M.F.Z. Daniel, S.A. Toledo-Filho, F. Foresti and O. Moreira-Filho. 1991. Estudos citogenéticos comparativos em *Sternopygus macrurus* (Gymnotoidei, Sternopygidae) das bacias Amazônica, São Francisco e Paraná. *Proceedings of the IX Encontro Brasileiro de Ictiologia*, p. 188.

Almeida-Toledo, L.F., M.F.Z. Daniel and S.A. Toledo-Filho. 1993. Nucleolar chromosome variants in *Sternopygus macrurus* (Pisces, Sternopygidae) from three Brazilian river basins. *Caryologia* 46: 53–61.

Almeida-Toledo, L.F., F. Foresti and S.A. Toledo-Filho. 1994. Nota sobre o cariótipo, padrão de bandas C e RONs de *Apteronotus* sp. (Pisces, Gymnotidae). *Proceedings of the V Simpósio de Citogenética Evolutiva e Aplicada de Peixes Neotropicais*, p. 20.

Almeida-Toledo, L.F., M.F.Z. Daniel, F. Foresti and S.A. Toledo-Filho. 1995. Sistema de cromossomos sexuais do tipo $X_1X_1X_2X_2$:X_1X_2Y em *Hypopomus* sp. (Gymnotoidei, Hypopomidae). *Revista Brasileira de Genética* 18 (3, Supplement): 465.

Almeida-Toledo, L.F., A.J. Stocker, F. Foresti and S.A. Toledo-Filho. 1996. Fluorescence *in situ* hybridization with rDNA probes on chromosomes of two nucleolus organizer region phenotypes of a species of *Eigenmannia* (Pisces, Gymnotoidei, Sternopygidae). *Chromosome Research* 4: 301–305.

Almeida-Toledo, L.F., M.F.Z. Daniel-Silva, C.E. Lopes and S.A. Toledo-Filho. 1998. Sondas teloméricas $(TTAGGG)_n$ e coloração por fluorocromos dos cromossomos sexuais de *Brachyhypopomus pinnicaudatus*. *Proceedings of the VII Simpósio de Citogenética Evolutiva e Aplicada de Peixes Neotropicais*, p. B24.

Almeida-Toledo, L.F., M.F.Z. Daniel-Silva and S.A. Toledo-Filho. 1999. Heterocromatina rica em GC em duas espécies do gênero *Eigenmannia* (Pisces, Gymnotiformes). *Genetics and Molecular Biology* 22 (3, Supplement): 39.

Almeida-Toledo, L.F., F. Foresti, M.F.Z. Daniel-Silva and S.A. Toledo-Filho. 2000a. Sex chromosome evolution in fish: The formation of the neo-Y chromosome in *Eigenmannia* (Gymnotiformes). *Chromosoma* 109: 197–200.

Almeida-Toledo, L.F., M.F.Z. Daniel-Silva, C.E. Lopes and S.A. Toledo-Filho. 2000b. Sex chromosome evolution in fish. Second occurrence of a X_1X_2Y sex chromosome system in Gymnotiformes. *Chromosome Research* 8: 335–340.

Almeida-Toledo, L.F., F. Foresti, E. Viegas-Péquignot and M.F.Z. Daniel-Silva. 2001a. XX:XY sex chromosome system with X heterochromatinization: an early stage of sex chromosome differentiation in the Neotropic electric eel *Eigenmannia virescens*. *Cytogenetics and Cell Genetics* 95: 73–78.

Almeida-Toledo, L.F., C.B. Moysés, M.F.Z. Daniel-Silva and C.E. Lopes. 2001b. Diversidade de espécies e distribuição geográfica do gênero *Eigenmannia* (Pisces, Gymnotiformes): Caracterização citogenética e molecular de três citótipos do Rio São Francisco, MG. *Proceedings of the 47° Congresso Nacional de Genética*, p. GA165.

Almeida-Toledo, L.F., M.F.Z. Daniel-Silva, C.B. Moysés, S.B.A. Fonteles, A. Akama and F. Foresti. 2002. Chromosome evolution in fish: Sex chromosome variability in *Eigenmannia virescens* (Gymnotiformes: Styernopygidae). *Cytogenetic and Genome Research* 99: 164–169.

Almeida-Toledo, L.F., C.B. Moysés, M.F.Z. Daniel-Silva and C.E. Lopes. 2004. Diversidade de espécies e distribuição geográfica do gênero *Eigenmannia*: caracterização citogenética e molecular de três citótipos do Rio São Francisco. *Proceedings of the X Simpósio de Citogenética Evolutiva e Aplicada de Peixes Neotropicais*, p. 15.

Bertollo, L.A.C., C.S. Takahashi, L.F. Almeida-Toledo, P.M. Galetti Jr, I. Ferrari, O. Moreira-Filho and F. Foresti. 1980. Estudos citogenéticos em peixes da Região Amazônica. I. Ordem Cypriniformes. *Ciência e Cultura* 32 (Supplement): p. 735.

Bordenave, S., P.A. Lopes, M.C. Pastori de Beletrami and A.S. Fenocchio. 1992. Levantamento citogenético no Rio Paraná (Chaco, Argentina). III. *Gymnotus carapo* (Pisces, Gymnotoidei). *Proceedings of the IV Simpósio de Citogenética Evolutiva e Aplicada de Peixes Neotropicais*, p. 42.

Borin, L.A. and H.F. Júlio-Junior. 1994. Ocorrência de um novo citótipo de *Gymnotus carapo* (Siluriformes, Gymnotoidei) no Alto Paraná. *Proceedings of the V Simpósio de Citogenética Evolutiva e Aplicada de Peixes Neotropicais*, p. 43.

Campos da Paz, R. 1997. *Sistemática e taxonomia de peixes elétricos das bacias dos rios Paraguai, Paraná e São Francisco, com notas sobre espécies presentes em rios costeiros do leste do Brasil (Teleostei: Ostariophysi: Gymnotiformes)*. Ph.D. Thesis. University of São Paulo, SP Brazil.

Campos da Paz, R. 2003. Family Gymnotidae. In: *Check List of the Freshwater Fishes of South America*, R.E. Reis, S.O. Kullander and C.J. Ferraris (eds.). EdiPUCs, Porto Alegre, pp. 483–486.

Cereali, S.C., M.C. Navarrete and O. Froelich. 2002. Análise cariotípica de duas espécies do gênero *Brachyhypopomus* (Pisces, Gymnotiformes) do Pantanal do Miranda-Abobral, Corumbá, MS. *IX Simpósio de Citogenética e Genética de Peixes*, p. 90.

Denton, T.E. 1973. *Fish Chromosome Methodology*. Charles C. Thomas, Chicago, USA.

Fernandes, C.C., J. Podos and J.G. Lundberg. 2004. Amazon Ecology: Tributaries enhance the diversity of electric fishes. *Science* 305: 1960–1962.

Fernandes, F.M.C., J.S. Albert, M.F.Z. Daniel-Silva, C.E. Lopes, W.G.R. Crampton and L.F. Almeida-Toledo. 2005. A new *Gymnotus* (Teleostei: Gymnotiformes: Gymnotidae) from the Pantanal Matogrossense of Brazil and adjacent drainages: continued documentation of a cryptic fauna. *Zootaxa* 933: 1–14.

Fernandes-Matioli, F.M.C. 1999. *Evolução e estrutura de populações no gênero Gymnotus (Pisces: Gymnotiformes)*. Ph. D. Thesis. University of São Paulo, SP.

Fernandes-Matioli, F.M.C. and L.F. Almeida-Toledo. 2001. A molecular phylogenetic analysis in *Gymnotus* species (Pisces: Gymnotiformes) with inferences on chromosome evolution. *Caryologia* 54: 23–30.

Fernandes-Matioli, F.M.C., L.F. Almeida-Toledo and S.A. Toledo-Filho. 1997. Extensive nucleolus organizer polymorphism in *Gymnotus carapo* (Gymnotoidei, Gymnotidae). *Cytogenetics and Cell Genetics* 78: 236–239.

Fernandes-Matioli, F.M.C., M.C.M. Marchetto, L.F. Almeida-Toledo and S.A. Toledo-Filho. 1998. High intraspecific karyological conservation in four species of *Gymnotus* (Pisces: Gymnotiformes) from Southeastern Brazilian basins. *Caryologia* 51: 221–234.

Fernandes-Matioli, F.M.C., S.R. Matioli and L.F. Almeida-Toledo. 2000. Species diversity and geographic distribution of *Gymnotus* (Pisces: Gymnotiformes) by nuclear (GGAC)$_n$ microsatellite analysis. *Genetics and Molecular Biology* 23: 803–807.

Ferraris Jr., C.J. 2003. Family Rhamphichthyidae. Family Gymnotidae. In: *Check List of the Freshwater Fishes of South America*, R.E. Reis, S.O. Kullander and C.J. Ferraris (eds.). EdiPUCs, Porto Alegre, pp. 483–486.

Foresti, F. 1987. *Estudos Cromossômicos em Gymnotiformes (Pisces, Ostariophysi)*. Tese de Livre Docência. Universidade Estadual Paulista, Botucatu, SP.

Foresti, F., L.F. Almeida-Toledo and S.A. Toledo-Filho. 1984. Chromosome studies in *Gymnotus carapo* and *Gymnotus* sp. (Pisces, Gymnotidae). *Caryologia* 37: 141–143.

Gupta, M., Y.S. Chyi, J. Romero-Severson and J.L. Owen. 1994. Amplification of DNA markers from evolutionarily diverse genomes using single primers of simple-sequence repeats. *Theoretical Applied Genetics* 89: 998–1006.

Hinegardner, R. and D.E. Rosen. 1972. Cellular DNA content and the evolution of teleostean fishes. *American Naturalist* 106: 621–644.

Howell, W.M. 1972. Somatic chromosomes of the black goast knifefish *Apteronotus albifrons* (Pisces: Apteronotidae). *Copeia* 1972: 191–193.

Knudsen, E.I. 1974. Behavioral thresholds to electric signals in high frequency electric fish. *Journal of Comparative Physiology* 91: 333–353.

Lundberg, J.G., W.M. Lewis, J.F. Saunders and F. Mago-Leccia. 1987. A major food web component in the Orinoco river channel: evidence from planktivorous electric fishes. *Science* 237: 81–83.

Mago-Leccia, F. 1994. Electric fishes of the continental waters of America. *Biblioteca de la Academia de Ciencias Fisicas Matematicas y Naturales*, Caracas, Venezuela, 29: 1–206.

Marchetto, M.C.N., F.M.C. Fernandes-Matioli, L.F. Almeida-Toledo. 1998. Estudos citogenéticos comparativos em populações de *Gymnotus pantherinus* (Pisces, Gymnotidae). *Proceedings of the VII Simpósio de Citogenética Evolutiva e Aplicada de Peixes Neotropicais*, p. B23.

Medaglia, A., M.M. Cestari and A. Fenocchio. 1995. Análises citogenéticas preliminares em *Electrophorus electricus* (Pisces, Electrophoridae). *Revista Brasileira de Genética* 18 (3 Supplement): 460.

Nakayama, C.M., J.I.R. Porto and E. Feldberg. 2000. Comparação cariotípica entre *Apteronotus anas* e *Apteronotus hasemani* (Apteronotidae, Gymnotiformes) da Bacia Amazônica Central. *Proceedings of the VIII Simpósio de Citogenética e Genética de Peixes*, p. 77.

Nelson, J.S.. 1994. *Fishes of the World*. 3rd Edition. John Wiley & Sons, Inc, New York.

Reis, R.E., S.O. Kullander, C.J. Ferraris Jr (Org). 2003. *Check List of Freshwater Fishes of South America*. EdiPUCs, Porto Alegre.

Rolón, A. 2002. Estudios citogenéticos de *Gymnotus carapo* (Pisces, Gymnotidae) del río Paraná médio (Argentina). *Proceedings of the IX Simpósio de Citogenética e Genética de Peixes*, p. 94.

Silva, E.B. and V.P. Margarido. 2001. Descrição de três citótipos em simpatria de *Gymnotus* (Pisces, Gymnotiformes) coletados no Rio Paraná, região de Guaíra, PR. *Proceedings of the 47° Congresso Nacional de Genética*, GA 176.

Silva, E.B., O. Moreira-Filho and V.P. Margarido. 2002a. Sistema de cromossomos sexuais múltiplos $X_1X_1X_2X_2$:X_1X_2Y em *Gymnotus* sp. (Pisces, Gymnotiformes) coletados no Rio Paraná, região de Guaíra, PR. *Proceedings of the IX Simpósio de Citogenética e Genética de Peixes*, p. 92.

Silva, E.B., O. Moreira-Filho and V.P. Margarido. 2002b. Ocorrência de três espécies de *Gymnotus* em simpatria, no Rio Paraná, na região de Guaíra, PR. *Proceedings of the IX Simpósio de Citogenética e Genética de Peixes*, p. 93.

Souza, L.O., C. Oliveira and F. Foresti. 1997. Estudo citogenético e avaliação do DNA nuclear em *Gymnotus carapo* (Pisces, Gymnotiformes). *Revista Brasileira de Genética* 20 (3, Supplement)A166, p. 102.

Vênere, P.C. 1998. *Diversificação cariotípica em peixes do médio Rio Araguaia, com ênfase em Characiformes e Siluriformes (Teleostei, Ostariophysi)*. Ph.D. Thesis. Universidade Federal de São Carlos, São Carlos, SP Brazil.

The Evolutionary Role of Hybridization and Polyploidy in an Iberian Cyprinid Fish—A Cytogenetic Review

Marta Gromicho and Maria João Collares-Pereira

INTRODUCTION

The development of cytogenetics, in particular of modern molecular techniques, has allowed a deeper knowledge on genomes structure and organization, providing a useful tool for understanding the mechanisms of evolution. However, the continuum of molecular-based events that have shaped the evolution of eukaryotic chromosomes, namely the nature and pattern of the frequent high amount of repetitive DNA, continue to pose a challenge for evolutionary biologists (Eichler and Sankoff, 2003).

In systematic studies, the classic typological concept of karyotype, prevalent for some decades, has been replaced by a dynamic one, which

Address for Correspondence: Universidade de Lisboa, Faculdade de Ciências, Departamento de Biologia Animal/Centro de Biologia Ambiental, Campo Grande, 1749-016 Lisboa, Portugal. E-mail: mlsilva@fc.ul.pt; mjpereira@fc.ul.pt

acknowledges the variation at both inter-individual and interpopulational levels. However, when the available data sets on fish karyotypes are analyzed, it is clear that they are still very incomplete (only ≈10-15% of all taxonomically known species were karyotyped) and frequently based on either a single or very few specimens (e.g., Klinkhardt *et al.*, 1995). Therefore, cytotaxonomists are now requested to increase their sampling procedure, if possible using non-destructive and higher resolution methods, in order to characterize the existing variability within species at the chromosome level.

In doing so, the possibility of finding either chromosome rearrangements or genome mutations like polyploidy in distinct animal groups, namely in those more plastic (*sensu* Venkatesh, 2003) with numerical variations, will also certainly increase. It is the case of fish and some other lower vertebrates, much less constrained in general than birds and mammals by their reproductive and ontogenetic strategies, i.e., external fertilization and development processes. As pointed out by Otto and Whitton (2000), polyploidy in higher vertebrates may generate severe abnormalities and early mortalities due to developmental disruptions probably related to changes in cell size and/or disruption of the balance between both maternally and paternally imprinted genes.

Moreover, these specific strategies are likely to favour the occurrence of hybridization between closely and even distantly related taxa (Arnold, 1997; Yakovlev *et al.*, 2000). Also, it is now well documented that as postulated by Schultz (1969), hybrids may overcome the meiotic barrier by altering the biparental mode of reproduction to a non-sexual mode, or even asexual manner, i.e., non-recombinant (Beukeboom and Vrijenhoek, 1998; revised by Dawley, 1989). Therefore, experimental crosses need to be performed, by applying molecular markers, in order to confirm whether both paternal and maternal genomes were indeed transmitted to progenies and the extent of genetic recombination produced. They will allow to evidence eventual exceptions to both classical bisexual reproduction and euploidy (e.g., Stöck *et al.*, 2002), which may be much more common than presently recognized.

The Iberian *Squalius alburnoides* (Steindachner) complex discovered by Collares-Pereira (1983), mainly based on karyological data, is a good example of what was previously stated as regards the need of both a wider population screening in karyotype analysis and of performing crossing experiments under controlled conditions. Indeed, combining both the use of cytogenetic and genetic markers—namely mini- and microsatellites with alleles diagnostic for parental genomes—contributed to the identification of both the mode of reproduction and the genomic composition of the different ploidies (2n-3n-4n) and to establish genome inter-relationships within this hybridogenetic complex.

Before presenting the state-of-the-art of the cytogenetics of this case of evolution-in-action by hybridization and polyploidy, it is important to briefly mention the problems related to nomenclature faced by taxonomists when dealing with non-Mendelian organisms. Major controversies concerning their nomenclature start from their hybrid origin and frequent dependence upon mating with Mendelian species (sympatric bisexual relatives which they 'parasite') (revised by Echelle, 1990). Theoretically, the lack of a gonochoristic reproduction precludes the use of the so-called biological-species-concept, with the exception of true clones (parthenogenetically produced). These are self-perpetuating entities (genetically autonomous, even with a hybrid multiple origin), with a specific evolutionary history and fate (reviewed by Cole, 1985), whereas gynogens, androgens and hybridogens involve constant and recurrent genetic parasitism.

Hotz *et al.* (1996) revised some of the proposals forwarded to formally designate hybrid clonal taxa and they recommended the use of the existing species category instead of introducing new categories or concepts in the International Code of Zoological Nomenclature, as suggested by several authors. In a pragmatic manner, they proposed to ensure stability by using the former designations (e.g., *Poecilia formosa* and *Rana esculenta*, which were described with a binomial status, or the hyphenated biotypes option—*Poeciliopsis monacha-2lucida* preferred by geneticists). More recently, Turner (1999) included 'unisexual species' in his revision on the nature of fish species and further discussed the application of the several species concepts. The adoption of the Mayrian concept was also proposed for such cases, once it is based on a testable hypothesis (demonstrable reproductive isolation) if molecular techniques are used on a population framework. However, to facilitate communication, a way to formally designate such apparently transient situations (in particular gynogenetics and hybridogenetics with multiple hybrid origins) needs to be adopted until these reproductive modes, their inter-relationships as well as the role of sex in their maintenance, will be fully understood. Therefore, at this stage, the use of the term 'complex' as done for the Iberian hybridogenetic fish by Collares-Pereira (1983, 1984) or of the expression 'species complex' as suggested by Echelle (1990)—when it is known that the complex comprises different species—should be recommended. They will mainly point out to the remaining (non-specialist) scientific community, the existence of distinct hybrids and ploidy forms, with atypical reproductive modes, under the designation of an apparently single 'biological entity'.

THE *SQUALIUS ALBURNOIDES* COMPLEX

General Description

Squalius alburnoides complex is an endemic minnow to the Iberian Peninsula and comprises diploid (2n=50), triploid (3n=75) and tetraploid (4n=100) females and males with different genome compositions. Studies on its origin revealed that S. *alburnoides* originated by way of interspecific hybridization (Alves *et al.*, 1997a; Carmona *et al.*, 1997) between *Squalius pyrenaicus* (Günther) (P genome) and an unknown species (A genome). Hybridization events were multiple and occurred over a long period of time—which may be traced back to the Upper Pliocene—but were unidirectional, once the mitochondrial DNA (mtDNA) of S. *alburnoides* is closely related to S. *pyrenaicus* mtDNA (Cunha *et al.*, 2004).

Presently, this hybrid minnow incorporates genomes from both the sympatric bisexual S. *pyrenaicus* in the southern river basins and from S. *carolitertii* (Doadrio) (C genome) in northern river basins. All specimens of the complex carry S. *pyrenaicus* mtDNA, indicating that this species was most likely the maternal ancestor of the complex (Alves *et al.*, 1997b). Diploid males, presenting allozyme genotypes (AA) predicted for the 'missing' ancestor, were collected from some localities of the southern basins. However, their mtDNA was also identical to those of sympatric S. *pyrenaicus*, suggesting that these nuclear nonhybrid males were reconstituted from the non-sexual hybrids (Alves *et al.*, 2002).

Therefore, S. *alburnoides* complex comprises two distinct forms: the most common with diploid and polyploid hybrid females and males; and another form constituted by the all-male diploid nuclear nonhybrid lineage. These two genetically distinct forms are readily identifiable through a few morphological characters, corresponding to the two forms defined by Collares-Pereira (1983, 1984). Form A includes specimens with hybrid genomes independently of their ploidy level, whereas form B comprises diploid nuclear nonhybrid specimens. The sex-ratio of the nonhybrid form observed until now is fully consistent with the data of the author, which found no female among 192 fish of form B captured in the Tejo, Guadiana and Sado drainages.

Triploid females predominate in most populations (in average 75%-85%), but some differences concerning the frequencies of genome combinations have also been found among drainages (reviewed in Alves *et al.*, 2001). Ploidy mosaicism (revised in Lamatsch *et al.*, 2002) was never detected among the hundreds of S. *alburnoides* specimens analyzed until now by way of flow cytometry methodology (Próspero and Collares-Pereira, 2000, unpublished data).

Hybrid females and males (with the exception of triploid males whose fertility could not be tested yet due to their rarity in natural populations) are fertile and exhibit distinct reproductive modes according to their ploidy level being involved in mechanisms of non-sexual reproduction (Fig. 1.3.1). Diploid hybrid females and males (PA) clonally transmit their genomes and produce diploid clonal eggs and sperm (PA). Rarely, some PA females (less than 3%) may reproduce through gynogenesis. Triploid females (PAA) reproduce by 'meiotic hybridogenesis' *sensu* Alves *et al.* (1998), where the genome of S. *pyrenaicus* (P) is discarded in each generation without recombination and the remaining two genomes (AA) undergo meiosis, in the process, producing haploid eggs and occasionally

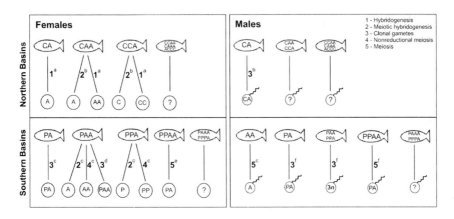

Fig. 1.3.1 Described reproductive features and ploidy of gametes of S. *alburnoides* complex in northern and southern populations (C = S. *carolitertii* genome; P = S. *pyrenaicus* genome; A = genome derived from the missing ancestor); a — Carmona *et al.* (1997); b — Pala and Coelho (2005); c — Alves *et al.* (1998); d — Alves *et al.* (2004); e — Gromicho and Collares-Pereira (2004); f — Alves *et al.* (1999) (see text for explanation).

also diploid eggs by an altered nonreductional meiosis. Symmetric tetraploid females and males (PPAA) produce normal diploid eggs and sperm (PA) via the usual meiosis. Nuclear diploid nonhybrid males (AA) exhibit normal meiosis producing haploid sperm (reviewed in Alves *et al.*, 2001). Recently, a new reproductive mechanism was found in a triploid female that generated both large triploid and small haploid eggs simultaneously, which after being fertilized with *S. pyrenaicus* sperm, originated diploid and tetraploid progeny, respectively (Alves *et al.*, 2004). According to Carmona *et al.* (1997) in the northern Douro drainage, allozyme patterns of mature primary oocytes indicated that diploid (CA) and triploid (CAA) females reproduce via hybridogenesis, discarding the *carolitertii* genome during oogenesis. However, the recent microsatellite analysis performed by Pala and Coelho (2005) of both natural populations and some experimental crosses from another northern basin (Mondego) revealed that triploid females represented 85% of the sampled fish and transmitted only one A genome to the offspring, evidencing that they also reproduce by meiotic hybridogenesis. These authors have also demonstrated the fact that diploid hybrid males (CA) transmit their genome clonally to offspring as the PA males from southern basins. However, since diploid nonhybrid males (AA) were not found, CA males may have a more important role regarding the complex maintenance in northern populations than PA in southern ones. Besides, diploid hybrid males were much more abundant than diploid hybrid females (CA) which were rare, conversely to the situation described in the south for the PA females.

The above described modes of reproduction imply continuous shifting between forms, where genomes derived from both parental ancestors are cyclically lost, gained or replaced (Alves *et al.*, 2001), and evidence the flexibility of this non-sexual system. Although sexually reproducing populations of the paternal ancestor may still exist, namely in some unexplored independent small drainages, in the three mainly surveyed basins (Tejo, Guadiana and Sado), the paternal species is most likely extinct. In these basins, the all-male lineage is apparently a significant piece in the dynamics of these hybrid populations, and has allowed the preservation of the ancestral paternal nuclear genome, which is likely to be closely related to *Anaecypris hispanica* (Steindachner), based on sequence analysis of nuclear genes (see Alves *et al.*, 2001; Robalo *et al.*, 2006).

Cytogenetic Data

The present review is based on the analysis of both fish from southern natural populations and the parents and progenies of numerous experimental crosses, using fibroblast culture procedure to ensure attaining higher-resolution metaphases. The genome composition of the fish was acquired through the use of genetic markers, namely variable microsatellite loci with some alleles diagnostic of each parental genome, as described in Gromicho and Collares-Pereira (2004).

Standard Karyotype

S. *alburnoides* complex and its maternal ancestor S. *pyrenaicus* have similar karyotypes composed mainly of biarmed chromosomes (meta- and submetacentric) compared to uniarmed (subtelo- and acrocentric) chromosomes. The biggest st-a chromosome pair, a chromosome marker for the Leuciscinae subfamily (Rab and Collares-Pereira, 1995) is one of the few pairs clearly distinguishable, as the remaining are mainly biarmed chromosomes, slightly decreasing in size with centromeres varying from median to almost terminal position.

Assigning the chromosomes into the main categories was difficult prior to obtaining high quality metaphases by fibroblast cell culture, which allowed some minor adjustments to the previously reported karyotypes of S. *alburnoides* complex by Collares-Pereira (1985) and of S. *pyrenaicus* by Collares-Pereira *et al.* (1998) (Gromicho and Collares-Pereira, 2004). As regards S. *pyrenaicus*, chromosome counts were invariably 2n=50, with 8 pairs of metacentric (m), 14 pairs of submetacentric (sm) and 3 pairs of subtelocentric to acrocentric (st-a) chromosomes, the largest pair of the set being st-a. Diploid nonhybrid males (AA) had a karyotype composed of 8 m pairs, 15 sm pairs and 2 st-a pairs (2n=50, NF=96) (Fig. 1.3.2a, b). Flow cytometry measurements of erythrocyte DNA content of S. *pyrenaicus* (PP), diploid nonhybrid males (AA) and diploid hybrids (PA) revealed no significant differences between them: 2.41±0.02, 2.43±0.06, and 2.44±0.08 pg/cell, respectively, supporting the similarity between the genomes of the bisexual ancestors (Collares-Pereira and Moreira da Costa, 1999; Próspero and Collares-Pereira, 2000).

Despite the observed similarity in karyotype structure between S. *pyrenaicus* and diploid nuclear nonhybrid males, the number of st-a

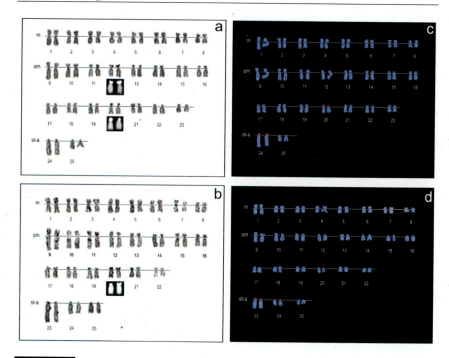

Fig. 1.3.2 Silver stained karyotypes of (a) *S. alburnoides* diploid nonhybrid male (AA) and (b) *S. pyrenaicus* male (PP) with detail of the four and two chromosomes with CMA$_3$$^+$/Ag$^+$ sites, respectively; karyotypes of (c) *S. alburnoides* diploid nonhybrid male (AA) and (d) *S. pyrenaicus* male (PP), both with six 5S rDNA sites.

pairs is apparently distinct. Although Collares-Pereira *et al.* (1998) hypothesized the existence of slightly differentiated sex chromosomes in *S. pyrenaicus* with the females being ZW heterogametic, no clearly heteromorphic chromosomes could be easily recognized in the karyotypes obtained as well as in *S. alburnoides* specimens.

Silver (Ag) and Chromomycin (CMA$_3$) Staining

The well-known pitfall in producing banding patterns on fish chromosomes has turned the nucleolus organizer regions (NORs) visualized indirectly through silver staining (Ag-NORs) as the most studied region on fish chromosomes. Ag-based techniques detect NORs by staining a complex of residual acidic protein associated with the fibrillar centre of the nucleolus and nascent pre-RNA, such as nucleolin (Jordan, 1987). It is commonly accepted that NORs on fish chromosomes can also be identified by fluorescent staining with chromomycin antibiotic

(CMA_3) that binds preferentially to GC-rich chromatin segments characteristic of rDNA sites of teleostean fishes (e.g., Mayr *et al.*, 1985; Amemiya and Gold, 1986; Schmid and Guttenbach, 1988). Correlations between Ag-NORs and CMA_3 positive sites have been reported for several fish species (e.g., Sola *et al.*, 2000; Jankun *et al.*, 2001, 2003; Rabova *et al.*, 2001, 2003), and their number and location have been among the few chromosome markers used in fish cytotaxonomy and phylogenetic analysis to date (e.g., Amemiya and Gold, 1990; Boron, 2001; Rabova *et al.*, 2001, 2003).

Since a full correspondence between CMA_3- and Ag-NOR positive marks and sites of major ribosomal genes was not completely established for our specimens, as further discussed below, we have designated them as Ag and CMA_3 positive (CMA_3^+/Ag^+) sites.

CMA_3^+ and Ag^+ sites were investigated in the *S. alburnoides* complex and in *S. pyrenaicus* through sequential staining, enabling us to evidence their correspondence in most metaphases.

In all the 52 specimens analyzed (9 diploid PA hybrids, 14 triploid PAA hybrids, 3 triploid PPA hybrids, 15 diploid AA nonhybrid males, 7 tetraploid PPAA hybrids and 4 *S. pyrenaicus* males), the CMA_3^+ and Ag^+ sites covered nearly the entire short arms of medium- to small-sized submetacentric chromosomes. While *S. pyrenaicus* presents only one pair of CMA_3^+/Ag^+ sites, both forms of the *S. alburnoides* complex are characterized by multiple CMA_3^+/Ag^+ sites.

Stability, variability and inheritance of CMA_3^+ and Ag^+ sites were studied by means of crossing experiments involving most genome compositions of the complex present in southern populations (Table 1.3.1). Cross analysis showed that the distribution pattern of multiple CMA_3^+ and Ag^+ sites in *S. alburnoides* complex can be explained by a Mendelian model of inheritance and includes at least three chromosome pairs, one being constant for the presence of CMA_3^+ and Ag^+ sites and the two other pairs presenting a CMA_3^+ and Ag^+ site polymorphism of presence/absence type (Fig. 1.3.3) (Gromicho and Collares-Pereira, 2004).

Crossing experiments revealed to be of major importance in understanding the banding patterns and disclosing the structure and dynamics of hybrid genomes. Together with microsatellite markers, CMA_3 and Ag patterns were also used to infer the inheritance of the additional reproductive mode of a PAA triploid female by Alves *et al.* (2004), which

Table 1.3.1   Summary of cytogenetic data from crossing experiments. G — Guadiana, T — Tejo, P — S. pyrenaicus genome, A — paternal ancestor genome, f — female, m — male.

River basin	FEMALES	Number of		Genomes (Gametes)	Number of crosses	MALES	Genomes (Gametes)	Number of		OFFSPRING Genomes	Number of	
		Ag^+ sites	CMA_3^+ sites					Ag^+ sites	CMA_3^+ sites		Ag^+ sites	CMA_3^+ sites
T	S. alb. 2n (n=1)	2-3	3	PA (PA)	1	S. alb. 2n (n=1)	PA (PA)	2-3	3	PPAA (n=3)	3-5	6
G	S. alb. 2n (n=2)	2-4	2 - 4	PA (PA)	4	S. alb. 2n (n=1)	AA (A)	3-5	4-5	PAA (n=8f)	3-6	4-6
G	S. alb. 3n (n=1)	4-5	5	PAA (A)	1	S. pyr. (n=1)	PP (P)	2	2	PA (n=5, 3f+2m)	1 - 3	3
T	S. alb. 4n (n=1)	5-6	6	PPAA (PA)	1	S. pyr. (n=1)	PP (P)	2	2	PPA (n=3)	3-5	4-5
G	S. alb. 3n (n=3)	3-5	5	PAA (A)	4	S. alb. 2n (n=0)	AA (A)	—	—	AA (n=9m)	3-5	4-5
T	S. alb. 4n (n=0)	—	—	PPAA (PA)	1	S. alb. 4n (n=1)	PPAA (PA)	4-5	6	PPAA (n=1)	5-6	6
T	S. alb. 3n (n=0)	—	—	PAA (A)	1	S. alb. 4n (n=1)	PPAA (PA)	5-6	6	PAA (n=2, 1f+1m)	4-5	4-5
G						S. alb. 2n (n=5)	AA (A)	2-5	4-5			
G						S. pyr. (n=2)	PP (P)	2	2			

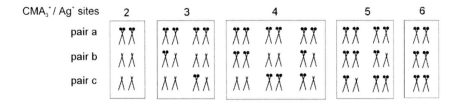

Fig. 1.3.3 Schematic representation of possible locations and CMA_3^+/Ag^+ numbers associated with AA genome, considering that CMA_3^+/Ag^+ sites are spread in three chromosome pairs: one chromosome pair with stable CMA_3^+/Ag^+ sites (pair a) and the other two chromosome pairs with CMA_3^+/Ag^+ site polymorphism of presence/absence type (pairs b and c) (from Gromicho and Collares-Pereira, 2004).

generated tetraploid (PPAA) and diploid (PA) progenies when fertilized with *S. pyrenaicus* sperm. The female and male parents exhibited 5 and 2 NOR-bearing chromosomes, respectively. The two diploid offspring analyzed cytogenetically exhibited four CMA_3^+ and Ag^+ sites, while the three tetraploid offspring exhibited 6 CMA_3^+ and Ag^+ sites. These numbers agree with the two reproductive mechanisms associated to this female: haploid eggs were probably produced by meiotic hybridogenesis as described by Alves *et al.* (1998), where the elimination of the unmatched genome allowed formation of bivalents and two meiotic divisions, while tetraploids were produced by normal fertilization of clonal triploid eggs.

Major and Minor Ribosomal Genes

The application of DNA markers as probes for fluorescence *in situ* hybridization (FISH) provides a new approach with wide applicability to chromosome studies allowing the comparison of genomes from different species. Moderately repeated DNA sequences such as the ribosomal genes have been the most used probes for *in situ* hybridization on fish chromosomes.

In higher eukaryotes, ribosomal genes (rDNAs) are organized as two distinct multigene families (major and minor ribosomal genes) comprising tandemly arrayed repeats composed of hundreds to thousands of copies. The major ribosomal genes (45S rDNA), coding for the 18S, 5.8S and 28S rRNAs, are related to NORs. The other family (5S rDNA) codes for the 5S rRNA and has extranucleolar location (reviewed in Long and David, 1980).

Simultaneous chromosomal mapping of the two gene families onto the karyotype of *S. alburnoides* complex and *S. pyrenaicus* was performed with double-colour FISH (db FISH) using 28S and 5S rDNA sequences as probes, in order to investigate their mutual relationship, which has proven to be an important source of information for karyological analysis.

No variation in the number of FISH signals was observed with the 5S rDNA probe that mapped constantly to 3 chromosomes per haploid genome, i.e., 2n=6, 3n=9 and 4n=12 signals (Fig. 1.3.2 c, d). 5S rDNA genes appeared to have a stable location as they were observed in a similar position on morphologically identical chromosomes, including the big acrocentric pair, which was always labeled.

Instead, in all the specimens studied, prominent intraindividual variations were detected in number, size and location of FISH signals with the 28S rDNA probe. Variation in the location of the major ribosomal genes within individuals was detected from that in the number of syntenic sites of the two genes combined with the information of those located in the big acrocentric pair (one of the few chromosome pairs from the complement that can be clearly distinguished).

Although quite variable in number and location, and not being the most frequent condition found in fishes (Martins and Wasko, 2004), syntenic location of the major and the 5S rDNA clusters was found in all the specimens studied, with a minimum of one syntenic site in AA and PAA fish and a maximum of ten sites in PPAA fish (Table 1.3.2). When both ribosomal genes share the same chromosome, they apparently overlap (Fig. 1.3.4) (Gromicho et al., 2006).

What was the Income of Recent Cytogenetic Data?

Although Collares-Pereira et al. (1998) hypothesized a ZW/ZZ sex chromosome heteromorphism for the sympatric species in northern drainages (S. carolitertii) along with even less conspicuousness for the species involved in the origin of the complex (S. pyrenaicus, the maternal ancestor, PP), no well-differentiated heteromorphic sex-chromosomes have been found on revisiting the previous descriptions of the different ploidy forms including AA and PP karyotypes. Moreover, in none of the distinct ploidies analyzed by the reported banding procedures sex-specific patterns were observed in number or site, either in parents or progenies.

Results from crossing experiments showed that the mechanism of sex determination in S. alburnoides cannot be fully explained by both simple sex-systems, i.e., neither by a female nor a male heterogamety (reviewed by Alves et al., 2001). Whereas the exclusive female offspring in crosses

Table 1.3.2 Summary of the number of major (28S) and minor (5S) ribosomal genes loci assigned by double FISH in the S. *alburnoides* complex and in S. *pyrenaicus*.

	28S			Syntenic sites	
	Range	*Average*	*5S*	*Range*	*Average*
AA	4 – 9	6.2	6	1 – 6	3.8
PA	4 – 8	6.4	6	3 – 6	4.0
PAA	6 – 12	8.3	9	1 – 9	4.6
PPA	5 – 11	7.5	9	3 – 8	4.8
PPAA	10 – 16	13.3	12	5 – 10	7.4
S. *pyr.*	4 – 9	6.1	6	2 – 6	3.6

between PA females and AA males could be in accordance with the former model, the offspring sex-ratio from PAA females seems to be dependent on the genome they receive from their father. In other words, while AA males only produce males, if the male parent is a S. *pyrenaicus*, then mixed broods of females and males are produced. Therefore, Alves et al. (1998, 1999) hypothesized that sex determination within the complex could be dependent on the existence of non-W-linked genes expressed differently, depending on the parents and population to which they belong.

However, the recent detection of a small 5S rDNA cluster near the centromeric/pericentromeric region of a single m chromosome in some metaphases from two S. *pyrenaicus* females (Gromicho et al., 2006) may suggest the existence of a primitive stage of sex chromosomes system differentiation towards a ZW/ZZ system in the maternal ancestor (PP). This system may have reached a higher differentiation in S. *carolitertii* according to Collares-Pereira et al. (1998) observations. Although more research is necessary with sex-specific probes to confirm the proposed model, the autosomic dependence of sex determination in the paternal ancestor (AA) coupled with the female heterogamety of S. *carolitertii* might even help to explain the paucity of forms (CAA, CCA and CA) and corresponding sex-ratios recently found by Pala and Coelho (2005) in Mondego drainage.

FISH with a 28S rDNA probe followed by sequential CMA_3 and Ag staining demonstrated a general lack of correspondence between classical and molecular cytogenetic techniques in all the specimens studied. Despite the positive correlation between CMA_3 and Ag, we clearly evidenced for the first time in fishes that not just other regions besides ribosomal genes were stained with these techniques, but also that the

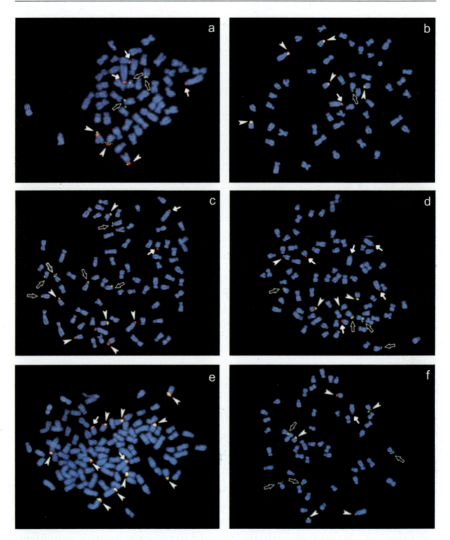

Fig. 1.3.4 Results of double-color FISH using biotin-labeled 28S rDNA (green) and digoxigenin-labeled 5S rDNA (red) probes on chromosomes counterstained with DAPI (blue) from (a) *S. pyrenaicus* and (b–f): *S. alburnoides* complex with different genotypes: (b) diploid nonhybrid male (AA); (c) triploid hybrid female (PAA); (d) triploid hybrid (PPA); (e) tetraploid hybrid (PPAA) and (f) diploid hybrid female (PA). Empty arrows: major ribosomal genes sites; Arrows: 5S rDNA sites; Arrowheads: syntenic major and minor ribosomal genes sites.

majority of the 28S rDNA sites were not detected (Fig. 1.3.5) (Gromicho *et al.*, 2005).

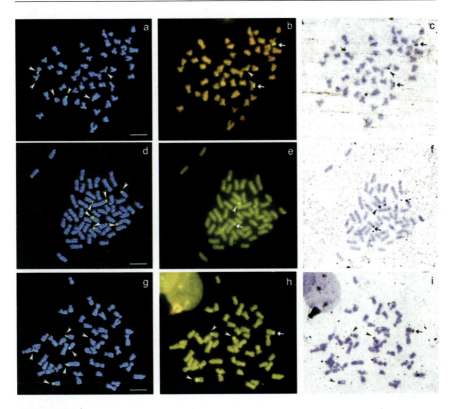

Fig. 1.3.5 Sequentially stained metaphases with a 28S rDNA probe by FISH (a, d, g), CMA$_3$-staining (b, e, h) and Ag-NOR staining (c, f, i). Metaphases with CMA$_3^+$/Ag$^+$ sites without correspondence with the major rDNA sites from (a–c): *S. alburnoides* diploid nonhybrid male (AA); (d–f): *S. pyrenaicus* male; (g–i): *S. alburnoides* diploid hybrid female (PA); Arrowheads: major ribosomal genes sites and correspondent CMA$_3^+$/Ag$^+$ sites; Arrows: CMA$_3^+$ / Ag$^+$ sites non-correspondent to major ribosomal genes sites. Scale bar = 5 μm.

It is well known that hybridization and polyploidy are associated with extensive genome plasticity and that these mechanisms introduce genome reorganizations that may also be linked with chromosomal changes and rearrangements (Soltis and Soltis, 2000). But if this could help to explain the results obtained for *S. alburnoides*, it does not apply to *S. pyrenaicus*, which still contributes to the complex persistence but is a 'normal' bisexual species. Peculiar heterochromatin characteristics associated with the high polymorphic nature of major ribosomal genes might be responsible for the obtained results. Thus, biochemical composition of heterochromatin and other possible factors that lead to these odd patterns need to be further investigated.

Moreover, with the growing outcome of results that doubt upon the reliability of CMA$_3$ and Ag to stain NORs (e.g., Vitturi *et al.*, 1999; Margarido and Galetti Jr., 2000; Dobigny *et al.*, 2002), care is advisable in considering Ag-NORs and CMA$_3$-stained sites as true indicators of major ribosomal gene sites. Whenever chromosomes cannot be clearly identified, such correspondence should always be tested through sequential banding on the same metaphases. The application of this methodology would also be important in determining whether staining of rDNA-free regions by CMA$_3$ and Ag is more frequent than originally predicted (Gromicho *et al.*, 2005). Nevertheless, CMA$_3$ and Ag staining proved to be useful chromosome markers and in spite of their high polymorphic nature in this fish complex, cross analyses evidenced that we have no reason to reject the hypothesis of Mendelian segregation based on inheritance patterns.

Notwithstanding, the CMA$_3$ and Ag banding pattern observed suggests that the karyotype of the missing paternal species of the *S. alburnoides* complex should have a multiple pattern supporting the hypothesized ancestry from *Anaecypris hispanica*-like individuals by Alves *et al.* (2001) and Robalo *et al.* (2006) M. Gromicho and M.J. Collares-Pereira, unpublished data) and emphasized the role of the all-male diploid lineage in preserving the missing ancestor genome, as previously suggested by Alves *et al.* (2002).

It is difficult to ascertain whether any of the chromosome pairs with CMA$_3{}^+$ and Ag$^+$ sites present in *S. alburnoides* AA males is homologous of the pair of *S. pyrenaicus* and, if so, to determine if it is the homomorphic pair or one of the polymorphic pairs (Fig. 1.3.2 a, b). The question also arises in inferring whether this potentially homologous pair is a consequence of the parental species of the *S. alburnoides* complex being closely related (common ancestry) or a clue for the existence of introgression through recombination between the two genomes.

Regarding the location of 5S rDNA in similar *sm* and *st-a* chromosomes in both *S. alburnoides* and *S. pyrenaicus*, it is in accordance with several results which have pointed out the conserved location of 5S rDNA sequences in closely related fish species (reviewed in Martins and Wasko, 2004).

Despite the great variability observed with 28S rDNA probe, it is possible to notice the similarity between *S. alburnoides* complex and *S. pyrenaicus* in terms of location and number of both ribosomal genes families, conversely to what was found with silver and CMA$_3$ staining (Gromicho and Collares-Pereira, 2004).

However, we must keep in mind that such interpretations are limited by the difficulty to clearly recognize each chromosome pair due to their morphological similarity and lack of evident banding patterns (precluded by the lack of genome compartmentalization in fish—Sumner, 1990). Thereafter, all morphology-based chromosome homologies remain ambiguous. The application of colour karyotyping to fish species will allow rapid and easier karyotype comparisons in species notoriously difficult to analyze. Chromosome painting (Multi-FISH) can be used to identify homologous chromosome segments in hybrid genomes and determine genome constitution in polyploids. The introduction of this technique into comparative cytogenetics has added significantly to the understanding of chromosome changes that occurred during the evolution of species (Wienberg et al., 1990). Future work will obviously also be aimed at increasing the resolution of S. alburnoides karyotype analyses by integrating chromosome arm and chromosome band-specific painting, and locus-specific, single-copy probe sets (see Ried et al., 1998).

To our knowledge, no other leuciscin fishes were mapped for 5S rRNA gene localization and, therefore, comparisons are precluded. Despite Cyprinidae being the most speciose fish family, only five other cyprinids were studied vis-a-vis their 5S localization (reviewed in Martins and Wasko, 2004). The described numbers of 5S rDNA loci were either two or four, mainly with interstitial position. This is considerably different from what we have observed since none had such a high number of 5S RNA loci, all with a terminal position as in S. alburnoides complex and in S. pyrenaicus.

Variation in the location of the major ribosomal genes within an individual is an uncommon discovery difficult to interpret but one might speculate about its eventual causes (see Gromicho et al., 2006). In the fish Leporinus friderici, intra-individual variations in the number of NORs were also identified—both with CMA$_3$-staining and FISH—what would not be expected unless independent structural changes occurred from cell to cell (Galetti et al., 1995). Aberrant transpositions events or an extremely high frequency of transposable elements may explain this phenomenon due to their ability to generate mutations and to influence genomic organization and gene expression (Kidwell, 2002; Lönnig and Saedler, 2002). These mobile elements can directly inactivate genes by insertions in their coding regions or serve as a substrate for DNA recombination due to their repeated nature. This can induce mainly interchromosomal rearrangements such as duplications, deletions, amplifications, inversions, or translocations, which may participate in the genome evolution (see Jaillon *et al.*, 2004). Inversions and translocations can also place a gene into a new genomic context and consequently modify its expression (Volff *et al.*, 1999).

Although major ribosomal genes have been insufficiently analyzed in Eurasian cyprinids, one single small *sm* chromosome pair bearing the major rDNA sites in its short arms is the most common situation for the studied Leuciscinae species (Rabova *et al.*, 2003 and references therein) and has been considered as a plesiomorphic condition by the authors. Multichromosomal locations, considered a derived condition, were already described for *Eupallasella percnurus* (Boron *et al.*, 1997), *Chondrostoma lusitanicum* (Rodrigues and Collares-Pereira, 1996; Collares-Pereira and Rab, 1999) and *Phoxinus phoxinus* (Boron, 2001). Even though it should be pointed out that none have displayed such a high number of major ribosomal gene sites (the 28S rDNA probe mapped to short-arms of in average six *sm* and *st-a* chromosomes in diploid specimens), and that it has never been found any site on the largest *st-a* chromosome pair (Rabova *et al.*, 2003 and references therein). Therefore, the studied Iberian cyprinids, although sharing the general pattern of European Leuciscinae in terms of chromosome number and morphology, may be exceptions to the assumed typical NOR phenotype (number and location of major rDNA sites).

These cytogenetic approaches revealed the major importance of crossing experiments and sequential banding treatments to understand banding patterns and to disclose the structure and dynamics of hybrid genomes, being highly recommended that such methodologies be adopted in future analyses together with more powerful cytogenetic molecular tools.

THE EVOLUTIONARY ROLE OF HYBRIDIZATION AND POLYPLOIDY

Evolutionary biology is now facing some important challenges which include, among others, the demonstration of the actual role and extension of hybridization and polyploidy processes in past and present animal evolution and diversification. Combining fields as diverse as paleontology, developmental and molecular biology with comparative genome analysis have opened new insights into these old paradigms, still almost only recognized with a primary role in the evolutionary history of plants (revised in Mable, 2003, 2004).

In fishes, heterospecific insemination may directly produce new taxa, with or without introgression (for revision see Dowling and Secor, 1997; Turner, 1999; Allendorf et al., 2001) or produce polyploid animal lineages, mainly through non-sexual modes of reproduction, which allow the removal of the meiotic barrier (Schultz, 1969). In fact, all described cases with altered gametogenesis have a hybrid origin and almost all of them form complexes with polyploid genomes (Vrijenhoek, 1998). However, the distinction between such allopolyploids and autopolyploids (in which the multiple sets of chromosomes are derived from the same ancestral species) is not a simple question for cytogeneticists.

This is due to the fact that both types become functionally diploid over time (e.g., White, 1978; Ohno, 1999; Wolfe, 2001), through a restoration process of diploid-like chromosomal behaviour, with extensive chromosomal rearrangement and gene silencing and loss. Genome reshaping in such 'degenerate polyploids' (Soltis and Soltis, 1999) involving intra and intergenomic reorganizations may be so rapid and dynamic (especially in allopolyploids, as experimentally evidenced in *Brassica* after some few generations—references in Soltis and Soltis, 1999) that only the most recent events are likely to be easily recognized as evolutionary polyploids. A paradigmatic example is the maize, which only recently has been proved to be an ancient allotetraploid (Gaut and Doebley, 1997) instead of a 'diploid' species. Indeed, the detection of such situations relies on the possibility of identifying the intergenomic chromosomal rearrangements by chromosome painting and gene mapping.

Although non-sexual lineages were previously considered as evolutionary dead-ends (White, 1978) and 'from no-hopers to hopeful monsters' (Vrijenhoek, 1989), this view has been questioned by the finding of very ancient lineages using molecular clocks both in invertebrates and vertebrates (see references in Judson and Normark, 1996; e.g., Spolsky et al., 1992; Schartl et al., 1995; Little and Herbert, 1997). Their evolutionary success depends on the ability to generate stable and reproductively isolated lineages, which may recover the bisexuality once the allotriploid (ABB) eventually becomes a symmetrical tetraploid (AABB). More than this indirect route to polyploidy a direct and quicker route may exist, if an unreduced hybrid oocyte (AB) is fertilized by an unreduced sperm (AB). In most non-sexual complexes, gene exchange among bisexual forms occurs and polyploidization acts as a mechanism for generating whole-genome redundancy, which has the potential for increasing gene divergence and acquisition of new functions (e.g., Holland, 1999; Ohno, 1999; Wendel, 2000; Venkatesh, 2003; Zhang, 2003), and may well purge the mutational 'melt-down' predicted by the Muller's ratchet mechanism.

These aspects have been well documented for decades in plant evolution, mainly in angiosperms and pterydophytes, once synthetic auto- and allopolyploids are easy to obtain and analyze in long-term series for life-history attributes and ecological parameters. Whereas, in animals, they have been overlooked mainly due to longstanding negative attitudes (revised in Dowling and Secor, 1997; Mable, 2004). However, after the general review made by Otto and Whitton (2000), developments in this field have attracted greater attention and two recent revisions were devoted in particular to polyploidy in fish, although with distinct perspectives (Leggatt and Iwama, 2003; Le Comber and Smith, 2004).

Thus, polyploidy is presently hypothesized for Petromyzontiformes (namely Petromyzontidae and Geotriidae) and confirmed in Chondrichthyes, Sarcopterygii and Actinopterygii in a total of 12 distinct orders (Leggatt and Iwama, 2003). Quite recently, Jaillon *et al.* (2004) found evidences of genome duplication followed by a massive loss of duplicated genes in the smallest known vertebrate genome *Tetraodon nigroviridis*, and of an ancient whole-genome-duplication in the teleost fish lineage subsequent to its divergence from mammals, deducing that the extinct ancestor had 24 chromosomes. Moreover, in certain Osteichthyes families (e.g., Cyprinidae, Cobitidae, and Characidae), polyploidy is clearly a recurrent process, with multiple origins (e.g., Tsigenopoulos *et al.*, 2002; Janko *et al.*, 2003; Cunha *et al.*, 2004).

As concerns the evolutionary potential of the Iberian fish complex, the presence of tetraploid females and males in natural populations follows the hypothesis made by Schultz (1969) that the restoration of an even ploidy value in tetraploids would allow a return to normal meiosis, serving as a stepping-stone to biparental reproduction. Although tetraploid individuals of S. *alburnoides* complex seem to be still uncommon, they may re-establish sexual reproduction, leading eventually to the formation of a new polyploid species if the isolating mechanisms required for speciation arise. The two alternative routes to tetraploidization may exist: although the most common route might be the syngamy of diploid clonal eggs and diploid clonal sperm produced by diploid hybrid males (Alves *et al.*, 1999), this route may be precluded from some southern basins, where diploid hybrid males have not yet been collected (see Alves *et al.*, 2001). In these basins, the possible route to tetraploidy may involve the normal fertilization of triploid clonal eggs, as described in Alves *et al.* (2004) following the most common process in non-sexual vertebrates (Dawley, 1989). This cross produced apparently all females, with a symmetric genome (PPAA). Such females should produce PA eggs (Alves *et al.*, 2001) although in natural populations, a return to triploidy in the next generation may also occur, through fertilization with either *S. pyrenaicus* (PP) or nonhybrid (AA) males.

There are now evidences pointing out that the *S. alburnoides* complex is likely to have been through different evolutionary constraints in southern versus northern drainages, since nuclear nonhybrid males have not yet been found in the later (see Pala and Coelho, 2005). Although distinct evolutionary scenarios may be hypothesized, the complex cannot be considered evolutionary condemned, especially in southern basins where the genetic diversity is much higher, as formerly predicted by Collares-Pereira (1989). On the contrary, as stated earlier, hybridization seems to have started an evolutionary process that may ultimately lead to a new sexually reproducing species (2n=100).

The *Squalius alburnoides* complex certainly constitutes a good example on how hybridization and subsequent genome multiplication processes coupled with mechanisms that prevent the loss of genetic variability during the transient stages can play a significant role in animal speciation. It should be used as a reference case that justifies the demand for more and wider-scope cytogenetic studies. Doing so, we might expect that, especially in lower vertebrates, similar cases that have been just ignored due to their difficult detection would be found. Moreover, this will certainly be a sound contribution for acknowledging the extension of the evolutionary role of hybridization and polyploidy in animal diversification.

Acknowledgements

We are especially grateful to M.J. Alves for helping with the crossing experiments, P. Ráb, M. Rábová and C. Ozouf-Costaz for receiving one of us (M.G.) in their laboratories to improve technical aspects of banding techniques and image analysis, M.M. Coelho for unpublished information regarding specimens genotypes, C. Bonillo for probe preparation and J.P. Coutanceau for technical assistance with FISH. We also thank Direcção Geral das Florestas for permission to collect specimens. This work was supported by Fundação para a Ciência e a Tecnologia (Portugal) and Fundo Social Europeu (UE) through the project POCTI/BSE/40868/2001 and by a grant to M.G. (SFRH/BD/1028/2000).

References

Allendorf, F.W., R.F. Leary, P. Spruell and J.K. Wenburg. 2001. The problems with hybrids: Setting conservation guidelines. *Trends in Ecology and Evolution* 16: 613–622.

Alves, M.J., M.M. Coelho and M.J. Collares-Pereira. 1997a. The *Rutilus alburnoides* complex (Cyprinidae): Evidence for a hybrid origin. *Journal of Zoological Systematics and Evolutionary Research* 35: 1–10.

Alves, M.J., M.M. Coelho, M.J. Collares-Pereira and T.E. Dowling. 1997b. Maternal ancestry of the *Rutilus alburnoides* complex (Teleostei, Cyprinidae) as determined by analysis of cytochrome *b* sequences. *Evolution* 51: 1584–1592.

Alves, M.J., M.M. Coelho and M.J. Collares-Pereira. 1998. Diversity in the reproductive modes of females of the *Rutilus alburnoides* complex (Teleostei, Cyprinidae): A way to avoid the genetic constraints of uniparentalism. *Molecular Biology and Evolution* 15: 1233–1242.

Alves, M.J., M.M Coelho, M.I. Próspero and M.J. Collares-Pereira. 1999. Production of fertile unreduced sperm by hybrid males of the *Rutilus alburnoides* complex (Teleostei, Cyprinidae): An alternative route to genome tetraploidization in unisexuals. *Genetics* 151: 277–283.

Alves, M.J., M.M. Coelho and M.J. Collares-Pereira. 2001. Evolution in action through hybridisation and polyploidy in an Iberian freshwater fish: A genetic review. *Genetica* 111: 375–385.

Alves, M.J., M.J. Collares-Pereira, T.E. Dowling and M.M. Coelho. 2002. The genetics of maintenance of an all-male lineage in the *Squalius alburnoides* complex. *Journal of Fish Biology* 60: 649–662.

Alves, M.J., M. Gromicho, M.J. Collares-Pereira, E. Crespo-López and M.M. Coelho. 2004. Simultaneous production of triploid and haploid eggs by triploid *Squalius alburnoides* (Teleostei: Cyprinidae). *Journal of Experimental Zoology* 301A: 552–558.

Amemiya, T. and J.R. Gold. 1986. Chromomycin A$_3$ stains nucleolus organizer regions of fish chromosomes. *Copeia* 1986: 226–231.

Amemiya, T. and J.R. Gold. 1990. Cytogenetic studies in North American minnows (Cyprinidae). XVII. Chromosomal NORs phenotypes of 12 species, with comments on cytosystematic relationships among 50 species. *Hereditas* 112: 231–247.

Arnold, M.L. 1997. *Natural Hybridization and Evolution*. Oxford University Press, Oxford.

Beukeboom, L.W. and R.C. Vrijenhoek. 1998. Evolutionary genetics and ecology of sperm-dependent parthenogenesis. *Journal of Evolutionary Biology* 11: 755–782.

Boron, A. 2001. Comparative chromosomal studies on two minnow fish, *Phoxinus phoxinus* (Linnaeus, 1758) and *Eupallasella perenurus* (Pallas, 1814), and associated cytogenetic-taxonomic considerations. *Genetica* 111: 387–395.

Boron, A., M. Jankun and J. Kusznierz. 1997. Chromosome study of swamp minnow *Eupallasella perunurus* (Dybowski, 1916) from Poland. *Caryologia* 50: 85–90.

Carmona, J.A., O.I. Sanjur, I. Doadrio, A. Machordom and R.C. Vrijenhoek. 1997. Hybridogenetic reproduction and maternal ancestry of polyploid Iberian fish: The *Tropidophoxinellus alburnoides* complex. *Genetics* 146: 983–993.

Cole, C.J. 1985. Taxonomy of parthenogenetic species of hybrid origin. *Systematic Zoology* 34: 359–363.

Collares-Pereira, M.J. 1983. *Estudo sistemático e citogenético dos pequenos ciprinídeos ibéricos pertencentes aos géneros Chondrostoma Agassiz,1835, Rutilus Rafinesque,1820 e Anaecypris Collares-Pereira,1983.* Ph.D. Thesis University of Lisbon (In Portuguese).

Collares-Pereira, M.J. 1984. The '*Rutilus alburnoides* (Steind, 1866) complex' (Pisces, Cyprinidae). I. Biometrical analysis of some Portuguese populations. *Arquivos do Museu Bocage* Sér. A 2: 111–143.

Collares-Pereira, M.J. 1985. The '*Rutilus alburnoides* (Steind, 1866) complex' (Pisces, Cyprinidae). II. First data on the karyology of a well-established diploid-triploid group. *Arquivos do Museu Bocage* Sér. A 3: 69–89.

Collares-Pereira, M.J. 1989. Hybridization in European cyprinids: evolutionary potential of unisexual populations. In: *Evolution and Ecology of Unisexual Vertebrates*, R.M. Dawley and J.P. Bogart (eds.). *New York State Museum Bulletin No.466*, New York, pp. 281–288.

Collares-Pereira, M.J. and L. Moreira da Costa. 1999. Intraspecific and interspecific genome size variation in Iberian Cyprinidae and the problem of diploidy and polyploidy, with review of genome sizes within the family. *Folia Zoologica* 48: 61–76.

Collares-Pereira, M.J. and P. Rab. 1999. NOR polymorphism in the Iberian species *Chondrostoma lusitanicum* (Pisces: Cyprinidae)—Re-examination by FISH. *Genetica* 105: 301–303.

Collares-Pereira, M.J., M.I. Próspero, R.I. Biléu and E.M. Rodrigues. 1998. *Leuciscus* (Pisces, Cyprinidae) karyotypes: Transect of Portuguese populations. *Genetics and Molecular Biology* 21: 63–69.

Cunha, C., M.M. Coelho, J.A. Carmona and I. Doadrio. 2004. Phylogeographical insights into the origins of the *Squalius alburnoides* complex via multiple hybridization events. *Molecular Ecology* 13: 2807–2817.

Dawley, R.M. 1989. An introduction to unisexual vertebrates. In: *Evolution and Ecology of Unisexual Vertebrates*, R.M. Dawley and J.P. Bogart (eds.). *New York State Museum Bulletin* No. 466, New York, pp. 1–18.

Dobigny, G., C. Ozouf-Costaz, C. Bonillo and V. Volobouev. 2002. 'Ag-NORs' are not always true NORs: New evidence in mammals. *Cytogenetic and Genome Research* 98: 75–77.

Dowling, T.E. and C.L. Secor. 1997. The role of hybridization and introgression in the diversification of animals. *Annual Review of Ecology and Systematics* 28: 593–619.

Echelle, A.A. 1990. Nomenclature and non-mendelian ('clonal') vertebrates. *Systematic Zoology* 39: 70–78.

Eichler, E.E. and D. Sankoff. 2003. Structural dynamics of eukaryotic chromosome evolution. *Science* 301: 793–797.

Galetti, Jr. P.M., C.A. Mestriner, P.J. Monaco and E.M. Rasch. 1995. Post-zygotic modifications and intra- and inter-individual nucleolar organizing region variations in fish: Report of a case involving *Leporinus friderici*. *Chromosome Research* 3: 285–290.

Gaut, B.S. and J.F. Doebley. 1997. DNA sequence evidence for the segmental allopolyploid origin of maize. *Proceedings of the National Academy of Sciences of the United States of America* 94: 6809–6814.

Gromicho, M. and M.J. Collares-Pereira. 2004. Polymorphism of major ribosomal gene chromosomal sites (NOR-phenotypes) in the hybridogenetic fish *Squalius alburnoides* complex (Cyprinidae) assessed through crossing experiments. *Genetica* 122: 291–302.

Gromicho, M., C. Ozouf-Costaz and M.J. Collares-Pereira. 2005. Lack of correspondence between CMA3-, Ag-positive signals and 28S rDNA loci in two Iberian minnows (Teleostei, Cyprinidae) evidenced by sequential banding. *Cytogenetic and Genome Research* 109: 507-511.

Gromicho, M., J.P. Coutanceau, C. Ozouf-Costaz and M.J. Collares-Pereira. 2006. Contrast between extensive variation of 28S rDNA and stability of 5S rDNA and telomeric repeats in the diploid-polyploid *Squalius alburnoides* complex and in its maternal ancestor *Squalius pyrenaicus* (Teleostei, Cyprinidae). *Chromosome Research* 14: 297-306.

Holland, P.W.H. 1999. Gene duplication: Past, present and future. *Seminars in Cell & Developmental Biology* 10: 541–547.

Hotz, H., T. Uzzell, P. Beerli, and G.D. Guex. 1996. Are hybrid clonal species? A case for enlightened anarchy? *Amphibia-Reptilia* 17: 315–320.

Jaillon, O., M. Aury, F. Brunet, J.L. Petit, N. Stange-Thomann, E. Mauceli, L. Bouneau, C. Fischer, C. Ozouf-Costaz, A. Bernot, S. Nicaud, D. Jaffe, S. Fisher, G. Lutfalla, C. Dossat, B. Segurens, C. Dasilva, M. Salanoubat, M. Levy, N. Boudet, S. Castellano, V. Anthouard, C. Jubin, V. Castelli, M. Katinka, B. Vacherie, C. Biémont, Z. Skalli, L. Cattolico, J. Poulain, V. de Berardinis, C. Cruaud, S. Duprat, P. Brottier, J.P. Coutanceau, J. Gouzy, G. Parra, G. Lardier, C. Chapple, K.J. McKernan, P. McEwan, S. Bosak, M. Kellis, J.N. Volff, R. Guigó, M.C. Zody, J. Mesirov, K. Lindblad-Toh, B. Birren, C. Nusbaum, D. Kahn, M. Robinson-Rechavi, V. Laudet, V. Schachter, F. Quétier, W. Saurin, C. Scarpelli, P. Wincker, E.S. Lander, J. Weissenbach and H.R. Crollius. 2004. Genome duplication in the teleost fish *Tetraodon nigroviridis* reveals the early vertebrate proto-karyotype. *Nature* (London) 431: 946–957.

Janko, K., P. Kotlik and P. Rab. 2003. Evolutionary history of asexual hybrid loaches (Cobitis, Teleostei) inferred from phylogenetic analysis of mitochondrial DNA variation. *Journal of Evolutionary Biology* 16: 1280–1287.

Jankun, M., P. Martinez, B.G. Pardo, L. Kirtiklis, P. Rab, M. Rabova and L. Sanchez. 2001. Ribosomal genes in Coregonid fishes (*Coregonus lavaretus*, *C. albula* and *C. peled*) (Salmonidae): Single and multiple nucleolus organizer regions. *Heredity* 87: 672–679.

Jankun, M., K. Ocalewicz, B.G. Pardo, P. Martinez, P. Woznicki and L. Sanchez. 2003. Chromosomal characteristics of rDNA in European grayling *Thymallus thymallus* (Salmonidae). *Genetica* 119: 219–224.

Jordan, G. 1987. At the heart of the nucleolus. *Nature (Lond.)* 329: 489–499.

Judson, O.P. and B.B. Normark. 1996. Ancient asexuals scandals. *Trends in Ecology and Evolution* 11: 41–46.

Kidwell, M.G. 2002. Transposable elements and the evolution of genome size in eukaryotes. *Genetica* 115: 49–63.

Klinkhardt, M., M. Tesche and H. Greven. 1995. *Database of Fish Chromosomes*. Westarp-Wissenchaften, Magdeburg.

Lamatsch, D.K., M. Schmid and M. Schartl. 2002. A somatic mosaic of the gynogenetic Amazon molly. *Journal of Fish Biology* 60: 1417–1422.

Le Comber, S.C. and C. Smith. 2004. Polyploidy in fishes: Patterns and processes. *Biological Journal of the Linnean Society* 82: 431–442.

Leggatt, R.A. and G.K. Iwama. 2003. Occurrence of polyploidy in fishes. *Reviews in Fish Biology and Fisheries* 13: 237–246.

Little, T.J. and P.D.N. Hebert. 1997. Clonal diversity in high arctic ostracodes. *Journal of Evolutionary Biology* 10: 233–252.

Long, E.O. and I.D. David. 1980. Repeated genes in eukaryotes. *Annual Review of Biochemistry* 49: 727–764.

Lönnig, W.E. and H. Saedler. 2002. Chromosome rearrangements and transposable elements. *Annual Review of Genetics* 36: 389–410.

Mable, B.K. 2003. Breaking down taxonomic barriers in polyploidy research. *Trends in Plant Science* 8: 582–590.

Mable, B.K. 2004. 'Why polyploidy is rarer in animals than in plants': Myths and mechanisms. *Biological Journal of the Linnean Society* 82: 453–466.

Margarido, V.P. and P.M. Galetti Jr. 2000. Amplification of a GC-rich heterochromatin in the freshwater fish *Leporinus desmotes* (Characiformes, Anostomidae). *Genetics and Molecular Biology* 23: 569–573.

Martins, C. and A.P. Wasko. 2004. Organization and evolution of 5S ribosomal DNA in the fish genome. In: *Focus on Genome Research*. C.R. Williams (ed.). Nova Science Publishers, Hauppauge, New York, USA, pp. 289–318.

Mayr, B., P. Rab and M. Kalat. 1985. Localizaton of NORs and counterstain-enhanced fluorescence studies in *Perca fluviatilis* (Pisces, Percidae). *Genetica* 67: 51–56.

Ohno, S. 1999. Gene duplication and the uniqueness of vertebrate genomes circa 1970-1999. *Seminars in Cell & Developmental Biology* 10: 517–522.

Otto, S.P. and J. Whitton. 2000. Polyploid incidence and evolution. *Annual Review of Genetics* 34: 401–437.

Pala, I. and M.M. Coelho. 2005. Contrasting views over a hybrid complex: Between speciation and evolutionary 'dead-end'. *Gene* (In press).

Próspero, M.I. and M.J. Collares-Pereira. 2000. Nuclear DNA content variation in the diploid-polyploid *Leuciscus alburnoides* complex (Teleostei, Cyprinidae) assessed by flow cytometry. *Folia Zoologica* 49: 53–58.

Rab, P. and M.J. Collares-Pereira. 1995. Chromosomes of European cyprinid fishes (Cyprinidae: Cypriniformes): A review. *Folia Zoologica* 44: 193–214.

Rabova, M., P. Rab and C. Ozouf-Costaz. 2001. Extensive polymorphism and chromosomal characteristics of ribosomal DNA in a loach fish, *Cobitis vardarensis* (Ostariophysi, Cobitidae) detected by different banding techniques and fluorescent *in situ* hybridization (FISH). *Genetica* 111: 413–422.

Rabova, M., P. Rab, C. Ozouf-Costaz, C. Ene and J. Wanzebock. 2003. Comparative cytogenetics and chromosomal characteristics of ribosomal DNA in the fish genus *Vimba* (Cyprinidae). *Genetica* 118: 83–91.

Ried, T., E. Schröck, Y. Ning and J. Wienberg. 1998. Chromosome painting: A useful art. *Human Molecular Genetics* 10: 1619–1626.

Robalo, J.I., Sousa Santos, C., Levy, A. and V.C. Almada. 2006. Molecular insights on the taxonomic position of the paternal ancestor of the *Squalius alburnoides* hybridogenetic complex. *Molecular Phylogenetics and Evolution* 39: 276-281.

Rodrigues, E.M. and M.J. Collares-Pereira. 1996. NOR polymorphism in the Iberian species *Chondrostoma lusitanicum* (Pisces, Cyprinidae). *Genetica* 98: 59–63.

Schartl, M., B. Wilde, I. Schlupp and J. Parzefall. 1995. Evolutionary origin of a parthenoform, the Amazon molly *Poecilia formosa*, on the basis of a molecular genealogy. *Evolution* 49: 827–835.

Schmid, M. and M. Guttenbach. 1988. Evolutionary diversity of reverse (R) fluorescent bands in vertebrates. *Chromosoma* 97: 327–344.

Schultz, R.J. 1969. Hybridization, unisexuality and polyploidy in the teleost *Poeciliopsis* (Poeciliidae) and other vertebrates. *American Naturalist* 103: 605–619.

Sola, L., S. De Innocentiis, E. Gornung, S. Papalia, A.R. Rossi, G. Marino, P. De Marco and S. Cataudella. 2000. Cytogenetic analysis of *Epinephelus marginatus* (Pisces: Serranidae), with the chromosome localization of the 18S and 5S rRNA genes and of the $(TTAGGG)_n$ telomeric sequence. *Marine Biology* 137: 47–51.

Soltis, D.E. and P.S. Soltis. 1999. Polyploidy: recurrent formation and genome evolution. *Trends in Ecology and Evolution* 14: 348–352.

Soltis, P.S. and D.E. Soltis. 2000. The role of genetic and genomic attributes in the success of polyploids. *Proceedings of the National Academy of Sciences of the United States of America* 97: 7051–7057.

Spolsky, C.M., C.A. Philips and T. Uzzell. 1992. Antiquity of clonal salamander lineages revealed by mitochondrial DNA. *Nature (London)* 356: 706–708.

Stöck, M., D.K. Lamatsch, C. Steinlein, J. Epplen, W-R. Grosse, R. Hock, T. Klapperstück, K. Lampert, U. Scheer, M. Schmid and M. Schartl. 2002. A bisexually reproducing all-triploid vertebrate. *Nature Genetics* 30: 325–328.

Tsigenopoulos, C.S., P. Rab, D. Naran and P. Berrebi. 2002. Multiple origins of polyploidy in the phylogeny of southern African barbs (Cyprinidae) inferred from mtDNA markers. *Heredity* 88: 466–473.

Turner, G.F. 1999. What is a fish species? *Reviews in Fish Biology and Fisheries* 9: 281–297.

Venkatesh, B. 2003. Evolution and diversity of fish genomes. *Current Opinion in Genetics & Development* 13: 588–592.

Vitturi, R., M.S. Colomba, R. Barbieri and M. Zunino. 1999. Ribosomal DNA location in the scarab beetle *Thorectes intermedius* (Costa) (Coleoptera: Geotrupidae) using banding and fluorescent *in-situ* hybridization. *Chromosome Research* 7: 255–260.

Volff, J.N., C. Körting, K. Sweeney and M. Schartl. 1999. The non-LTR retrotransposon *Rex3* from the fish *Xiphophorus* is widespread among teleosteans. *Molecular Biology and Evolution* 16: 1427–1438.

Vrijenhoek, R.C. 1989. Genetic and ecological constraints on the origins and establishment of unisexual vertebrates. In: *Evolution and Ecology of Unisexual Vertebrates*, R.M. Dawley and J.P. Bogart (eds.). *New York State Museum Bulletin* No. 466, New York, pp. 24–31.

Vrijenhoek, R.C. 1998. Clonal organisms and the benefits of sex. In: *Advances in Molecular Ecology*, G.R. Carvalho (ed.). IOS Press, Amsterdam, pp. 151–172.

Wendel, J.F. 2000. Genome evolution in polyploids. *Plant Molecular Biology* 42: 225–249.

Wienberg, J., A. Jauch, R. Stanyon and T. Cremer. 1990. Molecular cytotaxonomy of primates by chromosomal *in situ* suppression hybridization. *Genomics* 8: 347–350.

White, M.J.D. 1978. *Modes of Speciation*. WH Freeman & Co., San Francisco.

Wolfe, K.H. 2001. Yesterday's polyploids and the mystery of diploidization. *Nature Reviews/Genetics* 2: 333–341.

Yakovlev, V.N., Yu.V. Slyn'ko, I.G. Grechanov and E.Yu. Krysanov. 2000. Distant hybridization in fish. *Journal of Ichthyology* 40: 298–311.

Zhang, J. 2003. Evolution by gene duplication: an update. *Trends in Ecology and Evolution* 18: 292–298.

1.4

Chromosomal Changes and Stasis in Marine Fish Groups

Wagner Franco Molina

INTRODUCTION

In spite of the important role of fish in the marine environment, where they are important members of the community acting on its structure through processes of predation, competition, and territoriality (Sale, 1991), little is known about the genetic evolutionary processes leading to such diversity (Nelson, 1994). For instance, absence of biological information makes it difficult to handle the conservation of a significant biodiversity reserve (Norse, 1993). The fauna inhabiting the reefs is very specialized, with intricate evolutionary patterns of predator-prey (Floeter *et al.*, 2004) and competitive interactions among the species (Robertson, 1984; Harrington, 1993). These characteristics are reflected in an exuberant diversity of forms and colors, as well as, in some cases, diverse dynamics of karyotype differentiation patterns.

Address for Correspondence: Departamento de Biologia Celular e Genética, Universidade Federal do Rio Grande do Norte, Campus Universitário, 59078-970 Natal, RN, Brazil.
E-mail: molinawf@yahoo.com.br

Genetic analyses in marine populations have been developed to address delimitation of the areas (Feral, 2002), species distribution (Carvalho, 1993), speciation processes (Rocha et al., 2005), the identification of cryptic species (Pfeiler et al., 2002), the measures of the inter-population gene flow (Taylor and Hellberg, 2003) and the determination of genetic variability (DeWoody and Avise, 1999). The information generated from these research activities has had direct applications for biological conservation.

Physical factors such as active currents, temperature and salinity contribute to the definition of biogeographic boundaries and ecological patterns of the formation and maintenance of the particular faunas. Approximately 60% of the living fishes are marine, with a larger diversity occurring in the tropics, where more than 4,000 species can be found in reefs of the Indo-Pacific Ocean and about 400 varieties in the Caribbean (Nelson, 1994). In reef environments, Perciformes have a significant abundance and play an ecological role. Most of the representatives of this group are the results of extensive radiation that occurred during the Tertiary age, along with the radiation of the coral reefs, of which Labridae, Scaridae, Pomacentridae, Acanthuridae, Siganidae, Chaetodontidae, and Pomacanthidae are among the most characteristic groups (Sale, 1991).

The analysis of the genetic variation in marine teleosts in relation to several biological parameters (habitat, depth, type of egg, distribution area and size), indicates the existence of a larger level of heterozygosis between generalists than specialists, but that this is associated with a great number of other factors influencing genetic variation (Smith and Fujio, 1982). The combined action of certain parameters such as larval behavior, spawning time, and ocean currents in meso- or microscales can make it difficult to identify any single factor as responsible for genetic (Shulman and Bermingham, 1995) and/or cytogenetic variability (Molina and Galetti, 2004).

Among all fish orders, the Perciformes constitute a potentially useful model in understanding the genetic structure of marine populations, as well as the establishment of their evolutionary patterns. In this group can be found examples of extensive chromosome conservativeness (stasis) with symplesiomorphic patterns shared by a great number of species (Klinkhardt et al.,1995), as well as tendencies for karyotype diversification along the evolutionary history of some groups (Brum and Galetti, 1997; Klinkhardt et al., 1998). Other favorable factors that should be considered are its diversity and wide distribution on reefs and rocky substrata of oceanic islands as well as over the continental shelf. The origin and/or cause of these observed cytogenetical patterns are not always clear, but they can sometimes be correlated with biological aspects of the groups.

Speciation in Marine Environments: A Chromosomal Approach

Understanding the processes of speciation is one of the aspects derived from the genetic studies in marine fishes, which only now have begun to be investigated more thoroughly. Speciation frequently can be promoted (stasipatric model; White, 1973) or is followed by karyotype rearrangements (King, 1987), although this is not always observed (Molina *et al.*, 2002). According to various authors (e.g. John, 1981; King, 1993), the chromosome rearrangements (such as inversions, translocations and heterochromatinization processes) frequently involving sex chromosomes can act as an effective post-mating isolating mechanism if they prevent or negatively affect the meiotic pairing and segregation of the hybrids.

When comparing the degree of karyotype diversification and variation between fish groups with representatives in the freshwater and marine environment, more karyotype derivatives are observed in association with the former environment than with the latter, often related to the FN (Fundamental Number) (Brum, 1995). These modifications are mainly associated with the largest amount of partitions in the continental environments, occasioned by a higher number of obstacles to interpopulational gene flow (Moreira-Filho and Bertollo, 1991). Despite the apparent lack of physical barriers in the marine environment, strict behavioral and ecological factors can represent effective population gene-flow barriers (Rocha *et al.*, 2005). Different from single-dimensional conditions of lotic or lentic continental systems, associated with historical conditions that allow inferences on the distribution of a species, multiple and complex factors act within spatial-temporal conditions in order to define the final patterns of formation and the maintenance of populations and species in the marine environment.

It is almost a rule that species with wide geographic distribution exhibit morphologic character modifications. Reef fishes distributed through wide geographical areas can present subtle color differentiations

(Planes and Doherty, 1997), even sensible modifications of their corporal patterns (Bell *et al.*, 1982). Such differentiations can be detected through morphometric analyses, useful in the identification of minimum variations among the stocks, with the possibility of providing a supplemental source of evidence of genetic differentiation and stratification. For instance, Molina (2000) has identified marked inter-population variations of the pomacentrid *Abudefduf saxatilis*, through the multivariate morphometry analyses of populations from coastal regions of Brazil (NE and SE of Brazil) and those from the Saint Paul's Rocks (SPR) and the Rocas Atoll (respectively, about 960 and 300 km from the NE coast of Brazil). Clines involving some serial elements have been observed between the populations from the NE and SE of the Brazilian coast, with larger values found in the northern populations and oceanic islands, suggesting that temperature is a principal factor of this gradient (Molina, 2000). Interpopulational analyses among coastal regions and oceanic islands NE of Brazil, using RAPD markers, offer similar results to the morphologic data. The coastal populations, although separated by a distance of approximately 1.000 km, present greater genetic similarity to each other than to the insular populations, though these are distributed within an equivalent geographical distance (Rodrigues, 2003).

The genus *Abudefduf* has karyotypes dominated by a large amount of acrocentric chromosomes (Arai and Inoue, 1976). Cytogenetic analyses carried out in *A. saxatilis* populations of the SE, NE of Brazil (Aguilar *et al.*, 1998; Molina and Galetti, 2004a) and Saint Paul's Rocks, separated by distances of 3,000 km, did not demonstrate identifiable karyotype differences (Fig. 1.4.1). Results indicate that despite morphological changes, some gene flow continues to propitiate genetic cohesion for this species or is possibly accompanied by unrecognizable chromosome rearrangements.

Marine species inhabiting different environments seem to have developed a mechanism of *divergence with gene flow* through local adaptations (Beheregaray and Sunnucks, 2001). Historical events or geomorphological aspects of the environment are in a position to influence this condition (Pampoulie *et al.*, 2004). In marine organisms, the high potential for dispersion can delay the establishment of reproductive barriers, allowing for the maintenance of connections between widely dispersed populations (Rocha *et al.*, 2002). Controlled by the intensity and direction of marine currents, the dispersion of many organisms in the larval phase reaches great distances, which may influence the

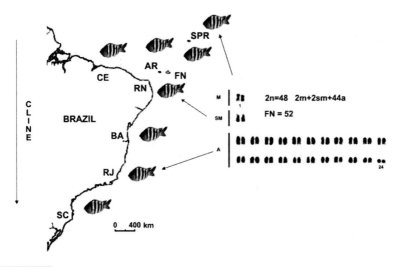

Fig. 1.4.1 Chromosomal conservativeness in morphologically differentiated populations of *Abudefduf saxatilis*, from coastal and insular regions of the Western Atlantic. RN — Rio Grande do Norte; SPR — the St. Paul's Rocks; AR — Atol das Rocas; FN — Fernando de Noronha Archipelago; CE — Ceará; BA — Bahia; RJ — Rio de Janeiro; SC — Santa Catarina.

biogeographical distribution of a species as well as its genetic patterns (Lacson, 1992; Molina and Galetti, 2004; Galetti *et al.*, 2005). In a general way, the absence of physical barriers frequently allows an extensive gene flow among the coastal populations of the Western Atlantic, contributing to a lesser degree of polymorphic chromosome/karyotype patterns in different stocks.

Among environments favorable to speciation processes, the oceanic islands and the atolls are outstanding and, in some cases, can act as a stepping-stone type model of dispersion. In the Atlantic Ocean, Briggs (1974) pointed out the importance of the Meso-Atlantic deep-water ridge as a strong obstacle for transatlantic dispersion. Only 9% of the total number of described species in this ocean are shown to have transatlantic dispersion, and none of them belong to species typical for the Eastern Atlantic. Similar to other vertebrate groups (Robbins *et al.*, 1983), the geographic isolation tends to favor the establishment of chromosome rearrangements during the speciation process.

Bellwood (1996) believes that the temporal and spatial continuity of reef environments could in part explain the conservative nature of the evolution of the coral reef fishes. However, in the case of more specialized species strictly associated with the reefs or inhabitants of shallow waters, the bio-ecological changes that occurred during the Pleistocene age (Planes et al., 1996) certainly promoted significant alterations in the historical processes related to the fauna. So, it is probable that fluctuations in the sea level have contributed to the speciation processes along the coast, as well as to the colonizing and endemism in the mid-Atlantic islands (Rocas Atoll's and St. Paul's Rocks). Climatic fluctuations due to changes in the shoreline in the last million years are an important source of geographical isolation that—in addition to particular current patterns—could provide transitory population isolation (Vermeij, 1978; Stepien and Rosenblatt, 1996), fomenting an efficient allopatric model for the formation of new species.

An obvious example of allopatry seems to have occurred in the insular speciation of *Stegastes sanctipauli* (2n=48; 22m+22sm+4a; FN=92) that are endemic to St. Paul's Rocks, about 1.000 km from the NE coast of Brazil, with one of the two coastal forms such as *S. variabilis* (2n=48; 18m+22sm+8a; FN=88) or *S. fuscus* (2n=48; 20m+22sm+6a; FN=90) (Molina and Galetti, 2004b) as probable ancestors. Besides sharing the same number diploid (2n=48), all these species had a homologous pair of NOR-bearing chromosomes (1st sm). The observed alterations suggest *S. fuscus* as the common ancestor for the insular species *S. sanctipauli*, due to the lower number of rearrangements involved in its differentiation (a pericentric inversion), as opposed to *S. variabilis*, from which a larger number of events would be necessary (two inversions).

In some groups, the karyotype structure seems to have resisted speciation when compared with morphological changes. A numerical and structural karyotype conservativeness (2n=48, FN=48), as extensively observed in some families of Perciformes such as Serranidae (Aguilar and Galetti, 1997; Sola et al., 2000), contrasts with remarkable morphological differentiations observed among the species. This asynchrony between endo-phenotype (karyotype pattern) and exo-phenotype (morphological traits) suggests that the speciation process in the karyotypically conservative marine groups may have been primarily established through mechanisms of pre-zygotic reproductive isolation (Molina et al., 2002).

The role of the chromosome rearrangements in speciation and adaptation in the marine environment is still debated in the face of a number of reports regarding balanced polymorphisms that were a result of structural heterozygotes without any apparent fitness reduction (Molina and Galetti, 2002). Some data stress the marginality of the role of chromosome rearrangements in marine teleost speciation (Caputo et al., 1996, 1997). The rearrangements that can arise in some populations may thus either remain in the form of balanced polymorphisms or become fixed by genetic drift as an allopatric speciation by-product without playing a causal role in the evolution of marine specific diversity (Caputo et al., 1999).

The existence of natural hybrids often provides an indication of still incipient speciation or the breakup of pre-zygotic reproductive isolation mechanisms.

Hybridization events have been reported for 56 fish families, mainly from observations in freshwater groups (Pyle and Randall, 1994). Interspecific crossings in the marine environment seem to be disfavored, as much by intense behavioral patterns as by great aggregations favoring crossings with individuals of the same species. In a general way, the majority of marine fish hybrids are extremely rare in nature and are probably the result of unusual circumstances (Pyle and Randall, 1994).

Examples of hybridization among several species of Perciformes (Allen, 1979) indicate that post-zygotic barriers are not completely effective between diverse groups and that karyotype micro-rearrangements of these species do not efficiently ensure genetic isolation from phylogenetically related forms. The failure of post-zygotic isolation suggests that the fixation of peculiar group behavioral patterns or ecological restrictions play an important role in marine fish speciation.

Among Serranidae, a marine group with conserved cytogenetic characteristics, the diversification in the genus *Hypoplectus* (hamlets) has been a subject of controversy (Fischer, 1980). In this genus exist several extensively sympatric and syntopic marine morphospecies, morphologically similar in all the diagnostic characters, differing only in the coloration patterns. Some evidence suggests that *hamlets* represent a species complex—a great complex of incipient sympatric species with a restricted exchange of genes between some morphospecies. The *hamlets* complex could be an example of speciation in progress as a result of pre-zygotic reproductive barrier reinforcement with origins in environmental fragmentations.

Complex intraspecific behavior patterns observed in some groups of reef fishes such as aggressive behavior, courting, signaling and morphological patterns among the individuals of a population can also favor increasing speciation rates (Vermeij, 1978). Hubbs (1970) related the great potential for hybridization in fishes—among other factors—to the dependence the break of weak ethologic mechanisms of reproductive isolation (Campton, 1987). However, in spite of this development, reef fishes seem to have significantly more complex ecological and ethological patterns than other fish groups (Sale, 1991). The partition of resources for the occupation of different marine microhabitats has been postulated in this environment to explain peculiarities in the habitat use that may be responsible in sustaining the diversity among the coral reef fish (Ormond et al., 1996).

Thus, the ethologic isolation of populations may constitute an efficient mechanism for the division of species in the absence of significant differentiation, as is suggested by the evolutionary conserved karyotypes of several perciform reef groups (Molina et al., 2002).

Geographic barriers, however, are responsible for most of the known speciation events. Isolated populations tend to fix chromosome rearrangements that can lead to a complete blockage of the gene flow. An example can be observed in two species of Eleotridae: *Dormitator maculatus* and *Eleotris pisonis*. Populations distributed along estuarine regions of the South Atlantic have 40m-sm+6st-a (2n=46; *D. maculatus*) and 46a (2n=46; *E. pisonis*), while on the Mexican coast of Central America, they are characterized by 34m-sm+12st-a (2n=46) and 2m-sm+42st-a (2n=44), respectively (Montes-Perez, 1981; Uribe-Alcocer and Ramirez-Scamilla 1989). The cytogenetic data indicated that the karyotype differences between the populations of northeastern Brazil and the Mexican coasts must be derived via pericentric inversions associated with translocations and fusions. The differences observed in the karyotype structure of the forms from South and Central America suggest the existence of sibling species (Molina, 2005). It is possible that physical barriers such as the delta of the Amazon River have restricted the gene flow between the Brazilian faunas and that of the Caribbean, and that such deterrents play an effective role in the formation of the species of both areas (Floeter and Gasparini, 2000; Rocha et al., 2002).

The diversity of species in marine environments has been explained as much by natural processes of allopatry that occur during the evolutionary history of some coastal areas (Molina, 2005), as by parapatric ecological speciation (Rocha et al., 2005; Sena and Molina, in press). However, the ethological complexity among some reef groups (Sale,

1991), in an effort to adapt to the dynamic coral reef environment allied to biological and oceanographic factors may have contributed to karyotypic stasis (i.e., a state of stability due to resistance to karyotype changes). Therefore, karyotypes with little differentiation would be reflexes of a scenario of intrinsic evolutionary dynamism (*sensu* Hawthorne and Via, 2001) mediated by environmental homogeneity, in a significant number of species.

Dispersion and Karyotype Changes

Majority of the fishes inhabiting coral reefs have a pelagic, dispersing larval stage, followed by a relatively sedentary adult life. Due to the sedentary condition of the adult stage, it has been accepted that the pelagic stage is responsible for most patterns of geographical distribution (Bonhomme and Planes, 2000). Most species endemic to the Brazilian Province (74%) as well as half of the endemic fish species of the tropical oceanic islands reproduce through demersal spawning with a short pelagic period and, consequently, restricted dispersion (Floeter and Gasparini, 2000).

The distances reached during the dispersal phases effect: (1) the gene flow and genetic differentiation of the species; (2) the extension and size of an intercrossing population; (3) the number of effective barriers to the gene flow and the potential for speciation; (4) the physical and biotic adaptive potential of local populations to environmental conditions; (5) the spatial scale of ecological communities; and (6) the geographical distribution of the species (Shulman and Bermingham, 1995). Populational genetic differentiations in marine fishes have been demonstrated using nuclear and mitochondrial DNA markers (Shulman and Bermingham, 1995) and allozymes (Doherty *et al.*, 1994). In contrast, several groups of fish maintain the same karyotype structure even over great geographic distances (Brum and Galetti, 1997).

Two main factors determine the genetic association of separate populations of coral reef fish through planktonic larval dispersion: the oceanographic factors and the life cycle of the species. In some Perciformes of the Caribbean region, no correlation was found between biological characteristics of the species (egg types, pelagic or not, and pelagic phase length) and the degree of genetic divergence among populations (Shulman and Bermingham, 1995), suggesting that while these factors play basic roles in the establishment of interpopulational gene flow, they may be masked by other factors, depending on environmental conditions.

A possible association between the larval pelagic period and chromosome rearrangements can be tested in some characteristic reef fish groups such as Pomacentridae and Labridae (Perciformes) (largely present in coral reef environments), through comparative cytogenetic analyses of species with differentiated dispersal abilities. Among the rearrangements that are more easily identifiable from ancestral karyotype with exclusively acrocentric (a) chromosomes (2n=48; FN=48), the events of pericentric inversion can be measured either by particular FN or by the average of a phylogenetically related group. In Pomacentridae, the karyotype composition among three species and different subfamilies has been correlated to dispersion capacities (Molina and Galetti, 2004).

The circumtropical distribution of *Abudefduf saxatilis* represents an extreme example of successful dispersion within the family (Allen, 1975). Although it has a relatively short pelagic phase of 23 days (Brothers *et al.*, 1983), this species possesses the capacity of metamorphosis close to floating debris in immature specimens, allowing an extended pelagic period and increasing its potential to disperse (Wellington and Victor, 1989). Another species, *Microspathodon chrysurus*, displays pronounced territorialism (Robertson, 1984), with a pelagic larval phase lasting 21-27 days (Wellington and Victor, 1989) and a distribution area ranging from the Caribbean to the north of South America (Robins *et al.*, 1986). Its karyotype contains a great number of subtelocentric (st) chromosomes and FN with intermediate values (48<FN<92). *Amphiprion* is considered one of the most cytogenetically derived groups amongst Pomacentridae (Alvarez *et al.*, 1980), with karyotypes dominated by meta-submetacentric (m-sm) chromosomes. It is likely that this kind of karyotype reflects isolation and small populations (Bell *et al.*, 1982). Regarded as a specialized organism due to its close association with a specific host organism (anemones), this species has one of the shortest pelagic periods among reef fishes (Wellington and Victor, 1989) and has been used in other studies as a model for dispersion parameters (Doherty *et al.*, 1994) and genetic variation (Smith and Fujio, 1982).

A negative correlation between specialization levels, larval pelagic periods, and the FN (Table 1.4.1)—an indicator of the amount of pericentric inversions, the main event involved in the karyotype differentiation of the Pomacentridae—has been demonstrated (Fig. 1.4.2).

Table 1.4.1 Association of the FN average obtained from the karyotype data of Pomacentridae and Labridae species and the pelagic larval period average (PLP) (from Victor, 1986; Wellington and Victor, 1989).

Pomacentridae	N	2n	NF	PPl
Amphiprioninae	3	48	84.6	16.0
Pomacentrinae	26	48	72.3	21.0
Pomacentrinae-Abudefduf	5	48	52.8	23.6
Chrominae	10	<48	49.6	24.8

Labridae	N	2n	NF	PPL
Bodianinae	5	48	80.0	31.8
Cheilininae	20	<48	60.4	35.6
Corinae	29	48	53.2	37.6

Therefore, the karyotypes of representatives of the subfamily Chrominae with longer pelagic periods possess the lowest FN per group. Amphiprioninae, a specialized and monophyletic group, has short larval periods and karyotypes with a more extensive accumulation of inversions than observed in other pomacentrid groups. The results in *Microspathodon chrysurus*, a typical representative of the Pomacentrinae, indicate intermediate values of FN and larval phase, corresponding to the general model. *A. saxatilis*, whose average larval phase period is much longer than that of other members of the subfamily, possesses a considerably lower FN. These analyses extended to different subfamilies through average values of the pelagic phase. FN have confirmed that this trend is disseminated through the entire family (Molina and Galetti, 2004a). Cytogenetic

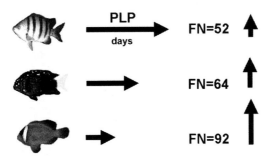

Fig. 1.4.2 Inverse correlation between the dispersive ability (PLP = pelagic larval period) and the number of chromosome arms (FN), as indication of pericentric inversions in an acrocentric ancestral karyotype. The arrows represent the parameter significance.

analyses already carried out in this family have identified the possible polyphyletic groups, as is the case of the genus *Abudefduf* within Pomacentrinae.

Among fishes of the family Labridae, the association of larval periods (Victor, 1986), with available cytogenetic data indicates a trend similar to that observed in the Pomacentridae (Sena and Molina, 2005). In this group, the correlations were established for each of the three subfamilies, including three species of Bodianinae, 14 of Corinae and five of Cheilininae, supporting the model proposed by Molina and Galetti (2004a). While it is not possible to generalize these results to all of the situations present in the marine environment, the results clearly identify the role of the larval dispersion capacity in genetic homogenization on a wide scale.

Chromosomal versus Morphological Diversification

The differentiation of marine species in many groups seems not to have been followed by significant karyotype diversification. The available karyotypic information in marine fishes has demonstrated a frequent karyotype 2n=48, composed by scrocentric chromosomes (Klinkhardt *et al.*, 1995; Brum and Galetti, 1997). This pattern has been identified mainly in marine representatives of Perciformes compared with freshwater families Cichlidae and Percidae (Brum, 1996). However, although initially considered to be a plesiomorphic character for all teleosteans further subdivided into five major basal lineages, it seems to reflect, instead, an apomorphic condition of Clupeocephala (Brum and Galetti, 1997).

Some specialized groups such as species of the Bovichtidae family have 2n=48 (Pisano *et al.*, 1995), a pattern considered ancestral for the Antarctic suborder Notothenioidei (Ozouf-Costaz *et al.*, 1991). The karyotype mostly persists, without modification, in marine species of Percomorpha (Brum, 1995).

In some families of Perciformes, cytogenetic features characterize the different populations. In the case of Gobiidae (Giles *et al.*, 1985; Caputo *et al.*, 1999) where polymorphism has originated from centric fusions and fissions, it suggests a great instability of its karyotypic pattern. In other families, however, cytogenetic comparative analysis did not provide evidence of any intraspecific (or interspecific) differentiation, even considering vast areas of distribution (Rossi *et al.*, 1996).

Several species with major morphological modifications possess indistinguishable karyotypes using conventional banding techniques, as has been observed for eight species of Haemulidae (Molina et al., 2005), or even after molecular cytogenetic analysis as well (Affonso, 2000). It is probable that karyotypes with similar macrostructure maintain internal changes within the lineage due to the micro-rearrangements, or undetectable rearrangements. It is not known whether the absence of greater differentiation rearrangements has been substituted by internal changes in the linkage groups sufficiently as effective as the first ones in the establishment of post-zygotic barriers in the speciation process (Hawthorne and Via, 2001).

Some groups studied in greater detail, such as Serranidae, have demonstrated a relative conservativeness of karyotype structure (Aguilar and Galetti, 1997; Sola et al., 2000). However, the presence of cryptic modifications has been found in the *Epinephelus adscensionis* (Molina et al., 2002), among other analyzed serranids. These cryptic karyotypes are apparently not as effective in the establishment of post-zygotic barriers as suggested by the significant rate of inter-specific hybridization in the Perciformes (Hubbs, 1970; Pyle and Randall, 1994).

Among Serranidae, hybrid cases have been observed, as much through morphological and meristic characteristics as through those detected with the analysis of nuclear and mitochondrial DNA sequences (Herwerden et al., 2002). It seems to be particularly true when it is taken into account that the same groups considered to be highly derived, such as the Pomacanthidae and Chaetodontidae, are conserved at karyotype level (Affonso et al., 2001) and exhibit greater inter-specific and inter-generic hybridization in reef areas (Pyle and Randall, 1994).

However, biological complexity and the marine environment favor exceptions. Such a case was identified in the pomacanthid *Centropyge aurantonotus* (2n=48; 4m+14sm+26st+4a), which is divergent in respect to the characteristic karyotype of the group. This species stands out for its karyotype structure (FN>48 due to the pericentric inversions) and other chromosomal features (multiple Ag-NORs sites, large heterochromatinic regions) that are strongly different from the rest of the family members, such as *C. ferrugatus* (2n=48a) (Affonso and Galetti, 2005).

Karyotype variability observed in the freshwater species *Astyanax scabripinnis* has been attributed to its biological characteristics. Isolation of populations in small streams results in interpopulational variation, which supports the proposition of species complex to this fish (Moreira-Filho and Bertollo, 1991). In contrast, chromosomal stability is compatible with low levels of interpopulational genetic variation. This fact could be related to the population structure of marine species, which have large reproductive groups, with dispersal (larval stage) or migratory (adults) abilities, associated with few effective physical barriers, which lead to genetic homogeneity through high gene flow (Lacson, 1992; Ward, 1995). In addition, behavioral factors such as aggregation of individuals in restricted geographical areas during the spawning period—such as that found in serranids (Bolden, 2000)—could favor the maintenance of undifferentiated karyotypes. It is possible that certain cytogenetic characteristics (low amount of heterochromatin, interstitial NOR sites) are associated with peculiar ethological and biological aspects (such as hermaphroditism and reproductive migrations) (Sale, 1991) acting to promote sufficient interpopulational gene flow (Molina *et al.*, 2002).

The chromosomal location and number of NOR sites, which in Characiformes and other groups are effective cytotaxonomic markers (Galetti *et al.*, 1984; Amemiya and Gold, 1986), are frequently less successful among the karyotypes of several Perciformes (Klinkhardt *et al.*, 1995; Sola *et al.*, 1997; 2000).

Interpopulational cytogenetic studies of freshwater species, in contrast to their salt-water counterparts, have demonstrated a number of cryptic forms within a single nominal species (Bertollo *et al.*, 2000, 2004; Torres *et al.*, 2005). Morphometric analyses have corroborated the interpopulational karyotype divergences in some species (Moreira-Filho and Bertollo, 1991; Maistro *et al.*, 1998). On the other hand, most of the marine groups have displayed conserved karyotypic features in their populations as observed among populations of the pomacentrid *Abudefduf saxatilis* even while significant differences in relation to the meristic and morphometric parameters were detected (Molina, 2000).

Karyotypic Orthoselection and Trends in Chromosome Evolution

The extensive diversification of forms and a common karyotype in Perciformes (2n=48; FN=48) reveals a favorable background for the identification of possible karyotype trends among the groups. The range of chromosome numbers in other orders is highly significant, with diploid numbers as low as 2n=33/34 in the monacanthid *Stephanolepis hispidus* (Tetraodontiformes; Sá-Gabriel and Molina, 2004) or 2n=40 in the balistid *Melichthys niger* (Tetraodontiformes; Sá-Gabriel and Molina, 2005) to 2n=100-102 in the triglid *Prionotus punctatus* (Scorpaeniformes; Corrêa, 1995) or 2n=240-270 in some anadromous Acipenseridae (Fontana *et al.*, 2001).

Qumsiyeh (1994) suggests that chromosome orthoselective processes would be selectively advantageous rearrangements in mammals due to their effect on gene recombination. Thus, while an increase in the diploid or FN values increase the recombination and variation possibilities that are important for adaptability in changeable environments, a decrease in number or FN values would reduce the recombination possibilities favorable for adaptation to specialized and stable environments. This would further increase the probability for new mutations to become fixed. Phillips and Ráb (2001) have supported the same idea as an explanation for the chromosome evolution in some groups of Salmonidae which exhibit numerous fusion and centric fission events.

When exposed to conditions of change, different karyotypes are predisposed to certain types of reorganization and not to others (King, 1981). Therefore, there is evidence that orthoselection (*sensu* White, 1973) takes place in a karyotypic format in some lineages and particular formats in others, and could be associated with specific adaptive values that have obvious evolutionary connotations. Tendencies for similar structural changes in karyotype have been observed in certain fish groups like the occurrence of Robertsonian translocations (Caputo *et al.*, 1999; Molina and Galetti, 2002), multiple pericentric inversions (Cano *et al.*, 1982; Ojima, 1983; Molina and Galetti, 2004b), or heterochromatinization events (Galetti *et al.*, 1991; Ueno *et al.*, 2001).

Pericentric Inversions and Karyotype Diversification

Pericentric inversions have been widely observed as one of the most frequent mechanisms associated with karyotype diversification in Perciformes (Galetti *et al.*, 2000, 2005; Molina and Galetti, 2004b). A reasonable estimate of the occurrence of this type of rearrangement is the change in the total number of chromosome arms (Fundamental Number, FN) in a fish group. Perciformes with karyotypes exclusively made of (a) chromosomes make up one of the best models for the observation of pericentric inversions (Molina and Galetti, 2004a). This fish group includes species with FN ranging from 22 to 96 (Klinkhardt et al., 1995).

Among the various examples of Perciformes possessing a derived karyotype pattern, in which FN is significantly larger than 48, can be considered the family Pomacentridae, since a significant amount of information is available on the basic cytogenetics of this group (about 34 species analyzed) primarily from the Pacific regions (Arai and Inoue, 1976; Arai et al., 1976; Ojima and Kashiwagi, 1981; Ojima, 1983; Takai and Ojima, 1995; Molina and Galetti, 2004b). Studies carried out by Arai and Inoue (1976) have generally demonstrated an increase in the fundamental number of karyotypes with 2n=48, due to pericentric inversions. Some species have exhibited conspicuous polymorphic chromosome numbers with the maintenance of FN, and variations in the diploid number, revealing Robertsonian events and associated pericentric inversions in the karyotypic evolution of the group (Ojima, 1983).

Karyotypes with 2n=48 (FN=48) were observed in a few species such as *Pomacentrus coelestis* (Arai and Inoue, 1976), *Chromis chromis* (Alvarez et al., 1980) and *C. multilineata* (Molina and Galetti, 2002). The data indicate that the main event of karyotype change was pericentric inversions (Arai and Inoue, 1976; Ojima, 1983). In some Pomacentridae of the genus *Stegastes*, high FN values have been detected in coastal Atlantic species, such as in *Stegastes variabilis* (FN=88), *S. fuscus* (FN=90), and the insular endemic species *S. sanctipauli* (FN=92) (Fig. 1.4.3). Karyotype evidence suggests extensive pericentric inversion events, acting in concert, accompanying the speciation process in the subfamily Pomacentrinae (Molina and Galetti, 2004b). In this group, biological characteristics that regulate gene flow—adhesive eggs, territorialism, and mutualism with sessil organisms (Smith and Fujio, 1982)—contributed to the partition of the marine environment in conspicuous micro-regions, favoring the establishment of such rearrangements. Due to the high phylogenetic kinship among these species, the increase of the chromosome numbers seems to reflect different degrees of phylogenetic derivation. In contrast to single rearrangements, multiple events of great magnitude could constitute important mechanisms of post-zygotic reproductive isolation (King, 1992) in the reef species speciation process.

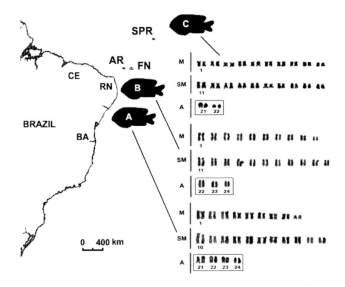

Fig. 1.4.3 Karyotype diversification by pericentric inversions in Pomacentridae species of the genus *Stegastes* with different geographical distribution in the Atlantic (A) *Stegastes variabilis* (2n=48; FN=88) (B) *S. fuscus* (2n=48; FN=90) and (C) *S. sanctipauli* (2n=48; FN=92). RN — Rio Grande do Norte; SPR — the St. Paul's Rocks.

The diploid number and karyotype formula in the holocentrids *Holocentrus ascensionis* (2n=50; 2m+6sm+16st+26a; FN=74) and *Myripristis jacobus* (2n=48; FN=48) demonstrate the karyotype diversity found in the Beryciformes. The first species presents 2n=50, with a karyotype showing all of the chromosome types and multiple NORs; the other has 2n=48a and FN=48, with single NORs and a peculiar pattern of heterochromatin. While stressing the active role of the pericentric inversions in the karyotype diversification in this family, the marked cytogenetic intra-familial differences could also indicate the need of a taxonomic revision of these taxa (Bacurau and Molina, 2004), already suggested by molecular phylogenetic reports (Colgan *et al.*, 2000).

Heterochromatin, Genomic Change and Karyotype dynamics

The constitutive heterochromatin seems to play an important role in the karyotype diversification in many fish groups. Some authors consider that the amount of heterochromatin (Molina and Galetti, 2002; Molina *et al.*,

2002) or its compositional aspects (Caputo *et al.*, 1997; Canapa *et al.*, 2002) carry out an active role in karyotype rearrangements.

Some fish genomes have a high amount of heterochromatin, as in the Characiformes groups (Souza and Moreira-Filho, 1995), showing large amplified heterochromatic regions in the autosomes, as well as heterochromatinization processes involved in the differentiation of heteromorphic sex chromosomes (Galetti *et al.*, 1991; Moreira-Filho *et al.*, 1993). This fact has been attributed to different causes such as heterochromatinization of the euchromatin process (*sensu* King, 1980) or the Rabl model (Schweizer and Loidl, 1987) for equilocal distributions of heterochromatin (Mantovani *et al.*, 2000).

A majority of representatives of Perciformes possessed a small heterochromatin content, which appears preferentially restricted to centromeric and pericentromeric regions (Mandrioli *et al.*, 2000; Ene, 2003). In general form, it could exert a little influence on the process of karyotype differentiation. A lower amount of heterochromatin could reduce the dynamics occasioned for the presence of large regions of repetitive sequences. This is particularly visible by the significant absence/reduction of some karyotype events in Perciformes such as B-chromosomes, heterochromatic polymorphisms, and by karyotypic structural change variation in numerous groups with extant karyotypes. Reduced heterochromatin amount is frequent in a large number of the species and seems to determine a gamma of common karyotype characteristics. For instance, the presence of single NOR sites, widely found in perciform species, could be related to the small amount of heterochromatin which could constrain the dispersion of ribosomal genes sequences in the karyotype.

Although the importance of the heterochromatin addition during the speciation processes is questioned (King, 1987), it could facilitate the occurrence of greater dynamism in the origin or establishment of rearrangements. In marine species, heterochromatinization processes have apparently played a relatively minor role than in freshwater groups.

A rare case of heterochromatin addition was identified in the karyotype of pomacanthid *Centropyge aurantonotus*, which, apart from having a diversified karyotype (4m+14sm+26st+4a) different from the *C. ferrugatus* (48a), has an uncommon amount of heterochromatin (GC-rich) in the centromeric regions, as well as over the short arms of nine chromosome pairs (Affonso and Galetti, 2005). It is quite probable that the diffusion of heterochromatin segments for other chromosomes has occurred through transposition, unequal exchanges, favored by the constant radial associations between acrocentric chromosomes (Aguilar *et al.*, 1998; Mantovani *et al.*, 2000).

Other groups such as the Tetraodontiformes exhibit a tendency to genome reduction (Brainerd *et al.*, 2001). This group, considered to be specialized, presents some lineages with a clear trend toward genomic reduction, possibly involving the loss of repetitive sequences (Neafsey and Palumbi, 2003).

Reduced genome content has been hypothesized as a result of the independent development of some groups of Tetraodontiformes (Tetraodontidae and Balistoidea), which is a remarkable evolutionary trait of Tetraodontidae, probably established after their split from Diodontidae (Brainerd *et al.*, 2001). At present, the smallest known vertebrate genome is that of the freshwater puffer fish *Tetraodon fluviatilis* (0.70pg) (Lamatsch *et al.*, 2000).

It is not entirely clear whether fitness processes motivate the causes of the genomic reduction in Tetraodontiformes or if it takes place because of a fixed neutral characteristic. However, evidently, it is the result of an imbalance between processes that produce insertions and those that promote deletions, resulting in the decrease of the genome size (Brenner *et al.*, 1993). Studies conducted on *Tetraodon nigroviridis* (Jaillon *et al.*, 2004) and *Takifugu rubripes* genomes (Christoffels *et al.*, 2004) suggest a whole-genome duplication event early (350 MYA.) during the evolution of ray-finned fishes, probably before the origin of teleosts, followed by a massive loss of duplicated genes. The anchoring of some sequences genomic clones by FISH in the chromosomes of *T. nigroviridis* show significant synteny, whereas others suggest reorganizations by fusions or fragmentations. Apparently, a relatively high degree of intrachromosomal rearrangement associated with a relatively low degree of interchromosomal exchange takes place in the *Tetraodon* lineage (Jaillon *et al.*, 2004). Pufferfish have approximately the same complement of genes as do other vertebrates (Brenner *et al.*, 1993). Thus this small size has presumably resulted from a loss of repetitive or other nonfunctional, noncoding DNA (Neafsey and Palumbi, 2003). A gradual imbalance in favor of small losses of DNA, preferentially heterochromatin segments, could explain the low content of DNA observed in the Tetraodontidae. Grützner *et al.* (1999) demonstrated a large number of repetitive sequences at the centromere of all *T. nigroviridis* and *T. rubripes* chromosomes, suggesting a loss of specific sequence classes rather than a miniaturization of the entire genome of the Tetraodontidae.

Different karyotype trends have been assumed in the Tetraodontiformes since different families have complex chromosome differentiation patterns with diploid numbers varying from 2n= 33/34 in the monachantid *Stephanolepis hispidus* (Sá-Gabriel and Molina, 2004) to 2n=52 in the diodontid *Chilomycterus spinosus* (Brum, 1995) and FN from 33 to 72 (Sá-Gabriel and Molina, 2005). The main chromosomal rearrangements involved in the karyotype diversification within the Tetraodontiformes are summarized in Figure 1.4.4.

The Triacanthidae, considered to be a less derived group, exhibit a karyotype of 48 acrocentric chromosomes, which is characteristic of a great number of Perciformes (Choudhury *et al.*, 1982). In Balistidae (2n=34-46; modal value 44) and Monacanthidae (2n=34), differentiation seems to have involved tandem fusions or centric fusions followed by pericentric inversions. The increased chromosome numbers in Diodontidae (2n=46-52; modal value 46) suggest that centric fissions associated with pericentric inversions have played an important role in the differentiation of the karyotype in this group (Arai and Nagaiwa, 1976;

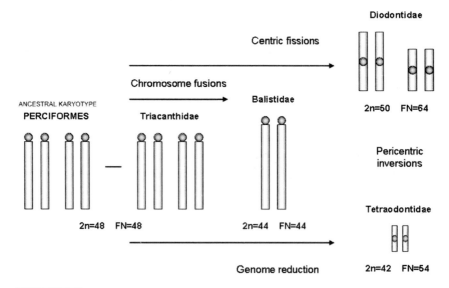

Fig. 1.4.4 The main chromosomal rearrangements involved in the karyotype diversification in Tetraodontiformes.

Brum, 2000). Neafsey and Palumbi (2003), analyzing the genome of spiny puffer *Diodon hystrix* (Diodontidae), identified that 22% of it is composed of repetitive DNA, which is almost seven times the proportion of repetitive DNA detected in *Takifugu rubripes* (Tetraodontidae). The genome size of *Diodon hystrix* (0.82pg) is twice as large as *Takifugu rubripes* (0.42pg).

The difference between spiny and smooth puffers could be explained by the increase in the rate or size of large deletions relative to large insertions, or by the decrease in the rate or size of large insertions relative to large deletions, contributing to a change in the equilibrium resulting from a change in the insertion/deletion profile in the tetraodontid lineage at a larger scale. The pufferfish genome (approximately 400Mb) possesses the structural complexity of the highest vertebrates, but it exhibits small numbers of unusual introns and the lack of repetitive sequence clusters and pseudogenes (Crnogorac-Jucevic *et al.*, 1997; Jaillon *et al.*, 2004).

The karyotype diversity in Tetraodontiformes is compatible with a scenario of extensive changes established during their adaptive radiation. Among the Tetraodontidae (2n=28-46; modal values 42, 44, 46), small chromosomes and diversified karyotypes have been observed as a consequence of genome reduction, like those that occur in the species of the genus *Sphoeroides* (2n=46) (Brum, 2000; Brum and Mota, 2002; Sá-Gabriel and Molina, 2005) and *Tetraodon* (2n=42) and *Takifugu* (2n=44) (Grützner *et al.*, 1999; Mandrioli, 2000; Mandrioli and Manicardi, 2001).

Several animal groups such as ciliates, nematodes, copepods, insects, hagfish, and marsupials have specific heterochromatin regulatory processes of elimination in post-somatic cellular lines, but retain this DNA in the nucleus of primordial germ line cells. Different causes have been identified with respect to this phenomenon, such as gene balancing for dose compensation in groups that display partial duplication of the genome (Kloc and Zagrodzinska, 2002). In hagfish, the elimination of part and/or whole chromosomes was verified during the beginning of the embryogenesis in different species (Nabeyama *et al.*, 2000). Other species present a loss of heterochromatin corresponding up to 55% of the total chromatin eliminated in the pre-somatic cellular lineages (Kubota *et al.*, 1997). These DNA sequences are formed highly and tandemly repeated, and it is predicted that they might be amplified by saltatory replication and may have evolved in a concerted manner (Kubota *et al.*, 2001). If similar mechanisms of selective heterochromatin elimination in evolution were established, this would result in fast drastic reductions of the genome of the species.

Contrary to the several examples of genome increase (i.e., poliploidization) (Comber and Smith, 2004) or evolutionary addition of repetitive sequences (i.e., heterochromatinization processes) (Galetti *et al.*, 1991; Affonso and Galetti, 2005), the mention of a higher number of cases of evolutionary reduction of genome in fishes is scarce.

Chromosomal Variation and Robertsonian Rearrangements

Among other events, centric fissions and fusions have been identified as important mechanisms of karyotype differentiation in many fish groups (Hartley and Horne, 1984; Amores *et al.*, 1990; Phillips and Rab, 2001). In Salmoniformes, they are among the major mechanisms of karyotype evolution (see Phillips and Ráb, 2001). In Perciformes, numerical variability derived from fusion and fission processes has been reported in numerous families, e.g., Pomacentridae (Ojima and Kashiwagi, 1981; Ojima, 1983; Molina and Galetti, 2002); Gobiidae (Thode *et al.*, 1985; Amores *et al.*, 1990, Caputo *et al.*, 1997; Ene, 2003), and Carangidae (Vitturi *et al.*, 1986), among other groups (i.e., Syngnathiformes; Vitturi *et al.*, 1998).

The presence of Robertsonian rearrangements necessarily leads to some degree of transitory polymorphisms, where under endogenous (meiotic selection, disruption of linkage groups) or exogenous (environmental selection for more adapted genotypes) influences (Amores *et al.*, 1990) some cytotypes become fixed.

In Gobiidae, the karyotype differentiation has been followed by frequent and complex modifications with a high tendency toward Robertsonian fusions (Thode *et al.*, 1988; Vasilev and Grigoryan, 1993; Caputo *et al.*, 1997; Ene, 2003). In the goby *Aphia minuta* (Adriatic Sea), this tendency is evident where up to 5 cytotypes have been observed, with $2n=44$, 43, 40 (2 cytotypes), and 38 chromosomes, with a constant $FN=44$ (Caputo *et al.*, 1999). It is possible that the variety of cytotypes reflects a recent junction of previously isolated populations. The analysis of *Gobius niger jozo* reveals the presence of four distinct cytotypes, $2n=52$ (8m-sm+10st+34a), 51 (9m-sm+10st+32a), 50 (10m-sm+10st+30a), and 49 (11m-sm+10st+28a), all with FN=60 (Vitturi and Catalano, 1989), demonstrating the presence of numerical and structural intrapopulational variation in addition to other characteristics of this family (Thode *et al.*, 1985). Specific types of centromeric or pericentromeric heterochromatin in acrocentric pairs may facilitate centric fusions (Caputo *et al.*, 1997; Ene, 2003), associated with frequent radial associations involving a great number of scrocentric chromosomes (Aguilar *et al.*, 1998; Molina and Galetti, 2002; Affonso and Galetti, 2005).

Robertsonian events are frequently at the origin of multiple sex chromosome systems (Almeida Toledo *et al.*, 1984; Brum *et al.*, 1992), which contribute, in some cases, to the establishment of genetically isolated populations.

In the Pomacentridae, almost all species show a conserved karyotype of 2n=48, but with significant differences in chromosome formulas (48=FN=92), indicating that the pericentric inversions have been the major type of chromosome change (Molina and Galetti, 2004b). The subfamily Chrominae, composed of the genera *Acanthochromis, Azurina, Chromis,* and *Dascyllus*, seems to be particularly inclined to Robertsonian polymorphism (Molina and Galetti, 2002). In this subfamily, centric fusions have been observed in heterozygous combinations in three *Dascyllus* species (2n=47-48, *Dascyllus trimaculatus*; 2n=34-37, *D. reticulates*; 2n=27-33, *D. aruanus*; Ojima and Kashiwagi, 1981) and in homozygous, likely fixed combinations in two other *Chrysiptera* species (Takai and Ojima, 1995). In *Chromis*, several species share a conserved karyotypic pattern (*C. chromis*; 2n=48/FN=48, Alvarez *et al.*, 1980; *C. chrysura*, 2n=48/FN=50; and *C. caerulea*, 2n=48/FN=48; Ojima and Kashiwagi, 1981; Ojima, 1983; *C. multilineata*, 2n=48/FN=48; Molina and Galetti, 2002). However, the reduction of the chromosome number and the presence of two cytotypes associated with the presence of large metacentric chromosomes, in *C. insolata* (2n=46-47, 4-3m+6sm+36-38a, FN=56) and a karyotype with nine large metacentrics in males of *C. flavicauda* (2n=39, 9m+6sm+24a; FN=54), also indicated the occurrence of centric fusions in this genus. Ribosomal 5S genes have been detected in the pericentromeric regions of two large metacentric pairs in the karyotype of *C. flavicauda* and of two acrocentric pairs in that of *C. insolata*. The karyotype structure, number and the location of the 5S sequences in these two species have demonstrated that the chromosomes bearing the 5S sequences are involved in the origin of this polymorphism.

Such polymorphisms, in fixed or transitory combinations, seem to constitute an evolutionary process of disseminated karyotypic orthoselection and are responsible for the modeling of the extant karyotypes in the subfamily Chrominae (Molina and Galetti, 2002).

In Labridae, while *Xyrichthys dea* and *X. pavo* have 44 acrocentric chromosomes, *X. twistii* (Cheilininae) shows a reduced diploid number (2n=22; 18m-sm+4a), indicating the occurrence of ample centric fusion events (Ueno and Takai, 2000).

Some groups of coral reef fish such as the parrotfish are among the more colored and overly dominant inhabitants of this habitat (Sale, 1991). Despite their importance, cytogenetic data of the Scaridae are very scarce. The first data for this group were provided by *Sparisoma rubripinne* (2n=46), which presented a differentiated diploid value, revealing a numeric reduction in relation to the *Scarus coelestinus* (2n=48). The presence of a large metacentric pair (1st pair) in the karyotype of *S. rubripinne* suggests a Robertsonian fusion event associated with pericentric inversions, which possibly moulded the karyotype of this species (Sena and Molina, 2005).

Some authors believe that transitory polymorphisms do not play any role in the speciation process (King, 1987; Caputo *et al.*, 1997). However, they could favor the occurrence of subsequent rearrangements that could determine an initial step for the establishment of post-zygotic barriers.

The origins of Robertsonian translocations have been clarified (Perry *et al.*, 2004). Fluorescence *in situ* hybridization with telomeric probes has become a useful tool for identification of such events since internalized ectopic telomeric sequences $(TTAGGG)_n$ are often maintained in the site of fusion. However, in several cases such telomeric sequences are lost (Meyne *et al.*, 1990). Thus, even in evident examples of centric fusions such as those in the long metacentric chromosomes of the species *Chromis flavicauda* and *C. insolata* (Molina and Galetti, 2002), no remnants of ectopic telomeric sequences have been found.

Studies for the detection of telomeric sequences in representatives of Scorpaeniformes indicate that centric or *tandem* (centromere-centromere or centromere-telomere) fusions have occurred, followed by losses of telomeric sequences as well (Caputo *et al.*, 1998), although in some cases they can be maintained. Prospective studies carried out in species with particular chromosome trends—such as the puffer fishes *Tetraodon fluviatilis* and *T. nigroviridis*—have not shown any signs of ectopic telomeric sequences.

Sex Chromosomes in Marine Groups

An increasing number of systems of differentiated sex chromosomes have been described for marine species—for instance, XX/XY in Ariidae (Gomes *et al.*, 1994), $X_1X_1X_2X_2/X_1X_2Y$ in Clupeidae (Brum *et al.*, 1992) or ZZ/ZW and ZZ/ZW_1W_2 in Synodontidae (Ueno *et al.*, 2001). The XX-X0 or ZZ-Z0 systems described for several species of marine deep fishes (e.g., see Chen, 1969) has not gained general acceptance (Ohno, 1974; Chourrout, 1986).

Processes of heterochromatinization in some elements seem to be less frequent mechanisms in the differentiation of heteromorphic sex chromosomes in marine species. Examples of such differentiation have been observed in the simple XX-XY system discovered in catfish *Netuma barba* (Ariidae, Siluriformes) (Gomes *et al.*, 1994) or in *Dicentrarchus labrax*, where an early stage of differentiation was hypothesized (Cano *et al.*, 1996). In the eel species *Conger myriaster* (Congridae), a ZZ-ZW system was described where the W had a specific heterochromatic pattern (Ojima and Ueda, 1982). There is an increase of interest in the analyses of the satellite DNA fractions of sex chromosome in fishes (Nanda *et al.*, 1990; Nakayama *et al.*, 1994; Devlin *et al.*, 1998; Vicente *et al.*, 2003).

In some cases, the review of karyotypes has been useful for clarifying the validity of some proposed sex systems. Thus, after an intense search for the identification of sex chromosomes in the labrid *Coris julis* (Cano *et al.*, 1982; Duchac *et al.*, 1982; Vitturi *et al.*, 1988), later studies of the three sexual types (female; primary and secondary males) did not identify sex chromosome heteromorphisms (Mandrioli *et al.*, 2000).

The cases of female heteromorphic sex chromosome systems in marine species appear less frequent, although the predominance of the male heteromorphic sex chromosome systems in marine fish may be due to the study of a limited number of groups (Galetti *et al.*, 2000).

Although only 5.4% of all freshwater fish species analyzed in the Neotropical region have heteromorphism involving sex chromosomes (Oliveira and Foresti, 1994), this represents a large number of species. Reviews by Moreira-Filho *et al.*, (1993) and Almeida-Toledo and Foresti (2001) point out five kinds of sex chromosome systems in Neotropic fish fauna: ZZ-ZW, XX-XY, $X_1X_1X_2X_2$-X_1X_2Y, XX-XY_1Y_2, and ZZ-ZW_1W_2.

These sex systems can be present in all the species of a group, such as ZZ-ZW in *Triportheus* karyotypes (Characiformes, Characidae) (Bertollo and Cavallaro, 1992) or in some species of a group, such as the XX-XY system in *Hoplias* (Characiformes, Erythrinidae) (Bertollo *et al.*, 2000).

In marine groups, the sporadic occurrence of sex chromosome systems originating from independent evolutionary events represents the most common situation. However, the sex chromosome distribution in some fish groups can express a monophyletic condition, shared by a part of the group. This is the case with the female heterogameity, which is widely spread among (Synodontidae). In *Saurida elongata, S. ulae, S. hoshinonis* (2n=48), a simple system was identified, with the males presenting two large acrocentric chromosomes (ZZ), while the females had only one heterochromatic microchromosome (W) (Ueno *et al.*, 2001). In *Trachinocephalus myops*, a reduced diploid number 2n=26 (FN=50) was observed for males and 27 (FN=51) for females, documenting an unusual multiple system $ZZ\text{-}ZW_1W_2$, with the presence of a heterochromatic microchromosome (Ueno *et al.*, 2001). *S. undosquamis* showed males with 2n=48 and females with 2n=47, with a microchromosome (Nishikawa and Sakamoto, 1978), possibly indicating another case of a $ZZ\text{-}ZW_1W_2$ system.

Multiple systems $X_1X_1X_2X_2X_1X_2Y$ originated by centric fusions have been observed in representatives of different families, namely Monacanthidae (Murofushi *et al.*, 1980; Sá-Gabriel and Molina, 2004), Ophichthidae (Murofushi and Yosida, 1984), Blenniidae (Carbone *et al.*, 1987), Clupeidae (Brum *et al.*, 1992) and Antarctic Notothenioidei (Morescalchi *et al.*, 1992; Pisano and Ozouf-Costaz, 2003), among others.

The occurrence of heteromorphic sex chromosomes favors the establishment of genetically isolated populations. It is possible that differentiated sex chromosome systems play an important role in the speciation process in marine environments, where geographical isolation is certainly less effective than in continental habitats.

Distribution of Specific DNA Sequences in the Karyotypes of Marine Species

Major rDNA sequences

Cytogenetic markers are being added to the morphological and morphometric characters in cytotaxonomic and phylogenetic studies

(Bertollo et al., 2004). Precise markers, obtained by means of structural analyses of the chromosomes, allow excluding the homoplasies, making phylogenetic hypotheses more reliable. The sites of major ribosomal genes, corresponding to NORs, have been most frequently used as chromosomal markers in comparative cytogenetics.

In some groups the NORs play the effective role of cytotaxonomic markers (Galetti et al., 1984). However, in the Perciformes, they have proven to be useless due to the extensive structural conservativeness found in the NOR-bearing chromosomes (Molina and Galetti, 2004b).

The location of the NORs in the pericentromeric position of acrocentric pairs of chromosomes has been claimed as an ancestral condition for many groups of Perciformes (Pomacanthidae; Affonso et al., 2001; Nothothenioidei, Ozouf-Costaz et al., 1997; Scianidae, LeGrande, 1988; Feldberg et al., 1999; Serranidae, Aguilar and Galetti, 1997). Therefore, variations in that pattern should be considered as derived (Brum, 1995; Aguilar and Galetti, 1997; Ozouf-Costaz et al., 1997; Affonso et al., 2002).

In groups with multiple tendencies of karyotype diversification, the major ribosomal sites demonstrate weak phylogenetic value. In more derived groups (for example, Tetraodontidae), the NORs exhibit variable frequency, being observed in one or two chromosome pairs (Mandrioli et al., 2000; Brum and Mota, 2002).

NOR sites from three geographical areas have been indicated as useful populational markers in Halichoeres poeyi (Labridae) along the Brazilian coast (Sena and Molina, 2005). Pisano and Ozouf-Costaz (2000) point out the importance of NOR sites to inferences about karyotypic diversification in the families Bovichtidae, Nototheniidae, and Channichthyidae (Notothenioidei).

The identification of ribosomal genes with 18S and 28S rDNA probes in fishes through fluorescence in situ hybridization has generally demonstrated similar results to those obtained by a silver nitrate staining technique (Ag-NORs), also showing in a number of groups that the size of polymorphisms of the NOR regions is due to structural variation in the number of ribosomal genes and not to molecular processes of gene regulation (Rossi et al., 1997).

Instead, ribosomal sites can be diversified in some families of Perciformes such as Labridae where single sites can be observed in some species of *Halichoeres* (Sena and Molina, 2005), together with examples of multiple NORs as in *Coris julis*, which shows four NORs-bearing chromosome pairs (Mandrioli *et al.*, 2000).

Marine groups with relatively undifferentiated karyotype show NORs-bearing likely homologous among species of different families (Pomacentridae, Scaridae, Labridae, Haemulidae, Serranidae, Pomacanthidae, among others), which indicates that besides the karyotype structures, large syntenic segments should possibly be maintained in these groups.

5S rDNA sequences

The chromosome mapping of ribosomal 5S rDNA sequences has provided relevant information in studies of karyotype evolution in fishes (Martins, present volume). Studies of 5S RNA gene localization comprises the orders Acipenseriformes, Anguilliformes, Characiformes, Perciformes, Salmoniformes, Tetraodontiformes (Martins and Galetti, 2001), among others.

For instance, the investigations by FISH based on the gene 5S RNA in pomacentrids *Chromis flavicauda* and *C. insolata* revealed the occurrence of centric fusions, despite the absence of internalized telomeric sites. Moreover, the location of these sequences was decisive in the identification of the chromosomes involved in the rearrangements. Whether these genes (or of heterochromatins associated with them) had an active role in promoting centric fusions is still unknown (Molina and Galetti, 2002).

The sequences of the ribosomal subunit 5S often occur in two chromosome pairs in an interstitial position, as observed in the pomacentrids (Molina and Galetti, 2002, 2004b). However, some species such as the serranid *Epinephelus marginatus* (Sola *et al.*, 2000) or the puffer species (Fischer *et al.*, 2000; Mandrioli and Manicardi, 2001) have a unique bearing-pair. Its occurrence in the karyotypes of representatives of subfamilies of Pomacentrinae and Chrominae seems to indicate the presence of two loci as a plesiomorphic condition for the family. The detection of one 5S sequence bearing-pair in karyotype of *Abudefduf saxatilis* (Molina and Galetti, 2004a), cytogenetically less derived, could reflect a different ancestral position in that species, although the presence of small, undetectable, or secondary sites cannot be excluded.

Even though the labrid *Coris julis* presents eight sites of major ribosomal genes, it only possesses four 5S ribosomal gene sites (Mandrioli *et al.*, 2000). The diversification of the 28S rDNA and the stability of the 5S sequences in karyotypes reinforce a hypothesis about a completely independent evolution between these rRNA genes (Martins and Galetti, 1999). The absence of synteny between the genes for rRNA 28S and 5S (Fischer *et al.*, 2000; Mandrioli and Manicardi, 2001) was also observed in *T. nigroviridis*.

In the chromosomes of six species of Channichthyidae, 5S rDNA sequences in a syntenic position with major rDNA sites were identified. In these species the 5S rDNA sequences revealed a lower discriminative value because similar patterns of distribution were shared by four of these species. However, in *Chionodraco hamatus*, the finding of 5S rDNA at telomeric position of the longest arm on the Y chromosome indicated homology between the distal part of the large Y chromosome with the unpaired acrocentric chromosomes possessing 5S rDNA sequences (Mazzei *et al.*, 2004).

In Pomacanthidae, 5S rDNA sites vary in number and position. While in the genus *Pomacanthus* they are observed in only one chromosomal pair, only in the karyotypes of the *Holacanthus* species were signals observed in four to five chromosomes. An extreme example of diversification of these sequences occurs in the karyotype of *Centropyge aurantonotus*, where 5S sites were identified in up to 18 chromosomes, preferentially associated with large heterochromatic blocks present on the short arms (Affonso and Galetti, 2005).

Conclusive Remarks

The species inhabiting the marine coastal waters are subjected to environmental conditions and biotic peculiarities, ranging from stable to extremely dynamic, that can be modified in regular (day, month, year) or irregular (event) periods, with a direct influence on the reproductive isolation of the populations. Such conditions provide an ideal background for the development of a variety of evolutionary strategies that partially explain the cytogenetic patterns observed in different groups of marine fish.

Among all fish orders, the Perciformes constitute a potentially useful model for the understanding of the genetic structure of marine populations, as well as the establishment of their evolutionary patterns, due to its diversity and wide distribution on reefs and rocky substrata of oceanic islands and over the continental shelf. In this group, examples of extensive chromosome conservativeness with symplesiomorphic patterns shared by a great number of species can be found, as well as tendencies for karyotypic diversification along the evolutionary history of some groups (Galetti et al., 2006). The origin or cause of these tendencies is not always clear, but they can sometimes be correlated with biological characteristics of the groups.

Primitive and common karyotypes, composed only by acrocentric elements, allow for determining and even quantifying rearrangements in perciform karyotypes. Therefore, extensive analyses in particular marine fishes have presented evidence that karyotypes of phylogenetically related groups demonstrate a particular tendency that can be submitted to the same type of chromosome rearrangements during their evolutionary history (orthoselective process). Robertsonian rearrangements can be found frequently in such large groups as Gobiidae (Amores et al., 1990) or in more restricted ones such as Chrominae (Pomacentridae) (Ojima and Kashiwagi, 1981).

Other sources of karyotypic evolutionary differentiation, such as pericentric inversions, are spread throughout diverse groups. Pericentric inversions are the most frequent event in karyotypes of many marine fishes (Takai and Ojima, 1995; Molina and Galetti, 2004b). Despite several examples of genome increase (i.e., poliploidization) or increase of highly repetitive sequences (i.e., heterochromatinization processes), the evolutionary trend towards reduction of genome in Tetraodontidae is peculiar. The constitutive heterochromatin seems to play an important role in the karyotype diversification in many fish groups. Some authors believe that the heterochromatin amount (Molina and Galetti, 2002; Molina et al., 2002) or its compositional aspects carry out an active role in karyotype rearrangements (Caputo et al., 1997).

Chromosomally mediated speciation is a process whereby fixation of chromosomal rearrangement initiates and contributes to divergence and reproductive isolation between populations (White, 1968; John, 1981, King, 1993). Although this is not a rule, an effective post-mating isolating mechanism could be reached by the presence of sex chromosome systems or wide karyotype changes. At present, five sex chromosome systems in marine fishes (ZZ-ZW, XX-XY, $X_1X_1X_2X_2$-X_1X_2Y, XX-XY_1Y_2, and ZZ-ZW_1W_2) have been reported. Advances in marine fish cytogenetics on new species and groups will be able to find new systems.

The lack of physical barriers in marine environments can facilitate the karyotype stability in some lineages due to effective interpopulational gene flow. Several examples demonstrate the absence of karyotype differentiation in specimens geographically distant (Molina and Galetti, 2004). Despite this, notable examples of allopatric differentiation occurred in the insular speciation of some forms (e.g., endemic pomacentrid *Stegastes sanctipauli*) or between South and Central Atlantic sibling species (e.g., eleotrids *Dormitator* and *Eleotris*), separated by the distance to Amazon river plume (Molina, 2005). The failure of reproductive isolation between several marine perciform species points out to an as yet incipient speciation, and/or to relatively similar karyotypes, suggested by stasis hypothesis.

The association between the larval pelagic period and chromosome rearrangements was well established in some reef families such as Pomacentridae and Labridae. Although this should not be the rule for other marine fish groups, it allows to characterize an unsuspected relation between biological patterns and chromosome differentiation in marine fishes (Molina and Galetti, 2004; Sena and Molina, 2005).

Several cytogenetic approaches have been used in the karyotype characterizations of marine teleosts species. Fluorochromes GC and AT-specific (Sola *et al.*, 2000; Ene, 2003; among others), replication bands (e.g., Salvadori *et al.*, 2003, among others), and fluorescence *in situ* hybridization with probes of telomeric sequences (Mandrioli *et al.*, 2000; Molina and Galetti, 2002), major rDNA (Ozouf-Costaz *et al.* 1996; Pisano *et al.*, 2001, among others), and 5S rDNA sites (Mazzei *et al.*, 2004, among others) have been utilized. These data have provided new insights into aspects of fish chromosome diversification.

Acknowledgements

I express my sincere thanks to Drs. Pedro Manoel Galetti Jr., Eva Pisano, Catherine Ozouf-Costaz, Paulo R. A. de Melo Affonso, Luis A. Carlos Bertollo, and the anonymous referee for their valuable comments and suggestions on the draft; and also to Allyson S. Souza for assistance with the preparation of the figures.

References

Affonso, P.R.A.M. 2000. *Caracterização citogenética de peixes de recifes de corais da Família Pomacanthidae (Perciformes)*. Masters' Thesis. UFSCar, SP, Brazil.

Affonso, P.R.A.M. and P.M. Galetti Jr. 2005. Chromosomal diversification of reef fishes from genus *Centropyge* (Perciformes, Pomacanthidae). *Genetica* 123: 227–233.

Affonso, P.R.A.M., W. Guedes, E. Pauls and P.M. Galetti Jr. 2001. Cytogenetic analysis of coral reefs fishes from Brazil (Families Pomacanthidae and Chaetodontidae). *Cytologia* 66: 379–384.

Aguilar, C.T. and P.M. Galetti Jr. 1997. Chromosomal studies in South Atlantic serranids (Pisces, Perciformes). *Cytobios* 89: 105–114.

Aguilar, C.T., M.M.O. Corrêa and P.M. Galetti Jr. 1998. Chromosome associations by centromeric heterochromatin in marine fishes. *Chromosome Science* 2: 73–76.

Allen, G.R. 1975. *Damselfishes of the South Seas*. T.F.H. Publications, Hong Kong.

Allen, G.R. 1979. *Butterfly and Angelfishes of the World*. Vol. 2. John Wiley & Sons, New York.

Almeida Toledo, L.F. and F. Foresti. 2001. Morphologically differentiated sex chromosomes in Neotropical freshwater fish. *Genetica* 111: 91–100.

Almeida Toledo, L.F., F. Foresti and S.A. Toledo Filho. 1984. Complex sex chromosome system in *Eigenmannia* sp. (Pisces, Gymnotiformes). *Genetica* 64: 165–169.

Alvarez, M.C., J. Cano and G. Thode. 1980. DNA content and chromosome complement of *Chromis chromis* (Pomacentridae, Perciformes). *Caryologia* 33: 267–274.

Amemiya, C.T. and J.R. Gold. 1986. Chromomycin A_3 stains nucleolus organizer regions of fish chromosomes. *Copeia* 1986: 226–231.

Amores, A., V. Giles, G. Thode and M.C. Alvarez. 1990. Adaptative character of a Robertsonian fusion in chromosomes of the fish *Gobius paganellus* (Pisces, Perciformes). *Heredity* 65: 150–155.

Arai, R. and K. Nagaiwa. 1976. Chromosomes of Tetraodontiform fishes from Japan. *Bulletin of the National Science Museum* 2: 59–72.

Arai, R., M. Inoue and H. Ida. 1976. Chromosomes of four species of coral fishes from Japan. *Bulletin of the National Science Museum* 2: 137.

Bacurau, T.O.F and W.F. Molina. 2004. Karyotypic diversification in two Atlantic species of Holocentridae (Pisces, Beryciformes). *Caryologia* 57: 300–304.

Beheregaray, L.B. and P. Sunnucks. 2001. Fine-scale structure, estuarine colonization and incipient speciation in the marine silverside fish *Odontesthes argentinensis*. *Molecular Ecology* 10: 2849–2866.

Bell, L.J., J.T. Moyer and K. Numachi. 1982. Morphological and genetic variation in Japanese populations of the anemonefish *Amphiprion clarkii*. *Marine Biology* 72: 99–108.

Bellwood, D.R. 1996. The Eocene fishes of Monte Bolca: the earliest coral reef fish assemblage. *Coral Reefs* 15: 11–19.

Bertollo, L.A.C. and Z.I. Cavallaro. 1992. A highly differentiated ZZ/ZW sex chromosome system in a Characidae fish *Triportheus guentheri*. *Cytogenetics and Cell Genetics* 60: 60–63.

Bertollo, L.A.C., G.G. Born, J.A. Dergam, A.S. Fenocchio and O.M. Moreira-Filho. 2000. A biodiversity approach in the Neotropical Erythrinidae fish, *Hoplias malabaricus*. Karyotypic survey, geographic distribution of cytotypes and cytotaxonomic considerations. *Chromosome Research* 8: 603–613.

Bertollo, L.A.C., C. Oliveira, W.F. Molina, V.P. Margarido, M.S. Fontes, M.C. Pastori, J.N. Falcão and A.S. Fenocchio. 2004. Chromosome evolution in the erythrinid fish, *Erythrinus erythrinus* (Teleostei, Characiformes). *Heredity* 93: 228–233.

Bolden, S.K. 2000. Long-distance movement of a Nassau grouper (*Epinephelus striatus*) to a spawning aggregation in the central Bahamas. *Fishery Bulletin* 98: 642–645.

Bonhomme, F. and S. Planes. 2000. Some evolutionary arguments about what maintains the pelagic interval in reef fishes. *Environmental Biology of Fishes* 59: 365–383.

Brainerd, E.I., S.S. Slutz, E.K. Hall and R.W. Phillis. 2001. Patterns of genome size evolution in Tetraodontiform fishes. *Evolution* 55: 2363–2368.

Brenner, S., G. Elgar, R. Sandford, A. Macrae, B. Venkatesh and S. Aparicio. 1993. Characterization of the pufferfish (*Fugu*) genome as a compact model vertebrate genome. *Nature (Lond.)* 336: 265–268.

Briggs, J.C. 1974. *Marine Zoogeography*. McGraw-Hill Book Co. New York.

Brothers, E.B., D.McB. Williams and P.F. Sale. 1983. Lenght of larval life in twelve families of fishes at 'One Tree Lagoon', Great Barrier Reef, Australia. *Marine Biology* 76: 319–324.

Brum, M.J.I. 1995. *Correlação entre a filogenia e a citogenética dos peixes teleósteos*. *Sociedade Brasileira de Genética—Série Monografias* 2: 5–42.

Brum, M.J.I. 1996. Cytogenetic studies of Brazilian marine fish. *Brazilian Journal of Genetics* 19: 421–427.

Brum, M.J.I. 2000. Cytogenetic studies in tetraodontiforms *Sphoeroides tyleri* (Tetraodontiformes) and *Chiloycterus spinosus* (Diodontidae) from Rio de Janeiro, Brazil. *Chromosome Science* 4: 103–105.

Brum, M.J.I. and P.M. Galetti Jr. 1997. Teleostei plan ground karyotype. *Journal of Computational Biology* 2: 91–102.

Brum, M.J.I. and L.C.G. Mota. 2002. C-Banding and Nucleolar Organizer Regions of *Sphoeroides greeleyi* (Tetraodontidae, Tetraodontiformes). *Caryologia* 55: 171–174.

Brum, M.J.I., P.M. Galetti Jr., M.M.O. Corrêa and C.T. Aguilar. 1992. Multiple sex chromosomes in South Atlantic fish *Brevoortia aurea*, Clupeidae. *Brazilian Journal of Genetics* 15: 547–533.

Canapa, A., P.N. Cerioni, M. Barucca, E. Olmo and V. Caputo. 2002. A centromeric satellite DNA may be involved in heterochromatin compactness in gobiid fishes. *Chromosome Research* 10: 297–304.

Campton, D.E. 1987. Natural hybridization and introgression in fishes. Methods of detection and genetic interpretations. In: *Population Genetics in Fishery Management*, N. Ryman and F. Utter (eds.). University of Washington, Academic Press, Seattle, pp. 161–192.

Cano, J., G. Thode and M.C. Alvarez. 1982. Karyoevolutive considerations in 29 Mediterranean Teleost fishes. *Vie et Milieu* 32: 21–24.

Cano, J., A. Pretel, S. Melendez, F. Garcia, V. Caputo, A.S. Fenocchio and L.A.C. Bertollo. 1996. Determination of early stages of sex chromosome differentiation in the sea bass *Dicentrarchus labrax* L. (Pisces: Perciformes). *Cytobios* 87: 45–59.

Caputo, V., F. Marchegiani and E. Olmo. 1996. Karyotype differentiation between two species of carangid fishes, genus *Trachurus* Rafinesque, 1810 (Perciformes, Carangidae). *Marine Biology* 127: 193–199.

Caputo, V., F. Marchegiani, M. Sorice and E. Olmo. 1997. Heterochromatin heterogeneity and chromosome variability in four species of gobiid fishes (Perciformes: Gobiidae). *Cytogenetics and Cell Genetics* 79: 266–271.

Caputo, V., M. Sorice, R. Vitturi, R. Magistrelli and E. Olmo. 1998. Cytogenetic studies in some species of Scorpaeniformes (Teleostei: Percomorpha). *Chromosome Research* 6: 255–262.

Caputo, V., M.L. Caniglia and N. Machella. 1999. The chromosomal complement of *Aphia minuta*, a paedomorphic goby. *Journal of Fish Biology* 55: 455–458.

Carbone, P., R. Vitturi, E. Catalano and M. Macaluso. 1987. Chromosome sex determination and Y-autosome fusion in *Blennius tentacularis* Brunnich, 1765 (Pisces, Blennidae). *Journal of Fish Biology* 31: 597–602.

Carvalho, G.R. 1993. Evolutionary aspects of fish distribution: Genetic variability and adaptation. *Journal of Fish Biology* 47: 103–126.

Chen, T.R. 1969. Karyological heterogamety of deep-sea fishes. *Postilla* (Yale University) 130: 1–29.

Choudhury, R.C., R. Prasad and C.C. Das. 1982. Karyological studies in five tetraodontiform fishes from the Indian Ocean. *Copeia* 1982: 728–732.

Chourrout, D. 1986. Revue sur le dèterminisme genètique du sexe des poissons tèlèostèens. *Bulletin de la Societé Zoologique de France* 113: 123–144.

Christoffels, A. *et al.* 2004. *Fugu* genome analysis provides evidence for a whole-genome duplication early during the evolution of ray-finned fishes. *Molecular Biology and Evolution* 21: 1146–1151.

Colgan, D.J., C.G. Zhang and J.R. Paxtonm. 2000. Phylogenetic investigations of the Stephanoberyciformes and Beryciformes, particularly Whalefishes (Euteleostei: Cetomimidae), based on partial 12S rDNA and 16S rDNA sequences. *Molecular Phylogenetics and Evolution* 17: 15–25.

Corrêa, M.M.O. 1995. *Contribuição à citotaxonomia dos Scorpaeniformes (Osteichthyes – Teleostei): Estudos Citogenéticos em espécies do litoral do Rio de Janeiro, Brasil.* Dissertação de Mestrado, Universidade Federal do Rio de Janeiro, Brazil.

Crnogorac-Jurcevic, T., J.R. Brown, H. Leghach and L.C. Schalkwyk. 1997. *Tetraodon fluviatilis*, a new pufferfish model for genome studies. *Genomics* 41: 177–184.

Devlin, R.H., G.W. Stone and D.E. Smailus. 1998. Extensive direct tandem organization of a long repeat DNA sequence on the Y chromosome of chinook salmon (*Oncorhynchus tshawytscha*). *Journal of Molecular Evolution* 46: 277–287.

DeWoody, J.A. and J.C. Avise. 2000. Microsatellite variation in marine, freshwater and anadromous fishes compared with other animals. *Journal of Fish Biology* 56: 461–473

Doherty, P.J., P. Mather and S. Planes. 1994. *Acanthochromis polyacanthus*, a fish lacking larval dispersal, has genetically differenciated populations at local and regional scales on the Great Barrier Reef. *Marine Biology* 121: 11–21.

Duchac, B., F. Huber, H. Muller and D. Senn. 1982. Mating behaviour and cytogenetical aspects of sex-inversions in the fish *Coris julis* (Labridae, Teleostei). *Experientia* 38: 809–811.

Ene, A.C. 2003. Chromosomal polymorphism in the goby *Neogobius eurycephalus* (Perciformes: Gobiidae). *Marine Biology* 142: 583–588.

Feldberg, E., J.I.R. Porto, E.P. Santos and F.C. Valentim. 1999. Cytogenetic studies of two freshwater scianids of the genus *Plagioscion* (Perciformes, Scianidae) from the Central Amazon. *Genetics and Molecular Biology* 22: 351–356.

Feral, J.P. 2002. How useful are the genetic markers in attempts to understand and manage marine biodiversity? *Journal of Experimental Marine Biology and Ecology* 268: 121–145.

Fischer, C., C. Ozouf-Costaz, H. Roest Crollius, C. Dasilva, O. Jaillon, L. Bouneau, C. Bonillo, J. Weissenbach and A. Bernot. 2000. Karyotype and chromosome location of characteristic tandem repeats in the pufferfish *Tetraodon nigroviridis*. *Cytogenetics and Cell Genetics* 88: 50–55.

Fischer, E.A. 1980. Speciation in the Hamlets (*Hypoplectrus*: Serranidae)—A continuing enigma. *Copeia* 1980: 649–659.

Floeter, S.R. and J.L. Gasparini. 2000. The southwestern Atlantic reef fish fauna: composition and zoogeographic patterns. *Journal of Fish Biology* 56: 1099–1114.

Floeter, S.R., C.E.L. Ferreira, A. Dominici-Arosemena and I.R. Zalmon. 2004. Latitudinal gradients in Atlantic reef fish communities: Trophic structure and spatial use patterns. *Journal of Fish Biology* 64: 1680–1699

Fontana, F., J. Tagliavini and L. Congiu, 2001. Sturgeon genetics and cytogenetics: Recent advancements and perspectives. *Genetica* 111: 359–373.

Galetti Jr., P.M., F. Foresti, L.A.C. Bertollo and O. Moreira-Filho. 1984. Characterization of eight species of Anostomidae (Cypriniformes) fish on the basis of the nucleolar organizing region. *Caryologia* 37: 401–406.

Galetti Jr., P.M., C.A. Mestriner, P.C. Vênere and F. Foresti. 1991. Heterochromatin and karyotypic reorganization in fish of the family Anostomidae (Characiformes). *Cytogenetics and Cell Genetics* 56: 116–121.

Galetti Jr., P.M., C.T. Aguilar and W.F. Molina. 2000. An overview on marine fish cytogenetics. *Hydrobiologia* 420: 55–62.

Galetti Jr., P.M., W.F. Molina, P.R.A.M. Affonso and C.T. Aguilar. 2006. Assessing Genetic Diversity of Brazilian Reef Fishes by Chromosomal and DNA Markers. *Genetica* 126(1–2): 161–177.

Giles, V., G. Thode and M.C. Alvarez. 1985. A new Robertsonian fusion in the multiple chromosome polymorphism of a Mediterranean population of *Gobius paganellus* (Gobiidae, Perciformes). *Heredity* 55: 255–260.

Gold, J.R. 1979. Cytogenetics. In: *Fish Physiology.*, W.S. Hoar, D.J. Randall and J.R. Brett (eds.), Academic Press, New York, Vol. 8, pp. 353–405.

Gomes, V., V.N. Phan and M.J.A.C. Passo. 1994. Karyotypes of three species of marine catfishes from Brazil. *Boletim do Instituto Oceanográfico de São Paulo* 42: 55–61.

Grützner, F., G. Lütjens, C. Rovira, D.W. Barnes, H.H. Ropers and T. Haaf. 1999. Classical and molecular cytogenetics of the pufferfish *Tetraodon nigroviridis*. *Chromosome Research* 7: 655–662.

Harrington, M.E. 1993. Aggression in damselfish: Adult-juvenile interactions. *Copeia* 1993: 67–74.

Hartley, S.E. and M.T. Horne. 1984. Chromosome relationships in the genus *Salmo*. *Chromosoma* 90: 229–237.

Hawthorne, D.J. and S. Via. 2001. Genetic linkage of ecological specialization and reproductive isolation in pea aphids. *Nature* (*Lond.*) 412: 904–907.

Herwerden, V.L., C.R. Davies and J.H. Choat. 2002. Phylogenetic and evolutionary perspectives of the Indo-Pacific grouper *Plectropomus* species on the Great Barrier Reef, Australia. *Journal of Fish Biology* 60: 1591–1596.

Hubbs, C. 1970. Teleost hybridization studies. *Proceedings of the California Academy Sciences* 38: 289-297.

Jaillon, O. *et al.* 2004. Genome duplication in the teleost fish *Tetrodon nigroviridis* reveals the early vertebrate proto-karyotype. *Nature* (*Lond.*) 431: 946–957.

John, B. 1981. Chromosome change and evolutionary change: A critique. In: *Evolution and Speciation*, Atchley, W.R. and D.S. Woodruff (eds.). Cambridge University Press, Cambridge, pp. 23–51.

King, M. 1980. C-banding studies on Australian hylid frogs: secondary constriction structure and the concept of euchromatin transformation. *Chromosoma* 80: 191–217.

King, M. 1981. Chromosome change and speciation in lizards. In: *Evolution and Speciation. Essay in Honor of M.J.D. White*. W.R. Atchley and D.S. Woodruff (eds.), Cambridge University Press, Cambridge, pp. 262–285.

King, M. 1987. Chromosomal rearrangements, speciation and the theorical approach. *Heredity* 59: 1–6.

King, M. 1992. A dual level model for speciation by multiple pericentric inversions. *Heredity* 68: 437–440.

King, M. 1993. *Species Evolution: The Role of Chromosome Change*. Cambridge University Press, Cambridge.

Klinkhardt, M., M. Tesche and H. Greven. 1995. *Database of Fish Chromosomes*. Westarp Wissenschaften, Magdeburg.

Klinkhardt, M.B. 1998. Some aspects of karyoevolution in fishes. *Animal Research and Development* 47: 7–36.

Kloc, M. and B. Zagrodzinska. 2002. Chromatin elimination: An oddity or a common mechanism in differentiation and development? *Differentiation* 68: 84–91.

Kubota, S., T. Ishibashi and S. Kohno. 1997. A germline restricted, highly repetitive DNA sequence in *Paramyxine atami*: An interspecifically conserved but somatically eliminated element. *Molecular and General Genetics* 256: 252–256.

Kubota, S., J. Takano, R. Tsuneishi, S. Kobayakawa, N. Fujikawa, M. Nabeyama and S. Kohno. 2001. Highly repetitive DNA families restricted to germ cells in a Japanese hagfish (*Eptatretus burgeri*): A hierarchical and mosaic structure in eliminated chromosomes. *Genetica* 111: 319–328.

Lacson, J.M. 1992. Minimal genetic variation among samples of six species of coral reef fishes collected at La Parguera, Puerto Rico, and Discovery Bay, Jamaica. *Marine Biology* 112: 327–331.

Lamatsch, D.K., C. Steinlein, M. Schmid and M. Schartl. 2000. Noninvasive determination of genome size and ploidy level in fishes by flow cytometry: detection of triploid *Poecilia formosa. Cytometry* 39: 91–95.

Le Comber, S.C.L. and C. Smith. 2004. Polyploidy in fishes: patterns and processes. *Biological Journal of the Linnean Society* 82: 431–442.

LeGrande, W.H. 1988. Chromosome numbers of some gulf coast Scianid fishes. *Copeia* 1988: 91–493.

Maistro, E.L., C. Oliveira and F. Foresti. 1998. Comparative cytogenetic and morphological analysis of *Astyanax scabripinnis paranae* (Pisces, Characidae, Tetragonopterinae). *Genetics and Molecular Biology* 21: 201–206.

Mandrioli, M. 2000. Cytogenetic analysis of the pufferfish *Tetraodon fluviatilis* (Osteichthyes). *Chromosome Research* 8: 237–242.

Mandrioli, M. and G.C. Manicardi. 2001. Cytogenetic and molecular analysis of the pufferfish *Tetraodon fluviatilis* (Osteichthyes). *Genetica* 111: 433–438.

Mandrioli, M., M.S. Colomba and R. Vitturi. 2000. Chromosomal analysis of repeated DNAs in the rainbow wrasse *Coris julis* (Pisces, Labridae). *Genetica* 108: 191–195.

Mantovani, M., L.D.S. Abel, C.A. Mestriner and O. Moreira-Filho. 2000. Accentuated polymorphism of heterochromatin and nucleolar organizer regions in *Astyanax scabripinnis* (Pisces, Characidae): Tools for understanding karyotypic evolution. *Genetica* 109: 161–168,

Martins, C. and P.M. Galetti Jr. 1999. Chromosomal localization of 5S rDNA genes in *Leporinus* fish (Anostomidae, Characiformes). *Chromosome Research* 7: 363–367.

Martins, C. and P.M. Galetti Jr. 2001. Two 5S rDNA arrays in neotropical fish species: Is it a general rule for fishes? *Genetica* 111: 439–446.

Mazzei, F., L. Ghigliotti, C. Bonillo, J.P. Coutanceau, C. Ozouf-Costaz and E. Pisano. 2004. Chromosomal patterns of major and 5S ribosomal DNA in six icefish species (Perciformes, Notothenioidei, Channichthyidae). *Polar Biology* 28: 47–55.

Meyne, J., R.J. Baker, H.H. Hobart, T.C. Hsu, A.O. Ryder, O.G. Ward, J.E. Wiley, D.H. Wurster-Hill, T.L. Yates and R.K. Moyzis. 1990. Distribution of non-telomeric sites of the $(TTAGGG)_n$ telomeric sequence in vertebrate chromosomes. *Chromosoma* 99: 3–10.

Molina, W.F. 2000. *Análise da diversidade genética em Pomacentridae (Perciformes), através do uso combinado da citogenética, marcadores moleculares e morfometria multivariada.* Ph.D. Thesis. Universidade Federal de São Carlos, São Paulo, Brazil.

Molina, W.F. 2005. Intraspecific karyotypical diversity in brackish water fishes of the Eleotridae family (Pisces, Perciformes). *Cytologia* 70: 39–45.

Molina, W.F. and P.M. Galetti Jr. 2002. Robertsonian rearrangements in the reef fish *Chromis* (Perciformes, Pomacentridae) involving chromosomes bearing 5S rRNA genes. *Genetic and Molecular Biology* 25: 373–377.

Molina, W.F., P.M. Galetti Jr. 2004a. Karyotypic changes associated to the dispersive potential on Pomacentridae (Pisces, Perciformes). *Journal of Experimental Marine Biology and Ecology* 309: 109–119.

Molina, W.F. and P.M. Galetti Jr. 2004b. Multiple pericentric inversions and chromosome divergence in the reef fishes *Stegastes* (Perciformes, Pomacentridae). *Genetics and Molecular Biology* 27: 543–548.

Molina, W.F., F.A. Maia-Lima and P.R.A.M. Affonso. 2002. Divergence between karyotypical pattern and speciation events in Serranidae fish (Perciformes). *Caryologia* 55: 299–305.

Molina, W.F., L.C.B. Lima, A.J.M. Vasconcelos and I. Accioly. 2005. Extensive karyotypic similarity in marine fishes of the family Haemulidae. *Caryologia* (In press).

Montes-Pérez, R. 1981. *Estudios citogéneticos em Eleotris pisonis (Gobiidae, Perciformes)*. Tesis professional, Faculdad de Ciências. Univ. Nal. Autom. México.

Moreira-Filho, O. and L.A.C. Bertollo. 1991. *Astyanax scabripinnis* (Pisces, Characidae): a species complex. *Revista Brasileira de Genética* 14: 331–357.

Moreira-Filho, O., L.A.C. Bertollo and P.M. Galetti Jr. 1993. Distribution of sex chromosome mechanisms in Neotropical fish and description of a ZZ/ZW system in *Parodon hilarii* (Parodontidae). *Caryologia* 46: 115–125.

Morescalchi, A., J.C. Hureau, E. Olmo, C. Ozouf-Costaz, E. Pisano and R. Stanyon. 1992. A multiple sex-chromosome system in Antartic ice-fishes. *Polar Biology* 11: 655–661.

Murofushi, M. and T.H. Yoshida. 1984. Cytogenetical studies on fishes. VIII. XX-Y sex chromosome mechanism newly found in the snake eel, *Muraenichthys gymnotus* (Anguiliformes, Pisces). *Proceedings of the Japan Academy* 60B: 21–23.

Murofushi, M., S. Oikawa, S. Nishikawa and T.H. Yoshida. 1980. Cytogenetical studies in fishes. III. Multiple sex chromosome mechanism on the filefish *Stephanolepis cirrifer*. *Japanese Journal of Genetics* 55: 127–132.

Nabeyama, M., S. Kubota and S. Kohno. 2000. Concerted evolution of a highly repetitive DNA family in Eptatretidae (Cyclostomata, Agnatha) implies specifically differential homogenization and amplification events in their germ cells. *Journal of Molecular Evolution* 50: 154–169.

Nakayama, I., F. Foresti, R. Tewari, M. Schartl and D. Chourrout. 1994. Sex chromosome polymorphism and heterogametic males revealed by two cloned DNA probe in the ZW/ZZ fish *Leporinus elongatus*. *Chromosoma* 103: 31–39.

Nanda, I., W. Feichtinger, M. Schmid, J.H. Schröder, H. Zischler and J.T. Epplen. 1990. Simple repetitive sequences are associated with differentiation of the sex chromosomes in the guppy fish. *Journal of Molecular Evolution* 30: 456–462.

Neafsey, D.E. and S.R. Palumbi. 2003. Genome size evolution in pufferfish: A comparative analysis of diodontid and tetraodontid pufferfish genomes. *Genome Research* 13: 821–830

Nelson, J.S. 1994. *Fishes of the World*. 3rd Edition John Wiley & Sons, New York.

Nishikawa, S. and K. Sakamoto. 1978. Comparative studies on the chromosomes in Japanese fishes. IV. Somatic chromosomes of two lizardfishes. *Journal of Shimonoseki University of Fisheries* 27: 113–117.

Norse, E.A. 1993. *Global Marine Biological Diversity. A strategy for building conservation into decision making*. Island Press, Washington, 383 p.

Ohno, S. 1974. Sex chromosomes and sex determining mechanisms. In: *Animal Cytogenetics*, Vol. 4. Gebrüder Borntraeger, Berlin, pp. 46–63.

Ojima, Y. and E. Kashiwagi. 1981. Chromosomal evolution associated with Robertsonian fusion in the genus *Dascyllus* (Chromininae, Pisces). *Proceedings of the Japan Academy* 57B, 368.

Ojima, Y. and H. Ueda, 1982. A karyological study of the conger eel (*Conger myriaster*) *in vitro* cells, with special regard to the identification of the sex chromosome. *Proceedings of the Japan Academy* 58: 56–59.

Ojima, Y. 1983. Fish Cytogenetics. In: *Chromosomes in Evolution of Eukaryotic Groups*. A.K. Sharma, A. Sharma (eds.), Vol. 1, Chapter 2, CRC Press, Boca Raton, 254 p.

Oliveira, C. and F. Foresti. 1994. Revisão dos estudos citogenéticos em peixes Neotropicais de águas continentais. *Proceedings of IV Simpósio de Citogenética Evolutiva e Aplicada de Peixes Neotropicais*, Botucatu, Brazil.

Ormond, R.F.G., J.M. Roberts and R.Q. Jan. 1996. Behavioural differences in microhabitat use by damselfishes (Pomacentridae): Implications for ref. fish biodiversity. *Journal of Experimental Marine Biology and Ecology* 202: 85–95.

Ozouf-Costaz, C., J.C. Hureau and M. Beaunier. 1991. Chromosome studies of fish of the suborder Notothenioidei collected in the Weddell Sea during EPOS 3 cruise. *Cybium* 15: 271–289.

Ozouf-Costaz, C., E. Pisano, C. Bonillo and R. Williams. 1996. Ribosomal DNA location in the Antarctic fish *Champsocephalus gunnari* (Notothenioidei, Channichthyidae) by banding and fluorescence *in situ* hybridization. *Chromosome Research* 4: 557–561.

Ozouf-Costaz, C., E. Pisano, C. Thaeron and J. Hureau, 1997. Antarctic fish chromosome banding: Significance for evolutionary studies. *Cybium* 21: 399–409.

Pampoulie, C., E.S. Gysels, G.E. Maes, B. Hellemans, V. Leentjes, A.G. Jones and F.A.M. Volckaert. 2004. Evidence for fine-scale genetic structure and estuarine colonization in a potential high gene flow marine goby (*Pomatoschistus minutus*). *Heredity* 92: 434–445.

Perry, J., H.R. Slater and K.H.A. Choo. 2004. Centric fission: simple and complex mechanisms. *Chromosome Research* 12: 627–640.

Pfeiler, E., J. Colborn, M.R. Douglas and M.E. Douglas. 2002. Systematic status of bonefishes (*Albula* spp.) from the eastern Pacific Ocean inferred from analyses of allozymes and mitochondrial DNA. *Environmental Biology of Fishes* 63: 151–159.

Phillips, R. and P. Ráb. 2001. Chromosome evolution in the Salmonidae (Pisces): An update. *Biological Reviews* 76: 1–26.

Pisano, E. and C. Ozouf-Costaz. 2000. Chromosome change and the evolution in the Antarctic fish suborder Notothenioidei. *Antarctic Science* 72: 334–342.

Pisano, E. and C. Ozouf-Costaz. 2003. Cytogenetics and evolution in extreme environment: The case of Antarctic Fishes. In: *Fish Adaptations*, A.L. Val and B.G. Kapoor (eds.), Science Publishers, Inc., Enfield (NH), USA, pp. 309–336.

Pisano, E., C. Ozouf-Costaz, J.C. Hureau and R. Williams. 1995. Chromosome differentiation in the subantarctic Bovichtidae species *Cottoperca gobio* (Gunther, 1861) and *Pseudaphritis urvillii* (Valenciennes, 1832) (Pisces, Perciformes). *Antarctic Science* 7: 381–386.

Pisano, E., F. Mazzei, N. Derome, C. Ozouf-Costaz, J.C. Hureau and G. di Prisco. 2001. Cytogenetics of the bathydraconid fish *Gymnodraco acuticeps* (Perciformes, Notothenioidei) from Terra Nova Bay, Ross Sea. *Polar Biology* 24: 846–852.

Planes, S. and P.J. Doherty. 1997. Genetic and color interactions at a contact zone of *Acanthochromis polyacanthus*: A marine fish lacking pelagic larvae. *Evolution* 51: 1232–1243.

Planes, S., F. Bonhomme and R. Galzin. 1996. Genetic structure of *Dascyllus aruanus* populations in French Polynesia. *Marine Biology* 117: 665–674.

Pyle, R.L. and J.E. Randall. 1994. A review of hybridization in marine angelfishes (Perciformes: Pomacanthidae). *Environmental Biology of Fishes* 41: 127–145.

Qumsiyeh, M.B. 1994. Evolution of number and morphology of mammalian chromosomes. *Journal of Heredity* 85: 455–465.

Robbins, L.W., M.P. Moulton and R.J. Baker. 1983. Extent of geographic range and magnitude of chromosomal evolution. *Journal of Biogeography* 10: 533–541.

Robertson, D.R. 1984. Cohabitation of competing territorial damselfishes of a Caribbean coral reef. *Ecology* 65: 1121–1135.

Robins, C.R., G.C. Ray, J. Douglass and E. Freund. 1986. *A Field Guide to Atlantic Coast Fishes of North America*. The Petersen Field Guide Series, Houghton Mifflin, Boston.

Rocha, L.A., A.L. Bass, R. Robertson and B.W. Bowen. 2002. Adult habitat preferences, larval dispersal, and the comparative phylogeography of three Atlantic surgeonfishes (Teleostei: Acanthuridae). *Molecular Ecology* 11: 243–252.

Rocha, L.A., D.R. Robertson, J. Roman and B.W. Bowen. 2005. Ecological speciation in tropical reef fishes. *Proceedings of the Royal Society* B. 1-7, Published online. doi:10.1098/2004.3005.

Rodrigues, M.C. 2003. *Análise genética por marcadores RAPD em Abudefduf saxatilis (Pisces, Pomacentridae) no litoral nordeste do Brasil e arquipélago São Pedro e São Paulo*. Monografia. Universidade Federal do Rio Grande do Norte, Brazil.

Rossi, A.R., D. Crosetti, E. Gornung and L. Sola. 1996. Cytogenetic analysis of global populations of *Mugil cephalus* (striped mullet) by different staining techniques and fluorescent *in situ* hybridization. *Heredity* 76: 77–82.

Rossi, A.R., E. Gornung and D. Crosetti, 1997. Cytogenetic analysis of *Liza ramada* (Pisces, Perciformes) by different staining techniques and fluorescent *in situ* hybridization. *Heredity* 79: 83–87.

Sá-Gabriel, L.G. and W.F. Molina. 2004. Sex chromosomes in *Stephanolepis hispidus* (Monacanthidae, Tetraodontiformes). *Cytologia* 69: 447–452.

Sá-Gabriel, L.G. and W.F. Molina. 2005. Karyotype diversification in fishes of the Balistidae, Diodontidaae and Tetraodontidae (Tetraodontiformes). *Caryologia* 58: 229–237.

Sale, P.F. (ed.) 1991. *The Ecology of Coral Reef Fishes*. Academic Press, San Diego.

Salvadori, S., E. Coluccia, R. Cannas, A. Cau and A.M. Deiana. 2003. Replication banding in two Mediterranean moray eels: chromosomal characterization and comparison. *Genetica* 119: 253–258.

Schweizer, D. and J. Loidl. 1987. A model for heterochromatin dispersion and the evolution of C-band patterns. *Chromosome Today* 9: 61–74.

Sena, D.C.S. and W.F. Molina. 2005. Chromosomal rearrangements associated with the larval dispersal ability in Labridae (Perciformes, Labridae). *Neotropical Ichthyology* (In press).

Sena, D.C.S. and W.F. Molina. 2005. Robertsonian rearrangements and pericentric inversions in fishes of the Scaridae (Perciformes) family. *Caryologia*. (In press).

Shulman, M.A. and E. Bermingham. 1995. Early life histories, ocean currents, and the population genetics of Caribbean reef fishes. *Evolution* 49: 897–910.

Smith, P.J. and Y. Fujio. 1982. Genetic variation in marine teleosts: high variability in habitat specialists and low variability in habitat generalists. *Marine Biology* 69: 7–20.

Sola, L., O. Cipelli, E. Gornung, A.R. Rossi, F. Andaloro and D. Crosetti. 1997. Cytogenetic characterization of the greater amberjack, *Seriola dumerili* (Pisces: Carangidae), by different staining techniques and fluorescence *in situ* hybridization. *Marine Biology* 128: 573–577.

Sola, L., S. De Innocentiis, E. Gornung, S. Papalia, A.R. Rossi, G. Marino, P. De Marco and S. Cataudella. 2000. Cytogenetic analysis of *Epinephelus marginatus* (Pisces, Serranidae) with the chromosome localization of 18S and 5S rRNA genes and of the (TTAGGG)n telomeric sequence. *Marine Biology* 137: 47–51.

Souza, I.L., O. Moreira-Filho and L.A.C. Bertollo. 1995. Cytogenetic diversity in the *Astyanax scabripinnis* species complex (Pisces, Characidae). II. Different cytotypes living in sympatry. *Cytologia* 60: 273–281.

Stepien, C.A. and R.H. Rosenblatt. 1996. Genetic divergence in antitropical pelagic marine fishes (*Trachurus*, *Merluccius*, and *Scomber*) between North and South America. *Copeia* 1996: 586–598.

Takai, A. and Y. Ojima. 1995. Chromosome evolution associated with Robertsonian rearrangements in Pomacentrid fish (Perciformes). *Cytobios* 84: 103–110.

Taylor, M.S. and M.E. Hellberg. 2003. Genetic evidence for local retention of pelagic larvae in a Caribbean reef fish. *Science* 299: 107–108.

Thode, G., G. Martinez, J.L. Ruiz and J.R. Lopez. 1988. A complex chromosomal polymorphism in *Gobius fallax* (Gobiidae, Perciformes). *Genetica* 76: 65–71.

Thode, G., V. Giles and M.C. Alvarez. 1985. Multiple chromosome polymorphism in *Gobius paganellus* (Teleostei, Perciformes). *Heredity* 54: 3–7.

Torres, R.A., J.J. Roper, F. Foresti and C. Oliveira. 2005. Surprising genomic diversity in the Neotropical fish *Synbranchus marmoratus* (Teleostei, Synbranchidae): How many species? *Neotropic Ichthyology* 3: 277–284.

Ueno, K. and A. Takai. 2000. Chromosome evolution involving Robertsonian rearrangements in *Xirichthys* fish (Labridae, Perciformes). *Cytobios* 103: 7–15.

Ueno, K., K. Ota and T. Kobayashi. 2001. Heteromorphic sex chromosomes of lizardfish (Synodontidae): Focus on the ZZ-ZW1W2 system in *Trachinocephalus myops*. *Genetica* 111: 133–142.

Uribe-Alcocer, M. and A. Ramírez-Escamilla. 1989. Comparación citogenética entre las especies del genero *Dormitator* (Pisces, Gobiidae). *An. Instituto de Ciencias del Mar y Limnologie, Universad Nacional Autónomas de México* 16: 75–80.

Vasil'ev, V.P. and K.A. Grigoryan. 1993. Karyology of the Gobiidae. *Journal of Ichthyology* 33: 1–16.

Vermeij, G.J. 1978. *Biogeography and Adaptation. Patterns of Marine Life.* Harvard University Press, Cambridge, USA.

Vicente, V.E., L.A.C. Bertollo, S.R. Valentini and O. Moreira-Filho. 2003. Origin and differentiation of a sex chromosome system in *Parodon hilarii* (Pisces, Parodontidae). Satellite DNA, G- and C-banding. *Genetica* 119: 115–120

Victor, B.C. 1986. Duration of the planktonic larval stages of one hundred species of Pacific and Atlantic wrasses (family Labridae). *Marine Biology* 90: 317–326.

Vitturi, R. and E. Catalano. 1989. Multiple chromosome polymorphism in the Gobiid fish *Gobius niger jozo* L. 1758 (Pisces, Gobiidae). *Cytologia* 54: 231–235.

Vitturi, R., A. Mazzola, M. Macaluso and E. Catalano. 1986. Chromosomal polymorphism associated with Robertsonian fusion in *Seriola dumerilli* (Risso, 1980) (Pisces, Carangidae). *Journal of Fish Biology* 26: 529–534.

Vitturi, R., E. Catalano, A. Maiorca and T. Carollo. 1988. Karyological studies in *Coris julis* (Pisces, Labridae). *Genetica* 76: 219–223.

Vitturi, R., A. Libertini, M. Campolmi, F. Calderazzo and A. Mazzola. 1998. Conventional karyotype, nucleolar organizer regions and genome size in five Mediterranean species of Syngnathidae (Pisces, Syngnathiformes). *Journal of Fish Biology* 52: 677–687.

Ward, R.D. 1995. Population genetics of tunas. *Journal of Fish Biology* 47: 259–280.

Wellington, G.M and B.C. Victor. 1989. Planktonic larval duration of one hundred species of Pacific and Atlantic damselfishes (Pomacentridae). *Marine Biology* 101: 557–567.

White, M.J.D. 1968. Models of speciation. *Science* 159: 1065–1070.

White, M.J.D. 1973. *Animal Cytology and Evolution.* 3rd Edition. Cambridge University Press, Cambridge.

Karyotypic Evolution in Neotropical Fishes

Claudio Oliveira[1*], Lurdes Foresti de Almeida-Toledo[2] and Fausto Foresti[1]

INTRODUCTION

The Neotropical freshwater fish fauna is extremely rich, including 71 families and 4,475 species described, according to the most recent revision (Reis *et al.*, 2003). Moreover, an estimate in the same study suggested the existence of about 6,000 species in the rivers and lakes of the Neotropical regions. According to Schaefer (1998), about 8,000 freshwater fish species could exist in the Neotropics, which corresponds to about 25% of all fish species of the world. Vari and Malabarba (1998) agreed with this number and showed that all this diversity may occur in only about 0.003% of the planet's water. These data are very impressive considering the fact that the amount of freshwater species in the entire world is estimated to be about 13,000 species. The difference between the number of species described

Address for Correspondence: [1*]Departamento de Morfologia, Instituto de Biociências, Universidade Estadual Paulista, 18618-000, Botucatu, SP, Brazil.
E-mail: claudio@ibb.unesp.br
[2]Departamento de Genética Biologia Evolutiva Instituto de Biociências, Universidade de São Paulo, São Paulo, SP, Brazil.

and the actual number of species makes most research, including cytogenetic studies, difficult since many species still do not have the reference of a scientific name. Thus, several species possess a karyotype but have been identified only until the genus level.

The composition of the freshwater fish fauna of the Neotropics at high taxonomic level is singular since 3,192 species (71%) belong to the superorder Ostariophysi and 1,283 to other fish groups (Reis et al., 2003).

Only scattered information is available on Neotropical marine fishes. In Brazil, at least 1,298 marine species, being four of Myxini (hagfishes), 139 Chondrichthyes (sharks and rays), and 1,155 Actinopterygii (ray-finned fishes) have been recorded (Menezes et al., 2003).

Although some works investigating the cytogenetic aspects of Neotropical fish have been conducted during the 1940s (Wickbom, 1943), only in the 1960s, with the publication of the Post paper (1965), a series of regular publications by Europeans, Americans and Japanese about the karyotype of Neotropical fishes occurred (Ohno and Atkin, 1966; Muramoto et al., 1968; Fontana et al., 1970; Gyldenholm and Scheel, 1971; Hinegardner and Rosen, 1972; Howell, 1972; Scheel et al., 1972; Scheel, 1973; Chiarelli and Capanna, 1973; Denton, 1973; Hudson, 1976; Ojima et al., 1976; Thompson, 1979). Most of these data included only haploid or diploid numbers and since the fishes studied were mostly obtained from aquarists, the collect sites are, in general, not cited.

The first species studied by a researcher in the Neotropical region was *Synbranchus marmoratus* (Andrea, 1971), but the data was only presented in a Brazilian scientific meeting and never published. Then, several other fish groups began to be analyzed such as Poeciliidae and Curimatidae (Foresti, 1974; Foresti et al., 1974), Pimelodidae (Toledo, 1975), Loricariidae (Michelle, 1975), Erythrinidae (Bertollo, 1978) and Gymnotiformes (Almeida-Toledo, 1978). After these initial studies, several others have been developed with a wide range of groups, resulting in a considerable amount of knowledge available today. The cytogenetic studies are restricted to freshwater fishes mostly due to the fact that the freshwater fish fauna is among the richest in terms of number of species in the world (Nelson, 1994) and most researchers working in this area act in public universities inland of Brazil.

The first review on fish chromosome data about Neotropical species showed that until 1978, haploid and/or diploid numbers were known for about 252 species, belonging to 88 genera and 23 families (Almeida-Toledo, 1978). A second review showed that haploid and/or diploid numbers were known for about 433 species of freshwater fishes belonging to 145 genera and 33 families (Oliveira *et al.*, 1988). The first published review of chromosome data about Neotropical marine fishes showed that until 1996, diploid numbers were known for 44 species belonging to 38 genera and 23 families (Brum, 1996).

The present report is based on two databases (one with the freshwater species and the other with the marine species) created in 1988 and updated periodically. These databases are becoming larger each year and now they together contain about 3,000 data covering information about specific names and classifications, collecting places, haploid and diploid numbers, fundamental numbers, karyotypic formulae, number of nucleolus organizer regions (Ag-NORs), chromosome banding, B chromosomes, sex chromosomes and DNA content of 1,156 species. Unfortunately, the large size of these databases makes their publication in the traditional printed form impossible. The complete databases are available to the scientific community through the World Wide Web at the addresses: http://www.ibb.unesp.br/laboratorios/Lista%20Neotropicais.xls (freshwater fishes) and http://www.ibb.unesp.br/laboratorios/Lista%20Neotropicais%20Marinhos.xls (marine fishes) or upon direct request to the corresponding author. Studies using data within these databases should cite the present paper as the primary reference.

A general analysis of the available cytogenetic data clearly indicates that these studies are mainly concentrated in freshwater Neotropical fishes. Thus, today, cytogenetic information is available for 475 species of Characiformes (Table 1.5.1), 318 species of Siluriformes (Table 1.5.2), 48 species of Gymnotiformes (Table 1.5.3), 199 freshwater species that do not belong to the superorder Ostariophysi (Table 1.5.4), and 109 species of marine fishes (Table 1.5.5). The current databases also show that supernumerary chromosomes were described for 38 species (Table 1.5.6), sex chromosomes with female heterogamety were described for 40 species (Table 1.5.7) and sex chromosomes with male heterogamety were described for 22 species (Table 1.5.8).

In this chapter, we shall review the results obtained so far with fishes from the Neotropical region. All cytogenetic descriptions and discussions are based on the databases cited above. References taken from abstracts of scientific meetings are not cited here.

Table 1.5.1   Summary of the diploid number (2n) described for the order Characiformes. Numbers of known genera and species from Reis *et al.* (2003).

		Number of genera known/studied	Number of species known/studied	2n Range	Modal 2n
Acestrorhynchidae		1/1	15/5	50	50
Anostomidae		12/8	138/57	54	54
Characidae	Species Incertae Sedis	3/2	9/3	50-52	50-52
	Genera *Incertae* Sedis	85/33	637/189	28-52	50
	Aphyocharacinae	1/1	10/3	50	50
	Bryconinae	3/2	41/15	50	50
	Characinae	12/5	70/15	46-52	52
	Cheirodontinae	15/2	46/11	52	52
	Glandulocaudinae	19/3	50/6	52-54	52
	Iguanodectinae	2/1	11/1	50	50
	Serrasalminae	15/11	80/39	54-64	60
	Stethaprioninae	4/3	12/4	50	50
	Tetragonopterinae	1/1	2/2	50-52	52
Chilodontidae		2/2	7/2	54	54
Crenuchidae		12/2	73/11	38-50	50
Ctenoluciidae		2/2	7/2	36	36
Curimatidae		8/7	97/46	46-102	54
Cynodontidae		5/3	14/3	54	54
Erythrinidae		3/3	15/6	40-54	50
Gasteropelecidae		3/3	9/4	48-54	52-54
Hemiodontidae		5/4	28/9	54	54
Lebiasinidae		7/4	61/19	22-46	42
Parodontidae		3/2	23/9	54	54
Prochilodontidae		3/2	21/13	54	54
Total		226/107	1476/474		

Chromosome Evolution in Characiformes

The order Characiformes represents a wide fish group distributed in the tropical and Neotropical regions (Nelson, 1994). In spite of the ecological and economical importance of many characiform species, the phylogenetic relationships between many groups (particularly within the family Characidae) still remain unsolved. Molecular phylogenetic analyses by Ortí (1997) reach to a phylogeny very different from that published by Buckup (1998) based on morphological characters. Some studies

Table 1.5.2 Summary of the diploid number (2n) described for the order Siluriformes. Numbers of known genera and species from Reis *et al.* (2003).

	Number of genera known/studied	Number of species known/studied	2n Range	Modal 2n
Genus *Incertae Sedis*	1/1	1/1	60	60
Aspredinidae	12/1	36/4	50-64	50
Auchenipteridae	20/8	91/13	56-58	58
Callichthyidae	8/7	177/60	38-134	46
Diplomystidae	2/2	6/3	56	56
Doradidae	29/12	74/20	56-66	58
Heptapteridae	24/6	186/38	42-58	58
Loricariidae	94/30	673/124	36-96	54
Nematogenyidae	1/1	1/1	94	94
Pimelodidadae	31/13	83/30	50-58	56
Pseudopimelodidae	5/3	26/4	54	54
Trichomycteridae	41/5	171/20	48-64	54
Total	268/89	1525/318		

Table 1.5.3 Summary of the diploid number (2n) described for the order Gymnotiformes. Numbers of known genera and species from Reis *et al.* (2003).

	Number of genera known/studied	Number of species known/studied	2n Range	Modal 2n
Apteronotidae	12/1	52/4	22-52	52
Gymnotidae	2/2	19/15	38-54	52
Hypopomidae	7/3	25/6	36-50	42
Rhamphichthyidae	3/1	13/1	52	52
Sternopygidae	5/2	27/22	28-48	32
Total	29/9	136/48		

attempted to solve the relationships between smaller groups in the last few years, but the position of most groups of species needs to be investigated to a greater extent.

Table 1.5.1 shows that the cytogenetic data are known for 474 species and 107 genera of characiforms representing 32% of known species and 48% of known genera. The diploid modal number is 2n=54 chromosomes, but values below and above this modal number are also observed among fishes of this order.

The family Acestrorhynchidae is composed of one genus and 15 species (Menezes, 2003). Cytogenetic studies carried out in five species showed 2n=50 chromosomes (Table 1.5.1), many biarmed chromosomes,

Table 1.5.4 Summary of the diploid number (2n) described for Neotropical freshwater fishes that do not belong to the superorder Ostariophysi. Numbers of known genera and species from Reis *et al.* (2003).

	Number of genera known/studied	Number of species known/studied	2n range	Modal 2n
Elasmobranchii				
Rajiformes				
Potamotrygonidae	3/1	23/2	66	66
Actinopterygii				
Osteoglossiformes				
Arapaimidae	1/1	1/1	56	56
Osteoglossidae	1/1	2/2	54-56	56
Clupeiformes				
Pristigasteridae	3/1	5/1	48	48
Atheriniformes				
Atherinopsidae	9/1	94/2	46-48	46-48
Cyprinodontiformes				
Anablepidae	3/1	15/1	46	46
Cyprinodontidae	5/1	58/2	48	48
Poeciliidae	28/10	216/25	36-50	48
Rivulidae	27/13	235/41	20-54	48
Beloniformes				
Belonidae	4/1	9/1	52	52
Synbranchiformes				
Synbranchidae	2/1	4/1	42-46	42-46
Perciformes				
Sciaenidae	5/1	21/2	48	48
Polycentridae	2/1	2/1	46	46
Cichlidae	53/32	406/111	38-60	48
Gobiidae	12/3	40/4	46-48	46
Pleuronectiformes				
Achiridae	8/2	20/2	38-40	38-40
Sarcopterygii				
Lepidosireniformes				
Lepidosirenidae	1/1	1/1	38	38
Total	167/72	1152/200		

a single Ag-NOR-bearing chromosome pair, with the exception of *Acestrorhynchus pantaneiro* which displays two Ag-NOR-bearing chromosome pairs. The available cytogenetic data suggest a conserved macrokaryotypic structure. However, A. *pantaneiro* has 1.70 ± 0.04 pg of DNA/nucleus, while *Acestrorhynchus* sp. from Acre River has 3.10 ± 0.06 (Carvalho *et al.*, 2002b), pointing towards the need for further cytogenetic and DNA content studies, in order to investigate this difference in DNA content and even a possible diploid/tetraploid relationship.

Table 1.5.5 Summary of the diploid number (2n) described for the Neotropical marine fishes. Numbers of known genera and species from Menezes *et al.* (2003).

	Number of genera known/studied	Number of species known/studied	2n Range	Modal 2n
Muraenidae	7/1	16/1	42	42
Belonidae	7/1	12/1	48	48
Holocentridae	6/2	7/2	48-50	48-50
Clupeidae	8/1	12/2	45-48	46
Engraulidae	9/1	26/1	48	48
Poeciliidae	1/1	1/1	46	46
Ariidae	8/5	21/5	54-56	56
Batrachoididae	5/1	13/2	44	44
Atherinopsidae	4/3	14/4	48-50	48
Dactylopteridae	1/1	1/1	48	48
Scorpaenidae	7/1	20/2	40-46	40-46
Triglidae	2/1	5/1	100-102	100-102
Acanthuridae	1/1	4/3	34-48	34
Blenniidae	7/2	9/2	48	48
Carangidae	17/6	35/10	46-48	48
Centropomidae	1/1	4/3	48	48
Gerreidae	5/2	10/2	48	48
Gobiidae	21/1	40/1	48	48
Haemulidae	7/5	19/8	48	48
Labridae	7/2	16/5	48	48
Labrisomidae	4/1	10/1	48	48
Mugilidae	1/1	7/5	24-48	48
Mullidae	4/1	4/1	44	44
Pomacentridae	4/2	14/6	48	48
Pomacanthidae	3/3	5/5	48	48
Pomatomidae	1/1	1/1	48	48
Priachanthidae	3/1	4/1	50	50
Scaridae	4/2	10/2	46-48	46-48
Scianidae	21/4	54/5	46-48	48
Serranidae	18/4	45/8	46-48	48
Sparidae	4/2	9/2	48	48
Sphyraenidae	1/1	4/1	48	48
Achiridae	6/2	19/2	40-42	40-42
Bothidae	4/1	7/1	32	32
Cynoglossidae	1/1	10/1	46	46
Paralichthyidae	7/1	18/1	46	46
Diodontidae	3/2	7/2	46-52	46-52
Monacanthidae	4/3	8/3	33-40	33-40
Tetraodontidae	4/1	11/4	46	46
Total	228/73	532/109	24-102	48

Table 1.5.6 Summary of the known distribution of B chromosomes in Neotropical fishes.

	Species	Number of		
		2n	Bs	B size
Anostomidae				
Leporinus	L. friderici	54	0-1	small
	L. sp.	54	0-1	small
Characidae				
Genera Incertae Sedis				
Astyanax	A. eigenmanniorum	50	0-1	large
		48	0-1	large
	A. fasciatus	46	0-1	large
	A. scabripinnis	50	0-4	small
		50	0-2	large
	A. cf. schubarti	36	0-1	large
	A. sp.	50/51	0-1	
	A. sp. B	50	0-2	micro to large
Moenkhausia	M. intermedia	50	0-1	micro
	M. sanctaefilomenae	50	1-8	micro
Oligosarcus	O. pintoi	50-51	0-1	large
Serrasalminae				
Metynnis	M. lippincottianus	62	0-2	
Crenuchidae				
Characidium	C. cf. fasciatum	50	0-4	small
	C. cf. zebra	50	0-1	small
	C. sp.	50	0-1	small
Curimatidae				
Cyphocharax	C. modestus	54	0-4	micro
Steindachnerina	S. insculpta	54	0-2	micro
Erythrinidae				
Erythrinus	E. erythrinus	54/53	0-3	
		54	0-4	micro
Parodontidae				
Apareiodon	A. piracicabae	54	0-1	large
Prochilodontidae				
Prochilodus	P. brevis	54	0-2	micro
	P. lineatus	54	0-7	micro
	P. mariae	54	0-3	micro
	P. nigricans	54	0-2	micro
Callichthyidae				
Callichthys	C. callichthys	58	0-16	micro
Corydoras	C. aeneus	60-63	0-3	small

(Table 1.5.6 contd.)

(Table 1.5.6 contd.)

Heptapteridae				
Pimelodella	P. kronei	58	0-1	micro
Rhamdia	R. quelen	58	0-5	small
Loricariidae				
Hisonotus	H. leucofrenatus	54	0-2	large
Loricaria	L. prolixa	62	0-5	micro
Pimelodidadae				
Bergiaria	B. westermanni	56	0-5	small
Iheringichthys	I. labrosus	56	0-3	micro
Pimelodus	P. ortmanni	56	0-4	micro
	P. sp.	56	0-4	micro
Trichomycteridae				
Trichomycterus	T. davisi	54	0-2	micro
Belonidae				
Strongylura	S. cf. microps	52	1-7	micro
Synbranchidae				
Synbranchus	S. marmoratus	44	0-2	small
Cichlidae				
Gymnogeophagus	G. balzanii	48	0-4	micro
Satanoperca	S. jurupari	48	0-3	small

The family Anostomidae has 12 genera and 138 species (Garavello and Britski, 2003), *Leporinus* being the genus most speciose and widely distributed. Cytogenetic studies in 8 genera and 57 out 87 species showed that all species have 2n=54 chromosomes (Table 1.5.1). Two other conserved characteristics of this family are the presence of a very high number of biarmed chromosomes and a single Ag-NOR-bearing chromosome pair. Nuclear DNA content ranges from 2.57 ± 0.14 to 3.68 ± 0.06 pg DNA/nucleus (Carvalho *et al.*, 2002b). Two reports of supernumerary chromosomes were done for representatives of this family (Table 1.5.6). The most interesting feature of Anostomidae karyotypes is the presence of nine species with a ZZ/ZW sex chromosome system (Table 1.5.7). The sex chromosome systems in the genus *Leporinus* have recently reviewed (Vênere *et al.*, 2004).

The family Characidae is the largest freshwater fish family of the Neotropical region. The relationships inside the group are still poorly known and the family is possibly polyphyletic. Currently, 952 species are described, some of them grouped into 12 subfamilies but most of them not assigned to any subfamily. In a recent review by Reis *et al.* (2003), 3 species

Table 1.5.7 Summary of the known distribution of sex chromosome systems involving female heterogamety in Neotropical fishes.

	Species	2n	Sex chromosome system
Anostomidae			
Leporinus	L. aff. brunneus	54	ZW
	L. conirostris	54	ZW
	L. elongatus	54	ZW
	L. macrocephalus	54	ZW
	L. obtusidens	54	ZW
	L. reinhardti	54	ZW
	L. trifasciatus	54	ZW
	L. sp.	54	ZW
	L. sp.	54	ZW
Characidae			
Triportheus	T. albus	52	ZW
	T. angulatus	52	ZW
	T. elongatus	52	ZW
	T. cf. elongatus	52	ZW
	T. guentheri	52	ZW
	T. paranaensis	52	ZW
	T. signatus	52	ZW
	T. sp.	52	ZW
Cheirodontinae			
Odontostilbe	O. heterdon	52	ZW
	O. cf. microcephala	52	ZW
	O. paranaensis	52	ZW
Serrapinus	S. notomelas	52	ZW
Crenuchidae			
Characidium	C. cf. fasciatum	50	ZW
	C. sp. cf. C. alipionis	50	ZW
Gasteropelecidae			
Thoracocharax	T. cf. stellatus	52	ZW
Parodontidae			
Apareiodon	A. affinis	54-55	ZW1W2
Parodon	P. hilarii	54	ZW
	P. sp.	54	ZW
Prochilodontidae			
Semaprochilodus	S. taeniurus	54	ZW
Doradidae			
Opsodoras	O. sp	58	ZW
Heptapteridae			
Imparfinis	I. mirini	58	ZW

(Table 1.5.7 contd.)

(Table 1.5.7 contd.)

Loricariidae			
Hisonotus	*H. leucofrenatus*	54	ZW
	H. sp.	54	ZW
Otocinclus	*O.* aff. *vestitus*	72	ZW
Hemiancistrus	*H.* sp. 2	52	ZW
Hypostomus	*H.* sp. G	64	ZW
Sternopygidae			
Eigenmannia	*E. virescens*	38	ZW
	E. sp.	36	ZW
Poeciliidae			
Gambusia	*G. p. puncticulata*	48	ZW
	P. latipinna	46	ZW
	P. sphenops	46	ZW

and 85 genera (with 637 species) were considered *incertae sedis* to clearly show our present gap in the knowledge of this group. Cytogenetic data are available for 65 genera and 289 species (Table 1.5.1). The distribution of diploid number among the species and genera grouped as *incertae sedis* reflects their possible polyphyletic nature since the values range from $2n = 28$ to $2n = 52$ (Table 1.5.1). Among the *incertae sedis* genera, there are some of the most speciose characid genera: *Astyanax*, *Hyphessobrycon* and *Hemigrammus*.

The genus *Astyanax* is widely distributed in the Neotropical region, being found in almost all watercourses and in different types of environments (lagoons, rivers, streams, etc.) with high number of individuals in the populations. Although this group is one of the most studied cytogenetically, the discrimination among species is very difficult and many species complexes have been identified as single species. The most frequent diploid number in *Astyanax* is $2n = 50$ chromosomes but one species, *A. schubarti*, presents $2n = 36$ chromosomes (Morelli *et al.*, 1983).

The most studied species, *Astyanax scabripinnis*, exhibits individuals with three diploid numbers: $2n = 46$, $2n = 48$ and $2n = 50$, and in at least one location these three cytotypes were sympatric (Vieira, 2002). Moreover, several morphological and karyotypic differences have been found in this species, characterizing it as a species-complex (Moreira-Filho and Bertollo, 1991; Maistro *et al.*, 1992). Almost all samples analyzed of *A. scabripinnis* presented supernumerary macrochromosomes whose

Table 1.5.8 Summary of the known distribution of sex chromosome systems involving male heterogamety in Neotropical fishes.

	Species	2n	Sex chromosome system
Characidae			
Pselogrammus	P. kennedyi	50	XY
Serrasalminae			
Metynnis	M. mola	62	XY
Erythrinidae			
Erythrinus	E. erythrinus	52/51	X1X2Y
		54/53	X1X2Y
Hoplias	H. lacerdae	50	XY
	H. malabaricus	40-39	X1X2Y
		40-41	XY1Y2
		42	XY
Heptapteridae			
Pimelodella	P. sp.	46	XY
Loricariidae			
Pseudotocinclus	P. tietensis	54	XY
Ancistrus	A. sp.	39/40	X0
Hypostomus	H. ancistroides	68	XY
	H. macrops	68	XY
Pimelodidadae			
Steindachneridion	S. sp.	56	XY
Gymnotidae			
Gymnotus	G. carapo	54	X1X2Y
	G. sp.	40/39	X1X2Y
Hypopomidae			
Brachyhypopomus	B. pinnicaudatus		X1X2Y
Hypopomus	H. sp.	42/41	X1X2Y
Sternopygidae			
Eigenmannia	E. virescens	38	XY
	E. sp.	31-32	X1X2Y
Poeciliidae			
Gambusia	P. reticulata	46	XY
Rivulidae			
Gnatholebias	G. hoignei	46	XX/XY
Cichlidae			
Satanoperca	S. jurupari	48	XY
Gobiidae			
Awaous	A. strigatus	F 46	X1X2Y
Dormitator	D. maculatus	46	XY

number ranges from 0 to 2 among the specimens analyzed (Table 1.5.6) (Moreira-Filho *et al.*, 2004). However, samples from Jucu River exhibited up to four B chromosomes in males (Rocon-Stange and Almeida-Toledo, 1993). A molecular cytogenetic study by Mestriner et al. (2000) showed that the supernumerary macrochromosomes were almost all composed by a satellite DNA, made of a repeated sequence of only 51 base pairs also present in some other chromosomes of the A complement. A study of *A. scabripinnis* from different altitudes revealed significant difference in the frequency of B chromosomes, suggesting that the presence of B chromosomes could confer some selective advantage for the fish in higher altitudes (Porto-Foresti *et al.*, 1997).

The diploid number 2n=50 is also the most common found among the remaining genera of Characidae *Incertae sedis*, such as the well-studied genera *Moenkhausia* (8 species studied) and *Oligosarcus* (11 species studied). However, some groups are characterized by the presence of 2n=52 chromosomes such as the genera *Bryconamericus* (13 species studied) and *Triportheus* (11 species studied). In case of several genera of Characidae, only a reduced number of species was studied, making the discussion on their karyotypic evolution difficult.

The genus *Triportheus* includes 13 species (Malabarba and Lima, 2003). The cytogenetic study of 11 species of this genus showed 2n=52 chromosomes. In all species where males and females were karyotyped, a ZZ/ZW sex chromosome mechanism was found (Table 1.5.7). While the Z chromosome is apparently conserved, the W chromosome shows a differential evolution with morphological diversification not only between species, but also between local populations from the same hydrographic basin (Artoni *et al.*, 2001).

The genera *Hemigrammus* (18 species karyotyped) and *Hyphessobrycon* (39 species karyotyped) were mainly studied in the beginning of the decade of 70 and for many species only the haploid number was described. The values show a wide diploid number variation from 2n=28 to 2n=54 in *Hemigrammus* and a variation from 2n=42 to 2n=52 in *Hyphessobrycon*. New studies would be necessary for a better understanding of the patterns of karyotypic evolution in these genera, although the wide range found in diploid number may allow to anticipate that these genera are not natural groups.

A recent study of the genus *Oligosarcus* (Martinez *et al.*, 2004b) demonstrated that all karyotyped species have the first metacentric pair considerably larger than all other metacentric chromosomes in the karyotypes, confirming previous observations by Falcão and Bertollo (1985). This chromosomal feature has been observed in several species of the family Characidae, since its initial description by Scheel (1973). Daniel (1996) showed that this large chromosome pair presents similar R-banding patterns in five *Astyanax* species. Therefore, this chromosome characteristic may be considered as an important marker to further comparative cytogenetic analyses among Characidae.

The subfamily Aphyocharacinae includes one genus and 10 species (Lima, 2003). Cytogenetic information is available for three species, all of them showing 2n=50 chromosomes (Table 1.5.1). The number of Ag-NOR-bearing chromosome pairs ranges from 1 to 3. Nuclear DNA content studies for *Aphyocharax dentatus* showed 3.4 pg DNA/nucleus (Hinegardner and Rosen, 1972) and 2.45 ± 0.19 (Carvalho *et al.*, 2002b). Although these value differences could appear different due to the different techniques employed, we cannot exclude that they represent an actual difference between the samples due either to polymorphism among the samples analyzed or to problems in the species identification.

The subfamily Bryconinae comprises 3 genera and 41 species (Lima, 2003). Cytogenetic studies in 14 species of *Brycon* and the only known species of *Henochilus* showed in all of them 2n=50 chromosomes and a single Ag-NOR-bearing chromosome pair. A phylogenetic analysis of molecular data showed that the genus *Henochilus* is the sister group of *Brycon* and the genus *Salminus* the sister group of *Brycon* + *Henochilus* (Castro *et al.*, 2004). The already karyotyped species of the genus *Salminus* exhibited 2n=50 chromosomes, a diploid number which may corroborate the above phylogenetic hypothesis.

The subfamily Characinae presents 12 genera and 70 species (Lucena and Menezes, 2003). Cytogenetic studies in 5 genera and 15 species showed that almost all species have 2n=52 chromosomes, with the exception of *Phenacogaster aff. microstictus* (2n=50) and *P. cf. pectinatus* (2n=46) (Table 1.5.1). Comparative analysis of the literature data shows small differences between species, only related to the karyotypic formulae. Ag-NORs were observed in only one chromosome pair in almost all species of the subfamily, except for the species *Roeboides bonariensis*, which presents two chromosome pairs with Ag-NORs. Nuclear DNA content ranges from 2.18 ± 0.09 to 3.20 ± 0.11 pg DNA/nucleus in four species analyzed (Carvalho *et al.*, 2002b). The available cytogenetic data suggest that Characinae may constitute a natural group.

The subfamily Cheirodontinae presents 15 genera and 46 species (Malabarba, 2003). Species of this subfamily are commonly found but they generally present a very small size, which could explain the low number of cytogenetic studies conducted within the group. The available data show that all eleven species analyzed have $2n=52$ chromosomes (Table 1.5.1). In five species a ZZ/ZW sex chromosome system occurs (Table 1.5.7).

The subfamily Glandulocaudinae presents 19 genera and 50 species (Weitzman, 2003). Five species present $2n=52$ and one, *Glandulocauda melanopleura*, displays $2n=54$ chromosomes. The karyotypes comprise all types of chromosomes and usually two or three Ag-NOR-bearing chromosome pairs are present. Although one species presents a different diploid number, the subfamily may be considered karyotypically conserved. Nuclear DNA content was determined for two species with values ranging from 2.52 ± 0.09 to 3.06 ± 0.14 pg DNA/nucleus (Carvalho *et al.*, 2002b).

In the subfamily Iguanodectinae, which presents 2 genera and 11 species (Moreira, 2003), only the species *Iguanodectes spilurus* was studied showing the haploid chromosome number of $n=25$ (Table 1.5.1). Nuclear DNA content was determined for *Piabucus melanostomus* that presented 2.39 ± 0.04 pg DNA/nucleus, a value similar to those found in several other characiforms (Carvalho *et al.*, 2002b).

The subfamily Serrasalminae presents 15 genera and 80 species (Jégu, 2003). Cytogenetic studies were carried out in 11 genera and 39 species (Table 1.5.1). According to a molecular phylogeny proposed by Ortí et al. (1996), the subfamily Serrasalminae is monophyletic with the genera *Colossoma*, *Mylossoma* and *Piaractus* grouped in a monophyletic clade that is considered the sister group of all other Serrasalminae. The cytogenetic data are in agreement with this molecular phylogeny since all karyotyped species of the genera *Colossoma*, *Mylossoma*, and *Piaractus* present $2n=54$ chromosomes, while all species of the other genera present different diploid numbers with values ranging from $2n=58$ to $2n=64$. Species of this subfamily present a large number of Ag-NOR-bearing chromosome pairs (reaching up to 11 chromosome pairs). Nuclear DNA content was determined for 8 species with values ranging from 2.88 ± 0.10 to 3.58 ± 0.06 pg DNA/nucleus (Carvalho *et al.*, 2002b).

The subfamily Stethaprioninae presents 4 genera and 12 species (Reis, 2003a). Three genera and four species have been studied cytogenetically (Table 1.5.1). All species exhibit 2n=50 chromosomes, with differences in chromosome morphology between species. In all the species, only one submetacentric pair with Ag-NORs is present. The cytogenetic data available for this family reinforce the hypothesis proposed by Reis (1989) that the Stethaprioninae constitutes a monophyletic group, as also discussed by Carvalho *et al.* (2001).

After the research by Reis *et al.* (2003), the subfamily Tetragonopterinae was reduced to its type genus *Tetragonopterus* and two species. Both species exhibit 2n=52 chromosomes but in the species *T. argenteus* two cytotypes (one with 2n=50 and other with 2n=52) were found, suggesting the existence of cryptic species in the group. Nuclear DNA content of *T. argenteus* and *T. chalceus* are 2.99 ± 0.21 and 3.94 ± 0.09 pg DNA/nucleus, respectively (Carvalho *et al.*, 2002b).

Although numerous species and species groups of Characidae exhibit very conserved diploid numbers, suggesting a possible relationship, further studies with more species of this family are necessary in order to perform comparative analyses. Moreover, new phylogenetic studies between the characid groups are necessary for a better understanding of the patterns and the role of karyotypic changes in the evolution of the group.

The family Chilodontidae is composed of 2 genera and 7 species (Vari and Raredon, 2003). Cytogenetic studies carried out with 2 genera and 2 species of this family showed that both exhibited 2n=54 biarmed chromosomes (Table 1.5.1).

The family Crenuchidae is composed of 12 genera and 73 species (Buckup, 2003). Two genera and 11 species were studied. All species of the genus *Characidium* had 2n=50, some a ZZ/ZW sex chromosome system (Table 1.5.7) and some had supernumerary chromosomes (Table 1.5.6). The only species karyotyped of the genus *Crenuchus* showed 2n=38 chromosomes (Table 1.5.1).

The family Ctenoluciidae is composed of two genera and seven species (Vari, 2003a). Cytogenetic studies in one species of each genus showed in both 2n=36 chromosomes (Table 1.5.1). This diploid number is very low when compared with other Characiformes.

The family Curimatidae is composed of 8 genera and 97 species (Vari, 2003b). Cytogenetic studies carried out with seven genera and 46 species showed in 38 species 2n=54 chromosomes (almost all biarmed). Different diploid numbers were described for *Curimata ocellata* (2n≐56), *Curimatopsis* aff. *macrolepis* A (2n=52), C. aff. *macrolepis* B (2n=46), C. *myersi* (2n=46), *Cyphocharax* sp. (2n=58), *Potamorhina latior* (2n=56), *P. altamazonica* (2n=102), and *P. squamoralevis* (2n=102). The differences between the karyotypes of the species with 2n=54 are frequently related to the Ag-NOR position (Vênere and Galetti Jr, 1989). Instead, the presence of species with different diploid numbers is intriguing. Nuclear DNA content was determined for 8 species of this family that presented from 2.83 ± 0.26 (*Curimatella dorsalis*, 2n=54) to 3.80 ± 0.23 pg DNA/nucleus (*Potamorrhina squamoralevis*, 2n=102) (Carvalho *et al.*, 2002b). The large difference in diploid number associated with the small difference in nuclear DNA content reinforce the hypothesis of Feldberg *et al.* (1993) that chromosome fissions were the major mechanism of chromosome evolution in this family.

The family Cynodontidae is composed of 5 genera and 14 species (Toledo-Piza, 2003). Cytogenetic studies were conducted with 3 genera and 3 species (Table 1.5.1). All the species showed 2n=54 chromosomes and a single Ag-NOR-bearing chromosome pair. A recent study by Martinez *et al.* (2004 b) showed a predominance of biarmed chromosomes in all the species investigated. Only a few uniarmed chromosomes were observed in *Cynodon gibbus* and *Rhaphyodon vulpinus*, and acrocentric chromosomes were only observed in *R. vulpinus*. The nuclear DNA content of two species ranged from 2.02 ± 0.06 to 2.09 ± 0.28 pg DNA/nucleus (Carvalho *et al.*, 2002b). These close values reinforce the proposition of a very conserved karyotypic structure in the group.

The family Erythrinidae is composed of 3 genera and 15 species (Oyakawa, 2003). Cytogenetic studies carried out with 3 genera and 6 species (Table 1.5.1) show an impressive chromosome variability. The most remarkable cytogenetic characteristics of Erythrinidae are the low diploid number of many species identified with local samples and the presence of several sex chromosome systems, all of them involving male heterogameity (Table 1.5.8). In the genus *Hoplerythrinus*, the only karyotyped species, *H. unitaeniatus*, showed two diploid numbers 2n=48 and 2n=52. Additionally, a large variability in the karyotype of different samples has been observed (Giuliano-Caetano and Bertollo, 1988), suggesting that this species may represent a species complex. An extensive study conducted with *Hoplias malabaricus*, collected in different Brazilian localities, and some samples from Argentina, Uruguay and Surinam showed that seven general cytotypes might be clearly identified on the basis of their diploid number (2n=39 to 2n=42), chromosomal morphology and sex chromosome systems (Bertollo *et al.*, 2000). On the other hand, cytogenetic studies carried out with *H. lacerdae* showed the occurrence of the same diploid number (2n=50) in all sample studied. A cytogenetic study of six samples of *Erythrinus erythrinus* allowed the identification of four groups characterized on the basis of their chromosomal features (Bertollo *et al.*, 2004). A detailed analysis of chromosomes evolution in the family Erythrinidae is reported by L.A.C. Bertollo in the present volume, chapter 1.7.

The family Gasteropelecidae is composed of three genera and nine species (Weitzman and Palmer, 2003). Cytogenetic studies performed with three genera and four species (Table 1.5.1) exhibited a considerable karyotypic variability in the group with diploid numbers ranging from 2n=48 to 2n=54, including some intraspecific variability. According to Carvalho *et al.* (2002a), a common characteristic for the family is the presence of many subtelocentric and acrocentric chromosomes in their karyotypes. All species investigated presented a single Ag-NOR-bearing chromosome pair. Nuclear DNA content ranges from 2.18 ± 0.09 pg DNA/nucleus in *Thoracocharax* cf. *stellatus* to 2.8 pg in *Carnegiella strigata* and *Gasteropelecus levis* (Hinegardner and Rosen, 1972; Carvalho *et al.*, 2002b), suggesting that only small changes in the amount of nuclear DNA were fixed during the evolutionary process of the group.

The family Hemiodontidae is composed of 5 genera and 28 species (Langeani, 2003). Cytogenetic studies performed in 4 genera and 9 species (Table 1.5.1) showed the all of them presented 2n=54 chromosomes. A common characteristic was the presence on a single Ag-NOR-bearing chromosome pair and also a high number of biarmed chromosomes. Porto *et al.* (1993) showed that the seven species could be distinguished by small differences in their karyotype. The presence of 2n=54 chromosomes, most biarmed, in all the species analyzed of Hemiodontidae reinforce the idea that this family may be related with the clade composed by Prochilodontidae, Curimatidae, Anostomidae and Chilodontidae, as previously suggested by Porto *et al.* (1993).

The family Lebiasinidae is composed of 7 genera and 61 species (Weitzman and Weitzman, 2003). This group is composed of many species of very small size, which makes the cytogenetic studies difficult. Most data are restricted to the determination of the haploid number by Scheel (1973). An interesting characteristic of this family is the presence of reduced chromosome numbers with values ranging from n=11 in *Nannostomus unifasciatus* to 2n=46 in *N. trifasciatus* (Table 1.5.1). Both karyotyped species of the genus *Pyrrhulina* showed 2n=42 chromosomes and a high number of uniarmed chromosomes, a particular characteristic among Characiformes. An additional characteristic described for *Pyrrhulina* cf. *australis* was the presence of four chromosome pairs with Ag-NORs, one of them with NORs in both terminal regions (Oliveira *et al.*, 1991). Nuclear DNA content of *P.* cf. *australis* and *P. rachoviana* was 1.97 ± 0.15 and 2.4 pg DNA/nucleus, respectively (Hinegardner and Rosen, 1972; Carvalho *et al.*, 2002b).

The family Parodontidae is composed of 3 genera and 23 species (Pavanelli, 2003). Cytogenetic studies in 2 genera and 9 species showed 2n=54 chromosomes (Table 1.5.1). Additionally, in some local populations of *Apareiodon affinis*, the males display 2n=54 and females show 2n=55, which has been characterized as a ZZ/ZW1W2-type sex chromosome system (Table 1.5.7) (Moreira-Filho *et al.*, 1985; Jesus and Moreira-Filho, 2000). Other species of *Apareiodon* and several local populations of *A. affinis* do not have sex chromosomes. In the genus *Parodon* two species, *P. hilarii* and *Parodon* sp. exhibit a ZZ/ZW-type sex chromosome system (Centofante *et al.*, 2002). Almost all species of Parodontidae present one Ag-NOR-bearing chromosome pair. Nuclear DNA content analysis showed values ranging from 2.04 ± 0.12 to 2.53 ± 0.17 pg DNA/nucleus for two local samples of *A. affinis* (Carvalho *et al.*, 2002b).

The family Prochilodontidae is composed of 3 genera and 21 species (Castro and Vari, 2003). Cytogenetic studies of 2 genera and 13 species showed that all of them present 2n=54 biarmed chromosomes (Table 1.5.1). An interesting characteristic of the species of the genus *Prochilodus* is the presence of supernumerary microchromosomes, which were firstly reported by Pauls and Bertollo (1983) and then found in 4 species (Table 1.5.6) (Oliveira *et al.*, 2003). In the first study conducted with *P. lineatus* from the Mogi-Guaçu River, the B chromosomes presented a numeric variation among cells and among specimens (Pauls and Bertollo, 1983).

A further study in the same area showed that the intra-individual polymorphism disappeared and the fishes exhibited only inter-individual differences in the number of B chromosomes (Oliveira et al., 1997; Cavallaro et al., 2000). In P. lineatus, the B chromosomes have different morphology, are C-band positive (Cavallaro et al., 2000) and very rich in a satellite sequence that is also present in the A chromosomes (Jesus et al., 2003). The B-chromosomes of P. lineatus have a Mendelian inheritance pattern (Oliveira et al., 1997), which is very uncommon since B-chromosomes usually exhibit a non-Mendelian inheritance pattern (Beukeboom, 1994). In a species of the genus Semaprochilodus, S. taeniurus, a ZW-type sex chromosome system was described (Feldberg et al., 1987). Almost all species analyzed presented a single Ag-NOR-bearing chromosome pair. Nuclear DNA content analysis showed values ranging from 3.08 ± 0.07 to 3.72 ± 0.05 in P. argenteus and P. nigricans, respectively (Carvalho et al., 2002b). A sample of P. lineatus displayed a large variation in nuclear DNA content, 3.36 ± 0.32, possibly resulting from the presence of supernumerary chromosomes (Carvalho et al., 1998).

Although additional cytogenetic data are necessary for a better understanding of the relationships between members of the order Characiformes, the available data suggest that at least three evolutionary lines exist in the order. The first (the most representative) is composed of groups with diploid number close or equal to 2n=54; the second is composed of groups with diploid numbers around 2n=40 chromosomes; and the third one is composed of species with higher diploid number (around 2n=60), being represented by some species and genera of Serrasalminae.

The presence of 2n=54 biarmed chromosomes has been consistently found in the characiform families Anostomidae (Galetti Jr et al., 1995), Prochilodontidae (Oliveira et al., 2003), Curimatidae (Navarrete and Júlio Jr, 1997), and Chilodontidae (Vênere, 1998) (Table 1.5.1), four families that, according to Vari (1983), constitute a monophyletic group. The same diploid number was also found in all species analyzed of Hemiodontidae and Cynodontidae (Table 1.5.1) but the phylogenetic position of these families still remains unresolved.

Considering the fact that almost all species belonging to the sister group of Characiformes (the order Cypriniformes) present diploid numbers around 2n=50 (or 2n=100 in some tetraploid groups), Oliveira et al. (1988) suggested that the presence of 2n=54 chromosomes in

Characiformes could be a synapomorphy for this group. To test this hypothesis, the analysis of some primitive Characiformes should be necessary. At present, morphological studies point out that the African families Distichodontidae and Citharinidae are the putative primitive sister group of all Characiformes remained (Buckup, 1998). The karyotype was described only for *Distichodus affinis* that exhibit a diploid number of 2n=48 (34M+14SM), a single Ag-NOR-bearing chromosome pair and a very low amount of C-positive bands (Ráb *et al.*, 1998). The presence of 2n=48 in the Distichodontidae is not in agreement with the hypothesis that the primitive diploid number of characiforms is 2n=54. Instead, it can support the possibility that 2n=54 is a synapomorphy for the Neotropical characiforms only, although additional data are necessary to better test these hypotheses. Post (1965) reported the haploid chromosome numbers of two African species: *Alestes longipinnis* (n=24) and *Micralestes interruptus* (n=24) that belong to the family Alestiidae. This family is probably related to Characidae (Buckup, 1998), but the correct position of the families Alestiidae and Characidae among the Characiformes still remains unsolved.

Lucena (1993) and Buckup (1998) suggested that the family Ctenoluciidae is the sister group of the genus *Hepsetus* (an African characid) and that the clade Ctenoluciidae + *Hepsetus* was the sister group of the families Lebiasinidae and Erythrinidae. The consistent presence of low diploid numbers in Ctenoluciidae (2n=36), in Erythrinidae (2n=40 to 2n=54), and in Lebiasinidae (2n=22 to 2n=46) corroborates the possible close relationship between these groups. The hypothesis involving the relationship between Erythrinidae and Lebiasinidae was previously proposed by Oliveira *et al.* (1991) in a study of the karyotype of *Pyrrhulina* cf. *australis*.

Chromosome Evolution in Siluriformes

The order Siluriformes is worldwide distributed but a large number of families is endemic of the Neotropical regions (Nelson, 1994). Many species have high ecological and economical importance as commercial or aquarium fishes. The relationships between the catfish families have been studied for a long time, but only two hypotheses on the higher-level phylogeny of Siluriformes are available so far (de Pinna, 1998; Britto, 2003a).

Table 1.5.2 shows that the cytogenetic data are known for 318 species and 89 genera of Neotropical siluriforms, representing 21% of known species and 33% of known genera. Comparing siluriforms with characiforms, it is possible to observe that the first presents higher diploid numbers and higher variability between species.

In Reis et al. (2003), the species Conorhynchus conirostris, previously included in the family Pimelodidae, has been listed in siluriforms Incertae sedis and cannot be assigned to any known species of Siluriformes. Cytogenetic studies showed that C. conirostris presents 2n=60 chromosomes, which is an unusual number among siluriforms. The species presents a single Ag-NOR-bearing chromosome pair.

The family Diplomystidae, with 2 genus and 6 known species, is the most basal siluriform lineage (Fink and Fink, 1981; de Pinna, 1998; Ferraris, 2003a) and information on the group is fundamental for understanding the evolution of other siluriforms. The cytogenetic analysis of three species Diplomystes camposensis and D. nahuelbutaensis (Campos et al., 1997) and Olivaichthys mesembrinus (=D. mesembrinus) (Oliveira and Gosztonyi, 2000) showed that these three species present 2n=56 (Table 1.5.2). Considering that all karyotyped Diplomystidae species exhibit 2n=56 chromosomes, that this group is the sister group of all other Siluriformes and also that this diploid number is absent in Gymnotiformes (being very unusual among Characiformes) (Oliveira and Gosztonyi (2000) proposed that 2n=56 is a synapomorphy for the order Siluriformes. The mean diploid DNA content observed in erythrocyte nuclei of O. mesembrinus was 2.57 ± 0.15 pg (Oliveira and Gosztonyi, 2000). This value is higher than the mean value of 2.0 pg/diploid nuclei observed for teleosts (Hinegardner and Rosen, 1972). However, it is lower than the mean values observed in Characiformes (3.0 pg/diploid nucleus) (Carvalho et al., 1998) and Siluriformes (3.2 pg/diploid nucleus) (Fenerich et al., 2004). Since the available data about the DNA content of Siluriformes are limited to 43 species (Fenerich et al., 2004) and most of these data are from the family Callichthyidae (23 species), in which some species apparently evolved by polyploidy (Oliveira et al., 1993a), any conclusions about evolutionary changes in this parameter are tentative. Considering that the DNA content of Siluriformes ranges from 0.91 to 8.8 pg/diploid nucleus, it is possible to suggest that significant changes in DNA content have occurred during the evolution of this group without relevant modifications in diploid numbers.

The family Aspredinidae is composed of 12 genera and 36 species (Friel, 2003). Cytogenetic studies in one genera and four species showed that three species from Parana-Paraguay basin presented $2n=50$ chromosomes, and one species from Amazon basin presented $2n=64$ chromosomes (Table 1.5.2).

The family Auchenipteridae is composed of 20 genera and 91 species (Ferraris, 2003b). Studies in 8 genera and 13 species showed that almost all species have $2n=58$ chromosomes, but two species of the genus *Ageneiosus* present $2n=56$ chromosomes (Table 1.5.2). This difference in diploid number is interesting since until recently the genera *Ageneiosus* and *Tetranematichthys* were included separated family Ageneiosidae (Ferraris, 2003b).

The family Callichthyidae is composed of 8 genera and 177 species (Reis, 2003b). Cytogenetic studies in 7 genera and 60 species showed one of the higher variation in diploid number among fishes, with values ranging from $2n=38$ to $2n=134$ (Table 1.5.2). This family is the best investigated from the cytogenetic point of view and the available results have pointed to a very different pattern of chromosome evolution in the two subfamilies, Corydoradinae and Callichthyinae.

The subfamily Corydoradinae is now composed of three genera *Aspidoras*, *Scleromystax* and *Corydoras* (Britto, 2003b). The species of *Aspidoras* showed $2n=44$ (one species) and $2n=46$ (four species), with a high number of biarmed chromosomes and a single Ag-NOR-bearing chromosome pair, with interstitial marks (Shimabukuro-Dias *et al.*, 2004b). The similar characteristics found in the species of *Aspidoras* suggest that chromosome rearrangements, as fusions or translocations, may have occurred during the chromosome evolution of the species of this genus (Shimabukuro-Dias *et al.*, 2004b).

A general analysis of the available karyotypic data about *Corydoras* shows that there are common patterns of chromosome evolution in some groups of species and allows suggesting the occurrence of natural groups (Oliveira *et al.*, 1992; Shimabukuro-Dias *et al.*, 2004a).

The genus *Scleromystax* was recently proposed by Britto (2003b) in order to identify a particular group of *Corydoras* species, composed of C. *barbatus*, C. *macropterus*, C. *prionotos* and C. *lacerdae*. The main cytogenetic characteristics observed in the genus *Scleromystax*, such as high diploid number ($2n=64$ to $2n=86$), large number of biarmed chromosomes, multiple Ag-NORs at a terminal position in the short arms of biarmed chromosomes and the presence of C-band positive chromosome arms in a small number of chromosomes, suggested that these species belong to a natural group, which was identified as Group 1 by Oliveira et al. (1992).

The low nuclear DNA content observed in A. fuscoguttatus (Oliveira et al., 1993a) was also found in the species of the genus Scleromystax (cited as Corydoras), including S. barbatus, S. prionotos, and S. macropterus (Oliveira et al., 1993b). This finding allowed Oliveira et al. (1993a) to propose that these species of Scleromystax could be related to the species of Aspidoras. This hypothesis was corroborated by the phylogenetic studies (Britto, 2003b; Shimabukuro-Dias et al., 2004c). The presence of low diploid numbers in Aspidoras (2n=44-46), high diploid numbers in Scleromystax (2n=64-86) and other genera of Callichthyidae allow the authors suggesting that several terminal fusions may have occurred in the ancestor, which gave origin to the genus Aspidoras (Oliveira et al., 1993a).

Some species of Corydoras of the Group 2 exhibited high diploid numbers (2n=74 to 2n=120), a large number of uniarmed chromosomes, single Ag-NORs at a terminal position in the short arms, and very small C-band positive segments at a pericentromeric position in some short arms. Corydoras of the Group 3 exhibited 2n=56 to 2n=134 chromosomes, mainly biarmed elements, single Ag-NORs at interstitial position in the short arm, and a very small amount of C-band positive chromosome segments at pericentromeric position in a few chromosome pairs. Corydoras of the Group 4 show a small number of large chromosomes (2n=40 to 2n=52), mostly biarmed, and conspicuous C-band positive segments at centromeric and/or pericentromeric positions. Corydoras of the Group 5 exhibited a high diploid number (2n=92 to 2n=94), large chromosomes, mostly biarmed and multiple Ag-NORs (Oliveira et al., 1992; Shimabukuro-Dias et al., 2004a). Besides this high variation in diploid numbers, an eight times variation in nuclear DNA content was observed in Corydoras, with values ranging from 1.19 ± 0.13 in C. prionotos (2n=68) to 8.78 ± 1.50 pg DNA/nucleus in C. metae (2n=92) suggesting the occurrence of some events of polyploidy and/or gene duplication during the evolution of the group (Oliveira et al., 1992). Phylogenetic studies using morphological (Britto, 2003b) and molecular data (Shimabukuro-Dias et al., 2004c) are in general agreement with the above grouping but additional studies are necessary for a better understanding of chromosome evolution in the genus Corydoras.

The cytogenetic analyses conducted in the Callichthyinae subfamily showed the occurrence of two groups of species. The first group is composed of the genus *Callichthys*, with diploid numbers ranging from 2n=52 to 2n=60 and the presence of many biarmed chromosomes in the karyotypes. The second group includes the other genera, being characterized by diploid numbers ranging from 2n=60 to 2n=66 chromosomes and by high number of uniarmed chromosomes (Shimabukuro-Dias *et al.*, 2005).

The cytogenetic characteristics of the genus *Dianema* such as 2n=60-62 chromosomes, many uniarmed chromosomes, a single chromosome pair with Ag-NORs, the distribution pattern of C-band positive regions and its low amount of nuclear DNA (1.18 ± 0.07 pg of DNA/nucleus) are similar to those observed in *Hoplosternum littorale* (2n=60 and 1.36 ± 0.11 pg DNA/nucleus) and *Hoplosternum* sp. (2n=60), suggesting a close relationship between these genera, as previously demonstrated in the phylogenetic studies conducted in the family Callichthyidae (Reis, 1998; Shimabukuro-Dias *et al.*, 2004c). The cytogenetic characteristics showed to be similar to those found in the genus *Scleromystax* (one of the most primitive genera of Corydoradinae) reinforcing the hypothesis by Shimabukuro-Dias *et al.* (2004c) that *Dianema* is the most primitive genus of Callichthyinae.

In contrast to the extensive chromosome rearrangements observed in the Corydoradinae subfamily, a small number of chromosome changes seem to have occurred in the evolutionary history of the subfamily Callichthyinae. However, the karyotypic variability observed between the different samples of Callichthyinae analyzed could suggest the existence of several undescribed species in this subfamily or, alternatively, the existence of many intraspecific polymorphisms.

The family Doradidae is composed of 29 genera and 74 species (Sabaj and Ferraris, 2003). Cytogenetic studies carried out in 12 genera and 20 species showed a conserved diploid number of 2n=58 (Table 1.5.2). Only two species of *Anadoras* (2n=56), *Doras eigenmanni* (2n=66), and *Trachydoras paraguayensis* (2n=56) presented different diploid numbers. Almost all the species present a single Ag-NOR-bearing chromosome pair. A ZZ/ZW sex chromosome system was reported for *Opsodoras* sp. (Table 1.5.7). Nuclear DNA content ranges from 3.2 to 3.46 ± 0.13 pg DNA/nucleus in *Acanthodoras spinosissimus* and *Rhynodoras d'orbignyi*, respectively (Hinegardner and Rosen, 1972; Fenerich *et al.*, 2004).

The family Heptapteridae is composed of 24 genera and 186 species (Bockmann and Guazzelli, 2003). Cytogenetic studies carried out in 6 genera and 38 species showed diploid numbers ranging from 2n=42 to 2n=58 (Table 1.5.2), but most of the species present a diploid number of 2n=58. Among the species of several genera, more than one diploid number was described. Thus, in *Heptapterus*, one species presented 2n=42 and the other 2n=52; in *Imparfinis*, almost all the samples present 2n=58 but one species shows 2n=56; in *Pimelodella* four diploid numbers were found, 2n=46, 2n=52, 2n=56 and 2n-58; and in *Rhamdella* two species were present 2n=56 and one 2n=58. The most investigated genera, *Rhamdia*, exhibits a basic diploid number of 2n=58, with from 0 to 5 supernumerary chromosomes in some samples (Table 1.5.6). The wide difference in diploid number between the species analyzed suggests that many chromosome rearrangements occurred during the evolutionary history of the group. In a preliminary study conducted in *Imparfinis mirini*, the presence of a ZZ/ZW-type sex chromosome system was suggested to occur (Vissotto *et al.*, 1997). However, additional studies involving a large sample from several localities revealed a polymorphism involving a large C-band positive segment and the occurrence of sex chromosomes (Vissotto, 2000). The occurrence of a XX/XY sex chromosome system was reported for *Pimelodella* sp. by Dias and Foresti (1993) (Table 1.5.8). Almost all species of Heptapteridae exhibit a single Ag-NOR-bearing chromosome pair. Nuclear DNA content analysis in 15 samples of Heptapteridae showed a variation from 1.13 ± 0.09 to 2.38 ± 0.07 pg DNA/nucleus in *Pimelodella* sp. and *Imparfinis mirini*, respectively (Fenerich *et al.*, 2004).

The family Loricariidae is composed of 94 genera and 673 species traditionally divided into six subfamilies: Hypostominae, Ancistrinae, Loricariinae, Hypoptopomatinae, Neoplecostominae and Lithogeninae (Reis *et al.*, 2003). However, the subfamily Hypostominae was recently redefined with the exclusion of several genera, the incorporation of the older subfamily Ancistrinae, being now divided into five tribes: Corymbophanini, Rhinelepini, Hypostomini, Pterygoplichthini and Ancistrini (Armbruster, 2004).

Cytogenetic studies were carried out in 30 genera and 124 species of Loricariidae, showing a wide range of diploid numbers from 2n=36 to 2n=96 (Table 1.5.2).

The cytogenetic data for the subfamily Neoplecostominae showed that the species of *Neoplecostomus* (until recently considered as the only genus in the subfamily - Ferraris, 2003c) presented 2n=54 chromosomes. The karyotypic analysis in species of the genera *Hemipsilichthys*, *Kronichthys*, *Isbrueckerichthys* and *Pareiorhina* (Alves *et al.*, 2005) showed 2n=54 chromosomes and a very similar karyotypic structure in all the species, corroborating the hypothesis of Armbruster (2004) that all these genera could belong to a natural group. Considering that the subfamily Neoplecostominae represents the most basal clade of the family Loricariidae, excluding Lithogeninae (de Pinna, 1998; Armbruster, 2004), Alves *et al.* (2005) suggested that the presence of 2n=54 chromosomes and interstitial Ag-NORs could constitute a primitive character in Loricariidae.

Almost all species of the subfamily Hypoptopomatinae present 2n=54 chromosomes and a single Ag-NOR-bearing chromosome pair with NORs at interstitial position. Only two species in this group presented different diploid numbers, *Otocinclus* aff. *vestitus* (2n=72) and *Otocinclus gibbosus* (2n=58). An interesting cytogenetic characteristic only found in this subfamily is the presence of a XX/XY sex chromosome system in *Pseudotocinclus tietensis* (Andreata *et al.*, 1992) (Table 1.5.8) and a ZZ/ZW sex chromosome system in *Hisonotos leucofrenatus* (Andreata *et al.*, 1993) (Table 1.5.7). In both cases, the Y and W sex chromosomes are differentiated by the presence of C-band positive segments. In some specimens of *H. leucofrenatus*, a large supernumerary chromosome was also found (Andreata *et al.*, 1993) (Table 1.5.6).

Cytogenetic studies in the subfamily Loricariinae showed a great karyotypic complexity, with diploid number ranging from 2n=36 to 2n=74. A recent study in the genus *Harttia* showed that the species had from 2n=52 to 2n=58 chromosomes and differences in the number and position of the Ag-NORs (Alves *et al.*, 2003). Local samples of *Rineloricaria latirostris* presented polymorphism in diploid numbers (from 2n=36 to 2n=48), mainly attributed to the occurrence of chromosome fusions or fissions (Giuliano-Caetano, 1998). Alves *et al.* (2003) found 2n=70 in *Rineloricaria* n. sp., 2n=66 in *R. cadeae* and 2n=64 in *R. kronei*, with a high number of uniarmed chromosomes in all these species, suggesting either the occurrence of several different natural groups or a high rate of karyotypic evolution in the genus.

The diploid numbers in the subfamily Hypostominae (*sensu* Armbruster, 2004) range from 2n=38 to 2n=80. A detailed cytogenetic analysis showed that the large karyotypic complexity usually attributed to the family Loricariidae is mainly a characteristic of the tribes Ancistini and Hypostomini, once all Corymbophanini and Rhinelepini presented 2n=54 and all Pterygoplichthini 2n=52.

Only one species of *Corymbophanes*, a representative of the monotypic tribe Corymbophanini, the most basal clade in Hypostominae according to Armbruster (2004), was karyotyped and presented 2n=54 chromosomes and a single pair of chromosomes bearing interstitial Ag-NORs (Alves *et al.*, 2005). Members of the tribe Rhinelepini (genera *Rhinelepis* and *Pogonopoma*) also presented 2n=54 chromosomes. Cytogenetic studies in representatives of the tribe Pterygoplichthini (*Liposarcus* and *Glyptoperichthys*) showed that all of them have 2n=52 chromosomes that may represent a synapomorphy for this tribe (Alves *et al.*, unpublished data).

Diploid numbers range from 2n=38 to 2n=54 in the species of the tribe Ancistrini. However, this variation is restricted to the genus *Ancistrus*, since all karyotyped species of *Baryancistrus*, *Hemiancistrus*, *Megalancistrus*, and *Panaque* present 2n=52 chromosomes. The genus *Ancistrus* is the widest distributed in the family and may not form a natural group. A particular characteristic only recently found in one species of *Ancistrus* is the presence of a putative XX/X0 sex chromosome system (Alves *et al.*, unpublished data) (Table 1.5.8).

Cytogenetic studies of the genus *Hypostomus*, the only genus of the tribe Hypostomini according to Armbruster (2004), showed diploid number ranging from 2n=52 to 2n=80. Artoni and Bertollo (2001) suggested that centric fissions should be the main type of chromosome rearrangement occurring in the group. An additional peculiar characteristic of the tribe Hypostomini is the presence of multiple Ag-NORs, once in the other tribes and subfamilies of Loricariidae, the usual condition is the presence of a single Ag-NOR-bearing chromosome pair.

Nuclear DNA content studies in Loricariidae showed around 2.0 pg DNA/nucleus, in most species, including *Neoplecostomus paranensis* (2.26 ± 0.17 pg DNA/nucleus), representing the most primitive subfamily analyzed (Fenerich *et al.*, 2004). On the other hand, the analysis of species of the subfamily Hypostominae showed that in this group the nuclear DNA content ranges from 3.17 ± 0.17 to 4.90 ± 0.12 pg DNA/nucleus, with several species/populations having about 4.0 pg DNA/nucleus. Although additional studies are necessary, events of polyploidy could explain the high specific diversification of *Hypostomus*, similarly to what already suggested for the genus *Corydoras* (Oliveira *et al.*, 1992).

The family Nematogenyidae is made of only one species, *Nematogenys inermis* found exclusively in Southern Chile (de Pinna, 2003). The diploid number is 2n=94 chromosomes, which is an unusual even among siluriforms but unfortunately no additional cytogenetic information is available on the karyotype and the chromosomes in this species, which is seriously threatened due to environmental changes by men.

The family Pimelodidae is composed of 31 genera and 83 species (Lundberg and Littmann, 2003). Cytogenetic studies in 13 genera and 30 species showed diploid numbers ranging from 2n=50 to 2n=58 (Table 1.5.2), with the majority of species presenting 2n=56. Thus, this group may be considered as conservative in its diploid number. All species analyzed present a single Ag-NOR-bearing chromosome pair. Supernumerary microchromosomes were described to occur in the species *Bergiaria westermanni*, *Iheringichthys labrosus*, *Pimelodus ortmanni*, and *Pimelodus* sp. (Table 1.5.6). Recently, Swarça *et al.* (2003) suggested the presence of a XX/XY sex chromosome system in *Steindachneridium* sp. (Table 1.5.8). Nuclear DNA content ranges from 2.4 to 2.82 ± 0.20 pg DNA/nucleus (Fenerich *et al.*, 2004).

The family Pseudopimelodidae is composed of 5 genera and 26 species (Shibatta, 2003). Cytogenetic studies in 3 genera and 4 species showed a consistent 2n=54 diploid number (Table 1.5.2). This value contrasts with the modal diploid number of 2n=56 found in most other catfish families (Oliveira and Gosztonyi, 2000) and mainly with the diploid number found among representatives of the families Heptapteridae and Pimelodidae, previously considered as related to Pseudopimelodidae (Nelson, 1994), because in these last two families species with 2n=54 were not found. Thus, the presence of 2n=54 chromosomes may constitute an important characteristic to differentiate the species of Pseudopimelodidae from the species Heptapteridae and Pimelodidae (Martinez *et al.*, in press a). All species of this family present a single Ag-NOR-bearing chromosome pair. Nuclear DNA content determined for two species shows values of 2.23 ± 0.15 and 2.50 ± 0.18 pg DNA/nucleus (Fenerich *et al.*, 2004).

The family Trichomycteridae is composed of 41 genera and 171 species (de Pinna and Wosiacki, 2003). Cytogenetic studies in five genera and 20 species showed diploid number ranging from 2n=48 in *Trichomycterus punctulatus* to 2n=64 in *Vandellia cirrhosa*. However, most of the species exhibit 2n=54 chromosomes (Table 1.5.2), including the primitive species *Trichogenes longipinnis* (Lima and Galetti Jr, 1990).

A study of *Copionodon orthiocarinatus*, which belongs to the most primitive subfamily Copionodontinae, showed that this species also presents 2n=54 (Sato and Oliveira, unpublished data). An interesting characteristic of almost all *cis*-Andean species of *Trichomycterus* studied is the presence of 2n=54 chromosomes, even in widely distributed species such as *T. auroguttatus* (Sato *et al.*, 2004) and *T. spegazzini* (Gonzo *et al.*, 2000). The only exception is *T. diabolus* which presents 2n=56 chromosomes (Torres *et al.*, 2004). On the other hand, the trans-Andean species of *Trichomycterus* present different diploid numbers (Sato *et al.*, 2004), suggesting the presence of different cytogenetic units in the family Trichomycteridae. Supernumerary microchromosomes were described to occur in *Trichomycterus davisi* (Borin and Martins-Santos, 2000) (Table 1.5.6). Nuclear DNA content was described for two species that presented 2.30 ± 0.23 and 2.62 ± 0.19 pg DNA/nucleus (Fenerich *et al.*, 2004).

Chromosome Evolution in Gymnotiformes

Fishes of the order Gymnotiformes are exclusively found in the Neotropical regions where they are distributed in small and large rivers, lakes and shallow-water habitats (Ferraris, 1995). The most striking characteristic of gymnotoids is an organ system that allows these fishes to produce and receive electrical impulses (Ferraris, 1995). According to Albert and Campos-da-Paz (1998), the high taxonomic level of gymnotoids was recently re-evaluated by dividing the order into five families: Gymnotidae, Sternopygidae, Rhamphichthyidae, Hypopomidae and Apteronotidae. The existence of many undescribed species (Albert, 2003a,b; Albert and Crampton, 2003; Campos-da-Paz, 2003; Ferraris, 2003d) together with several taxonomic problems, including the existence of species theoretically distributed throughout South America (such as *Gymnotus carapo* and *Eigenmannia virescens*) have made cytogenetic studies in the order very difficult.

Table 1.5.3 shows that the cytogenetic data are known for 48 species and 9 genera of gymnotiforms representing 35% of the 136 known species and 31% of the 29 known genera. Comparing the diploid numbers found in gymnotoids with those found in siluriforms, it is possible to observe that the first one presents a much lower diploid numbers.

The family Apteronotidae is composed of 12 genera and 52 species (Albert, 2003b) and represents the most diversified gymnotoid family. Cytogenetic studies in 4 species of one genus showed diploid number ranging from 2n=22 or 2n=24 in A. *albifrons* to 2n=52 in A. *anas*, A. *hasemani*, and *Apteronotus sp.* (Table 1.5.3). The value of 2n=22 in A. *albifrons* by Hinegardner and Rosen (1972) may be incorrect since two other studies showed 2n=24 (Almeida-Toledo *et al.*, 1981). Considering the fact that three other species of the family exhibited 2n=52, several chromosome rearrangements should have occurred during specific differentiation.

The family Gymnotidae is composed of 2 genera and 19 species, but according to Campos-da-Paz (2003), many undescribed forms are present in museum collections awaiting formal description. Cytogenetic studies in 2 genera and 15 species showed that the diploid number ranges from 2n=38 to 2n=52 (Table 1.5.3). Two studies in *Electrophorus electricus* (previously the only member of the old family Electrophoridae) showed 2n=52 chromosomes. The genus *Gymnotus* showed a wide variation in diploid numbers: 2n=39/40, 2n=40, 2n=42, 2n=48, 2n=50, 2n=52, and 2n=54 (Foresti *et al.*, 1984; Fernandes-Matioli *et al.*, 1997). Much of this variation was due to different values of diploid number attributed to the species *G. carapo* and probably due to taxonomic problems in the species identification. An interesting feature found in a sample of *Gymnotus sp.* is the presence of an $X_1X_1X_2X_2/X_1X_2Y$-type sex chromosome systems. The same system was also suggested for a sample identified as G. *carapo* with females having 2n=54 chromosomes (Table 1.5.8).

The nuclear DNA content in three cytotypes of *Gymnotus carapo* showed very low variation. Thus, the cytotype with 2n=40 presented 1.25 ± 0.10 pg DNA/nucleus, the cytotype with 2n=50 presented 1.38 ± 0.08, and the cytotype with 2n=54 presented 1.28 ± 0.10. These data suggest that the differences between the cytotypes were due to chromosome rearrangements not associated with large changes in DNA amount.

The family Hypopomidae is composed of 7 genera and 25 species (Albert and Crampton, 2003). Cytogenetic studies in 3 genera and 6 species showed that the diploid number ranges from 2n=36 to 2n=50 (Table 1.5.3). Besides the large difference observed in the diploid number, there is also a significant difference in the number of biarmed and uniarmed chromosomes between the species, suggesting a frequent occurrence of pericentric inversions Two species presented $X_1X_1X_2X_2/X_1X_2Y$-type sex chromosomes system (Table 1.5.8).

The family Rhamphichthyidae is composed of 3 genera and 13 species (Ferraris, 2003d). The only genus and species studied, *Rhamphichthys* cf. *marmoratus* presented 2n=52 chromosomes (Table 1.5.3) (Almeida-Toledo, 1978).

The family Sternopygidae is composed of 5 genera and 27 species, but according to Albert (2003a), a large number of undescribed species are known in museum collections and many more are likely to be captured from continued field studies, particularly in Western Amazonia. Cytogenetic studies were carried out in two genera and 22 species (Table 1.5.3) showing karyotypic variability and the occurrence of different sex-chromosome systems (Tables 1.5.7 and 1.5.8). In the genus *Sternopygus*, only the karyotype of one species, *S. macrurus*, was described, showing two diploid numbers: 2n=48 and 2n=46 (more frequent). Several species of the genus *Eigenmannia* showed great karyotypic and chromosomal diversity, as it is reported in detail in the present book, chapter 1.2.

The high degree of intraspecific polymorphism, strongly indicates the necessity of extensive studies in order to better identify the large number of undescribed species in the family Sternopygidae and also to improve the knowledge on the karyotypic characteristics of this group (Almeida-Toledo *et al.*, 1984,1988, 2000, 2001; Almeida-Toledo and Foresti, 2001).

Chromosome Evolution in Cichlidae

Fishes of the family Cichlidae are the most important species-rich non-Ostariophysan fish family in freshwater, and one of the major vertebrate families (Kullander, 1998, 2003b). Cichlids are found in freshwaters of Africa, Asia, and Central and South America. In the Neotropical regions, the family Cichlidae is composed of 53 genera and 406 species. Phylogenetic analyses combining molecular and morphological characters (Farias *et al.*, 2000) showed that Asian cichlids are the most basal lineage for the family Cichlidae and that the Neotropical cichlids belong to a monophyletic clade, which is the sister group of the African cichlids.

Table 1.5.4 shows that the cytogenetic data are known for 32 genera and 111 species of cichlids representing 27% of the known species and 60% of the known genera. Diploid numbers range from 2n=38 to 2n=60 with a clear modal number of 2n=48. The presence of 2n=48 chromosomes in cichlids possibly reflects the same feature observed in the order Perciformes in which more than 65% of the species present 2n=48 and in more than 50% of them the karyotypes are composed of 48 uniarmed chromosomes (Klinkhardt, 1998). However, an interesting feature observed in the Neotropical cichlids is that only a few species present 2n=48 uniarmed chromosomes. Thus, for example, in *Apistogramma*, the species *A. agassizii* (2n=46) and *A. ortmanni* (2n=46) present 24 biarmed chromosomes and 22 uniarmed chromosomes while *A. borellii* (2n=38) presents 22 biarmed chromosomes and 16 uniarmed chromosomes (Thompson, 1979). In this case, chromosome evolution possibly involved the occurrence of chromosome fusions or translocations and also pericentric inversions.

Another interesting case of deviation of the diploid number 2n=48 is observed in the genus *Symphysodon*, popularly known by aquarists over the word as 'discus'. All the samples analyzed so far presented 2n=60 many biarmed chromosomes, and in some cases, several microchromosomes. The presence of gross differences between karyotypes of fishes of this group collected in different localities suggests that most local populations are presently isolated and may constitute different species.

In *Gymnogeophagus balzanii*, an interesting case of supernumerary chromosomes restrict to gonadal tissue was described (Feldberg and Bertollo, 1985) (Table 1.5.6). The presence of B chromosomes was also reported in *Satanoperca jurupari* (Table 1.5.6), which also displays an XX/XY-type sex chromosome system (Table 1.5.8). Nuclear DNA content ranged from 2.0 to 2.8 pg DNA/nucleus in the species analyzed indicating the occurrence of only small changes in the group (Hinegardner and Rosen, 1972). A general overview of the karyoevolution of cichlids from the different geographic sectors supported the hypothesis that Neotropical species have a higher rate of evolution than observed among cichlids from the Old World (Feldberg *et al.*, 2003).

Chromosome Evolution in Other Groups of Freshwater Fishes

A general analysis in Table 1.5.4 shows that the cytogenetic data are known for 200 species and 72 genera of non-Ostariophysi groups, which represent 17% of the known species of those groups and 43% of the known genera.

The family Potamotrygonidae, freshwater stingrays, is composed of 3 genera and 23 species found exclusively in the Neotropical region (Carvalho *et al.*, 2003). Cytogenetic studies in one genus and two species showed in both 2n=66 chromosomes (Table 1.5.4). The presence of high diploid numbers is a common feature among Chondrichthyes and the occurrence of 2n=66 chromosomes was reported for the marine families Torpedinidae and Mobulidae (Klinkhardt et al., 1995). As observed among other Chondrichthyes, the species of Potamotrygonidae present a high number of uniarmed chromosomes. An interesting characteristic described for this group is the presence of multiple Ag-NORs.

The order Osteoglossiformes is represented in the Neotropical region by two families Arapaimidae (one genus and one species) and Osteoglossidae (one genus and two species) (Ferraris, 2003e, f). *Arapaima gigas* and *Osteoglossum bicirrhosum* have 2n=56 chromosomes while *O. ferreirai* presents 2n=54 chromosomes (Table 1.5.4). A diploid number of 2n=54 was previously reported for *O. bicirrhosum* but it was not confirmed. Similar diploid numbers were described for two other osteoglossids (Klinkhardt et al., 1995), which also displayed high number of uniarmed chromosomes.

The order Clupeiformes is represented in the Neotropical region by the families Clupeidae, Engraulididae, and Pristigasteridae (de Pinna and Di Dario, 2003). Only 3 genera and 5 species of the family Pristigasteridae were studied (Table 1.5.4). *Pristigaster cayana* presents 2n=48 chromosomes, almost all uniarmed. Similar characteristics are observed in the marine species of Clupeiformes that present diploid numbers ranging from 2n=44 to 2n=52, with some exceptions (Klinkhardt et al., 1995).

The family Atherinopsidae, is composed of nine genera and 94 species (Dyer, 2003). *Basilichthys australis* and *B. microlepidotus* from Chile showed 2n=48 and 2n=46 chromosomes, respectively (Table 1.5.4). Among the order Atheriniformes, 12 species presented 2n=48 and only *B. microlepidotus* 2n=46 (Table 1.5.4). Additionally, only a few differences in karyotypic formulae were observed between the species (Klinkhardt et al., 1995).

The order Cyprinodontiformes is represented in Neotropical region by the families Anablepidae (15 species), Cyprinodontidae (58 species), Poeciliidae (216 species) and Rivulidae (235 species) (Reis et al., 2003). In spite of the large number of species present in Neotropical regions, cytogenetic data are available only for one species of Anablepidae, two Cyprinodontidae, 25 Poeciliidae and 41 Rivulidae (Table 1.5.4). Although modal diploid numbers found for these families is 2n=48 and most species present almost all uniarmed chromosomes, a conspicuous chromosome change seems to have occurred in Poeciliidae and mainly in Rivulidae with species presenting diploid number ranging from 2n=20 to 2n=54. A reduction in the diploid number apparently occurred in many species since diploid numbers below 2n=48 are not uncommon. Low diploid numbers were also found in many African Cyprinodontidae of the genus *Aphyosemion* (Scheel, 1973). On the other hand, almost all species of the genus *Fundulus* from North America present 2n=48 chromosomes (Klinkhardt *et al.*, 1995). ZZ/ZW-type sex chromosomes were described for *Gambusia p. puncticulata*, *Poecilia latipinna* and *P. sphenops* (Table 1.5.7) and XX/XY-type sex chromosomes were described for *P. reticulata* and *Gnatholebias hoignei* (Table 1.5.8). Nuclear DNA content ranges from 1.5 to 3.0 pg DNA/nucleus in this group (Hinegardner and Rosen, 1972).

The family Belonidae contains 10 genera and 33 species (Lovejoy and Collette, 2003) but of them, only 4 genera and 9 species occur in the Neotropical region. A cytogenetic study conducted on *Strongylura* cf. *microps* showed 2n=52 chromosomes (Table 1.5.4), which is unusual for a family with marine ancestors. Additionally, this species presented from 1 to 7 supernumerary microchromosomes (Table 1.5.6).

The family Synbranchidae is represented in the Neotropical region by 2 genera and 4 species (Kullander, 2003a) but the number of species is expected to increase considerably in the next years with the description of several unnamed forms of *Synbranchus*. All samples analyzed so far were identified as *Synbranchus marmoratus* (Table 1.5.4), and three different diploid numbers were found: 2n=42, 2n=44, 2n=46. Cytogenetic analysis by Melillo *et al.* (1996) revealed at least five cytotypes in 11 samples differing in diploid number (2n=42, 44, 46), number of metacentric-submetacentric pairs (two or three), size of the second metacentric-submetacentric pair (small or medium sized), type of the Ag-NOR-bearing chromosome (subtelocentric or acrocentric), and position of NORs in the acrocentric chromosomes (short or long arm). An additional cytotype for this species was recorded by Torres (2000). A particular characteristic of this group is the high values of nuclear DNA content that ranges from 5.56 ± 0.38 to 8.54 ± 0.73 pg of DNA/nucleus. The available cytogenetic data reinforce the hypothesis that the genus *Synbranchus* might include several additional species besides those already described.

The order Perciformes is mainly composed of marine species but some species are found in freshwater in the Neotropical region. The most important and representative group is Cichlidae, which was already discussed. The Scianidae (two species) present 2n=48, the Polycentridae analyzed (one species) presents 2n=46, and Gobiidae (four species) present 2n=46 and 2n=48 (Table 1.5.4). Species of Scianidae and Polycentridae present basically uniarmed chromosomes while the species of Gobiidae present more diversified karyotypes. Among the representatives of this family, an X1X2Y-type sex chromosome system was identified in *Awaous strigatus* and an XX/XY-type in *Dormitator maculatus* (Table 1.5.8).

Achiridae is an important family of the order Pleuronectiformes widely distributed in North, Central, and South America with freshwater and marine representatives. Cytogenetic studies conducted in two freshwater species showed 2n=38 (*Hypoclinemus mentalis*) and 2n=40 (*Catathyridium jenynsi*) (Table 1.5.4) and a large number of biarmed chromosomes, suggesting the occurrence of several chromosome fusions in the karyotypic history of this species. These values are very similar to those found in the marine representatives of the family (Azevedo *et al.*, 2005).

Among the Sarcopterygii, the only representative in Neotropical region is *Lepidosiren paradoxa* (Arratia, 2003) which presents 2n=38 biarmed chromosomes (Table 1.5.4) and a very high DNA value of 248 pg DNA/nucleus. In the Sarcopterygii species from Australia, *Neoceratodus forsteri*, a diploid number of 2n=38 was found (Rock *et al.*, 1996), while species from Africa showed 2n=34 and 80 pg DNA/nucleus (*Protopterus annectens* and *P. aethiopicus*) and 2n=68 and 163 pg DNA/nucleus in *P. dolloi* (Vervoort, 1980). According to Vervoort (1980), *P. dolloi* possibly represents a tetraploid form originated from an ancestor with the diploid number 2n=34, also found in *P. annectens* and in *P. aethiopicus*. All species so far analyzed in this group exhibit very large chromosomes, including the Neotropical *L. paradoxa* and also a very high nuclear DNA content, which is a particular characteristic of this lineage.

Chromosome Evolution in Marine Fishes

The cytogenetic studies of Neotropical marine fishes started only in 1983 with the first description of the karyotypes of *Microponogias furnieri* and *Macrodon ancylodon* (Gomes *et al.*, 1983a, b). Our current knowledge about marine fishes in the Neotropical region is quite limited, and for almost all families only a few species were studied (Table 1.5.5).

A low number of papers have been published in scientific journals and our present knowledge came from brief reports presented in scientific meetings, which are not easily found by most scientists. Due to these limitations, a detailed cytogenetic analysis of each family of marine fish of the Neotropical region will not afforded in the present report. However, we will provide a general overview and suggest to those interested in particular families, to check our database in the web and related studies known from other parts of the world (Klinkhardt *et al.*, 1995). According to available data on marine Neotropical fishes, the karyotypic formulae are known for most of the species analyzed (109). The diploid number ranges from $2n=24$ in a cytotype of *Mugil curema* to $2n=100/102$ in *Prionotus punctatus*. Among the species analyzed, 68 (62%) exhibited $2n=48$ chromosomes and among these, 35 (51%) present only uniarmed chromosomes. This high incidence of $2n=48$ chromosome is a characteristic particularly found among the fishes of the order Perciformes and may be due to the presence of $2n=48$ chromosomes in the ancestor of the group as discussed by Brum (1995). This conservative diploid number contrasts with the variability observed in some orders such as Scorpaeniformes with diploid numbers ranging from $2n=40$ to $2n=100/102$ and Pleuronectiformes with diploid numbers ranging from $2n=32$ to $2n=46$.

Karyotypic polymorphism among local populations was reported for a few species. *Micropogonias furnieri* presented $2n=48$ chromosomes in a sample from Cananeia and $2n=46$ in a sample from Rio de Janeiro. In *Mugil curema* two cytotypes: $2n=24$ and $2n=28$. The species *Diplectrum formosum* presented $2n=46$ in a sample from the Littoral Fluminense and $2n=48$ in a sample from Baia de Guanabara, both in Rio de Janeiro. Small differences in karyotypic formulae were reported to occur in *Orthopristis rubber* and *Sphoeroides greeleyi*.

Supernumerary chromosomes were described for one species (0.9% of the species karyotyped), *Sphoeroides greeleyi*, which presented from 0 to 2 additional microchromosomes (Table 1.5.6). Sex chromosomes were reported in three species (2.7% of the species karyotyped), including *Brevoortia aurea* (Clupeidae) and *Stephanolepis hispidus* (Monacanthidae) which present an $X_1X_1X_2X_2/X_1X_2Y$-type sex chromosome system and *Netuma barba* (Ariidae), which present an XX/XY-type sex chromosome system (Table 1.5.8). These numbers are smaller than those observed for freshwater fishes (about 6%), suggesting that such type of chromosome changes were either not frequent or were not positively selected in the marine groups.

Although the number of marine fish species so far studied has been small, the available cytogenetic data are in general more detailed due to the fact that they were conducted after 1980 when almost all current cytogenetic techniques had already developed. The number of nucleolus organizer regions was described for 70 (64%) species and most of them exhibited only one pair of Ag-NOR-bearing chromosomes. Other chromosome banding techniques such as C-banding were applied in 47 (43%) species. The nuclear DNA content is known for only two species (about 2% of the total number of species karyotyped), with 1.24 ± 0.01 pg DNA/nucleus in *Micropogonias furnieri* (2n=48) and 1.57 ± 0.03 in *Menticirrhus americanus* (2n=48) (Gomes *et al.*, 1983a, b).

Occurrence and Meaning of Chromosome Polymorphism

As specified in the present review, Neotropical fishes present a generally wide intraspecific and interspecific karyotypic variability. Some groups present large differences in diploid number and/or karyotypic formulae as has been observed in the genus *Corydoras* (Callichthyidae), whose karyotypic variability is among the most extensive described for fishes. Conversely, some groups appear extremely homogeneous in terms of diploid number and karyotypic formulae. In some cases, as it can be observed in the family Curimatidae, thirty-three species show the same diploid number (2n=54) but one has 2n=46, one 2n=48, two 2n=56, one 2n=58 and one 2n=102. In the family Anostomidae, all species analyzed have 2n=54 chromosomes, but display different mechanisms of sex determination.

These karyotypic characteristics are, in general, in accordance with theoretical models correlating population structure with probability of fixation of chromosome rearrangements. Thus, according to Oliveira *et al.* (1988), Neotropical species with high mobility and population composed of large number of individuals maintain stable chromosome constitutions. In contrast, groups characterized by reduced mobility and low population density usually show extensive interspecific and intraspecific karyotype variation. The existence of cytogenetic variability at the karyotype microstructure level, in species or groups with high mobility or with a large number of individuals in their populations may be explained by the ancient geological age of the large drainage systems in the Neotropical region, which made it possible for many groups to maintain isolation for millions of years, thus allowing the fixation of several rare chromosome rearrangements in these species. As a result, important chromosomal changes could have accumulated within the species despite maintaining the general macrostructure of the karyotypes.

This hypothesis is in accordance with the studies by Santos *et al.* (2002) on chromosome synapses at synaptonemal complex level of artificial hybrids between *Piaractus mesopotamicus* and *Colossoma macropomum* (Serrasalminae). Both parental species present 2n=54 metacentric and submetacentric chromosomes, and may be distinguished only by their characteristic C-band patterns. An analysis of the chromosome pairing of hybrids showed that there are only a few chromosome regions with homology between these species, revealing that many chromosomal changes occurred in their specific evolutionary history; most of them are unidentified by conventional karyotyping. These findings also suggest a possible selective pressure against chromosome rearrangements that could alter the gross karyotypic organization, keeping the same diploid number conserved.

Occurrence of Cryptic Species and Species Complexes

Among Neotropical fishes, there are many nominal species with large geographical distribution, which includes their occurrence in different basins. Cytogenetic studies with samples from different areas have shown that many local populations present sharp karyotypic characteristics, which is suggestive of the hypothesis that several unidentifiable species are actually observed instead of a single species widely distributed (Bertollo *et al.*, 1978, 1979, 2000, 2004; Morelli *et al.*, 1983; Foresti, 1987; Giuliano-Caetano and Bertollo, 1988; Oliveira *et al.*, 1990; Moreira-Filho and Bertollo, 1991; Cestari and Galetti Jr, 1992; Maistro *et al.*, 1992; Porto and Feldberg, 1992; Andreata *et al.*, 1993; Moreira-Filho *et al.*, 2004).

Until 1972, only one species of *Eigenmannia* had been assigned as occurring in rivers of the State of São Paulo, Brazil (Britski, 1972). Cytogenetic studies revealed that this genus was composed of a complex of at least four species initially characterized by their karyotypes: *Eigenmannia virescens* with 2n=38 chromosomes, *Eigenmannia* sp. 1 with 2n=31/32, *Eigenmannia* sp. 2 with 2n=28, and *Eigenmannia* sp. 3 with 2n=36 (Almeida-Toledo, 1978; Foresti, 1987). Cytogenetic analysis of species composed of many isolated local populations, as *Astyanax scabripinnis*, has demonstrated that almost all of them could be identified by their characteristic karyotypes, including those that were not discriminated by morphological data (Moreira-Filho and Bertollo, 1991; Maistro *et al.*, 1998). This data suggests that karyotype may have been evolved at a high rate in some groups of Neotropical fishes when compared with the modifications occurred in characters related to morphological traits (Maistro *et al.*, 1998).

Polyploidy and Evolution

It is largely known that several fish groups such as the families Salmonidae and Catostomidae have evolved by polyploidy (Allendorf and Thorgaard, 1984). Among Neotropical fishes, the analyses carried out in the family Callichthyidae showed that the DNA content ranged up to eight times among species of the genus *Corydoras* (Oliveira *et al.*, 1992), which allowed suggesting that for two species groups, the most important event in the process of speciation was polyploidy. In these groups, two different situations were found. In a group there was a close relation between DNA content and diploid number duplication; in the other group, this evidence was not found, indicating that different pathways of changes could have led to the reduction of diploid number in the tetraploid forms. These finding are possibly related with a different origin of these groups: the first one could be originated in a recent polyploidy event and the second would be the result of a more ancient one.

Among the Loricariidae, the study of different species of *Hypostomus* revealed a great variation in the diploid numbers, which ranged from 2n=64 to 2n=76. The analysis of nuclear DNA content of species in the Loricariidae family by Fenerich *et al.* (2004) showed that the variation in diploid number does not seem to be directly related to the genome size. Thus, *Hypostomus* sp. 1 from the Paranapanema river (2n=64) presented 3.89±0.22 pg of DNA, and Hypostomus sp. 3 from the Edgardia stream (2n=72) presented 3.67±0.24 pg of DNA. Another species of the subfamily Hypostominae, *Liposarcus anisitsi*, presented 2n=52 and 3.96±0.22 pg of DNA. However, *Neoplecostomus paranensis*, a species of the subfamily Neoplecostominae, considered as one of the most primitive species in the family Loricariidae, presented 2n=54 and 2.26±0.17 pg of DNA. Similarly, other basal species, *Hisonotus* sp. 1 (Hypoptopomatinae) presented 2n=54 and 2.66±0.12 pg of DNA, and *Hisonotus* sp. 2

presented 2n=54 and 1.78±0.10 pg of DNA. These results suggest that the differences found between the species of this group could be due not only to simple chromosome rearrangement events, but also to the occurrence of changes involving all the genome of the species, with the possible occurrence of polyploidy in the origin of some members of the subfamily Hypostominae.

Triploidy is a rare phenomenon in the higher groups of vertebrates, occurring only among fish, amphibians and reptiles, probably due to their incipient mechanisms of sex determination (Thorgaard and Gall, 1979). Natural triploidy occurs in some fish species, including Neotropical fishes, where this rare phenomenon has been reported mainly in the genus *Astyanax* (three cases described). An extreme example is the occurrence of triploid populations of *Poecilia* maintained by gynogenesis (Sola *et al.*, 1993).

Triploids are good models for the analysis of gene regulation process. For instance, the study of triploids of *Eigenmania* sp. (Almeida-Toledo *et al.*, 1985), *Astyanax scabripinnis* (Maistro *et al.*, 1995), and *Trichomycterus davisi* (Borin *et al.*, 2002) revealed a normal activity of ribosomal genes in all the three Ag-NOR-bearing chromosomes. The analysis of other genes could be useful in answering the following question: why are triploids almost identical to diploid individuals in terms of morphological characteristics?

CONCLUSION

Considering the great number of freshwater and marine fishes living in Neotropical regions, the number of species studied from the cytogenetic point of view remains limited. Moreover, we have to consider the necessity to re-analyze several species for which either only the basic haploid or diploid chromosomal numbers were described or the specimens were not identified at specific taxonomic level.

However, there is no doubt that the application of cytogenetic methods to the analysis of the Neotropical fishes has made an important contribution for a better knowledge of this fauna. These studies have provided valuable information for the understanding of the relationship between groups, the occurrence of cryptic species and species complexes, the mechanism of sex determination and sex chromosome evolution, nucleolus organizer regions distribution, supernumerary chromosomes and the relationship of polyploidy and evolution.

In future we should consider the necessity to use more elaborate molecular cytogenetic techniques such as *in situ* hybridization and chromosome painting, for a better understanding of the variability detected by traditional analysis. Moreover, studies integrating cytogenetics with populational, morphological and systematic data are also needed to arrive at a correct view of the importance of karyotypic changes in the species evolution.

References

Albert, J.S. 2003a. Sternopygidae. In: *Check List of the Freshwater Fishes of South America*, R.E. Reis, S.O. Kullander and C.J. Ferraris (eds.). Edipucrs, Porto Alegre, pp. 487–491.

Albert, J.S. 2003b. Apteronotidae. In: *Check List of the Freshwater Fishes of South America*, R.E. Reis, S.O. Kullander and C.J. Ferraris (eds.). Edipucrs, Porto Alegre, pp. 497–502.

Albert, J.S. and R. Campos da Paz. 1998. Phylogenetic systematics of Gymnotiformes with diagnosis of 58 clades: A review of available data. In: *Phylogeny and Classification of Neotropical Fishes*, L.R. Malabarba, R.E. Reis, R.P. Vari, Z.M.S. Lucena and C.A.S. Lucena (eds.). Edipucrs, Porto Alegre, pp. 419–446.

Albert, J.S. and W.G.R. Crampton. 2003. Hypopomidae. In: *Check List of the Freshwater Fishes of South America*, R.E. Reis, S.O. Kullander and C.J. Ferraris (eds.). Edipucrs, Porto Alegre, pp. 494–496.

Allendorf, F.W. and G.H. Thorgaard. 1984. Tetraploidy and the evolution of Salmonid fishes. In: *Evolutionary Genetics of Fishes*, B.J. Turner (ed.). Plenum Publishing Company, New York, pp. 1–53.

Almeida-Toledo, L.F. 1978. *Contribuição à citogenética dos Gymnotoidei (Pisces, Ostariophysi)*. Doctoral Thesis. Universidade de São Paulo, São Paulo.

Almeida-Toledo, L.F. and F. Foresti. 2001. Morphologically differentiated sex chromosomes in Neotropical freshwater fish. *Genetica* 111: 91–100.

Almeida-Toledo, L.F., F. Foresti and S.A. Toledo-Filho. 1981. Constitutive heterochromatin and nucleolus organizer region in the knifefish, *Apteronotus albifrons* (Pisces, Apteronotidae). *Experientia* 37: 953–954.

Almeida-Toledo, L.F., F. Foresti and S.A. Toledo-Filho. 1984. Complex sex chromosome system in *Eigenmannia* sp. (Pisces, Gymnotiformes). *Genetica* 64: 165–169.

Almeida-Toledo, L.F., F. Foresti and S.A. Toledo-Filho. 1985. Spontaneous triploidy and NOR activity in *Eigenmannia* sp. (Pisces, Sternopygidae) from the Amazon Basin. *Genetica* 66: 85–88.

Almeida-Toledo, L.F., E. Viegas-Pequignot, F. Foresti, S.A. Toledo Filho and B. Dutrillaux. 1988. BrdU replication patterns demonstrating chromosome homoeologies in two fish species, genus *Eigenmannia*. *Cytogenetics and Cell Genetics* 48: 117–120.

Almeida-Toledo, L.F., F. Foresti, M.F.Z. Daniel and S.A. Toledo-Filho. 2000. Sex chromosome evolution in fish: The formation of the neo-Y chromosome in *Eigenmannia* (Gymnotiformes). *Chromosoma* 109: 197–200.

Almeida-Toledo, L.F., F. Foresti, M.F.Z. Pequignot and M.F.Z. Danial-Silva. 2001. XX:YY sex chromosome system with X heterochromatinization: An early stage of sex chromosome and differentation in the Neotropic electric eel *Eigenmannia viresscens*. *Cytogenetics and Cell Genetics* 95: 73-78.

Alves, A.L., C. Oliveira and F. Foresti. 2003. Karyotype variability in eight species of the subfamilies Loricariinae and Ancistrinae (Teleostei, Siluriformes, Loricariidae). *Caryologia* 56: 57–63.

Alves, A.L., C. Oliveira and F. Foresti. 2005. Comparative cytogenetic analysis of eleven species of subfamilies Neoplecostominae and Hypostominae (Siluriformes: Loricariidae). *Genetica* 124: 124-127.

Andrea, M. 1971. Contribuição ao estudo da biologia e do cariótipo do mussum (*Synbranchus marmoratus*). *Ciência e Cultura* 23: 103–104.

Andreata, A.A., L.F. Almeida-Toledo, C. Oliveira and S.A. Toledo-Filho. 1992. Chromosome studies in Hypoptopomatinae (Pisces, Siluriformes, Loricariidae): XX/XY sex chromosome heteromorphism in *Pseudotocinclus tietensis*. *Cytologia* 57: 369–372.

Andreata, A.A., L.F. Almeida-Toledo, C. Oliveira and S.A. Toledo-Filho. 1993. Chromosome studies in Hypoptomatinae (Pisces, Siluriformes, Loricariidae). II. ZZ/ZW sex-chromosome system, B chromosomes, and constitutive heterochromatin differentiation in *Microlepdogaster leucofrenatus*. *Cytogenetics and Cell Genetics* 63: 215–220.

Armbruster, J. W. 2004. Phylogenetic relationships of the sucker-mouth armored catfishes (Loricariidae) with particular emphasis on the Hypostominae and the Ancistrinae. *Zoological Journal of the Linnean Society* 141: 1–80.

Arratia, G. 2003. Lepidosirenidae. In: *Check List of the Freshwater Fishes of South America*. R.E. Reis, S.O. Kullander and C.J. Ferraris (eds.). Edipucrs, Porto Alegre, pp. 671–672.

Artoni, R.F. and L.A.C. Bertollo. 2001. Trends in the karyotype evolution of Loricariidae fish (Siluriformes). *Hereditas* 134: 201–210.

Artoni, R.F., J.N. Falcão and O. Moreira-Filho. 2001. An uncommon condition for a sex chromosome system in Characidae fish. Distribution and differentiation of the ZZ/ZW system in *Triportheus*. *Chromosome Research* 9: 449–456.

Azevedo, M.F.C., C. Oliveira, B.G. Pardo, P. Martinez and F. Foresti. 2005. Cytogenetic analysis of seven species of the family Achiridae (Teleostei: Pleuronectiformes). *Genetica* 125: 125-132.

Bertollo, L.A.C. 1978. *Estudos citogenéticos no gênero Hoplias Gill, 1903* (Pisces, Eryhrinidae). Doctoral Thesis. Faculdade de Medicina de Ribeirão Preto, Ribeirão Preto.

Bertollo, L.A.C., C.S. Takahashi and O. Moreira-Filho. 1978. Cytotaxonomic considerations on *Hoplias lacerdae* (Pisces, Erytrinidae). *Revista Brasileira de Genética* 1: 103–120.

Bertollo, L.A.C., C.S. Takahashi and O. Moreira-Filho. 1979. Karyotypic studies of two allopatric populations of the genus *Hoplias* (Pisces, Erytrinidae). *Revista Brasileira de Genética* 2: 17–37.

Bertollo, L.A.C., G.G. Born, J.A. Dergam, A.S. Fenocchio and O. Moreira-Filho. 2000. A biodiversity approach in the neotropical Erythrinidae fish, *Hoplias malabaricus*. Karyotypic survey, geographical distribution of cytotypes and cytotaxonomic considerations. *Chromosome Research* 8: 603–613.

Bertollo, L.A.C., C. Oliveira, W.F. Molina, V.P. Margarido, M.S. Fontes, M.C. Pastori, J.N. Falcão and A.S. Fenocchio. 2004. Chromosome evolution in the erythrinid fish, *Erythrinus erythrinus* (Teleostei, Characiformes). *Heredity* 93: 228–233.

Beukeboom, L.W. 1994. Bewildering Bs: An impression of the 1st B-chromosome Conference. *Heredity* 73: 328–336.

Bockmann, F.A. and G.M. Guazzelli. 2003. Heptapteridae. In: *Check List of the Freshwater Fishes of South America*, R.E. Reis, S.O. Kullander and C.J. Ferraris (eds.). Edipucrs, Porto Alegre, pp. 406–431.

Borin, L.A. and I.C. Martins-Santos. 2000. Intra-individual numerical chromosomal polymorphism in *Trichomycterus davisi* (Siluriformes, Trichomycteridae) from Iguaçu river basin in Brazil. *Genetics and Molecular Biology* 23: 605–607.

Borin, L.A., I.C. Martins-Santos and C. Oliveira. 2002. Natural triploid in *Trichomycterus davisi* (Siluriformes, Trichomycteridae): mitotic and meiotic characterization by chromosome banding and synaptonemal complex analyses. *Genetica* 115: 253–258.

Britski, H.A. 1972. Peixes de água-doce do Estado de São Paulo: sistemática. In: *Poluição e Piscicultura*. Faculdade de Saúde Pública da Universidade de São Paulo, São Paulo, pp. 79–108.

Britto, M.R. 2003a. *Análise filogenética da ordem Siluriformes com ênfase nas relações da superfamília Loricarioidea (Teleostei: Ostariophysi)*. Doctoral Thesis. Instituto de Biociências, Universidade Estadual de São Paulo, São Paulo, Brazil.

Britto, M.R. 2003b. Phylogeny of the subfamily Corydoradinae Hoedeman, 1952 (Siluriformes: Callichthyidae), with a definition of its genera. *Proceedings of the Academy of Natural Science of Philadelphia* 153: 119–154.

Brum, M.J.I. 1995. *Correlações entre a filogenia e a citogenética dos peixes teleósteos*. Sociedade Brasileira de Genética, Série Monografias, Ribeirão Preto.

Brum, M.J.I. 1996. Cytogenetic studies of Brazilian marine fish. *Revista Brasileira de Genética* 19: 421–427.

Buckup, P.A. 1998. Relationships of the Characidiinae and phylogeny of characiform fishes (Teleostei: Ostariophysi). In: *Phylogeny and Classification of Neotropical Fishes*, L. R. Malabarba, R. E. Reis, R. P. Vari, Z. M. S. Lucena and C.A.S. Lucena (eds.). Edipucrs, Porto Alegre, pp. 123–144.

Buckup, P.A. 2003. Crenuchidae. In: *Check List of the Freshwater Fishes of South America*, R.E. Reis, S.O. Kullander and C.J. Ferraris (eds.). Edipucrs, Porto Alegre, pp. 87–95.

Campos, H., G. Arratia and C. Cuevas. 1997. Karyotypes of the most primitive catfishes (Teleostei: Siluriformes: Diplomystidae). *Journal of Zoology Systematic and Evolutionary Research* 35: 113–119.

Campos-da-Paz, R. 2003. Gymnotidae. In: *Check List of the Freshwater Fishes of South America*, R. E. Reis, S. O. Kullander and C. J. Ferraris (eds.). Edipucrs, Porto Alegre, pp. 483–486.

Carvalho, M.L., C. Oliveira and F. Foresti. 1998. Nuclear DNA content of thirty species of Neotropical fishes. *Genetics and Molecular Biology* 21: 47–54.

Carvalho, M.L., C. Oliveira and F. Foresti. 2001. Cytogenetic analysis of three species of the families Characidae and Curimatidae (Teleostei, Characiformes) from the Acre river. *Chromosome Science* 5: 91–96.

Carvalho, M.L., C. Oliveira and F. Foresti. 2002a. Description of a ZZ/ZW sex chromosome system in *Thoracocharax* cf. *stellatus* (Teleostei, Characiformes, Gasteropelecidae). *Genetics and Molecular Biology* 25: 299–303.

Carvalho, M.L., C. Oliveira, M.C. Navarrete, O. Froehlich and F. Foresti. 2002b. Nuclear DNA content determination in Characiformes fish (Teleostei, Ostariophysi) from the Neotropical region. *Genetics and Molecular Biology* 25: 49–55.

Carvalho, M.R., N.R. Lovejoy and R.S. Rosa. 2003. Potamotrygonidae. In: *Check List of the Freshwater Fishes of South America*, R.E. Reis, S.O. Kullander and C.J. Ferraris (eds.). Edipucrs, Porto Alegre, pp. 22–28.

Castro, R.M.C.C. and R.P. Vari. 2003. Prochilodontidae. In: *Check List of the Freshwater Fishes of South America*, R.E. Reis, S.O. Kullander and C.J. Ferraris (eds.). Edipucrs, Porto Alegre, pp. 65–70.

Castro, R.M.C., R.P. Vari, F. Vieira and C. Oliveira. 2004. Phylogenetic analysis and redescription of the genus *Henochilus* (Characiformes: Characidae). *Copeia* 2004: 496–506.

Cavallaro, Z.I., L.A.C. Bertollo, F. Ferfectti and J.P.M. Camacho. 2000. Frequency increase and mitotic stabilization of a B chromosome in the fish *Prochilodus lineatus*. *Chromosome Research* 8: 627–734.

Centofante, L., L.A.C. Bertollo and O. Moreira-Filho. 2002. A ZZ/ZW sex chromosome system in a new species of the genus *Parodon* (Pisces, Parodontidae). *Caryologia* 55: 139–150.

Cestari, M.M. and P.M. Galetti Jr. 1992. Chromosome studies of *Serrasalmus spilopleura* (Characidae, Serrasalminae) from the Parana-Paraguay rivers: Evolutionary and cytotaxonomic considerations. *Copeia* 1992: 108–112.

Chiarelli, A.B. and E. Capanna. 1973. Checklist of fish chromosomes. In: *Cytotaxonomy and Vertebrate Evolution*, A.B. Chiarelli and E. Capanna (eds.). Academic Press, London, pp. 205–232.

Daniel, M.F.Z. 1996. *Estudos citogenéticos comparativos em quatro espécies do gênero Astyanax (Pisces, Characidae)*. Masters' thesis. Instituto de Biociências, São Paulo.

de Pinna, M.C.C. 1998. Phylogenetic relationships of Neotropical Siluriformes (Teleostei: Ostariophysi): Historical overview and synthesis of hypotheses. In: *Phylogeny and Classification of Neotropical Fishes*, L.R. Malabarba, R.E. Reis, R.P. Vari, Z.M.S. Lucena and C.A.S. Lucena (eds.). Edipucrs, Porto Alegre, pp. 279–330.

de Pinna, M.C.C. 2003. Nematogenyidae. In: *Check List of the Freshwater Fishes of South America*, R.E. Reis, S.O. Kullander and C.J. Ferraris (eds.). Edipucrs, Porto Alegre, pp. 268–269.

de Pinna, M.C.C. and F. Di Dario. 2003. Pristigasteridae. In: *Check List of the Freshwater Fishes of South America*, R.E. Reis, S.O. Kullander and C.J. Ferraris (eds.). Edipucrs, Porto Alegre, pp. 43–45.

de Pinna, M.C.C. and W. Wosiacki. 2003. Trichomycteridae. In: *Check List of the Freshwater Fishes of South America*, R.E. Reis, S.O. Kullander and C.J. Ferraris (eds.). Edipucrs, Porto Alegre, pp. 270–290.

Denton, T.E. 1973. *Fish Chromosome Methodology*. Charles C. Thomas Publisher, Springfield, Illinois.

Dias, A.L. and F. Foresti. 1993. Cytogenetic studies on fishes of the family Pimelodidae (Siluroidei). *Revista Brasileira de Genética* 16: 585–600.

Dyer, B.S. 2003. Atherinopsidae. In: *Check List of the Freshwater Fishes of South America*, R.E. Reis, S.O. Kullander and C.J. Ferraris (eds.). Edipucrs, Porto Alegre, pp. 515–525.

Falcao, J.N. and L.A.C. Bertollo. 1985. Chromosome characterization in Acestrorhynchinae and Cynopotaminae (Pisces, Characidae). *Journal of Fish Biology* 27: 603–610.

Farias, I.P., G. Ortì, A. Meyer. 2000. Total evidence: molecules, morphology, and the phylogenetics of cyclid fishes. *Journal of Experimental Zoology* 288: 76–92.

Feldberg, E. and L.A.C. Bertollo. 1985. Discordance in chromosome number among somatic and gonadal tissue cells of *Gymnogeophagus balzanii* (Pisces, Cichlidae). *Revista Brasileira de Genética* 7: 639–645.

Feldberg, E., L.A.C. Bertollo, L.F. Almeida-Toledo, F. Foresti, O. Moreira-Filho and A.F. Santos. 1987. Biological aspects of Amazonian fishes IX. Cytogenetic studies in two species of the genus *Semaprochilodus* (Pisces, Prochilodontidae). *Genome* 29: 1–4.

Feldberg, E., J.I.R. Porto, C.M. Nakayama and L.A.C. Bertollo. 1993. Karyotype evolution in Curimatidae (Teleostei, Characiformes) from the Amazon region. II. Centric fissions in the genus *Potamorhina*. *Genome* 36: 372–376.

Feldberg, E., J.I.R. Porto and L.A.C. Bertollo. 2003. Chromosomal changes and adaptation of cichlid fishes during evolution. In: *Fish Adaptation*, A.L.Val and B.G. Kapoor (eds.). Science Publishers Inc., Enfield (NH), USA, pp. 285–308.

Fenerich, P.C., C. Oliveira and F. Foresti. 2004. Nuclear DNA content in 20 species of Siluriformes (Teleostei: Ostariophysi) from the Neotropical region. *Genetics and Molecular Biology* 27: 350–354.

Fernandes-Matioli, F.M.C., L.F. Almeida-Toledo and S.A. Toledo-Filho. 1997. Extensive Nucleolus organizer region polymorphism in *Gymnotus carapo* (Gymnotoidei, Gymnotidae). *Cytogenetics and Cell Genetics* 78: 236–239.

Ferraris, C.J. 1995. Catfishes and knifefishes. In: *Encyclopedia of Fishes*, J.R. Paxton and W. N. Eschmeyer (eds.). Academic Press, San Diego, pp. 106–112.

Ferraris, C.J. 2003a. Diplomystidae. In: *Check List of the Freshwater Fishes of South America*, R.E. Reis, S.O. Kullander and C.J. Ferraris (eds.). Edipucrs, Porto Alegre, pp. 255–256.

Ferraris, C.J. 2003b. Auchenipteridae. In: *Check List of the Freshwater Fishes of South America*, R.E. Reis, S.O. Kullander and C.J. Ferraris (eds.). Edipucrs, Porto Alegre, pp. 470–482.

Ferraris, C.J. 2003c. Neoplecostominae. In: *Check List of the Freshwater Fishes of South America*, R.E. Reis, S.O. Kullander and C.J. Ferraris (eds.). Edipucrs, Porto Alegre, pp. 319–320.

Ferraris, C.J. 2003d. Rhamphichthyidae. In: *Check List of the Freshwater Fishes of South America*, R.E. Reis, S.O. Kullander and C.J. Ferraris (eds.). Edipucrs, Porto Alegre, pp. 492–493.

Ferraris, C.J. 2003e. Arapaimidae. In: *Check List of the Freshwater Fishes of South America*, R.E. Reis, S.O. Kullander and C.J. Ferraris (eds.). Edipucrs, Porto Alegre, pp. 31–31.

Ferraris, C.J. 2003f. Osteoglossidae. In: *Check List of the Freshwater Fishes of South America*, R.E. Reis, S.O. Kullander and C.J. Ferraris (eds.). Edipucrs, Porto Alegre, pp. 30–30.

Fink, S. and W. Fink. 1981. Interrelationships of the ostariophysan fishes (Pisces, Teleostei). *Zoological Journal of Linnean Society* 72: 297–353.

Fontana, F., A.B. Chiarelli and A. Rossi. 1970. Il cariotipo di alcune specie di Cyprinidae, Centrarchidae, Characidae, studiate mediante colture *in vitro*. *Caryologia* 23: 549–564.

Foresti, F. 1974. Técnica para obtenção de cromossomos de pequenos peixes. *Ciência e Cultura* 26: 248.

Foresti, F. 1987. *Estudos cromossômicos em Gymnotiformes (Pisces, Ostariophysi).* Universidade Estadual Paulista, Botucatu.

Foresti, F., L.F. Almeida-Toledo and S.A. Toledo-Filho. 1984. Chromosome studies in *Gymnotus carapo* and *Gymnotus* sp. (Pisces, Gymnotidae). *Caryologia* 37: 141–146.

Foresti, F., L.M. Oliveira and W.A. Angeleli. 1974. Caracterização cromossômica de peixes do gênero Curimatus (Cypriniformes, Curimatidae). *Ciência e Cultura* 26: 249.

Friel, J.P. 2003. Aspredinidae. In: *Check List of the Freshwater Fishes of South America*, R.E. Reis, S.O. Kullander and C.J. Ferraris (eds.). Edipucrs, Porto Alegre, pp. 261–267.

Galetti Jr, P.M., C.A. Mestriner, P.J. Monaco and E.M. Rasch. 1995. Post-zygotic modifications and intra- and inter-individual nucleolar organizing region variations in fish: report of a case involving *Leporinus friderici*. *Chromosome Research* 3: 285–290.

Garavello, J.C. and H.A. Britski. 2003. Anostomidae. In: *Check List of the Freshwater Fishes of South America*, R.E. Reis, S.O. Kullander and C.J. Ferraris (eds.). Edipucrs, Porto Alegre, pp. 71–84.

Giuliano-Caetano, L. 1998. *Polimorfismo cromossomico Robertsoniano em populações de Rineloricaria latirostris (Pisces, Loricariinae).* Doctoral Thesis. Universidade Federal de São Carlos, São Carlos.

Giuliano-Caetano, L. and L.A.C. Bertollo. 1988. Karyotype variability in *Hoplerythrinus unitaeniatus* (Characiformes, Erythrinidae). I. Chromosome polymorphism in the Rio Negro population (Manaus, state of Amazonas). *Revista Brasileira de Genética* 11: 299–306.

Gomes, V., A.E.A.M. Vazzoler and P.V. Ngan. 1983a. Estudos cariotipicos de peixes da familia Sciaenidae (Teleostei, Perciformes) da regiao de Cananeia, SP, Brasil. 2. Sobre o cariotipo de *Menticirrhus americanus* (Linnaeus, 1758). *Boletim do Instituto Oceanográfico* 32: 187–191.

Gomes, V., A.E.A.M. Vazzoler and P.V. Ngan. 1983b. Estudos cariotipicos de peixes da familia Sciaenidae (Teleostei, Perciformes) da regiao da Cananeia, SP, Brasil. 1. Sobre o cariotipo de *Micropogonias furneiri* (Desmarest, 1823). *Boletim do Instituto Oceanográfico* 32: 137–142.

Gonzo, G.M., A.S. Fenocchio and C. Pastori. 2000. Chromosome characterization of *Trichomycterus spegazzini* (Siluriformes, Trichomycteridae) from three hydrographic basins of the northwest of Argentina. *Caryologia* 53: 39–43.

Gyldenholm, A.O. and J.J. Scheel. 1971. Chromosome number of fishes. *Journal of Fish Biology* 3: 479–486.

Hinegardner, R. and D.E. Rosen. 1972. Cellular DNA content and the evolution of Teleostean fishes. *American Naturalist* 106: 621–644.

Howell, W.M. 1972. Somatic chromosomes of the black ghost knifefish, *Apteronotus albifrons* (Pisces, Aptenorotidae). *Copeia* 1972: 191–193.

Hudson, R.G. 1976. A comparison of karyotypes and erythrocyte DNA quantities of catfish (Siluriformes) with phylogenetic implications. Ph.D. Thesis. North Carolina State University at Raleigh, North Carolina, USA.

Jégu, M. 2003. Serrasalminae. In: *Check List of the Freshwater Fishes of South America*, R.E. Reis, S.O. Kullander and C.J. Ferraris (eds.). Edipucrs, Porto Alegre, pp. 182–196.

Jesus, C.M. and O. Moreira-Filho (2000). Cytogenetic studies in some *Apareiodon* species (Piscies, Parodontiae). *Cytologia* 65: 397–402.

Jesus, C.M., P.M. Galetti Jr, S. Valentini and O. Moreira-Filho. 2003. Molecular characterization and chromosomal localization of two families of satellite DNA in *Prochilodus lineatus* (Pisces, Prochilodontidae), a species with B chromosomes. *Genetica* 118: 25–32.

Klinkhardt, M.B. 1998. Some aspects of karyoevolution in fishes. *Animal Research and Development* 47: 7–36.

Klinhkhardt, M.B., M. Tesche and H. Greven. 1995. *Database of Fish Chromosomes*. Westarp-Wiss, Magdeburg.

Kullander, S.O. 1998. A phylogeny and classification of the South American Cichlidae (Teleostei: Perciformes). In: *Phylogeny and Classification of Neotropical Fishes*, L. R. Malabarba, R. E. Reis, R. P. Vari, Z. M. S. Lucena and C. A. S. Lucena (eds.). Edipucrs, Porto Alegre, pp. 461–498.

Kullander, S.O. 2003a. Synbranchidae. In: *Check List of the Freshwater Fishes of South America*, R.E. Reis, S.O. Kullander and C.J. Ferraris (eds.). Edipucrs, Porto Alegre, pp. 594–595.

Kullander, S.O. 2003b. Cichlidae. In: *Check List of the Freshwater Fishes of South America*, R.E. Reis, S.O. Kullander and C.J. Ferraris (eds.). Edipucrs, Porto Alegre, pp. 605–654.

Langeani, F. 2003. Hemiodontidae. In: *Check List of the Freshwater Fishes of South America*, R.E. Reis, S.O. Kullander and C.J. Ferraris (eds.). Edipucrs, Porto Alegre, pp. 96–100.

Lima, N. R. W. and P. M. Galetti Jr. 1990. Chromosome characterization of the fish *Trichogenes longipinnis*. A possible basic karyotype of Trichomycteridae. *Revista Brasileira de Genética* 13: 239–245.

Lima, R.S. 2003. Aphyocharacinae. In: *Check List of the Freshwater Fishes of South America*, R.E. Reis, S.O. Kullander and C.J. Ferraris (eds.). Edipucrs, Porto Alegre, pp. 197–199.

Lima, F.C.T. 2003. Bryconinae. In: *Check List of the Freshwater Fishes of South America*, R.E. Reis, S.O. Kullander and C.J. Ferraris (eds.). Edipucrs, Porto Alegre, pp. 200–208.

Lovejoy, N.R. and B.B. Collette. 2003. Belonidae. In: *Check List of the Freshwater Fishes of South America*, R.E. Reis, S.O. Kullander and C.J. Ferraris (eds.). Edipucrs, Porto Alegre, pp. 586–588.

Lucena, C.A.S. 1993. *Estudo filogenético da família Characidae com uma discussão dos grupos naturais propostos (Teleostei, Ostariophysi, Characiformes)*. Doctoral Thesis. Instituto de Biociências, São Paulo.

Lucena, C.A.S. and N.A. Menezes. 2003. Characinae. In: *Check List of the Freshwater Fishes of South America*, R.E. Reis, S.O. Kullander and C.J. Ferraris (eds.). Edipucrs, Porto Alegre, pp. 200–208.

Lundberg, J.G. and M.W. Littmann. 2003. Pimelodidae. In: *Check List of the Freshwater Fishes of South America*, R.E. Reis, S.O. Kullander and C.J. Ferraris (eds.). Edipucrs, Porto Alegre, pp. 432–446.

Maistro, E.L., F. Foresti, C. Oliveira and L.F. Almeida-Toledo. 1992. Occurrence of macro B chromosome in *Astyanax scabripinnis paranae* (Pisces, Characiformes, Characidae). *Genetica* 87: 101–106.

Maistro, E.L., A.L. Dias, F. Foresti, C. Oliveira and O. Moreira-Filho. 1995. Natural triploidy in *Astyanax scabripinnis* (Pisces, Characidae) and simultaneous occurrence of macro B chromosomes. *Caryologia* 47: 233–239.

Maistro, E.L., C. Oliveira and F. Foresti. 1998. Comparative cytogenetic and morphological analysis of *Astyanax scabripinnis paranae* (Pisces, Characidae, Tetragonopterinae). *Genetics and Molecular Biology* 21: 201–206.

Malabarba, L.R. 2003. Cheirodontinae. In: *Check List of the Freshwater Fishes of South America*, R.E. Reis, S.O. Kullander and C.J. Ferraris (eds.). Edipucrs, Porto Alegre, pp. 215–221.

Malabarba, M.C.S.L. and F.C.T. Lima. 2003. *Triportheus*. In: *Check List of the Freshwater Fishes of South America*, R.E. Reis, S.O. Kullander and C.J. Ferraris (eds.). Edipucrs, Porto Alegre, pp. 157–158.

Martinez, E.R.M., C. Oliveira and F. Foresti. 2004a. Cytogenetic Analyses of *Pseudopimelodus mangurus* (Teleostei: Siluriformes: Pseudopimelodidae). *Cytologia*. 69: 419-424.

Martinez, E.R.M., C. Oliveira and H.F. Júlio Junior. 2004b. Cytogenetic analysis of species of the genera *Acestrorhynchus*, *Oligosarcus* and *Rhaphiodon* (Teleostei: Characiformes). *Caryologia*. 57(3): 294-299.

Melillo, M.I.F.M, F. Foresti and C. Oliveira. 1996. Additional cytogenetic studies on local populations of *Synbranchus marmoratus* (Pisces, Synbranchiformes, Synbranchidae). *Naturalia* 21: 201–208.

Menezes, N.A. 2003. Acestrorhynchidae. In: *Check List of the Freshwater Fishes of South America*, R.E. Reis, S.O. Kullander and C.J. Ferraris (eds.). Edipucrs, Porto Alegre, pp. 231–233.

Menezes, N.A., P.A. Buckup, J.L. Figueiredo and R.L. Moura. 2003. *Catálogo das espécies de peixes marinhos do Brasil*. Museu de Zoologia da Universidade de São Paulo, São Paulo.

Mestriner, C.A., P.M. Galetti Jr, S.R. Valentini, I.R.G. Ruiz, L.D.S. Abel, O. Moreira-Filho and J.P.M. Camacho. 2000. Structural and functional evidence that a B chromosome in the characid fish *Astyanax scabripinnis* is an isochromosome. *Heredity* 85: 1–9.

Michelle, J.L. 1975. *Estudo cariotípico de algumas espécies da Família Loricariidae (Pisces)*. Másters' Thesis. Universidade de São Paulo, Ribeirão Preto.

Moreira, C. 2003. Iguanodectinae. In: *Check List of the Freshwater Fishes of South America*, R.E. Reis, S.O. Kullander and C.J. Ferraris (eds.). Edipucrs, Porto Alegre, pp. 172–173.

Moreira-Filho, O. and L.A.C. Bertollo. 1991. *Astyanax scabripinnis* (Pisces, Characidae): a species complex. *Revista Brasileira de Genética* 14: 331–358.

Moreira-Filho, O., L.A.C. Bertollo and P.M. Galetti Jr. 1985. Karyotypic study of some species of family Parodontidae (Pisces, Cypriniformes). *Caryologia* 38: 47–55.

Moreira-Filho, O., L.A.C. Bertollo and P.M. Galetti Jr. 2004. B chromosomes in the fish *Astyanax scabripinnis* (Characidae, Tetragonopterinae): An overview in natural populations. *Cytogenetic and Genome Research* 106: 230–234.

Morelli, S., L.A.C. Bertollo, F. Foresti and O. Moreira-Filho. 1983. Cytogenetic considerations on the genus *Astyanax* (Pisces, Characidae). I. Karyotypic variability. *Caryologia* 36: 235–244.

Muramoto, J., S. Ohno and N.B. Atkin. 1968. On the diploid state of the fish Order Ostariophysi. *Chromosoma* 24: 59–66.

Navarrete, M.C. and H.F. Júlio Jr. 1997. Cytogenetic analysis of four curimatids from the Paraguay Basin, Brazil (Pisces: Characiformes: Curimatidae). *Cytologia* 62: 241–247.

Nelson, J.S. 1994. *Fishes of the World*. 3rd Edition. John Wiley & Sons, Inc., New York.

Ohno, S. and N.B. Atkin. 1966. Comparative DNA values and chromosome complements in eight species of fishes. *Chromosoma* 18: 455–456.

Ojima, Y., K. Ueno and M. Hayashi. 1976. A review of the chromosome numbers of fishes. *La Kromosomo* 2: 19–47.

Oliveira C. and A.E. Gosztonyi. 2000. A cytogenetic study of *Diplomystes mesembrinus* (Teleostei, Siluriformes, Diplomystidae) with a discussion of chromosome evolution in siluriforms. *Caryologia* 53: 31–37.

Oliveira, C., L.F. Almeida-Toledo, F. Foresti, H.A. Britski and S.A. Toledo-Filho. 1988. Chromosome formulae of Neotropical freshwater fishes. *Revista Brasileira de Genética* 11: 577–624.

Oliveira, C., L.F. Almeida-Toledo and S.A. Toledo-Filho. 1990. Comparative cytogenetic analysis in three cytotypes of *Corydoras nattereri* (Pisces, Siluriformes, Callichthyidae). *Cytologia* 55: 21–26.

Oliveira, C., A.A. Andreata, L.F. Almeida-Toledo and S.A. Toledo-Filho. 1991. Karyotype and nucleolus organizer regions of *Pyrrhulina* cf. *australis* (Pisces, Characiformes, Lebiasinidae). *Revista Brasileira de Genética* 14: 685–690.

Oliveira, C., L.F. Almeida-Toledo, L. Mori and S.A. Toledo-Filho. 1992. Extensive chromosome rearrangements and nuclear DNA content changes in the evolution of the armoured catfishes genus *Corydoras* (Pisces, Siluriformes, Callichthyidae). *Journal of Fish Biology* 40: 419–431.

Oliveira, C., L.F. Almeida-Toledo, L. Mori and S.A. Toledo-Filho. 1993a. Cytogenetic and DNA content in six genera of the family Callichthyidae (Pisces, Siluriformes). *Caryologia* 46: 171–188.

Oliveira, C., L.F. Almeida-Toledo, L. Mori and S.A. Toledo-Filho. 1993b. Cytogenetic and DNA content studies on armoured catfishes of the genus *Corydoras* (Pisces, Siluriformes, Callichthyidae) from the southeast coast of Brazil. *Revista Brasileira de Genética* 16: 617–629.

Oliveira, C., S.M.R. Saboya, F. Foresti, J.A. Senhorini and G. Bernardino. 1997. Increased B-chromosome frequency and absence of drive in the fish *Prochilodus lineatus*. *Heredity* 79: 473–476.

Oliveira, C., M. Nirchio, A. Granado and S. Levy. 2003. Karyotypic characterization of *Prochilodus mariae*, *Semaprochilodus kneri* and *S. laticeps* (Teleostei: Prochilodontidae) from Caicara Del Orinoco, Venezuela. *Neotropical Ichthyology* 1: 47–52.

Ortí, G. 1997. Radiation of Characiform fishes: evidence from mitochondrial and nuclear DNA sequences. In: *Molecular Systematics of Fishes*, T.D. Kocher and C.A. Stephien (eds.). Academic Press, London, pp. 219–243.

Ortí, G., P. Petry, J.I.R. Porto, M. Jegu and A. Meyer. 1996. Patterns of nucleotide change in mitochondrial ribosomal RNA genes and the phylogeny of piranhas. *Journal of Molecular Evolution* 42: 168–182.

Oyakawa, O.T. 2003. Erythrinidae. In: *Check List of the Freshwater Fishes of South America*, R.E. Reis, S.O. Kullander and C.J. Ferraris (eds.). Edipucrs, Porto Alegre, pp. 238–240.

Pauls, E. and L.A.C. Bertollo. 1983. Evidence for a system of supernumerary chromosomes in *Prochilodus scrofa* Steindachner, 1881 (Pisces, Prochilodontidae). *Caryologia* 36: 307–314.

Pavanelli, C.S. 2003. Parodontidae. In: *Check List of the Freshwater Fishes of South America*, R.E. Reis, S.O. Kullander and C.J. Ferraris (eds.). Edipucrs, Porto Alegre, pp. 46–50.

Porto, J.I.R. and E. Feldberg. 1992. Comparative cytogenetic study of the armored catfishes of the genus *Hoplosternum* (Siluriformes, Callichthyidae). *Revista Brasileira de Genética* 15: 359–367.

Porto, J.I.R., E. Feldberg, J.N. Falcão and C.M. Nakayama. 1993. Cytogenetic studies in Hemiodidae (Ostariophysi, Characiformes) fishes from the Central Amazon. *Cytologia* 58: 397–402.

Porto-Foresti, F., E.L. Maistro, F. Foresti and C. Oliveira. 1997. Estimated frequency of B-chromosomes and population density of *Astyanax scabripinnis paranae* in a small stream. *Revista Brasileira de Genética* 20: 377–380.

Post, A. 1965. Vergleichende Untersuchungen der Chromosomenzahlen dei Susswasser Teleosteen. Zeitschrift für zoologische *Systematik und Evolutionsforschung* 3: 47–93.

Ráb, P., M. Rabova and C. Ozouf-Costaz.1998. Karyotype analysis of the African citharinid fish *Distichodus affinis* (Osteichthyes, Characiformes) by different staining techniques. *Journal of African Zoology* 112: 185–191.

Reis, R.E. 1989. Systematic revision of the Neotropical characid subfamily Stethaprioninae (Pisces, Characiformes). *Comunicações do Museu de Ciências da PUCRS (Porto Alegre)* 2: 3–86.

Reis, R.E. 1998. Systematics, Biogeography, and the fossil Record of the Callichthyidae: a review of the available data. In: *Phylogeny and Classification of Neotropical Fishes*, L.R. Malabarba, R.E. Reis, R.P. Vari, Z.M.S. Lucena and C.A.S. Lucena (eds.). Edipucrs, Porto Alegre, pp. 351–362.

Reis, R.E. 2003a. Stethaprioninae. In: *Check List of the Freshwater Fishes of South America*, R.E. Reis, S.O. Kullander and C.J. Ferraris (eds.). Edipucrs, Porto Alegre, pp. 209–211.

Reis, R.E. 2003b. Callichthyidae. In: *Check List of the Freshwater Fishes of South America*, R.E. Reis, S.O. Kullander and C.J. Ferraris (eds.). Edipucrs, Porto Alegre, pp. 291–309.

Reis, R.E., Kullander, S.O. and C.J. Ferraris Jr., (eds.). 2003. *Check List of the Freshwater Fishes of South America*. Porto Alegre: Edipucrs.

Rock, J., M. Eldridge, A. Champion, P. Johnston and J. Joss. 1996. Karyotype and nuclear DNA content of the Australian lungfish, *Neoceratodus forsteri* (Ceratodidae: Dipnoi). *Cytogenetics and Cell Genetics* 73: 187–189.

Rocon-Stange, E.A. and L.F. Almeida-Toledo. 1993. Supernumerary B chromosomes restricted to males in *Astyanax scabripinnis* (Pisces, Characidae). *Revista Brasileira de Genética* 16: 601–615.

Sabaj, M.H. and C.J. Ferraris. 2003. Doradidae. In: *Check List of the Freshwater Fishes of South America*, R.E. Reis, S.O. Kullander and C.J. Ferraris (eds.). Edipucrs, Porto Alegre, pp. 456–469.

Santos, V.H., F. Foresti, C. Oliveira, L.F. Almeida-Toledo, S.A. Toledo-Filho and G. Bernardino. 2002. Synaptonemal complex analysis in the fish species *Piaractus mesopotamicus* and *Colossoma macropomum*, and in their interspecific hybrid. *Caryologia* 55: 73–79.

Sato, L.R., C. Oliveira and F. Foresti. 2004. Karyotype description of five species of *Trichomycterus* (Teleostei: Siluriformes: Trichomycteridae). *Genetics and Molecular Biology* 27: 45–50.

Schaefer, S.A. (1998). Conflict and resolution impact of new taxa on phylogenetic studies of the Neotropical cascudinhos (Siluroidei: Loricariidae). In: *Phylogeny and Classification of Neotropical Fishes*, L.R. Malabarba, R.E. Reis, R.P. Vari, Z.M.S. Lucena and C.A.S. Lucena (Eds.). Edipucrs, Porto Alegre, pp. 375–400.

Scheel, J.J. 1973. *Fish chromosomes and their evolution*. Internal Report of Danmarks Akvarium, Charlottenlund, Denmark.

Scheel, J.J., V. Simonsen and A.O. Gyldenholm. 1972. The karyotypes and some electrophoretic patterns of fourteen species of the genus *Corydoras*. Zeitschrift für zoologische *Systematik und Evolutionsforschung* 10: 144–152.

Shibatta, O.A. 2003. Pseudopimelodidae. In: *Check List of the Freshwater Fishes of South America*, R.E. Reis, S.O. Kullander and C.J. Ferraris (eds.). Edipucrs, Porto Alegre, pp. 401–405.

Shimabukuro-Dias, C.K., C. Oliveira and F. Foresti. 2004a. Karyotype variability in eleven species of the genus *Corydoras* (Siluriformes, Callichthyidae). *Ichthyological Exploration of Freshwaters* 15: 135–146.

Shimabukuro-Dias, C.K., C. Oliveira and F. Foresti. 2004b. Cytogenetic analysis of five species of the subfamily Corydoradinae (Teleostei: Siluriformes: Callichthyidae). *Genetics and Molecular Biology* 27: 549–554.

Shimabukuro-Dias, C.K., Oliveira, C. and F. Foresti. 2005. Comparative cytogenetic studies in species of the subfamily Callichthyinae (Teleostei: Siluriformes: Callichthyidae). *Caryologia.* 58(2): 102-111.

Shimabukuro-Dias, C.K., C. Oliveira, R.E. Reis and F. Foresti. 2004c. Molecular phylogeny of the armored catfish family Callichthyidae (Ostariophysi, Siluriformes). *Molecular Phylogenetics and Evolution* 32: 152–163.

Sola, L., A.R. Rossi, S. Bressanello, E.M. Rasch and P. J. Monaco. 1993. Cytogenetics of bisexual/unisexual species of *Poecilia*. V. unisexual poeciliids with anomalous karyotypes from northeastern Mexico. *Cytogenetics and Cell Genetics* 63: 189–191.

Swarça, A.C., A.S. Fenocchio, M.M. Cestari and A.L. Dias. 2003. Analysis of heterochromatin by combination of C-banding and CMA3 and DAPI staining in two fish species (Pimelodidae, Siluriformes). *Genetica* 119: 87–92.

Thompson, K.W. 1979. Cytotaxonomy of 41 species of Neotropical Cichlidae. *Copeia* 1979: 679–691.

Thorgaard, G.H. and G.A.E. Gall. 1979. Adult triploids in a rainbow trout family. *Genetics* 93: 961–973.

Toledo, V. 1975. *Contribuição ao estudo citogenético da família Pimelodidae (Pisces).* Master Thesis. Universidade de São Paulo, Ribeirão Preto.

Toledo-Piza, M. 2003. Cynodontidae. In: *Check List of the Freshwater Fishes of South America,* R.E. Reis, S.O. Kullander and C.J. Ferraris (eds.). Edipucrs, Porto Alegre, pp. 234–237.

Torres, R.A. 2000. *O gênero Synbranchus (Pisces, Synbranchinae, Synbranchidae): interrelações citotaxonômicas, evolutivas e natureza da variabilidade cariotípica.* Doctoral Thesis. Instituto de Biociências, Botucatu.

Torres, R.A., C. Oliveira and F. Foresti. 2004. Cytotaxonomic diagnosis of *Trichomycterus diabolus* (Teleostei: Trichomycteridae) with comments about its evolutionary relationships with co-generic species. *Neotropical Ichthyology* 2: 123–125.

Vari, R.P. 1983. Phylogenetic Relationships of the Families Curimatidae, Prochilodontidae, Anostomidae, and Chilodontidae (Pisces: Characiformes). *Smithsonian Contributions to Zoology* 378: 1–61.

Vari, R.P. 2003a. Ctenoluciidae. In: *Check List of the Freshwater Fishes of South America,* R.E. Reis, S.O. Kullander and C.J. Ferraris (eds.). Edipucrs, Porto Alegre, pp. 252–253.

Vari, R.P. 2003b. Curimatidae. In: *Check List of the Freshwater Fishes of South America,* R.E. Reis, S.O. Kullander and C.J. Ferraris (eds.). Edipucrs, Porto Alegre, pp. 51–64.

Vari, R.P. and L.R. Malabarba. 1998. Neotropical Ichthyology: An overview. In: *Phylogeny and Classification of Neotropical Fishes,* L.R. Malabarba, R.E. Reis, R.P. Vari, Z.M.S. Lucena and C.A.S. Lucena (eds.). Edipucrs, Porto Alegre, pp. 1–11.

Vari, R.P. and S.J. Raredon. 2003. Chilodontidae. In: *Check List of the Freshwater Fishes of South America,* R.E. Reis, S.O. Kullander and C.J. Ferraris (eds.). Edipucrs, Porto Alegre, pp. 85–86.

Vênere, P.C. 1998. *Diversificação cariotípica em peixes do médio rio Araguaia, com ênfase em Characiformes e Siluriformes (Teleostei, Ostariophysi)*. Doctoral Thesis. Universidade Federal de São Carlos, São Carlos, Brazil.

Vênere, P.C. and P.M. Galetti Jr. 1989. Chromosome evolution and phylogenetic relationships of some Neotropical Characiformes of the family Curimatidae. *Revista Brasileira de Genética* 12: 17–25.

Vênere, P.C., I.A. Ferreira, C. Martins and P.M. Galetti Jr. 2004. A novel ZZ/ZW sex chromosome system for the genus *Leporinus* (Pisces, Anostomidae, Characiformes). *Genetica* 121: 75–80.

Vervoort, A. 1980. Tetraploidy in *Protopterus* (Dipnoi). *Experientia* 36: 294–296.

Vieira, M.M.R. 2002. *Análise da evolução de populações locais de Astyanax scabripinnis (Pisces, Characiformes, Characidae) da região de Botucatu com base em caracteres cromossômicos e de DNA mitocondrial*. Doctoral Thesis. Instituto de Biociências, Botucatu.

Vissotto, P.C. 2000. *Análise citogenética no gênero Imparfinis (Pisces, Siluriformes, Pimelodidae)*. Doctoral Thesis. Instituto de Biociências, Botucatu, Brazil.

Vissotto, P.C., F. Foresti and C. Oliveira. 1997. A ZZ/ZW sex chromosome system in *Imparfinis mirini* (Pisces, Siluriformes). *Cytologia* 62: 61–66.

Weitzman, M. and S.H. Weitzman. 2003. Lebiasinidae. In: *Check List of the Freshwater Fishes of South America*, R.E. Reis, S.O. Kullander and C.J. Ferraris (eds.). Edipucrs, Porto Alegre, pp. 241–251.

Weitzman, S.H. 2003. Glandulocaudinae. In: *Check List of the Freshwater Fishes of South America*, R.E. Reis, S.O. Kullander and C.J. Ferraris (eds.). Edipucrs, Porto Alegre, pp. 222–230.

Weitzman, S.H. and L. Palmer. 2003. Gasteropelecidae. In: *Check List of the Freshwater Fishes of South America*, R.E. Reis, S.O. Kullander and C.J. Ferraris (eds.). Edipucrs, Porto Alegre, pp. 101–103.

Wickbom, T. 1943. Cytological studies on the family Cyprinodontidae. *Hereditas* 29: 1–24.

Chromosomal Evolution in Mugilidae, Mugilomorpha: An Overview

**Luciana Sola*, Ekaterina Gornung, Maria Elena Mannarelli
and Anna Rita Rossi**

INTRODUCTION

The Teleostei represent approximately 96% of all living 'fishes' and comprise 23.637 species, grouped in 426 families and 38 orders (Nelson, 1994). The cytotaxonomic knowledge of this division only covers approximately 2700 species, i.e., only 11.5% of the extant species, as indicated in the most recent review by Klinkhardt *et al.* (1995). This percentage has only slightly increased in the last decade. Thus, compared to other vertebrates, there are not enough data to reconstruct a general picture of the chromosome evolution in this taxon. Nevertheless, Brum and Galetti (1997), on re-examining the available karyological data, proposed that the karyotype of 2n=48 acrocentrics, originally considered

Address for Correspondence: *Department of Animal and Human Biology, University of Rome 1, Via A. Borelli 50, 00161 Rome, Italy. E-mail: luciana.sola@uniroma1.it

to be primitive for all teleosts (Ohno, 1974), has to be considered to be a synapomorphic condition to Clupeomorpha and Euteleostei only. Furthermore, they suggested that the Teleostei ground plan condition for the number of chromosomes is more likely 2n=60.

The Mugilidae family, the only representative of Mugilomorpha, also the most primitive among the three series of the superorder Acanthopterygii, can be considered to be one of the typical marine euteleostean families showing the highly conservative 48 uniarmed karyotype.

The karyological data regarding Mugilidae reviewed here are based on our own studies as also from existing literature. Our studies concerning Mediterranean grey mullets (Rossi *et al.*, 1996, 1997, 2000; Gornung *et al.*, 2001, 2004; unpublished data) were undertaken to investigate—in a group of systematically close species—whether the morphologically conservative karyotype revealed the occurrence of finer differentiated features when analyzed using either differential staining techniques or Fluorescence *In Situ* Hybridization (FISH) with several types of DNA probes. This was done because it was thought to lead to a more general understanding of the karyological evolutive trends in the different taxa of teleosts characterized by the 48-uniarmed-chromosome conservative karyotype. By framing our data into literature studies on grey mullets, for a total of 16 species analyzed, the chromosomal evolution in Mugilidae appears to be more complex than the picture suggested by the early descriptions of their conservative karyotype.

Systematics

The family Mugilidae groups fish species which are distributed in all the tropical and temperate coastal marine and brackish water of the world. In the past, the family was either considered to be a representative of a distinct order (Berg, 1947; Bertin and Arambourg, 1958) or as a suborder of Perciformes (Rosen, 1964; Greenwood *et al.*, 1966; Bini, 1968; Gosline, 1968). More recently, both Nelson (1994) and Thomson (1997) recognized the original classification and placed only the Mugilidae family back in the order of Mugiliformes, Mugilomorpha.

Due to the considerable uniformity of the external morphological characters, the family has been subjected to numerous systematic revisions at both the generic and specific level. The first detailed revision was by Schultz (1946), who identified 16 valid genera, and subsequently revised the genus *Chelon* (Schultz, 1953). In the following years, apart from some locally restricted systematic revisions (cf., Thomson, 1997), the taxonomic organization of the family was repeatedly defined by Thomson (1954, 1963, 1997). In his latest and most comprehensive work, Thomson (1997) recognizes 14 genera (*Agonostomus, Aldrichetta, Cestraeus, Chaenomugil, Chelon, Crenimugil, Joturus, Liza, Mugil, Myxus, Oedalechilus, Rhinomugil, Sicamugil, Valamugil*) and 62 valid species (compared to the 17 genera and 80 species considered valid by Nelson, 1994), leaving 18 species in the status of species *inquerenda*. The FishBase (Froese and Pauly, 2004) reports further changes, which include three more genera (*Moolgarda, Neomyxus* and *Xenomugil*), for a total of 75 species. In all these quoted revisions, approximately half of the species belong to the genera *Mugil* and *Liza* which, respectively, group 12-16 and 23 species, while most of the remaining genera include only one or two species.

According to Thomson (1997), the evolutionary series concerning the six grey mullet genera, namely *Chelon, Liza, Mugil, Oedalechilus, Rhinomugil* and *Valamugil*, whose representatives have been karyologically investigated, allocates *Mugil* to a basal position and *Liza, Chelon, Oedalechilus* to an apical position. More specifically, *Chelon* and *Oedalechilus*, which both contain only two species and display many features common to *Liza*, are considered as derived from this latter.

KARYOLOGICAL STUDIES

Karyological data are available for 16 species (Table 1.6.1). For many of them, only the chromosome number and morphology are known. However, more recent literature includes the cytogenetic characterization of several species, reporting data obtained through either classical or molecular cytogenetics (Table 1.6.2). Classical cytogenetic studies made use of differential staining techniques to investigate numbers and chromosomal locations of the nucleolus organizer regions (NORs) and the chromosomal distribution and composition of the constitutive heterochromatin. Molecular cytogenetic studies, on the other hand, aimed at the chromosomal localization of specific DNA sequences, multicopy genes or repeats, by FISH.

The Chromosome Complement

The karyotype of all grey mullet species studied till date (Table 1.6.1), with one notable exception in *Mugil curema*, has been found to be the

Table 1.6.1 Karyological studies in Mugilidae.

Species[1]	2n	karyotype		FN[2]	References
cytotype A					
Liza					
L. macrolepis (Smith, 1849)	48		48a	(48)	Choudhury et al., 1979
L. oligolepis (Bleeker, 1859)	48		48a	(48)	Choudhury et al., 1979
Mugil					
M. cephalus Linneo, 1758	48	48m[3]	48a	(48)	Cataudella and Capanna, 1973; Cataudella et al., 1974
" "	48			(48)	Natarajan and Subrahmanyam, 1974
" "	48		48a	(48)	Le Grande and Fitzsimons, 1976
" "	48		48a	48	Cano et al., 1982
" "	48		48a	(48)	Amemiya and Gold, 1986
" "	48	2st[4]	+46a	48	Arefyev, 1989
" "	48		48a	48	Crosetti et al., 1993; Rossi et al., 1996; Gornung et al., 2001, 2004
M. guimardianus Desmaret,1831	48		48a	(48)	Nirchio et al., 2003
M. liza Valenciennes, 1836	48		48a	(48)	Pauls and Coutinho, 1990;
" "	48		48a	48	Nirchio and Cequea, 1998; Nirchio et al., 2001; Rossi et al., 2005
M. parsia (Hamilton-Buchanan, 1822)	48		48a	(48)	Chatterjee and Majhi, 1973
" "	48		48a	(48)	Khuda-Bukhsh and Manna, 1974
M. platanus (Günther, 1880)	48		48a	(48)	Jordão et al., 1992
M. speigleri (Bleeker, 1858)	48		48a	(48)	Rishi and Singh, 1982
M. trichodon Poev, 1875	48		48a	(48)	Nirchio et al., 2005a

(Table 1.6.1. contd.)

(Table 1.6.1. contd.)

Rhinomugil

M. corsula (Hamilton-Buchanan, 1822)	48					Nayyar, 1966
" "	48					Khuda-Bukhsh and Manna, 1974
R. corsula (Swainson, 1820)	48					Chakrabarti and Khuda-Bukhsh, 2000

cytotype B

Chelon

C. labrosus (Risso, 1826)	48		2st	48a	(48)	Cataudella and Capanna, 1973, Cataudella *et al.,* 1974
"	48	2sm		48a	(48)	Delgado *et al.,* 1990, 1991, 1992
"	48		2st	48a	(48)	Gornung *et al.,* 2001, 2004

Liza

L. aurata (Risso, 1810)	48		2st	+46a	(48)	Cataudella *et al.,* 1974
"	48		2st	+46a	48	Cano *et al.,* 1982
"	48	2sm		+46a	50	Delgado *et al.,* 1990, 1991, 1992
"	48		2st	+46a	(48)	Gornung *et al.,* 2001, 2004
L. ramada (Risso, 1826)	48		2st	+46a	(48)	Cataudella and Capanna, 1973, Cataudella *et al.,* 1974
"	48	2sm		+46a	50	Delgado *et al.,* 1990, 1991, 1992
"	48		2st	+46a	(48)	Rossi *et al.,* 1997; Gornung *et al.,* 2001, 2004
L. saliens (Risso, 1810)	48		2st	+46a	(48)	Cataudella *et al.,* 1974
"	48	2sm		+46a	50	Arefyev, 1989
"	48		2st	+46a	(48)	Gornung *et al.,* 2001, 2004

Oedalechilus

O. labeo (Cuvier, 1829)	48		2st	+46a	(48)	Cataudella *et al.,* 1974
"	48		2st	+46a	(48)	Rossi *et al.,* 2000; Gornung *et al.,* 2001, 2004

(Table 1.6.1. contd.)

(Table 1.6.1. contd.)

cytotype C						
Mugil curema Valenciennes, 1836						
cytotype C1						
M. curema	(USA)	28	20m		+ 4st	+ 4a	(48)	Le Grande and Fitzsimons, 1976
" "	(Brazil)	28	20m		+ 4st	+ 4a	(48)	Cipriano *et al.*, 2002; Nirchio *et al.*, 2005b
cytotype C2						
M. curema	(Venezuela)	24	22m	+ 2sm			48	Nirchio and Cequea, 1998; Nirchio *et al.*, 2001, 2003, 2005b; Rossi *et al.*, 2005

[1]The specific name given in the original article is reported, even in case of changes of attribution.
[2]When in brackets number of arms calculated by present authors.
[3]Chromosomes are reported as biarmed in the text, but are clearly acrocentric in the figures shown.
[4]Two chromosomes are reported as subtelocentric in the text, but are clearly acrocentric in the figures shown.

morphologically conservative karyotype, corresponding to the 48 uni-armed-chromosome type considered to be ancestral for all teleosts by Ohno (1974), and more recently considered restricted to the ancestor of Euteleostei and Clupeomorpha by Brum and Galetti (1997).

Considering the minor or major differences among the karyotypes, three main cytotypes (Table 1.6.1 and Fig. 1.6.8) can be identified for the 16 grey mullet species studied, as proposed by Choudhury *et al.* (1979).

(1) **Cytotype A**: A karyotype composed of 48 exclusively acrocentric chromosomes, gradually decreasing in size, so that only the largest and the smallest chromosome pairs can be identified with any certainty. This karyotype (Fig. 1.6.1a) characterizes 10 species (see Table 1.6.1 for references): 2 species of *Liza* (*L. macrolepis* and *L. oligolepis*), 5 species of *Mugil* (*M. cephalus*, *M. gaimardianus*, *M. liza*, *M. platanus* and *M. trichodon*) and 3 species which were formerly attributed to the genus *Mugil*, *M. corsula*, *M. parsia*, *M. speigleri* and later, respectively, assigned to the genera *Rhinomugil*, *Liza* and *Valamugil* (Thomson, 1997). For these latter species, the specific name given in the original article is reported in Table 1.6.1.

(2) **Cytotype B**: A karyotype composed of 46 acrocentric plus two submeta/subtelocentric chromosomes. This karyotype chara-cterizes five species, *Liza aurata*, *L. ramada*, *L. saliens*, *Chelon labrosus* and *Oedalechilus labeo* (references in Table 1.6.1). The subtelocentric chromosome pair appears to be homeologous in the three *Liza* species and in *C. labrosus*, as it is the smallest one of the chromosome complement, chromosome pair number 24 (Fig. 1.6.1b), whereas it is a medium-sized chromosome pair, classified as number 9, in *O. labeo* (Fig. 1.6.1c).

(3) **Cytotype C**: A karyotype mainly or even exclusively composed of biarmed chromosomes, which characterizes only one species, *Mugil curema* (references in Table 1.6.1). In the specimens from Louisiana, USA, and from Paranà, Brazil, a karyotype (Fig. 1.6.1d) composed of 2n=28 chromosomes, 20 of which are metacentrics, 4 subtelocentrics and 4 acrocentrics (hereon referred as cytotype C1) has been described. In specimens from the Caribbean Sea, Venezuela, an all biarmed chromosome complement (Fig. 1.6.1e), 2n=24 (hereon referred as cytotype C2) has been found.

Cytotypes A and B appear to be very close to each other and to be both considered as all-uniarmed karyotypes, i.e., the plesiomorphic condition in the family. Indeed, though in the cytotype B species, some authors (Arefyev, 1989; Delgado *et al.*, 1992) have classified the

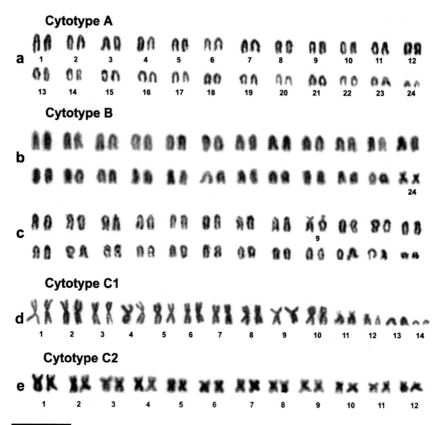

Cytotype A

Cytotype B

Cytotype C1

Cytotype C2

Fig. 1.6.1 Representative Giemsa-stained karyotypes of different cytotypes found in Mugilidae species. Cytotype A, (a) *Mugil cephalus*; Cytotype B, (b) *Liza aurata*, (c) *Oedalechilus labeo*; Cytoypes C (by the courtesy of M. Nirchio): (d) Cytotype C1, *Mugil curema* from Brazil, (e) Cytotype C2, *Mugil curema* from Venezuela.

chromosome pair that bears short arms as submetacentric, it has to be considered that NORs (see below) are located on them. It is well known that NORs are affected by a variable number of ribosomal repeats, as well as by late condensation, both of which are conditions that may produce variable sizes of chromosomal regions bearing ribosomal genes. Thus, when ribosomal genes are located on short arms, subtelocentric chromosomes may occasionally appear and be classified as submetacentric. Since the number of chromosomal arms—fundamental number (FN)—along with the diploid number, is commonly used either to analyze Robertsonian events (i.e., chromosomal fusion or fission) or pericentric inversions, to consider the chromosomes that bear NORs on their short arms in cytotype B as biarmed pair would be misleading in the study of chromosomal evolution in Mugilidae.

A fundamental number equal to 48 also characterizes M. *curema* cytotypes C, which are quite different and undoubtedly derived from cytotypes A and B. Le Grande and Fitzsimons (1976), in their first description of the M. *curema* karyotype from Louisiana (cytotype C1), proposed it had evolved from an ancestral group with an all-acrocentric chromosome complement similar to that of M. *cephalus*, as a result of extensive Robertsonian (Rb) fusions. In cytotype C2, found in specimens of M. *curema* from Venezuela, Robertsonian (Rb) fusions may have possible occurred for all the acrocentrics of the chromosome complement, resulting into the maximum reduction of the diploid number to 24, for a conserved FN of 48.

No morphologically differentiated sex chromosomes were found in any of the analyzed mugilid species. However, in L. *oligolepis* and L. *macrolepis*, Choudhury *et al.* (1979) referred to a possible process of morphological differentiation of one pair of autosomes into heterochromosomes, having observed the presence of 23 bivalents and 2 univalents in male meiosis.

Nucleolus Organizer Regions and Major Ribosomal Genes

The nucleolus organizer regions, NORs, represent the chromosomal regions where the clusters of the multigene family of major (18S, 5.8S and 28S) ribosomal genes (rDNAs) are located. The silver (Ag) staining (Howell and Black, 1980) is commonly used to detect the nucleolus organizer regions (NORs) which have been functionally active in the preceding interphase. Fluorochrome staining with chromomycin A_3 (CMA_3) in fish (Amemiya and Gold, 1986) produces positive signals on the NORs, which usually contain GC-enriched DNA, so that this fluorochrome is believed to show all the structural NORs in the chromosome complement. However, recent evidences in some species, included fish (Gromicho *et al.*, 2005), have put doubt upon the accuracy of Ag and CMA_3 in NORs detection, so that FISH, which can easily localize ribosomal genes on metaphase chromosomes, thanks to their repetitiveness, appears to be the most reliable method for this purpose.

Twelve species of Mugilidae have been analyzed for NORs (Table 1.6.2 and references therein included), four of which (M. *gaimardianus*, M. *platanus*, M. *trichodon*, R. *corsula*, plus M. *curema* cytotype C1) exclusively by Ag-staining. For the remaining species (included M. *curema* cytotype C2), data have been confirmed by CMA_3-staining or FISH.

Table 1.6.2   Differential staining techniques and FISH in karyological studies of Mugilidae

Species	AgNO$_3$	C-bands	CMA$_3$	DAPI	FISH				Reference
					18S	5S	telomeric	Sat-DNA	
cytotype A									
M. cephalus	+	–	+	–	–	–	–	–	Amemiya and Gold, 1986
"	+	+	+	+	–	–	–	–	Crosetti et al., 1993
"	+	+	+	+	+	–	–	–	Rossi et al., 1996
"	–	–	–	–	–	+	–	–	Gornung et al., 2001
"	–	–	–	–	–	–	+	–	Gornung et al., 2004
"	–	–	–	–	–	–	–	+	Present paper
M. gaimardianus	+	–	–	–	–	–	–	–	Nirchio, personal communication
M. liza	+	–	–	–	–	–	–	–	Nirchio et al., 2001
M. platanus	–	–	+	+	+	+	+	–	Rossi et al., 2005
M. trichodon	+	+	–	–	–	–	–	–	Jordão et al., 1992
"	+	+	–	–	–	–	–	–	Nirchio et al., 2005a
R. corsula	+	+	–	–	–	–	–	–	Chakrabarti and Khuda-Bukhsh, 2000
cytotype B									
C. labrosus	+	+	–	–	–	–	–	–	Delgado et al., 1990, 1991
"	–	–	–	–	+	+	–	–	Gornung et al., 2001
"	–	–	–	–	–	–	+	–	Gornung et al., 2004
"	–	–	–	–	–	–	–	+	Present paper
L. aurata	+	+	–	–	+	+	–	–	Delgado et al., 1990, 1991
"	–	–	–	–	–	–	+	–	Gornung et al., 2001
"	–	–	–	–	–	–	+	–	Gornung et al., 2004
"	–	–	–	–	–	–	–	+	Present paper

(Table 1.6.2 contd.)

(Table 1.6.2 contd.)

		1	2	3	4	5	6	7	8
L. ramada	Delgado et al., 1990, 1991	–	–	–	–	–	–	+	+
"	Rossi et al., 1997	–	–	–	+	+	+	+	+
"	Gornung et al., 2001	–	–	+	–	–	–	–	–
"	Gornung et al., 2004	–	+	–	–	–	–	–	–
"	Present paper	+	–	–	–	–	–	–	–
L. saliens	Gornung et al., 2001	–	–	+	+	+	+	+	+
"	Gornung et al., 2004	–	+	–	–	–	–	–	–
"	Present paper	+	–	–	–	–	–	–	–
O. labeo	Rossi et al., 2000	–	–	–	+	+	+	+	+
"	Gornung et al., 2001	–	–	+	–	–	–	–	–
"	Gornung et al., 2004	–	+	–	–	–	–	–	–
"	Present paper	+	–	–	–	–	–	–	–
cytotype C									
cytotype C1									
M. curema	Nirchio et al., 2005b	–	–	–	–	–	–	+	+
cytotype C2									
M. curema	Nirchio et al., 2001, 2005b	–	–	–	–	–	–	+	+
"	Rossi et al., 2005	–	+	+	+	+	+	–	–

+ : applied; – : not applied

In all species excluding *Rhinomugil corsula*, NORs are borne by a single chromosome pair. However, their location is different between and within the three main cytotypes.

In the six analyzed species with cytotype A, up to four different locations of NORs have been observed. In three species of *Mugil*, *M. cephalus* (Fig. 1.6.2a), *M. liza* and *M. platanus*, NORs are located at the terminal region of the long arms of the largest chromosome pair. In the other two species of the genus, *M. trichodon* and *M. gaimardianus* (Fig. 1.6.2b), NORs are interstitial and located on a medium-sized chromosome pair. Finally, in *Rhinomugil corsula*, Ag-NORs were detected in the terminal regions of five acrocentric chromosomes (Fig. 1.6.8). From the figures shown in the original paper (Chakrabarti and Khuda-Bukhsh, 2000), two of them appear to belong to chromosome pair number 1.

In all five species with cytotype B, NORs were localized on the short arms of the only subtelocentric chromosome pair, i.e., number 24 in *C. labrosus*, *L. aurata*, *L. ramada* and *L. saliens* (Fig. 1.6.2c), and number 9 in *O. labeo* (Fig. 1.6.2d). *L. saliens* and *O. labeo* also show additional and variable rDNA sites. Indeed, in three out of the eight analyzed specimens of *L. saliens* (Gornung *et al.*, 2001), additional small NOR-sites were detected, by both Ag- and CMA$_3$-staining and by FISH, in proximity to the centromeres of one or both homologues of chromosome pair 9 (Fig. 1.6.8). In *O. labeo* (Rossi *et al.*, 2000), two out of 16 specimens showed one small additional, Ag-negative but CMA$_3$- and FISH-positive, NOR-site in a paracentromeric region of one medium sized chromosome, numbered 10 (Fig. 1.6.8).

In *M. curema* populations with different cytotypes C, the NORs location identifies a further degree of differentiation among the 'mostly biarmed' and the 'all-biarmed' karyotypes. Indeed, in the Brazilian population of *M. curema* (cytotype C1, Fig. 1.6.2e), Ag-NORs have been observed on the short arms of the subtelocentric chromosome pair number 11, whereas in the Venezuelan population of *M. curema* (cytotype C2, Fig. 1.6.2f), NORs have been observed at the terminal region of the largest metacentric chromosome pair, which is compatible with the hypothesis of

Fig. 1.6.2 Representative Ag-stained metaphase plates of some of the different NORs locations found in Mugilidae species with different cytotypes. Cytotype A, (a) *Mugil cephalus*, (b) *Mugil trichodon* (by the courtesy of M. Nirchio); Cytotype B, (c) *Liza saliens*, (d) *Oedalechilus labeo*; Cytoypes C (by the courtesy of M. Nirchio): (d) Cytotype C1, *Mugil curema* from Brazil, (e) Cytotype C2, *Mugil curema* from Venezuela. Arrows indicate NORs.

Cytotype A

Cytotype B

Cytotype C1   **Cytotype C2**

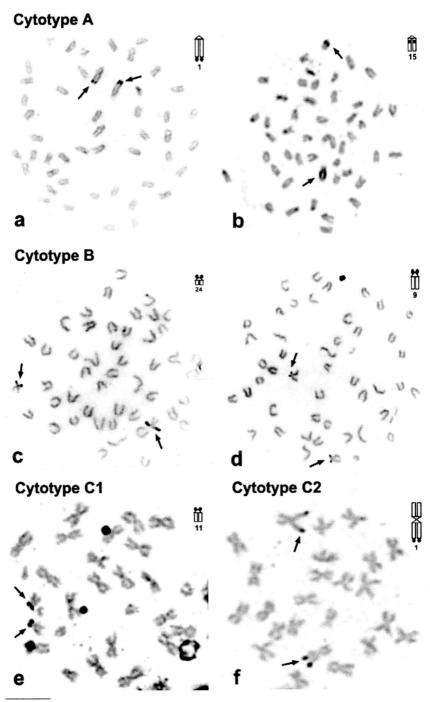

Fig. 1.6.2

Rb fusions from an ancestral group with an all-acrocentric chromosome complement similar to that of M. cephalus (Le Grande and Fitzsimons, 1976).

To summarize (Fig. 1.6.8), by pursuing a parsimonious criterion, at least five different species-specific NORs locations have been observed in the 11 Mugilidae species with cytotypes A and B, and two in the two M. curema populations with cytotypes C1 and C2.

Minor Ribosomal Genes (5S rDNA)

The minor ribosomal genes are also constituted by a multigene family organized in tandem arrays, with repeat units composed of a coding region and a non-transcribed spacer sequence. The 5S rRNA genes are generally detected in distinct areas of the genome from the major ribosomal genes and are often located on different chromosomes (Martins, present volume). As they are less variable than NORs, they are usually considered to be good chromosome markers and their chromosomal distribution to be a useful cytotaxonomic tool in the assessment of evolutionary relationships among closely related species. They can only be detected by FISH.

Eight species of Mugilidae have been analyzed for 5S rDNA (Table 1.6.2 and references included therein) and in all of them, with the exception of Liza saliens, the clusters of minor ribosomal genes are located on a single chromosome pair, which is different from the NOR-bearing chromosome pairs.

In the two species investigated with cytotype A, M. cephalus (Fig. 1.6.3a) and M. liza, the 5S ribosomal genes are located in a subcentromeric position of the smallest chromosome pair.

In four of the five species with cytotype B, L. ramada, L. aurata (Fig. 1.6.3b), C. labrosus, and O. labeo, the 5S ribosomal genes are located in a subcentromeric position of one medium-sized chromosome pair, considered to be homeologous in all species and classified as chromosome pair 8. The remaining species with cytotype B, L. saliens (Fig. 1.6.3c) shows a 5S rDNA site in a subtelomeric position of the long arms of chromosome pair 1, in addition to the 5S ribosomal genes on chromosome pair 8.

As far as the cytotype C is concerned, only the Venezuelan population of M. curema (Fig. 1.6.3d), with cytotype C2, has been investigated for minor ribosomal genes, and the 5S rDNA sites were found in a paracentromeric position of a medium-sized chromosome pair, whose arms

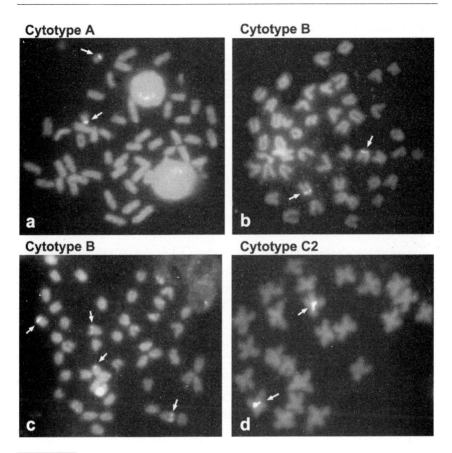

Fig. 1.6.3 Representative FISH metaphase plates of the different 5S rDNA locations found in Mugilidae species with different cytotypes. Cytotype A, (a) *Mugil cephalus*; Cytotype B, (b) *Liza aurata*, (c) *Liza saliens*; Cytoype C2, (e) *Mugil curema* from Venezuela. Arrows indicate 5S rDNA sites.

are evidently larger than those of the smallest acrocentric chromosome pair of *M. cephalus*. Thus, these data do not seem to support the hypothesis (Le Grande and Fitzsimons, 1976) of Rb fusions from an ancestral group with an all-acrocentric chromosome complement similar to the one of *M. cephalus*, unless we suppose that further chromosome rearrangements have occurred after Rb events.

To summarize (Fig. 1.6.8), three different chromosome pairs bear 5S rDNA sites in the seven species of Mugilidae with cytotypes A and B analyzed, and one pair in the population of *M. curema* with cytotype C2.

The (TTAGGG)n Telomeric Sequence

Telomeric DNA is widely conserved among vertebrates and consists of (TTAGGG)n sequence repeats. These sequences are mainly found at the ends of chromosomes, but the distribution at non-telomeric sites has been observed in a variety of vertebrate species. Though the origin and the evolutionary role of (TTAGGG)n repeats at non-telomeric sites is yet to be completely understood, the non-telomeric sites are undoubtedly involved in recombination events with great frequency (Ashley and Ward, 1993; Slijepevic et al., 1996; Kilburn et al., 2001), which supports the original hypothesis by Meyne et al. (1990) that the interstitial sites of telomeric sequences may provide a greater flexibility for karyotype changes.

The chromosomal distribution of the (TTAGGG)n repeats has been investigated in eight species of Mugilidae (Table 1.6.2 and references included therein), two with cytotype A (M. cephalus and M. liza), all the five species with cytotype B, and the Venezuelan population of M. curema, with cytotype C2.

In all species (Fig. 1.6.4a-c), the telomeric probe hybridized to the ends of all chromosomes and was also found to be interspersed with clusters of major ribosomal genes (Fig. 1.6.4d). No other additional, interstitial or pericentromeric, sites were detected in either species, including M. curema which has a karyotype composed of all-metacentric Robertsonian chromosomes.

Constitutive Heterochromatin and Satellite DNA

Constitutive heterochromatin is typically composed of repetitive DNA, commonly located at pericentromeric and telomeric regions of chromosomes, and in most fish species it is also associated to NORs (Phillips and Reed, 1996). Classical cytogenetic techniques produce information regarding its chromosomal location and reveal a possible differential enrichment in base content. Indeed, C-banding (Sumner, 1972) produces a Giemsa-positive heteropycnosis in the heterochromatin-rich chromosomal regions, due to a lower rate of DNA removal from heterochromatin, compared to euchromatin, during C-banding treatment (Sumner, 1990). On the other hand, in cold-blooded vertebrates fluorochrome staining with chromomycin A_3 (CMA_3) and 4',6-diamidino-2-phenylindole (DAPI) dyes specifically detects GC- or AT-enriched DNA, respectively, in the heterochromatic chromosomal

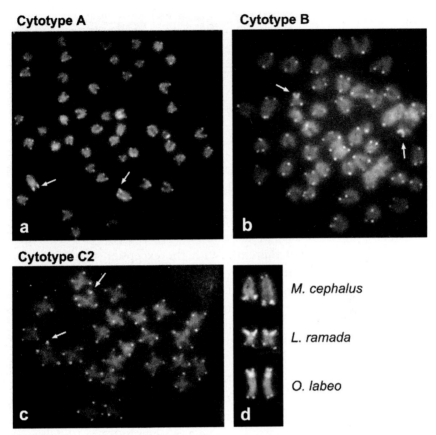

Fig. 1.6.4 Representative metaphase plates from species of Mugilidae with different cytotypes after FISH with a telomeric probe. Cytotype A, (a) *Mugil cephalus*; Cytotype B, (b) *Liza saliens*; Cytoype C2, (c) *Mugil curema* from Venezuela. Arrows indicate NORs. (d) selected samples of NOR-bearing chromosomes showing hybridization signals scattered along the NORs.

regions. As heterochromatin is mainly composed of satellite DNA, i.e., very highly repetitive non-coding DNA sequences arranged in huge clusters supposed to evolve in a concerted fashion (Ugarkovic and Phlol, 2002), the isolation of families of satellite DNA and their chromosomal localization constitute additional cytogenetic markers which are useful for investigating cytotaxonomic relationships among related species.

C-banding was applied to ten Mugilidae species (Table 1.6.2 and references included therein), and in all of them revealed a pericentromeric location of the heterochromatin in virtually all chromosomes, as well as its association to the major ribosomal genes at every different chromosomal location observed in different cytotypes. As an example, the C-banding pattern obtained in *O. labeo* is shown in Fig. 1.6.5a.

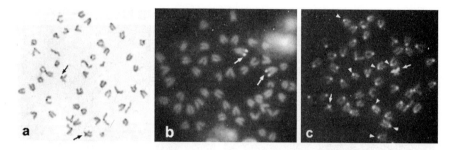

Fig. 1.6.5 Metaphase plates after C-banding (a) in *Oedalechilus labeo* and after CMA$_3$-staining in (b) *Mugil cephalus* and (c) *Liza ramada*. Arrows indicate NORs.

Fluorochrome staining was applied to six Mugilidae species, M. *cephalus*, M. *liza*, L. *ramada*, L. *saliens*, O. *labeo* and M. *curema* (Table 1.6.2 and references included therein). With the exception of the CMA$_3$-positive and the DAPI-negative NORs, generally no chromosome region with a differential increase of AT- or GC-rich DNA was detected. As an example, the CMA$_3$-staining pattern obtained in M. *cephalus* is shown in Fig. 1.6.5b. One notable exception to this general picture is constituted by L. *ramada* (Rossi *et al.*, 1997), which is characterized by a GC-richness of centromeric heterochromatin (Fig. 1.6.5c).

Such a difference in DNA composition of centromeric heterochromatin prompted us to investigate whether it reflects phylogenetic relationships among the Mediterranean mullets, namely M. *cephalus*, L. *saliens*, L. *aurata*, L. *ramada*, C. *labrosus* and O. *labeo*. A comparative study was carried out on the interspecific chromosomal distribution of two types/families of highly repetitive (satellite) DNA obtained in two species, L. *ramada* and M. *cephalus*. L. *ramada* was chosen because is the only species with GC rich centromeric heterochromatin (Rossi *et al.*, 1997), and M. *cephalus* is considered to be the closest to the ancestral Mugilidae form (Gornung *et al.*, 2001). The preliminary results of this study are reported below.

Electrophoresis of the total DNA of L. *ramada* and M. *cephalus* after complete digestion with a panel of 17 restriction endonucleases revealed a prominent band of approximately 250 bp in both species, which indicates the presence of highly repetitive genomic sequences (Singer, 1982). The finest bands were obtained with *Hind*III and *Kpn*I in L. *ramada*, and with *Kpn*I and *Pst*I in M. *cephalus*. After partial digestion, major bands corresponding to multiples of the 250 bp monomer were obtained, indicating the repetitive fragments are clustered in tandem arrays. The KpnI-fragments of both *L. ramada* and *M. cephalus* were extracted from bands, labelled and used directly as probes, named as LrK and McK, respectively, for FISH on chromosome preparations of different species.

In homologous hybridizations, each probe hybridized to the centromeric regions of all chromosomes in the source species (Figs. 1.6.6a, 1.6.7a). No cross hybridization was observed, even under low stringency conditions, when the LrK-fragments were used as a probe to metaphases of *M. cephalus* (Fig. 1.6.6e), or, conversely, when the McK-fragments were used as a probe to metaphases of *L. ramada* (Fig. 1.6.7c). These results indicate that the repeats, in spite of their similar size and similar chromosomal distribution in the source species, belong to rather diversified satellite families and they do not enclose, or are not significantly enriched by, restriction fragments of any conserved repeats, such as 5S or 45S rRNA genes, which might have produced confusing FISH patterns.

Subsequently, FISH was simultaneously applied with both probes on separate areas of the same slide in order to maintain identical hybridization conditions for the two probes, to chromosome preparations of the remaining four Mediterranean species. The probe LrK hybridized to the centromeres of all chromosomes in *L. aurata* (Fig. 1.6.6b) and *L. saliens* (Fig. 1.6.6c), and of a subset of 18-20 chromosomes in *C. labrosus* (Fig. 1.6.6d), while no signals were obtained in *O. labeo* chromosomes (Fig. 1.6.6f). On the other hand, concurrent hybridization with the probe McK produced no signals either in the other *Liza* species or in *C. labrosus* (negative patterns not shown), whereas positive signals were found at the centromeres of 2-3 chromosome pairs in *O. labeo* (Fig. 1.6.7b). The results are summarized in Table 1.6.3.

DISCUSSION AND CONCLUDING REMARKS

To compare all existing data, idiograms were drawn (Fig. 1.6.8) summarizing the karyotypes (Table 1.6.1) and the major and minor ribosomal genes locations known (Table 1.6.2) for the Mugilidae species that have been karyologically analyzed.

The number of analyzed species represents approximately 25% of the species belonging to the family and, thus, though a homogeneous and thorough comparison is not possible since not all the species were studied

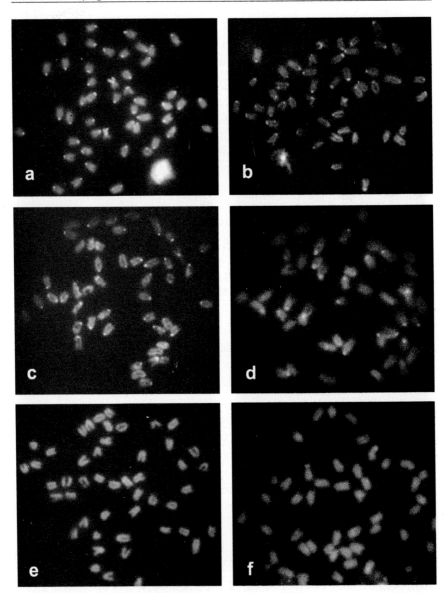

Fig. 1.6.6 Metaphase plates after FISH with the satellite LrK probe in (a) *Liza ramada*, (b) *Liza aurata*, (c) *Liza saliens*, (d) *Chelon labrosus*, (e) *Mugil cephalus*, (f) *Oedalechilus labeo*.

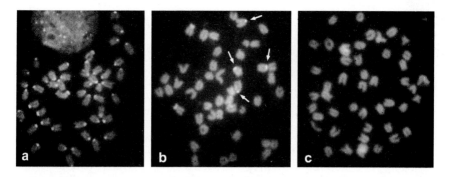

Fig. 1.6.7 Metaphase plates after FISH with the satellite McK probe in (a) *Mugil cephalus*, (b) *Oedalechilus labeo* (arrows indicate FISH-positive signals), (c) *Liza ramada*.

Table 1.6.3 Results of FISH-mapping with KpnI restriction endonuclease DNA fragments from *L. ramada* (LrK) and *M. cephalus* (McK).

FISH signals	Liza ramada		Liza aurata		Liza saliens		Chelon labrosus		Oedalechilus labeo		Mugil cephalus	
probe	LrK	McK	LrK	McK	LrK	McK	LrK	McK	LrK	McK	LrK	McK
occurrence	+	–	+	–	+	–	+	–	–	+	–	+
location	pc	–	pc	–	pc	–	pc	–	–	pc	–	pc
numbers	48	–	48	–	48	–	18-20	–	–	4-6	–	48

pc : pericentromeric

with the same techniques, some general considerations can be made on chromosome evolution in Mugilidae.

Among the studied genera, *Mugil* is considered to be the most primitive (Thomson, 1997) and this is in agreement with its conservative 48 acrocentric karyotype, cytotype A, shared by all the species of the genus investigated, with the exception of *M. curema*. The cytotype A is also shown by three more species belonging to the genera *Liza* and *Rhinomugil*, i.e., for a total of ten of the 16 analyzed species. Thus, cytotype A can reasonably be considered to be plesiomorphic for the family.

Obviously, the most important chromosome changes mark two lineages: the lineage including species with cytotypes A and B, on the one hand, and the clearly divergent *Mugil curema* lineage, with cytotypes C, on the other, which merit a separate discussion.

Considering the lineage including cytotypes A and B, cytotype B is also shared by species belonging to different genera, *Liza*, *Chelon*,

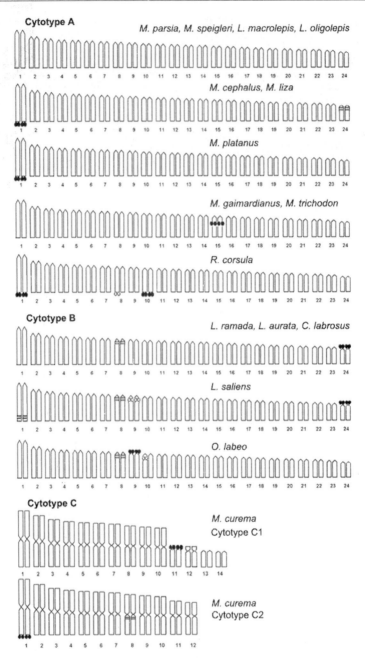

Fig. 1.6.8 Idiograms summarizing the karyotypes and the major and minor ribosomal genes locations reported in species of Mugilidae with different cytotypes. *Solid circles* represent constant and *open circles* represent variable locations of 18S rDNA. *Grey bars* represent locations of 5S rDNA.

Oedalechilus, which are closely related according to Thomson (1997). A hypothesis, previously proposed for serranids (Sola *et al.*, 1993), stated that the shift from the 48-acrocentric karyotype to a karyotype composed of 46 acro- plus 2 subtelocentrics (or 2 submetacentrics—depending on the authors' classification) is related to the 'appearance' of NORs over the short arms of the latter ones. By considering this hypothesis, it was previously speculated (Gornung *et al.*, 2001) that the derived cytotype B might have originated from an ancestral karyotype similar to the one found in M. *cephalus*, through a translocation of NORs from the terminal region of chromosomes 1 to the short arms of chromosomes 24 (*L. saliens, L. ramada, L. aurata, C. labrosus*) or chromosomes 9 (*O. labeo*). This possible shift might therefore explain the split among cytotypes A and B in Mugilidae.

Assuming the NOR location shown by M. *cephalus*, M. *platanus* and M. *liza*, i.e., by three of the five species of the genus investigated for NORs, to be plesiomorphic, under the most parsimonious hypothesis NORs may have been involved in at least three independent translocation events, the two reported above in species with cytotypes B, and one, followed by a paracentric inversion, in two species with cytotype A, M. *trichodon* (Nirchio *et al.*, 2005a) and M. *gaimardianus* (Nirchio, personal communication). The interspersion of ribosomal and telomeric sequences reported to date in a limited number of fish species (Salvadori *et al.*, 1995; Abuín *et al.*, 1996; Sola *et al.*, 2003), but shared by all the 8 grey mullets species examined (Table 1.6.2), might constitute the structural basis of the rearrangements involving these chromosome regions in this taxon. Nevertheless, at the intraspecific level, other factors seem to influence the 'stability' of NORs, given that, on the one hand, global samples of M. *cephalus* (Rossi *et al.*, 1996), genetically well differentiated and supposed to be under incipient speciation (Rossi *et al.*, 1998), share the same single NOR location, and, on the other, single populations of *O. labeo* (Rossi *et al.*, 2000), *L. saliens* (Gornung *et al.*, 2001) and *R. corsula* (Chakrabarti and Khuda-Bukhsh, 2000) present variable NORs sites, in addition to the species-specific NOR-bearing chromosome pair.

It is worth noting that the five *Liza* species analyzed exhibit both cytotypes: A (two species), and B (three species). Considering that the differences among and within cytotypes A and B are mainly related to the NORs location, the question arises as to whether NORs are too variable as chromosome markers to be useful in the reconstruction of phylogenies.

Other cytogenetical markers, i.e., 5S rDNA and satellite DNA, though the latter is only available for Mediterranean species, provide additional perspectives for studying the phylogenetic relationships within the A-B lineage.

If we assume that the 5S rDNA location (in a paracentromeric position of chromosome pair number 24) in M. cephalus and in M. liza is plesiomorphic, one could suppose that a single translocation event involving this site might have occurred in a common ancestor with a karyotype similar to cytotype A. This would, therefore, have resulted in a common 5S rDNA location on chromosome pair 8, which can be observed in all the species with cytotype B. On the other hand, the presence of a second 5S rDNA site on its chromosomes 1 clearly distinguishes L. saliens from the other species with cytotype B, and may indicate its higher relatedness to cytotype A. Indeed, by considering this second site, major and minor rRNA genes show alternative and reverse locations on chromosomes 1 and 24 in M. cephalus and L. saliens. Thus, an event of reciprocal translocation for these two sets of genes might also be hypothesized, which may signify that L. saliens (subgenus Protomugil, Popov, 1930) is the most primitive in the Liza-Chelon-Oedalechilus series (Gornung et al., 2001).

Satellite DNA is constituted by rapidly evolving repetitive sequences and is, therefore, considered to be a valuable tool for disclosing evolutionary relationships, since the sharing of the same satellite DNA families by two or more taxa is regarded as a signal of their relatedness (Arnason et al., 1992; Ugarković and Plohl, 2002). Thus, the similar chromosomal distribution of the KpnI-defined satellite DNA of L. ramada in all the three Liza species analyzed indicates their recent separation from a common ancestor, as well as their relatedness to C. labrosus. This fact supports the hypothesis of a non-monophyletic origin of Liza exclusive of Chelon, based on mt-rDNA and allozyme data (Rossi et al., 2004). The absence of this same satellite DNA from the genomes of species with the same, O. labeo, or a different cytotype, M. cephalus, indicates a more distant position of these latter from the Liza-Chelon group. On the other hand, the hybridization of the M. cephalus satellite DNA probe to chromosomes of O. labeo, though limited to only a few of them, suggests a closer relationship, so far never evidenced, among them. Thus, the phylogenetic reconstruction based on morphological features by Thomson (1997), which considers the genera Chelon and Oedalechilus 'doubtless derived from the genus Liza', is not supported by satellite DNA data.

As far as the *Mugil curema* lineage is concerned, cytotype(s) C are the most divergent among mugilid chromosome complements, the only ones where centric fusions occurred, giving rise to metacentric and submetacentric chromosomes.

The differences between cytotype C1 and C2 could be attributed to a different extent of centric fusions, which involved most (C1) or all (C2), the acrocentric chromosomes. The location of NORs in cytotype C2 observed in the Venezuelan population (the most derived and completely biarmed karyotype) is in agreement with a direct derivation from the plesiomorphic all acrocentric chromosome complement shared by *M. cephalus*, *M. liza* and *M. platanus*. However, the 5S rDNA location observed in the same population is not consistent with this, unless a translocation event involving this site occurred during phyletic evolution, after the separation of the *M. curema* lineage. On the other hand, the only chromosome marker studied in the cytotype C1 Brazilian population does not provide an overlapping picture, given the NORs location might only be explained by a (further and/or independent?) translocation event.

The pattern of telomeric repeat distribution identified in one (C2) of the two cytotypes investigated, indicates the absence of interstitial sites, which is considered to be a stabilizing factor for fusions (Slijepcevic, 1998). The structural basis that prompted Rb rearrangements in this lineage could be clarified by the analysis of centromeric heterochromatin DNA composition, which one would expect to be very differentiated not only respect to that of *Mugil cephalus*, but also to those of all other mullet genera. The analysis of satDNA might also disclose the relatedness among the two cytotypes.

The question arises on whether such important differences in the cytogenetic features indicate the existence of an intra-specific chromosomal Rb polymorphism, or reveal a higher taxonomic status of M. *curema* populations with different cytotypes. Recently, Nirchio *et al.* (2005b), after parallel cytogenetic, meristic and morphometric analyses of samples from Brazil and Venezuela, suggested that the observed differences might indicate the existence of two different species instead of the geographic variation of the same polytypic species. The analyses of other molecular markers, i.e., nuclear or mitochondrial sequences are needed to clarify the magnitude of the genetic divergence and thus the systematic positions of populations with different cytotypes.

To conclude, the cytogenetic features of the three cytotypes found in Mugilidae species are unable to identify unequivocal karyoevolutive trends within the family, especially within the all uni-armed cytotypes. Actually, the few available cytogenetic markers, 18 S and 5S rDNA, constitutive heterochromatin and satellite DNA, telomeric sequences, have different structural bases and roles, and are subjected to quite different evolutive rates and pressures. Therefore, they do not necessarily provide consistent indications of relatedness among species/genera. Moreover, although it is likely to be correct to pursue a parsimonious criterion, any established homeology (or not) is arbitrary and cannot be ascertained in the absence of other markers that make such morphologically similar chromosomes distinguishable. Nevertheless, on the one hand, these cytogenetic characters must be investigated and be considered in any phylogenetic reconstruction, and, on the other, they were able to disclose that even a conservative karyotype hides karyological changes. In this context, the similar patterns of NORs and 5S rDNA sites distribution in *L. ramada*, *L. aurata* and *C. labrosus*, as well as the partial conservation of satellite DNA, cannot be neglected in any way and altogether underlines a high affinity of *Chelon* to genus *Liza*. Similarly, within the general karyotype conservativeness in the family, the cytogenetic features of the M. *curema* lineage are striking and may have revealed that something evolutionarily important is currently happening within it.

The data here reported constitute a basic plot to which new data on other species of the family should hopefully continue to be added in order to provide a more general picture of karyoevolutive trends in the family, useful in the understanding of more general chromosomal evolutionary processes in fish.

References

Abuín, M., P. Martinez and L. Sanchez. 1996. Localization of the repetitive telomeric sequence (TTAGGG)n in four Salmonid species. *Genome* 39: 1035–1038.

Amemiya, C.T. and J.R. Gold. 1986. Chromomycin A₃ stains nucleolus organizer regions of fish chromosomes. *Copeia* 1986: 226–231.

Arefyev, V.A. 1989. The application of the method of colchicine baths to studies of karyotypes of the young of two mullet species (Mugilidae) from the Black Sea. In: *Early Life History of Marine Species.* Sb Nauchn Tr VNIRO 1989, Moscow, pp. 139–149.

Arnason, U., S. Gretarsdottir and B. Widegren. 1992. Mysticete (baleen whale) relationships based upon the sequence of the common cetacean DNA satellite. *Molecular Biology and Evolution* 9: 1018–1028.

Ashley, T. and D.C. Ward. 1993. A 'hot spot' of recombination coincides with an interstitial telomeric sequence in the Armenian hamster. *Cytogenetic and Cell Genetics* 62: 169–171.

Berg, L.S. 1947. *Classification of Fishes: Both Recent and Fossil.* J. W. Edwards. Ann Arbor, H.S. Michigan, USA.

Bertin, L. and C. Arambourg. 1958. Super-ordre des Téléostéens. In: *Traité de Zoologie,* XIII, P. P. Grassé (ed.). Masson et cie, Paris, pp. 2205–2500.

Bini, G. 1968. *Atlante dei Pesci delle coste italiane.* Mondo Sommerso Editrice, Vol 4. Osteitti, Milano.

Brum, M.J.I. and P. Galetti Jr. 1997. Teleostei ground plan karyotype. *Journal of Comparative Biology* 2: 91–102.

Cano, J., G. Thode and M.C. Alvarez. 1982. Karyoevolutive considerations in 29 Mediterranean teleost fishes. *Vie et Milieu* 32: 21–24.

Cataudella, S. and E. Capanna. 1973. Chromosome complements of three species of Mugilidae (Pisces, Perciformes). *Experientia* 29: 489–491.

Cataudella, S., M.V. Civitelli and E. Capanna. 1974. Chromosome complements of Mediterranean mullets (Pisces, Perciformes). *Caryologia* 27: 93–105.

Chakrabarti, J. and A.R. Khuda-Bukhsh. 2000. Chromosome banding studies in an Indian mullet: Evidence of structural rearrangments from NOR location. *Indian Journal of Experimental Biology* 5: 467–470.

Chatterjee, K. and A. Majhi. 1973. Chromosomes of *Mugil parsia* Hamilton (Teleostei, Mugiliformes, Mugilidae). *Genen Phaenen* 16: 51–54.

Choudhury, R.C., R. Prasad and C.C. Das. 1979. Chromosomes of six species of marine fishes. *Caryologia* 32: 15–21.

Cipriano, R.R., M.M. Cestari and A.S. Fenocchio. 2002. Levantamento citogenético de peixes marinhos do litoral do Paraná. *IX Simposio de Citogenética e Genetica de Peixes. Maringa, Brasil.*

Crosetti, D., J.C. Avise, F. Placidi, A.R. Rossi and L. Sola. 1993. Geographic variability in the grey mullet *Mugil cephalus*: Preliminary results of mtDNA and chromosome analyses. *Aquaculture* 111: 95–101.

Delgado, J.V., A. Molina, J. Lobillo, A. Alonso, D. Llanes and M.E. Camacho. 1990. Estudio citogénetico en tres especies de la familia Mugilidae, *Liza aurata, Liza ramada* y *Chelon labrosus*: Morfometria chromosomica, bandas C y distribucion de los NORs. *Actas III Congreso Nacional de Acuicultura:* 301–305.

Delgado, J.V., G. Thode, J. Lobillo, M.E. Camacho, A. Alonso and A. Rodero. 1991. Detection of the nucleolar organizer regions in the chromosomes of the family Mugiliidae (Perciformes): Technical improvements. *Archivos de Zootecnia* 40: 301–305.

Delgado, J.V., A. Molina, J. Lobillo, A. Alonso and M.E. Camacho. 1992. Morphometrical study on the chromosomes of three species of mullet (Teleostei, Mugilidae). *Caryologia* 45: 263–271.

Froese, R. and D. Pauly. 2004. FishBase. World Wide Web electronic publication. www.fishbase.org, version (03/2004).

Gornung, E., C.A. Cordisco, A.R. Rossi, S. De Innocentiis, D. Crosetti and L. Sola. 2001. Chromosomal evolution in Mugilidae: Karyotype characterization of *Liza saliens* and comparative localization of major and minor ribosomal genes in the six Mediterranean mullets. *Marine Biology* 139: 55–60.

Gornung, E., M.E. Mannarelli, A.R. Rossi and L. Sola. 2004. Chromosomal evolution in Mugilidae (Pisces, Mugiliformes): FISH mapping of the (TTAGGG)n telomeric repeat in the six Mediterranean mullets. *Hereditas* 140: 1–2.

Gosline, W.A. 1968. The suborder of Perciform fishes. *Proceedings of the United States National Museum* 124: 1–78.

Greenwood, P.H., D.E. Rosen, S.H. Weitzman and G.S. Myers. 1966. Phyletic studies of teleostean fishes with a provisional classification of living forms. *Bulletin of the American Museum of Natural History* 131: 339–456.

Gromicho, M., C. Ozouf-Costaz and M.J. Collares-Pereira. 2005. Lack of correspondence between CMA_3-, Ag-positive signals and 28S rDNA loci in two Iberian minnows (Teleostei, Cyprinidae) evidenced by sequential banding. *Cytogenetic and Genome Research* 109: 507-511.

Howell, W.M. and D.A. Black. 1980. Controlled silver-staining of nucleolus organizer regions with a protective colloidal developer: A 1 step-method. *Experientia* 36: 1014–1015.

Jordao, L.C., C. Oliveira, F. Foresti and H. Godinho. 1992. Caracterizacao citogenética da tainha, *Mugil platanus* (Pisces, Mugilidae). *Boletim do Instituto de Pesca Sao Paulo* 19: 63–66.

Khuda-Bukhsh, A.R. and G.K. Manna. 1974. Somatic chromosomes in seven species of teleostean fishes. *Chromosome Information Service* 17: 5–6.

Kilburn, A.E., M.J. Shea, R.G. Sargent and J.H. Wilson. 2001. Insertion of a telomere repeat sequences into a mammalian gene causes chromosome instability. *Molecular Cell Biology* 21: 126–135.

Klinkhardt, M., M. Tesche and H. Greven. 1995. *Database of Fish Chromosomes*. Westarp Wissenschaften, Magdeburg.

LeGrande, W.H. and J.M. Fitzsimons. 1976. Karyology of the mullets *Mugil curema* and *Mugil cephalus* (Perciformes: Mugilidae) from Louisiana. *Copeia* 1976: 388–391.

Meyne, J., R.J. Baker, H.H. Hobart, T.C. Hsu, O.A. Ryder, O.G. Ward, J.E. Wiley, D.H. Wursterhill, T.L. Yates and R.K. Moyzis. 1990. Distribution of non-telomeric sites of the (TTAGGG)n telomeric sequence in vertebrate chromosomes. *Chromosoma* 99: 3–10.

Natarajan, R. and K. Subrahmanyam. 1974. A karyotype study of some teleosts from Portonovo waters. *Proceedings of Indian Academy of Sciences.* 79B: 173–196.

Nayyar, R.P. 1966. Karyotype studies in thirteen species of fishes. *Genetica* 37: 78–92.

Nelson, J.S. 1994. *Fishes of the World*. John Wiley & Sons, New York.

Nirchio, M. and H. Cequea. 1998. Karyology of *Mugil liza* and M. *curema* from Venezuela. *Boletin de Investigaciones Marinas y Costeras* 27: 45–50.

Nirchio, M., D. González and J.E. Pérez. 2001. Estudio citogenético de *Mugil curema* y M. *liza* (Pisces: Mugilidae): Regiones organizadoras del nucleolo. *Boletino del Instituto Oceanografico, Universidad de Oriente Venezuela* 40: 3–7.

Nirchio, M., F. Cervigon, Rebelo J.I. Porto, J.E. Perez, J.A. Gomez and J. Villalaz. 2003. Karyotype supporting *Mugil curema* Valenciennes, 1836 and *Mugil gaimardianus* Desmarest, 1837 (Mugilidae: Teleostei) as two valid nominal species. *Scientia Marina* 67: 113–115.

Nirchio, M., E. Ron and A.R. Rossi. 2005a. Karyological characterization of *Mugil trichodon* Poey, 1876 (Pisces: Mugilidae). *Scientia Marina* 69: 525-530.

Nirchio, M., R.R. Cipriano, M.M. Cestari and A.S. Fenocchio. 2005b. Cytogenetical and morphological features reveal significant differences among Venezuelan and Brazilian samples of *Mugil curema*. *Neotropical Ichthyology* 3: 107-110.

Ohno, S. 1974. Protochordata, Cyclostomata and Pisces. In: *Animal Cytogenetics*, Vol. 4, Chordata 1. B. John (ed.), Gebrüder Borntraeger, Berlin.

Pauls, E. and I.A. Coutinho. 1990. Levantamento citogenético em peixes de maior valor econômico do litoral Fluminense, RJ (23° Lat/S). *Congresso Brasileiro de Zoologia 17. Universidade Estadual de Londrina* p. 325.

Phillips, R.B. and K.M. Reed. 1996. Application of fluorescence in situ hybridization (FISH) techniques to fish genetics: A review. *Aquaculture* 140: 197–216.

Rishi, K.K. and J. Singh. 1982. Karyological studies on five estuarine fishes. *The Nucleus* 25: 178–180.

Rosen, D.E. 1964. The relationships and taxonomic position of the halfbeaks, killifishes, silversides, and their relatives. *Bulletin of the American Museum of Natural History* 127: 217–268.

Rossi, A.R., D. Crosetti, E. Gornung and L. Sola. 1996. Cytogenetic analysis of global populations of *Mugil cephalus* (striped mullet) by different staining techniques and fluorescent in situ hybridization. *Heredity* 76: 77–82.

Rossi, A.R., E. Gornung and D. Crosetti. 1997. Cytogenetic analysis of *Liza ramada* (Pisces, Perciformes) by different staining techniques and fluorescent *in situ* hybridization. *Heredity* 79: 83–87.

Rossi, A.R., Capula, M., Crosetti, D., Sola, L. and D.E. Campton. 1998. Allozyme variation in global populations of striped mullet, *Mugil cephalus* (Pisces: Mugilidae). *Marine Biology* 131: 203–212.

Rossi, A.R., E. Gornung, D. Crosetti, S. De Innocentiis and L. Sola. 2000. Cytogenetic analysis of *Oedalechilus labeo* (Pisces: Mugilidae), with a report of NOR variability. *Marine Biology* 136: 159–162.

Rossi, A.R., A. Ungaro, S. De Innocentiis, D. Crosetti and L. Sola. 2004. Phylogenetic analysis of Mediterranean mugilids by allozymes and 16S mt rRNA genes investigation: is the genus *Liza* monophyletic? *Biochemical Genetics* 42: 301–315.

Rossi, A.R., E. Gornung, L. Sola and M. Nirchio. 2005. Comparative molecular cytogenetic analysis of two congeneric species, *Mugil curema* and M. *liza*, (Pisces, Mugiliformes), characterized by significant karyotype diversity. *Genetica* 125: 27–32.

Salvadori, S., A. Deiana, E. Coluccia, G. Floridia, E. Rossi, O. Zuffardi. 1995. Colocalization of (TTAGGG)n telomeric sequences and ribosomal genes in Atlantic eels. *Chromosome Research* 3: 54–58.

Schultz, L.P. 1946. A revision of the genera of mullets, fishes of the family Mugilidae, with descriptions of three new genera. *Proceedings of the United States National Museum* 96: 377–395.

Schultz, L.P. 1953. Mugilidae. In: *Fishes of the Marshall and Marianas Islands*, Vol 1. Smithsonian Institution United States National Museum, Bulletin 202, United States Government Printing Office, Washington.

Singer, M.F. 1982. Highly repeated sequences in mammalian genome. *International Review of Cytology* 76: 76–112.

Slijepcevic, P. 1998. Telomeres and mechanisms of Robertsonian fusion. *Chromosoma* 107: 136–140.

Slijepcevic, P., Y. Xiao, I. Dominguez and A.T. Natarajan. 1996. Spontaneous and radiation-induced chromosomal breakage at interstitial telomeric sites. *Chromosoma* 104: 596–604.

Sola, L., S. Bressanello, A.R. Rossi, V. Iaselli, D. Crosetti and S. Cataudella. 1993. A karyotype analysis of the genus *Dicentrarchus* Gill by different staining techniques. *Journal of Fish Biology* 43: 329–337.

Sola, L., E. Gornung, H. Naoi, R. Gunji, C. Sato, K. Kawamura, R. Arai and T. Ueda. 2003. FISH-mapping of 18S ribosomal RNA genes and telomeric sequences in the Japanese bitterlings *Rhodeus ocellatus kurumeus* and *Tanakia limbata* (Pisces, Cyprinidae) reveals significant cytogenetic differences in morphologically similar karyotypes. *Genetica* 119: 99–106.

Sumner, A.T. 1972. A simple technique for demonstrating centromeric heterochromatin. *Experimental Cell Research* 75: 304–306.

Sumner, A.T. 1990. *Chromosome Banding*. Unwin Hyman Ltd., London.

Thomson, J.M. 1954. The Mugilidae of Australia and adjacent seas. *Australian Journal of Marine and Freshwater Research* 5: 70–131.

Thomson, J.M. 1963. A bibliography of systematic references to the grey mullets (Mugilidae). Technical Paper. Division of Fisheries and Oceanography. CSIRO Australia, No 16.

Thomson, J.M. 1997. The Mugilidae of the world. *Memoirs of the Queensland Museum* 41: 457–562.

Ugarković, D. and M. Plohl. 2002. New EMBO Member's review: variation in satellite DNA profiles—Causes and effects. *EMBO Journal* 21: 5955–5959.

Chromosome Evolution in the Neotropical Erythrinidae Fish Family: An Overview

Luiz Antonio Carlos Bertollo

INTRODUCTION

Erythrinidae is a relatively small Characiformes family endemic to South America, with representatives broadly distributed throughout the Brazilian basins. Erythrinidae species are found in several habitats, right from small lakes and lagoons to large rivers, with a carnivorous diet (Oyakawa, 2003). Erythrinidae belongs to a major clade, composed of three other families: Lebiasinidae, Ctenoluciidae and the African Hepsetidae (Buckup, 1991, 1998; Vari, 1995). Indeed, Fink and Fink (1981) had already suggested Erythrinidae to have a narrow relationship with Ctenoluciidae and Hepsetidae.

Three genera are recognized for this family: *Erythrinus* Scopoli (1977), *Hoplerythrinus* Gill (1895) and *Hoplias* Gill (1903). While *Erythrinus* and

Address for Correspondence: Cytogenetics Laboratory, Department of Genetics and Evolution, Federal University of São Carlos, 13565-905 São Carlos, SP, Brazil.
E-mail: bertollo@power.ufscar.br

Hoplerythrinus species may reach up to nearly 40 cm in length, *Hoplias* species may present even larger sizes, some reaching almost 100 cm (Oyakawa, 2003).

Hoplias presents the broadest geographical distribution, with approximately nine described species (Oyakawa, 2003), *H. malabaricus* and *H. lacerdae* being the best known representatives of the group. However, Oyakawa (1990) considers each one of these two nominal species as a separate group, suggesting that *H. macrophthalmus* may also represent a third group. On the other hand, only three *Hoplerythrinus* species and only two *Erythrinus* species are recognized, among which the better known ones are *H. unitaeniatus* and *E. erythrinus*.

In fact, the taxonomy and systematics of the Erythrinidae family are still poorly defined, and what is currently known on this group certainly does not correspond to reality. A large number of yet to be described species have already been recognized for the *H. lacerdae* group from museum collections (Oyakawa, 1990). Unfortunately, no revision studies are yet available for the *H. malabaricus* group, as also for *Erythrinus* and *Hoplerythrinus*.

Genus *Hoplias*

Of the nine *Hoplias* species currently recognized, chromosomal studies are available only for *H. malabaricus* and representatives of the *lacerdae* group due to the broad distribution of the species.

Hoplias malabaricus

H. malabaricus is the most widespread erythrinid species and although it is usually considered as a single taxonomic entity, many evidences point to the existence of a group of species. Right from the first karyotypic studies (Bertollo, 1978), growing evidence has showed that *H. malabaricus* is a heterogeneous group with significant karyotype differences (Bertollo *et al.*, 1979, 1983, 1997a, b, 2000; Ferreira *et al.*, 1989; Dergam and Bertollo, 1990; Lopes and Fenocchio, 1994; Scavone *et al.*, 1994; Bertollo and Mestriner, 1998; Lopes *et al.*, 1998; Bertollo *et al.*, 2000; Born and Bertollo, 2000, 2001; Lemos *et al.*, 2002; Pazza *et al.*, 2003; Vicari *et al.*, 2003).

Seven main cytotypes (A-G) have been recognized for *H. malabaricus* (Bertollo *et al.*, 2000), which can be characterized by their diploid number, chromosome morphology and distinct sex chromosome systems

(Table 1.7.1; Fig. 1.7.2). Some of these cytotypes have a broad geographical distribution, across northern to southern Brazil, even reaching Uruguay and Argentina, as is the case of cytotypes A and C (Fig. 1.7.1). Comparative studies show small changes in the centromere position and the number and location of nucleolus-organizing regions as well as in heterochromatic regions among cytotype A populations, showing that some evolutionary divergences have already occurred within this group (Born and Bertollo, 2001; Vicari et al., 2005).

The seven general cytotypes of H. malabaricus can be grouped in two big clusters, the first including cytotypes A-D and the second cytotypes E-G (Bertollo et al., 2000). In the first cluster, despite the numeric chromosome differences, the four cytotypes share a general karyotypic macrostructure (Fig. 1.7.2, A-D). Cytotype B, with a simple XX/XY sex chromosome system, and cytotype D, with a multiple $X_1X_1X_2X_2/X_1X_2Y$ sex chromosome system, are probably derived forms of cytotypes similar to A and C, respectively (Bertollo et al., 2000). On the other hand, cytotypes E-G in the second cluster also share common karyotypic features. The larger size of the first metacentric chromosome pairs, as well as the occurrence of acrocentric chromosomes in cytotypes E and G are both outstanding (Fig. 1.7.2, E-G). It is also probable that cytotype F, homomorphic, and cytotype G, with a multiple XX/XY_1Y_2 sex chromosome system, are derived forms of a cytotype similar to E, particularly concerning the large metacentric chromosome pair 1 and the sex chromosomes (Bertollo et al., 2000).

Sex chromosomes. A relevant feature in H. malabaricus is the occurrence of three distinct sex chromosome systems, found in cytotypes B, D and G (Fig. 1.7.2 B, D, G). All these cytotypes show a more restricted geographical distribution compared to the cytotypes A, C and F, which do not possess differentiated sex chromosomes (Fig. 1.7.1), thus reinforcing their derived condition.

The XX/XY sex chromosomes, initially proposed by Bertollo et al. (1979), were further confirmed by C- and G-banding in a population from the Parque Florestal do Rio Doce (Minas Gerais State, Brazil), belonging to cytotype B (Born and Bertollo, 2000). In this study it was evidenced that an accumulation of heterochromatin is closely related to the differentiation of the long arm of the subtelocentric X chromosome (Fig. 1.7.3), probably from an ancestral cytotype similar to A.

Table 1.7.1 Karyotypic data for the Erythrinidae fish family.

Cytotype 2n number	Karyotypic formula	Sex	Sex system	Geographic distribution	Sympatric cytotypes
Hoplias malabaricus					
A 2n=42	42 M/SM	f/m	Not differentiated	Wide	AC/AD/ACD/AF
B 2n=42	40 M/SM + 2 ST	f	XX	Restricted	Not detected
	41 M/SM + 1 ST	m	XY		
C* 2n=40	40 M/SM	f/m	Not differentiated	Wide	CA/CAD/CF/CG
D 2n=40/39	40 M/SM	f	$X_1X_1X_2X_2$	Restricted	DA / DAC
	39 M/SM	m	X_1X_2Y		
E 2n=42	40 M/SM + 2 A	m	Not identified	Restricted	EG
F* 2n=40	40 M/SM	f/m	Not differentiated	Moderate	FA / FC
G 2n=40/41	40 M/SM	f	XX	Restricted	GC / GE
	40 M/SM + 1 A	m	XY_1Y_2		
Hoplias lacerdae group					
2n=50	50 M/SM	f/m	Not differentiated	Wide	
Hoplerythrinus uniteniatus					
A 2n=48	48 M/SM	f/m	Not differentiated	Wide	Not detected
B 2n=48	46 M/SM + 2 A	f/m	Not differentiated	Restricted	Not detected
C 2n=52	46 M/SM + 6 A	f/m	Not differentiated	Restricted	Not detected
D 2n=52	44M/SM+4ST+4A	f/m	Not differentiated	Restricted	Not detected
Erythrinus erythrinus					
A 2n=54	46A+2ST+6M	f/m	Not differentiated	Wide	Not detected
B 2n=54/53	46A+2ST+6M	f	$X_1X_1X_2X_2/$	Restricted	Not detected
	44A+2ST+7M	m	X_1X_2Y		
C 2n=52/51	38A+6ST+8M/SM	f	$X_1X_1X_2X_2/$	Restricted	Not detected
	36A+6ST+9M/SM	m	X_1X_2Y		
D 2n=52/51	44A+2ST+6M/SM	f	$X_1X_1X_2X_2/$	Restricted	Not detected
	42A+2ST+7M/SM	m	$X_1\underline{X_2}Y$		

2n=diploid chromosome number; M = metacentric, SM = submetacentric, ST = subtelocentric, A=acrocentric chromosomes; m = male; f = female; *cytotypes with the same diploid number and karyotypic formula, but differing in chromosome morphology.

Fig. 1.7.1 Distribution of the analyzed samples of the family Erythrinidae through Brazil and some other South America localities. Solid circles refer to *Hoplias malabaricus* cytotypes, solid squares to *Hoplias lacerdae* group, solid triangles to *Hoplerythrinus unitaeniatus* cytotypes and solid stars to *Erythrinus erythrinus* cytotypes. (1 = cytotype A; 2 = cytotype B; 3 = cytotype C; 4 = cytotype D; 5 = cytotype E; 6 = cytotype F; 7 = cytotype G). For convenience, the samples from two Brazilian states are shown in detail. The large open circles indicate sympatry among distinct *H. malabaricus* cytotypes.

Hoplias malabaricus

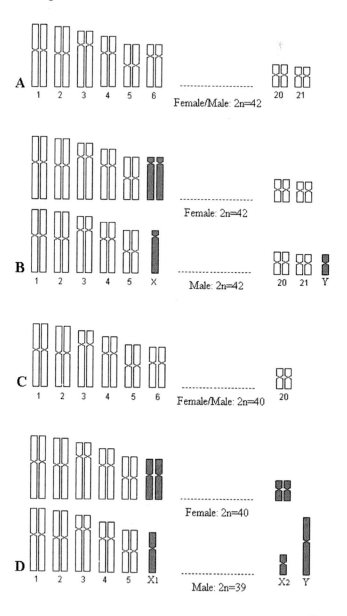

(Fig. 1.7.2 contd.)

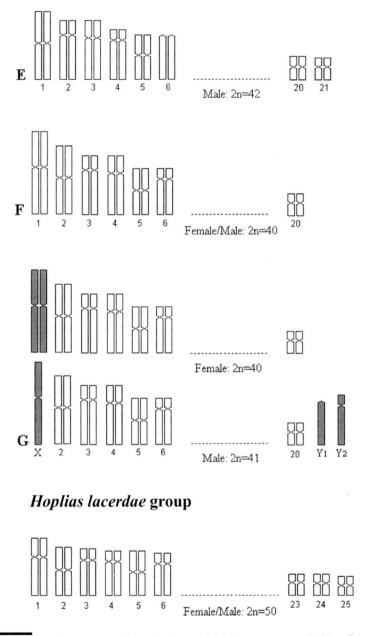

Hoplias lacerdae group

Fig. 1.7.2 Partial idiograms of the *Hoplias malabaricus* cytotypes (**A-G**) and *Hoplias lacerdae* group, showing some of their most remarkable features.

Hoplias malabaricus   *Hoplias malabaricus*

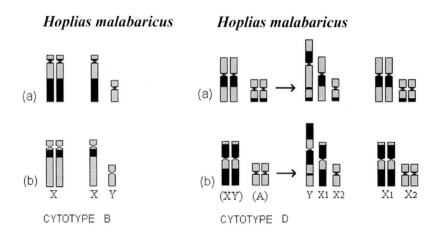

Hoplias malabaricus

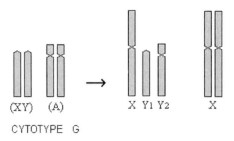

Erythrinus erythrinus

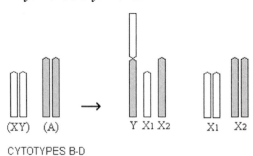

Fig. 1.7.3 Idiograms of the sex chromosome systems found in *Hoplias malabaricus* cytotypes B, D and G, and in *Erythrinus erythrinus* cytotypes B, C and D. (a) C-banded chromosomes; (b) G-banded chromosomes. Arrows indicate the proposed origin of the current sex chromosomes from a probable ancestral homomorphic (XY) chromosomes and an autosomal (A) pair (see text for description).

The multiple $X_1X_1X_2X_2/X_1X_2Y$ sex chromosome system in cytotype D was analyzed by Bertollo et al. (1983, 1997b) and Bertollo and Mestriner (1998). Structural rearrangements, probably from an ancestral cytotype similar to C, are involved in the origin of this sex system. Distinct banding data indicate that the long arm of the Y chromosome resulted from a translocation between an ancestral chromosome similar to the current X_2 and the short arm of an ancestral chromosome similar to the current X_1. Consequently, the short arm of the Y chromosome corresponds to the long arm of the current X_1 chromosome (Fig. 1.7.3). Conventional meiotic data and synaptonemal complexes analyses suggest a normal pattern for male meiosis, representing a stabilized multiple sex chromosome system in the species (female $2n=40$; male $2n=39$). Indeed, two types of balanced metaphase II cells are formed during spermatogenesis, showing 19 and 20 chromosomes, respectively. In addition, the meiotic trivalent, formed by the X_1, X_2 and Y chromosomes, presents a synaptic adjustment which may be important for normal meiotic development (Bertollo and Mestriner, 1998).

Regarding the origin of the XX/XY_1Y_2 sex chromosomes of cytotype G (Fig. 1.7.3), no accurate information is as yet available. However, a comparative analysis between cytotypes E-G (Fig. 1.7.2) suggests that ancestral chromosomes, similar to pairs 4 and 6 found in cytotype E, may be related to karyotypic rearrangements that resulted in the first large metacentric pair of cytotypes F and G, as well as in the sex chromosomes of cytotype G (Bertollo et al., 2000).

Sympatry areas. Distinct cytotypes living in sympatry have already been detected in several South American regions (Fig. 1.7.1). Two sympatric cytotypes have been the commonly observed cases, although in the Porto Rico region (Paraná State – Brazil) three co–existing cytotypes (A, C and D) were also detected. In all these cases, there exist no records of possible hybrid forms, suggesting a probable reproductive isolation between the cytotypes.

Hoplias lacerdae Group

This fish group was reviewed by Oyakawa (1990), where several species, including H. lacerdae and H. brasiliensis were recognized. However, the majority of these species are yet to be described. Based on this review, some cytogenetic data probably refer to species that fit the undescribed category, considering their origin from places that are not in agreement with the restricted distribution of *H. lacerdae* and *H. brasiliensis*, while some other samples from fish farms probably refer to *Hoplias* cf. *lacerdae*.

In contrast to *H. malabaricus*, the species of the *lacerdae* group show a relatively stable karyotype, with no significant variations between the distinct populations examined (Fig. 1.7.1), with 2n=50 meta-submetacentric chromosomes (Table 1.7.1; Fig. 1.7.2). Bertollo *et al.* (1978), suggested an XX/XY sex chromosome system for representatives of *H.* cf. *lacerdae*. However, this sex system was not found in other analyzed samples from fish farms or natural populations (Morelli, 1998). Therefore, even though belonging to the same genus, *H. malabaricus* and the species of the *lacerdae* group exhibit an extremely differentiated karyotypic evolution.

Genus *Hoplerythrinus*

Hoplerythrinus is a small erythrinid genus, represented by *H. cinereus* and *H. gronovii*, with restricted distribution, and *H. unitaeniatus*, with a broader distribution through South America (Oyakawa, 2003). Only *H. unitaeniatus* has been cytogenetically analyzed (Giuliano-Caetano, 1986; Giuliano-Caetano and Bertollo, 1988, 1990; Giuliano-Caetano *et al.*, 2001; Diniz and Bertollo, 2003), also appearing as a diversified species with at least four distinct cytotypes (Table 1.7.1).

Cytotype A is characterized by 2n=48 meta-submetacentric chromosomes. The other cytotypes present particular features such as a larger diploid number, 2n=52 (cytotypes C and D), distinct numbers and sizes of acrocentric chromosomes (cytotypes B-D), and the occurrence of subtelocentric chromosomes (cytotype D). Thus, the karyotypic evolution in this group was accomplished by an increase/reduction in the chromosome number, as well as by structural rearrangements modifying the chromosome forms (Fig. 1.7.4). While cytotype A shows a broader geographical distribution, cytotypes B-D have a more restricted one (Fig. 1.7.1). Parallel with those macrostructural alterations, karyotypic differences concerning the number and locations of the Ag-NORs, as well as of the 18S and 5S rDNA sites, have also been observed among cytotypes (Diniz and Bertollo, 2003).

Together with the interpopulational karyotypic differentiations, a chromosome polymorphism was also reported for a population from the Negro river, near the city of Manaus (Amazonas State – Brazil). In this population, which belongs to cytotype A, specimens with different

Hoplerythrinus unitaeniatus

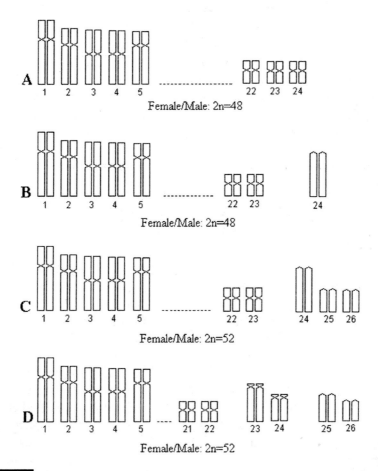

Fig. 1.7.4 Partial idiograms of the *Hoplerythrinus unitaeniatus* cytotypes (A-D), showing some of their most remarkable features.

karyotypes were found, probably due to pericentric inversions. Thus, specimens with 2n=48 M/SM (standard karyotype) were observed alongside specimens with 2n=46 M/SM + 2 A, and with 2n=47 M/SM + 1 A, with the two acrocentric chromosomes presenting similar or distinct sizes (Giuliano-Caetano and Bertollo, 1988). On the other hand, a pronounced polymorphism was also verified in the Prata river population (Minas Gerais State – Brazil), where cytotype D is the standard type. Variations in the diploid number (2n=50 and 2n=51) and in the morphology of several chromosomes were observed, in addition to chromosomes lacking homologues in the karyotype (Diniz, 2002). However, due to their complexity, such variations are yet to be properly interpreted.

No heteromorphic sex chromosomes were found in any of the *H. unitaeniatus* populations. Therefore, although *H. malabaricus* and *H. unitaeniatus* show a nonconservative karyotypic evolution, *H. unitaeniatus* present a particular evolutionary history for conserving the homomorphic condition between the sexes.

Genus *Erythrinus*

According to Oyakawa (2003), only two species are described for this group: *E. erythrinus* and *E. kessleri*, the latter possibly restricted to Bahia State (Brazil), while *E. erythrinus* presents a broader South American distribution.

Erythrinus is the least studied erythrinid genus. Indeed, only *E. erythrinus* was recently analyzed (Bertollo *et al.*, 2004), in which a karyotypic diversity was also found. Five Brazilian populations and one Argentinean population were studied (Fig. 1.7.1), comprising four (A, B, C, D) distinct cytotypes or karyotypic groups (Table 1.7.1; Fig. 1.7.5). A high number of acrocentric chromosomes stands out in all populations, which basically differentiates *Erythrinus* from the other Erythrinidae genera, where these chromosomes are rarely observed (Table 1.7.1). On the other hand, the four cytotypes may be mutually differentiated by the diploid number (2n=54 or 52) as well as by the number of A, ST and M/SM chromosomes (Table 1.7.1; Fig. 1.7.5).

Sex chromosomes. The most marking feature of cytotypes B-C-D is the presence of a multiple sex chromosome system of the $X_1X_1X_2X_2/X_1X_2Y$ type, absent in cytotype A. The pronounced size of the metacentric Y chromosome, the largest in the karyotype, is pointed out, while chromosomes X_1 and X_2 belong to the acrocentric group. It appears that a centric fusion may have created the Y chromosome from ancestral chromosomes similar to the current X_1 and X_2, with the consequent numerical reduction in male karyotypes (Figs. 1.7.3, 1.7.5). A comparative analysis between the cytotypes indicates that the mainly chromosomal rearrangements related to the karyotypic evolution in *E. erythrinus* are probably due to centric fusions and pericentric inversions (Bertollo *et al.*, 2004).

Erythrinus erythrinus

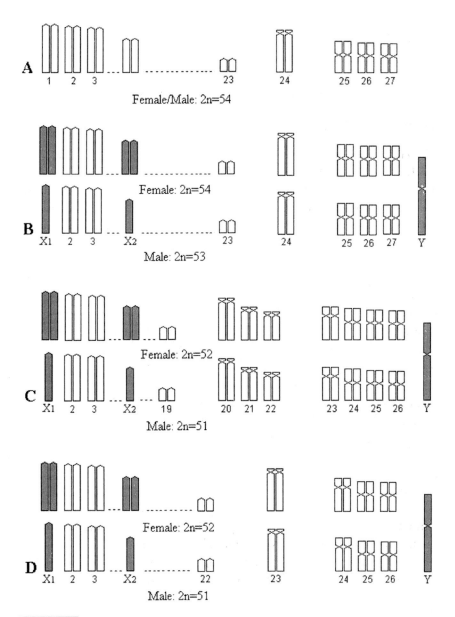

Fig. 1.7.5 Partial idiograms of the *Erythrinus erythrinus* cytotypes (A-D), showing some of their most remarkable features.

Supernumerary chromosomes. Supernumerary microchromosomes were also found in cytotypes A and B, varying from 0 to 4, with an apparent mitotic instability (Bertollo et al., 2004). It is interesting to point out that even though these two cytotypes differ in regard to sex chromosomes, their karyotype structure (Fig. 1.7.5 A, B), as well as the common occurrence of supernumerary chromosomes make them closer to each other than to the remained cytotype groups.

Conclusive Remarks

This chapter evidences the karyotypic diversity in the family Erythrinidae, particularly concerning *Hoplias malabaricus*, *Hoplerythrinus unitaeniatus* and *Erythrinus erythrinus*, which exhibit an extensive chromosome evolution associated with numerical and structural rearrangements. The cytotypes detected in these three species allow for a precise characterization of different local populations—many of them isolated in nature—where the chromosomal rearrangements are already well fixed. Even in the cases where the co-existence of distinct *H. malabaricus* cytotypes has been observed, hybrid forms do not seem to occur, which reinforces the possibility of reproductive isolation between the cytotypes. Thus, each one of those nominal species may include distinct biological species, thus representing species complexes. The need of a taxonomic revision in the family Erythrinidae as a whole becomes evident.

On the other hand, the presence of distinct sex chromosome systems among the Erythrinidae fish, which appear to have differentiated from separate events, such as heterochromatin accumulation, translocations and centric fusions on different chromosomes, is outstanding. The occurrence of any homology between the sex chromosomes of distinct cytotypes is an open question, requiring complementary analyses. Nevertheless, sex chromosomes in *H. malabaricus* as also in *E. erythrinus* is not a general feature, with cytotypes bearing heteromorphic sex chromosomes together with cytotypes where males and females display homomorphic karyotypes. Therefore, differentiated sex chromosomes do not represent a basal characteristic in any one of these two genera, corresponding to a derived trait for some cytotypes.

Acknowledgements

Research supported by the Conselho Nacional de Desenvolvimento Científico e Tecnológico (CNPq). I thank Dr Liano Centofante and Marcelo Ricardo Vicari for assistance in the organization of the figures.

References

Bertollo, L.A.C. 1978. *Estudos Citogenéticos no Gênero Hoplias Gill 1903 (Pisces, Erythrinidae)*. Ph.D. Thesis. Universidade de São Paulo, Ribeirão Preto, Brazil.

Bertollo, L.A.C. and C.A. Mestriner. 1998. The X_1X_2Y sex chromosome system in the fish *Hoplias malabaricus* (Pisces, Erythrinidae). II. Meiotic analyses. *Chromosome Research* 6: 141–147.

Bertollo, L.A.C., C.S. Takahashi and O. Moreira-Filho. 1978. Cytotaxonomic considerations on *Hoplias lacerdae* (Pisces, Erythrinidae). *Brazilian Journal of Genetics* 1: 103–120.

Bertollo, L.A.C., C.S. Takahashi and O. Moreira-Filho. 1979. Karyotypic studies of two allopatric populations of the genus *Hoplias* (Pisces, Erythrinidae). *Brazilian Journal of Genetics* 2: 17–37.

Bertollo, L.A.C., C.S. Takahashi and O. Moreira-Filho. 1983. Multiple sex chromosomes in the genus *Hoplias* (Pisces, Erythrinidae). *Cytologia* 48: 1–12.

Bertollo, L.A.C., O. Moreira-Filho and M.S. Fontes. 1997a. Karyotypic diversity and distribution in *Hoplias malabaricus* (Pisces, Erythrinidae): Cytotypes with 2n=40 chromosomes. *Brazilian Journal of Genetics* 20: 237–242.

Bertollo, L.A.C., M.S. Fontes, A.S. Fenocchio and J. Cano. 1997b. The X_1X_2Y sex chromosome system in the fish *Hoplias malabaricus*. I. G-, C- and chromosome replication banding. *Chromosome Research* 5: 493–499.

Bertollo, L.A.C., G.G. Born, J.A. Dergam, A.S. Fenocchio and O. Moreira-Filho. 2000. A biodiversity approach in the Neotropical Erythrinidae fish, *Hoplias malabaricus*. Karyotypic survey, geographic distribution of cytotypes and cytotaxonomic considerations. *Chromosome Research* 8: 603–613.

Bertollo, L.A.C., C. Oliveira, W.F. Molina, V.P. Margarido, M.S. Fontes, M.C. Pastori, J.N. Falcão and A.S. Fenocchio. 2004. Chromosome evolution in the erythrinid fish, *Erythrinus erythrinus* (Teleostei: Characiformes). *Heredity* 93: 228–233.

Born, G.G. 2000. *Estudo da Diversidade Cariotípica no Grupo Hoplias malabaricus (Pisces, Erythrinidae). Cariótipo 2n=42*. Ph.D. Thesis. Universidade Federal de São Carlos, São Carlos, Brazil.

Born, G.G. and L.A.C. Bertollo. 2000. An XX/XY sex chromosome system in a fish species, *Hoplias malabaricus*, with a polymorphic NOR-bearing X chromosome. *Chromosome Research* 8: 111–118.

Born, G.G. and L.A.C. Bertollo. 2001. Comparative cytogenetics among allopatric populations of the fish, *Hoplias malabaricus*. Cytotypes with 2n=42 chromosomes. *Genetica* 110: 1–9.

Buckup, P.A. 1991. *The Caracidiinae: A Phylogenetic Study of the South American Darters and their Relationships with other Characiform Fishes*. Ph.D. Thesis. The University of Michigan, Ann Arbor.

Buckup, P.A. 1998. Relationships of the Characidiinae and phylogeny of characiform fishes (Teleostei: Ostariophysi). In: *Phylogeny and Classification of Neotropical Fishes*, L.R. Malabarba, R.E. Reis, R.P. Vari, Z.M.S. Lucena and C.A. Lucena (eds.). Edipucrs, Porto Alegre, pp. 123–144.

Dergam, J.A. and L.A.C. Bertollo. 1990. Karyotypic diversification in *Hoplias malabaricus* (Osteichthyes, Erythrinidae) of the São Francisco and Alto Paraná basins, Brazil. *Brazilian Journal of Genetics* 13: 755–766.

Diniz, D. 2002. *Estudos Citogenéticos Populacionais em Hoplerythrinus unitaeniatus (Pisces, Erythrinidae). Análise da Biodiversidade*. M.Sc. Dissertation. Universidade Federal de São Carlos, São Carlos, Brazil.

Diniz, D. and L.A.C. Bertollo. 2003. Karyotypic studies on *Hoplerythrinus unitaeniatus* (Pisces, Erythrinidae) populations. A biodiversity analysis. *Caryologia* 56: 303–313.

Ferreira, R.H.R., C.G. Fonseca, L.A.C. Bertollo and F. Foresti. 1989. Cytogenetics of fishes from Parque Estadual do Rio Doce (MG). I. Preliminary study of "*Hoplias malabaricus*" (Pisces, Erythrinidae) from Lagoa Carioca and Lagoa dos Patos. *Brazilian Journal of Genetics* 12: 219–226.

Fink, S.V. and W.L. Fink. 1981. Interrelationships of the ostariophysan fishes (Teleostei). *Zoological Journal of the Linnean Society* 72: 297–353.

Giuliano-Caetano, L. 1986. *Estudo Citogenético em Hoplerythrinus unitaeniatus (Pisces, Erythrinidae) de Diferentes Bacias Hidrográficas do Brasil*. M.Sc. Dissertation. Universidade Federal de São Carlos, São Carlos, Brazil.

Giuliano-Caetano, L. and L.A.C. Bertollo. 1988. Karyotype variability in *Hoplerythrinus unitaeniatus* (Characiformes, Erythrinidae). I. Chromosome polymorphism in the rio Negro population (Manaus, State of Amazonas) Brazil. *Brazilian Journal of Genetics* 11: 299–306.

Giuliano-Caetano, L. and L.A.C. Bertollo. 1990. Karyotype variability in *Hoplerythrinus unitaeniatus* (Pisces, Characiformes, Erythrinidae). II. Occurrence of natural triploidy. *Brazilian Journal of Genetics* 13: 231–237.

Giuliano-Caetano, L., L.C. Jorge, O. Moreira-Filho and L.A.C. Bertollo. 2001. Comparative cytogenetic studies on *Hoplerythrinus unitaeniatus* populations (Pisces, Erythrinidae). *Cytologia* 66: 39–43.

Lemos, P.M.M., A.S. Fenochio, L.A.C. Bertollo and M.M. Cestari. 2002. Karyotypic studies on two *Hoplias malabaricus* populations (Characiformes, Erythrinidae) of the 2n=42 group, from the first plateau of the Iguaçu river basin (Paraná state, Brazil). *Caryologia* 55: 193–198.

Lopes, P.A. and A.S. Fenocchio.1994. Confirmation of two different cytotypes for the Neotropical fish *Hoplias malabaricus* Gill 1903 (Characiformes). *Cytobios* 80: 217–221.

Lopes, P.A., A.J. Alberdi, J.A. Dergam and A.S. Fenocchio. 1998. Cytotaxonomy of *Hoplias malabaricus* (Osteichthyes, Erythrinidae) in the Aguapey river (Province of Corrientes, Argentina). *Copeia* 1998: 485–487.

Morelli, S. 1998. *Citogenética Evolutiva em Espécies do Gênero Hoplias, Grupo lacerdae. Macroestrutura Cariotípica, Heterocromatina e Regiões Organizadoras de Nucléolo.* Ph.D. Thesis. Universidade Federal de São Carlos, São Carlos, Brazil.

Pazza, R. and H.F. Julio Jr. 2003. Occurrence of three sympatric cytotypes of *Hoplias malabaricus* (Pisces, Erythrinidae) in the upper Paraná river foodplain (Brazil). *Cytologia* 68: 159–163.

Oyakawa, O.T. 1990. *Revisão Sistemática das Espécies do Gênero Hoplias (Grupo lacerdae) da Amazônia Brasileira e Região Leste do Brasil (Teleostei: Erythrinidae).* M.Sc. Dissertation. Universidade de São Paulo, São Paulo, Brazil.

Oyakawa, O.T. 2003. Family Erythrinidae. In: *Check List of the Freshwater Fishes of South and Central America*, R.E. Reis, S.O. Kullander, C.J. Ferraris Jr. (eds.). Edipucrs, Porto Alegre, pp. 238–240.

Scavone, M.D.P., L.A.C. Bertollo and M.M. Cavallini. 1994. Sympatric occurrence of two karyotypic forms of *Hoplias malabaricus* (Pisces, Erythrinidae). *Cytobios* 80: 223–227.

Vari, R.P. 1995. The Neotropical fish family Ctenoluciidae (Teleostei: Ostariophysi: Characiformes): Supra and intrafamilial phylogenetic relationships, with a revisionary study. *Smithsonian Contributions to Zoology* 564: 1–97.

Vicari, M.R. 2003. *Citogenética Comparativa de Hoplias malabaricus (Pisces, Erythrinidae). Estudos em Região Divisora de Águas para as Bacias dos Rios Tibagi, Iguaçu, Ivaí e Ribeira (Ponta Grossa, PR).* M.Sc. Dissertation. Universidade Federal de São Carlos, São Carlos, Brazil.

Vicari, M.R., R.F. Artoni and L.A.C. Bertollo. 2003. Heterochromatin polymorphism associated with 18S rDNA: A differential pathway among *Hoplias malabaricus* fish populations. *Cytogenetic and Genome Research* 101: 24–28.

Vicari, M.R., R.F. Artoni and L.A.C. Bertollo. 2005. Comparative cytogenetics of *Hoplias malabaricus* (Pisces, Erythrinidae): a population analysis in adjacent hydrographic basins. *Genetics and Molecular Biology* 28: 103–110.

SECTION

2

Fish Cytogenetics and
Biodiversity Conservation

Cytogenetics as a Tool in Fish Conservation: The Present Situation in Europe

Petr Ráb[1,3*], Jörg Bohlen[1], Marie Rábová[1], Martin Flajšhans[2,3] and Lukáš Kalous[1, 4]

INTRODUCTION

Fishes are by far the most species-rich group of vertebrates. More than 30,000 of the 40,000 vertebrate species comprise fishes. The exploration of this enormous biodiversity is still a challenge for ichthyologists. This is

Address for Correspondence:
[1*]Laboratory of Fish Genetics, Section of Evolutionary Biology and Genetics, Institute of Animal Physiology and Genetics (IAPG), AS CR, 27721 Liběchov, Czech Republic. E-mail: rab@iapg.cas.cz
[2]Department of Genetics and Breeding, Research Institute of Fishery and Hydrobiology (RIFH), University of South Bohemia, Vodňany, Czech Republic.
[3]Joint Laboratory of Genetics, Physiology and Reproduction of Fishes of IAPG AS CR Liběchov and RIFH USB Vodňany, Czech Republic.
[4]Department of Zoology and Fisheries, Faculty of Agrobiology, Food and Natural Resources, Czech University of Agriculture, Praha 6 – Suchdol, Czech Republic.

true not only for remote areas in exotic and/or tropical regions of the world but—due to historical reasons—also holds for freshwater fish diversity in Europe.

European scientists initiated systematic research and played a leading role in fish research for many centuries. At the beginning of twentieth century, several opinions about the diversity of the European fishes had been published, and scientists started to focus more on exploring exotic countries. Their interest in European fish diversity started to wane, a situation that has continued until recently. In the 1980s, the diversity of European freshwater fishes, in particular, was much less understood than in that for any other continent.

In 1997, Kottelat presented the first critical review of the systematics and nomenclature of European (not including the former USSR) freshwater fishes in more than 100 years. This work was an inspiration to many European ichthyologists, and once again generated interest in the investigation of the present-day diversity of freshwater fishes of Europe. This increased interest in fish diversity was encouraged by many molecular genetic and cytogenetic studies, which revealed that the varieties in any investigated fish group were higher than previously assumed. The new wave of interest resulted in the publication of descriptions of many new taxa (*Aphanius lozanoi*, *Barbatula pindos*, *Barbus balcanicus*, *B. carpathicus*, *Chondrostoma almaca*, *Coregonus arenicolus*, *C. atterensis*, *C. fatioi*, *C. fontanae*, *Phoxinellus dalmaticus*, *P. jadovensis*, *P. krbavensis*, *Rhodeus colchicus*, *Romanogobio pentatrichus*, *Salmo aphelios*), to mention a few examples in the re-estimation of the taxonomic status of several other taxa (e.g., *Gobio* spp., *Romanogobio* spp., *Cobitis* spp., *Coregonus* spp., *Salmo* spp., *Cottus* spp.) and in detailed studies on the genetic diversity of some unusual taxa (e.g., forms of evolutionary hybrid origin in Central European *Cobitis* or Iberian *Squalius*). The increasing understanding that former widespread taxa, in fact, contain many overlooked or forgotten taxa, significantly increased the number of localized endemic species. These species often inhabit only a very small distributional range. This makes them vulnerable to negative impacts like introduction of alien species, water removal for irrigation purposes, monotonization of habitat structures and eutrophication. Therefore, many of the newly discovered/considered taxa are threatened today and the definition of conservation units and/or measures for such species has, therefore, become an urgent task for conservation biologists, ichthyologists and geneticists.

Another important problem for fish conservation in Europe is intentional and/or accidental introduction of alien fish species. These species can have a direct effect as predatory species (e.g., largemouth bass *Micropterus salmoides*, northern pike *Esox lucius*, Amur sleeper goby *Percottus glenii*), or an indirect effect by way of ecological competition (e.g., Prussian carp *Carassius 'gibelio'*), or transmission of diseases (e.g., stone moroko *Pseudorasbora parva*), especially in areas where the endemic fish faunas are restricted in distribution. Sound knowledge of the origins, phylogeography, population structure and dynamics, and many other biological factors are required for effective management of these invasive and harmful intruders.

To define conservation units and cope with alien species, it is necessary to have an understanding of the natural diversity of the animals under consideration , a task that is nowadays generally addressed in terms of genetic diversity. In fishes, such estimates have to include the study of cytogenetic diversity, because fishes are very tolerant in terms of genomic changes at level of their chromosome sets, a fact that leads to a much higher cytogenetic diversity than in other vertebrates. In fact, variability in certain chromosomal regions in fishes can be thought of as equivalent to differences in chromosome number, and even ploidy level. These differences may occur at the intra-individual, intrapopulational or intraspecific levels. In some cases, they can also be diagnostic for individual groups, and scale from population level to family level. Therefore, we feel that the time is ripe to use fish cytogenetics as genetic tool in conservation biology.

Cytogenetics as Compared to Other Tools Used in Study of Fish Diversity

The voluminous and ever-increasing body of recent literature on fish biodiversity clearly demonstrates the fact that molecule-based approaches using various genetic tools such as molecular phylogeny and phylogeography have greatly contributed to our understanding of fish diversity. Park and Moran (1994)—just to mention some from several others authors on this topic—have clearly summarized the set of applicable genetic markers, their characteristics (tissue requirement, number of loci, coverage of genome, relative rate of evolution) and their suitability for analysis of a particular objective (population genetics, phylogeny, phylogeography) or type of problem (natural hybridizations, pedigree analysis, etc.).

However, cytogenetics has practically never been listed among other genetic tools used in fish conservation, although Ráb and Bohlen (2000) have identified those problems that can be addressed using such an approach. The apparent absence of cytogenetics in the toolbox of genetic markers is most likely due to the misunderstanding or even lack of knowledge of the informative/resolution power of cytogenetic data.

Figure 2.1.1 provides a simplified explanation of the different levels of genome organization stressing the resolution power of cytogenetics as a tool in addressing fish diversity.

The various morphological characters can be compared to a city plan. Many cities differ from each other; some are huge while others are small. Others, however, were designed by the same architect and are very similar in form and layout. In the same way and by comparison, the pike (genus *Esox*) has a similar body plan (i.e., similar morphological characters) in all of its five extant species. To elaborate further, every city (of note) has a library that stores information, either classically in books, or more recently also in a variety of electronic media. The architecture of such a library building can be compared to chromosomal level of genome organization— the place where data are stored, processed, multiplied, found, lost, exchanged, stolen. For many studies, it is important to determine whether the architecture of the library building is new and modern (= highly rearranged chromosome complements in terms of evolution), is ancient (= highly evolutionary conserved chromosome set) or has two different architectural styles, inappropriately mixed during its construction (= hybrid situations). Our example can also easily demonstrate the following facts. (1) In different cities, the same architect could build the same library (= the morphologically similar karyotypes in evolutionary conserved groups, as in the morphologically differentiated leuciscin lineage of the family Cyprinidae). (2) Similar cities can have libraries designed by different architects (= morphologically very similar species can have highly different karyotypes, as in 'umbroids' of the order Esociformes). (3) Some cities can also have two or more libraries (= polyploid situations). (4) Some of the libraries are under extensive reconstruction (= adaptations of karyotype patterns driven by apparent correlation with some other factors, such as observed in anadromous salmonid fishes).

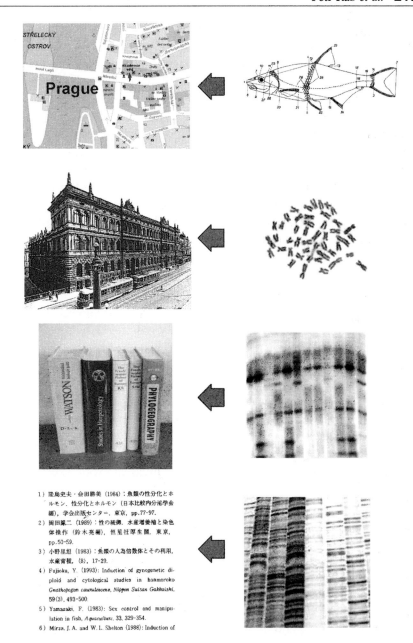

1）隆島史夫・会田勝美（1964）：魚類の性分化とホルモン．性分化とホルモン（日本比較内分泌学会編），学会出版センター，東京，pp.77-97.

2）岡田薫二（1989）：性の統御，水産増養殖と染色体操作（鈴木亮爾編），恒星社厚生閣，東京，pp.50-59.

3）小野里坦（1983）：魚類の人為倍数体とその利用，水産育種，(8), 17-29.

4）Fujioka, Y. (1993): Induction of gynogenetic diploid and cytological studies in honmoroko *Gnathopogon caurulescens*, *Nippon Suisan Gakkaishi*, 59(3), 493-500.

5）Yamazaki, F. (1983): Sex control and manipulation in fish, *Aquaculture*, 33, 329-354.

6）Mirza, J. A. and W. L. Shelton (1988): Induction of

Fig. 2.1.1 A comparison between the different levels of genome organization (right) and the different levels of information contained in city plan, library, books and letters in books (left) provides a simplified explanation of the resolution power of chromosomal data.

As pointed out above, the library usually contains books or some other information media units (=genes). Similarly and as at the previous level, we can determine that: (a) different libraries (= chromosome complement) in many cities (= organisms) have the same books and vice versa; (b) in any particular library, some books are present in several copies (= regional multiplication of particular genes), while in others certain books are missing. Many other possible comparisons are possible within the context of our example. When we open a book, we can see sentences composed of words and written in letters (= set of sequence-based characters). We can even take our comparisons further: the number of letters is limited by our alphabet (= genetic code of nucleotide triplets); same things are repeated in different books, and there are different sentences in the same books. We may partly understand the sense of many sentences (= progress in genomic projects), but in other sentences we have to make an effort to even read the letters themselves (exemplified here by Japanese letters in Fig. 2.1.1).

We hope that our oversimplified explanation has showed that each biological reality in addressing fish diversity requires different tools that describe the different various levels of genome organization, from morphology to sequences (and their combinations). This also means that cytogenetics (= study of library building architectures) studies and describes the morphology and other markers of cytological (nuclear) objects, which themselves are made up of component parts bearing genetic information. In this chapter, we shall provide several examples of the application of cytogenetic methodology to fish conservation issues in Europe, dealing with both endangered and invasive exotic species.

Case Studies

Hybrid polyploid fish complexes

Reproductive contact between distantly related species in lower vertebrates (fishes, amphibians and reptiles) may sometimes result in altered, otherwise evolutionarily highly conserved, gametogenetic mechanisms in their progeny. This is associated with changes in reproductive modes, and frequently also with polyploidy. Such alternative reproductive modes include parthenogenesis, gynogenesis and hybridogenesis, an equivalent of plant apomixis. They result in either clonal (parthenogenesis, gynogenesis) or hemiclonal (hybridogenesis) inheritances. The list of more than 90 documented cases of such hybrid complexes with altered mode of reproduction among lower vertebrates indicates that they are not rare (reviewed in Vrijenhoek *et al.*, 1989; Alves *et al.*, 2001). The asexual (also called unisexual and more recently non-sexual, asexual) vertebrates are of a cryptic nature and genetic methods have to be used in their discovery. In Europe, there are several known hybrid asexual fish, including the cyprinids *Squalius 'alburnoides'* in the Iberian Peninsula, *Carassius 'gibelio'*, an invasive Asian intruder, cobitid spiny loaches of the genus *Cobitis* in the entire Europe north of the Mediterranean drainage, and several other 'suspicious' cases. Their evolutionary significance is not well understood at present; some of these lineages seem to be evolutionary dead-ends, while in others evidence suggests that they may be transitory links in speciation processes. Apart from their discovery, the study of such asexual vertebrates includes identification of parental genomes using various molecular markers, experimental analyses of modes of reproduction, population dynamics analyses, that are usually very important for the programmes for conservation of endangered taxa (e.g., Alves *et al.*, 2001; Janko *et al.*, 2003). Cytogenetics is a an inseparable part of such research, especially because of very frequent occurrence of polyploidy in hybrid complexes.

Squalius Minnows (Cyprinidae)

Speciose and evolutionary diversified Eurasian leuciscine cyprinids, though not still well defined phylogenetically (Howes, 1991; Gilles *et al.*, 1998; Hänfling and Brandl, 2000) have remarkably and highly conserved karyotypes in terms of nearly invariable diploid chromosome number (2n = 50), karyotype structure with most frequently 8 pairs of metacentrics (m), 12 - 15 pairs of submetacentrics (sm) and 5 – 2 pairs of subtelo- (st) to acrocentrics (a), and NOR phenotype where NORs are typically situated at the shorter arms of one st chromosome pair (Ráb and Collares-Pereira, 1995; Rábová *et al.*, 2003; Bianco *et al.*, 2004). Such a karyotype is also characterized by the largest chromosome pair of the complement that is st – a 'leuciscine' cytotaxonomic marker. Its presence was noted by Vasil'ev (1985) and then repeatedly confirmed in all other subsequent studies dealing with the chromosomes of Eurasian cyprinids (Ráb and Collares-Pereira, 1995; Bianco *et al.*, 2004 and references therein) (Fig. 2.1.2).

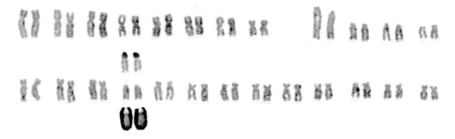

Fig. 2.1.2 Karyotype of male of *Leuciscus albus* (Trasimeno Lake, Italy) arranged from conventionally Giemsa-stained chromosomes; the pair of chromosomes carrying the NORs are also showed after sequential Ag- (above) and CMA3-staining (below).

Exceptions to this pattern include European representatives of the phoxinin lineage of leuciscins *Phoxinus phoxinus* and *Eupallasella perenurus* (Boroń, 2001), and no doubt a number of other leuciscine cyprinids from both inside and outside Europe, e.g., *Acanthobrama marmid* (Gaffaroglu, 2003) with NORs present in two pairs of chromosomes, or *Aspius aspius* with the position of NORs resulting probably from paracentric inversion (Ráb *et al.*, 1990). The genus *Chondrostoma* appears to include species with more differentiated NOR phenotypes while retaining the common leuciscine karyotype. *Chondrostoma lusitanicum* is characterized by a structural rearrangement of the translocation type, resulting in polymorphism of a number of NOR sites (Collares-Pereira and Ráb, 1999; Rodriguez and Collares-Pereira, 1999). Recently, Ganthe *et al.* (2004) described a double NOR phenotype in *C. macrolepidotum*, where both the sites are situated pericentromerically on one chromosome pair. This indicates a small inversion of a part of the original, simple NOR site around the centromere. Interestingly, Kalous *et al.* (2006) have discovered a rather complex NOR phenotype in the karyotype of the highly endangered Iberian species *C. arrigonis*. The nature of this is presently unknown and requires an in-depth investigation using molecular tools.

Accumulated data suggest that our present opinion about the conservative character of leuciscine karyotypes will be modified to some extent in the near future (wide discrepancy between localization of major ribosomal genes using the conventional and molecular methods described by Gromicho *et al.*, 2005; Gromicho and Collares-Pereira, Chapter 1.3, present volume). Such a conservative character of the cytogenetic/ cytotaxonomic data set is not effectively useful in identification of applicable conservation units.

However, one exception includes the hybrid polyploid complex of the Iberian minnow *Squalius alburnoides* . This asexual vertebrate form arose by way of multiple hybridization events between Iberian cyprinids S. *caroliterii* or *S. pyrenaicus* with a still-unidentified and unknown ancestor. The latter is most probably closely related to the endemic genus *Anaecypris*, where the only extant species of the genus is one of the most endangered European freshwater fishes (Collares-Pereira *et al.*, 2002) and the conservation significance of cytogenetics is obvious. An extensive review of the evolutionary role of hybridization and polyploidy in the endemic Iberian *Squalius ´alburnoides´* complex is provided by Gromicho and Collares-Pereira (Chapter 1.3 present volume).

Cobitis Loaches (Cobitidae)

Spiny loaches of the genus *Cobitis* (as recently recognized) apparently represent a polyphyletic, speciose assemblage (Nalbant *et al.*, 2001; Perdices and Doadrio, 2001). They are small, bottom-dwelling fishes extensively distributed in the freshwaters of Eurasia. Their karyotypes are highly evolutionarily diversified in terms of karyotype structures and other chromosomal markers. But they do retain almost the same diploid chromosome number 2n = 50 (reviewed in Ráb and Slavík, 1996). The exceptions include several species where 2n was decreased by chromosomal rearrangements (e.g., the European C. *taenia* and Asian C. *takatsuensis*, and C. *´sinensis´*) or increased via polyploidization events (large race of C. *´biwae´* and C. *´striata´* in Japan). The remarkable diversity of their karyotypes undoubtedly reflects different rates of chromosome change in particular lineages. In one recently recognized subgenus *Bicanestrinia* (in fact, there are four phylogenetic lineages, according to Bohlen *et al.*, 2006), the karyotype and NOR phenotype appears rather uniform (Rábová *et al.*, 2001), while in the subgenus *Cobitis* (or *Cobitis* s. str.), these cytotaxonomic markers are highly variable. This primarily western Palearctic lineage (Perdices *et al.*, 2004) is characterized by an invariable 2n = 50, except for 2n = 48 in C. *taenia* and karyotypes variously include from uni-armed to bi-armed chromosomes. In several European species of this lineage such as C. *vardarensis* (Rábová *et al.*, 2001), C. *taenia* (Boroń, 1999, 2003; Ráb *et al.*, 2000; Boroń *et al.*, 2004 and references therein), C. *elongatoides* (Ráb *et al.*, 2000), C. *tanaitica* and C. *taurica* (Rábová *et al.*, 2004; Janko *et al.*, 2005), an extensive structural polymorphism of major ribosomal genes sites of the presence/absence type was detected by means of conventional and/or molecular cytogenetic techniques. The nature of such unusual polymorphism is not understood well at present, but the evolutionary hybrid origin of these taxa might be hypothesized. Except for *C. vardarensis*, all of the other four species have been identified as parental species of hybrid diploid-polyploid complexes (Šlechtová *et al.*, 2000; Janko *et al.*, 2003, 2005 and references therein). Cytogenetics was and still is extensively used as a tool in discovering the enormous hybrid diversity of these loaches across Europe (Sofradzija and Berberovic, 1978; Vasil'ev and Vasil'eva, 1982; Boroń, 1992; reviewed by Bohlen and Ráb, 2001). The evolutionary scenario reconstructed by means of phylogenetic analyses of mtDNA and some nuclear markers suggested that during several of the last interglaciation cycles, all four species came into reproductive contact several times, resulting in reciprocal and polyphyletic origins of nearly—all female hybrids with a gynogenetic mode of reproduction (Janko *et al.*, 2003, 2005). Such hybrids co-occur with one of the parental species because they need bisexually reproducing diploid males for reproduction and act thus as sperm-dependent parasites. The review of hybrid diversity across Europe (Bohlen and Ráb, 2001; Boroń, 2003; Janko *et al.*, 2003) has documented the presence of hybrids with nearly all possible genome combinations (*C. taenia* = TT, *C. elongatoides* = EE, *C. tanaitica* = NN, *C. taurica* = CC) in diploid (TE, TN, EN, NC, TC), triploid (e.g., TTE, TEE, EEN, ENN, TTN, TNN, NNC, TTC, TCC, NCC, rarely also three-genome TEN or TEC) and tetraploid hybrids (combinations not shown). The karyotypes of all four species differ remarkably: *C. taenia* (2n = 48, karyotype pairs of m, 10 pairs of sm and 9 pairs of st to a chromosomes where one pair of m is characteristically larger than others) (Fig. 2.1.3a), *C. elongatoides* (2n = 50, 15 m+ 8 sm+1 st+1 a) (Fig. 2.1.3b), *C. tanaitica* (2n = 50, 6 m+13 sm+6 st-a) (Fig. 2.1.3c), *C. taurica* (2n = 50, 5 m+15 sm+4 st+1 a, where two pairs of m are distinctly larger) (Fig. 2.1.3d).

These karyotype differences, especially the ratio of m and st-a chromosomes, in combination with chromosome numbers, can be used practically for the unambiguous assessment of the genome compositions in diploid, triploid and tetraploid asexual hybrids. They even enable karyological dissection of haploid genomes within three-genome hybrids TEN or TEC (Fig. 2.1.4). Such an approach in the identification of various hybrid genomes, combined with the parallel application of

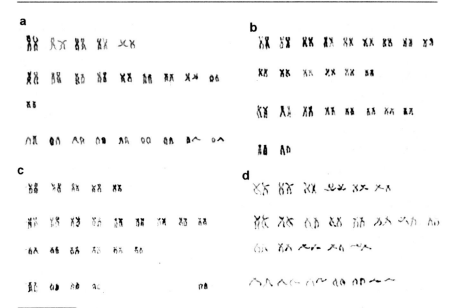

Fig. 2.1.3 Karyotype of male of *C. taenia* (TT) (lower Bug R., Ukraine) (a); female of *C. elongatoides* (EE) (upper Vltava R., Czech Republic) (b) male of *C. tanaitica* (NN) (lower Bug R., Ukraine) (c) and male of *C. taurica* (CC) (Black R., Crymea Peninsula) (d) all arranged from conventionally Giemsa-stained chromosomes.

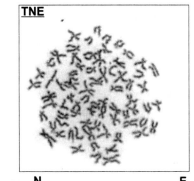

Fig. 2.1.4 Metaphase cell of triploid (TNE) hybrid female of *Cobitis* (Loire R., France) containing haploid genomes of *C. taenia* (T), *C. tanaitica* (N) and *C. elongatoides* (E).

mtDNA and nuclear markers in specimens under analysis, has actually led to the discovery of an even higher genetic diversity of hybrids. This is because the hybrids of the same genome combinations can possess mtDNA markers from different parental species due to the reciprocal origin of hybrids (Janko *et al.*, 2003).

Loaches of the genus *Cobitis* are protected in almost all European countries and are listed in varying degrees of endangered status. However, the present legislation and conservation texts in Europe do not reflect the enormous observed hybrid and ploidy diversity between the four parental loach species (as briefly reviewed above) nor the species diversity of *Cobitis* loaches occurring mainly in the Mediterranean drainage. Depth analysis by Bohlen and Ráb (2001) revealed the existing, and in fact paradoxical, situation where the species C. *taenia* is stringently protected in many European countries, although this fairly well-identifiable loach species may not even occur in this particular country. They also documented the fact that formal legislative acts do not follow the recent discoveries of *Cobitis* loach diversity in Europe. A very important problem arising from their analysis is how to treat hybrid biotypes in conservation and legislation. They are not the usual F1 hybrids, but as sperm-dependent hybrids reproductively dependent on one of the parental species. They persist over long time-scales; each locality contains its own unique combination of hybrid biotypes, and they pose a formal nomeclatural problem. In conclusion, European hybrid complexes of loaches should be treated and deserve protection as components of existing biodiversity just as other asexual vertebrates (e.g., Kraus, 1995). Cytogenetics, therefore, continues to play an important and critical role in the discovery, screening, evaluation and understanding of diversity in endangered fishes in Europe.

Misgurnus Weatherfishes (Cobitidae)

Species diversity in the cobitid genus *Misgurnus*, with its disjunct distribution in Europe and East Asia, is presently not well known. More species are bound to be discovered during faunal and taxonomic surveys (Vasil'eva, 2001). In Europe, only M. *fossilis* is considered to be an endangered or critically endangered species in many national and European legislative acts. The main reason for the dramatic decline of this fish species is undoubtedly habitat alteration. Weatherfish occur specifically in riverine sidearms, a habitat that disappears successively in any river management operation. The representatives of the genus exhibit diploid-polyploid relationships. Chromosome studies of the Asian species M. *anguillicaudatus* (in fact a 'catch-all' taxon that includes several species – Kottelat, 2001) revealed the presence of diploid, triploid and tetraploid specimens (Hitotsumachi *et al.*, 1969; Ojima and Takai, 1979; Li *et al.*, 1983; Lee *et al.*, 1987; Yu *et al.*, 1989; Arai *et al.*, 1991), while M. *mizolepis* contained diploid individuals only (Ueno *et al.*, 1985; Lee *et al.*, 1987). The diploid chromosome number of the European species M. *fossilis* remains unclear and available reports provide conflicting data. Post (1965) described a haploid number n = 24 for specimens from the lower Elbe area (northern Germany), and very similar to the 2n = 50 found by Vujošević *et al.* (1983) in individuals from Danube River in Serbia. On the other hand, Raicu and Taisescu (1972) (lower Danube River), Ene and Suciu (2000) (Danube Delta area) and Boron (2000) (Vistula and Bug river areas) exclusively found fishes with 2n = 100. Our unpublished data from the upper Danube River area in the Czech republic also showed fishes exclusively with 2n = 100 (Fig. 2.1.5). Although still insufficiently surveyed across Europe, these data indicate the presence of specimens with at least two different ploidy levels. This is a strong

Fig. 2.1.5 Metaphase cell of juvenile *Misgurnus fossilis* (Dyje R., middle Danube R.) with 100 chromosomes; evidence of a tetraploid form of European weatherfish.

indication of some type of hybrid asexually complex, and thus the likelihood for the existence of even more biological species than are currently recognized. Such a hypothesis is highlighted by the recent discovery of a cryptic clonal lineage in *M. anguillicaudatus* in Japan (Morishima *et al.*, 2002), indicating a hybridization event known for asexual vertebrates. The weatherfish case again documents the fact that cytogenetics can play a very important role in discovering actual fish diversity in Europe.

'Suspicious' Cases

The obvious lack of application of cytogenetics in biodiversity research and conservation programmes results in poor or only limited knowledge about the actual number and distribution of hybrid asexual complexes in fishes. Hence, the actual number of biological species that can hybridize with 'asexual consequence' is underestimated. As stated above, such hybrid asexual complexes are cryptic in nature and genetic methods have to be used for their disclosure. One of the first signs of such a complex is the apparent prevalence of females, together with the occurrence of triploid individuals in the population. Cytogenetic and/or flow cytometry screening can easily discover presence of such individuals (Collares-Pereira *et al.*, 1995). However, spontaneous triploidy arising via failure of second body extrusion during homospecific fertilization has long been considered as the mechanism responsible for triploidy (Benfey, 1989). Undoubtedly, a number of such triploid individuals can be found in populations. Collares-Pereira *et al.* (1995) have described a single triploid specimen of balitorid stone loach *Barbatula barbatula* from upper Vistula basin, Slovakia. Subsequently, a large survey of nearly 200 specimens of stone loach across water basins in Central Europe (unpublished results) did not reveal any other triploid specimens, suggesting an autopolyploid origin in the above specimen. Karyotype dissection of the triploid complement can really suggest such a conclusion, but in some groups like leuciscine cyprinids, the karyotypes of hybridizing taxa are so similar to each other that without application of other genetic methods, it is difficult to decide whether triploids arose via homospecific or heterospecific fertilization (= interspecific hybridization). However, allotriploidy (=at least two very similar genomes from different species) should be seriously considered (Fig. 2.1.6).

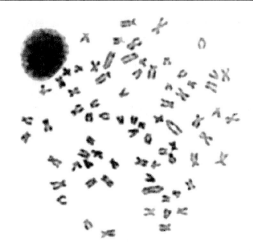

Fig. 2.1.6 Triploid metaphase cell 3n = 75 (above) and corresponding karyotype of *Leuciscus* aff. *leuciscus* (lower Dyje R., middle Danube R. basin) arranged from conventionally Giemsa-stained chromosomes. The three haploid genomes have the same karyotype structure, which prevents the assessment the auto- or allopolyploid origin.

Invasive and/or Exotic Fish Species—A Recent Problem in Fish Conservation

Carassius carps (Cyprinidae)

The genus *Carassius* includes the single autochthonous European species *C. carassius* and a still unknown number of Asian species. Although five species are currently recognized in Japan (Hosoya, 2000), on the Asian continent only the nominal species *C. auratus*, with some subspecies such as *C. a. gibelio*, is listed. Several biological species are obviously present. Such higher species diversity is indirectly confirmed by the fact that, as in Japan (Murukami *et al.*, 2001 and references therein), some hybrid asexual complexes do occur in China, as suggested by extensive genetic surveys of Chinese populations (e.g., Yang *et al.*, 2001). These species result from interspecific hybridizations associated with polyploidy. The populations of the Asian *Carassius*, therefore, contain diploid (2n = 100), triploid (3n ~ 150) (Fig. 2.1.7) and also (but rarely) tetraploid (4n ~ 200) individuals. Due to a number of both accidental and incidental introductions into Europe at the beginning of 1950s through the former USSR and Yugoslavia (Boroń, 1994; Kalous, 2005 and references therein), several still unidentified lineages of Asian origin have been treated taxonomically as *C. gibelio* (but also consult Kalous *et al.*, 2004). These species established themselves in European freshwaters (Kalous and Šlechtová, 2004). This exotic fish appeared to be a very successful competitor and harmful pest (Richardson *et al.*, 1995) to the autochthonous fish faunas, characterized by increasing population abundance and expanding distribution ranges (e.g., Holčík, 1980; Slavík and Bartoš, 2004 and references therein). Although the main reason for its expansion is undoubtedly accidental stocking in the aquaculture industry, its specific mode of reproduction is also a very important factor (reviewed in Várkonyi and Tóth, Chapter 2.2 present volume).

It is evident that in order to understand the apparently very complex biology of reproduction of this non-native, rapidly expanding pest fish, and to cope with its conservation problem, inclusion of cytogenetic screening of Prussian carp populations into research programmes is necessary.

Sturgeons and Their Natural Hybrids— Another Important Conservation Issue

Acipenseriform fishes (sturgeons, paddlefishes and shovelnoses of the genera *Acipenser, Huso, Scaphirhynchus, Pseudoscapirhynchus, Polyodon* and

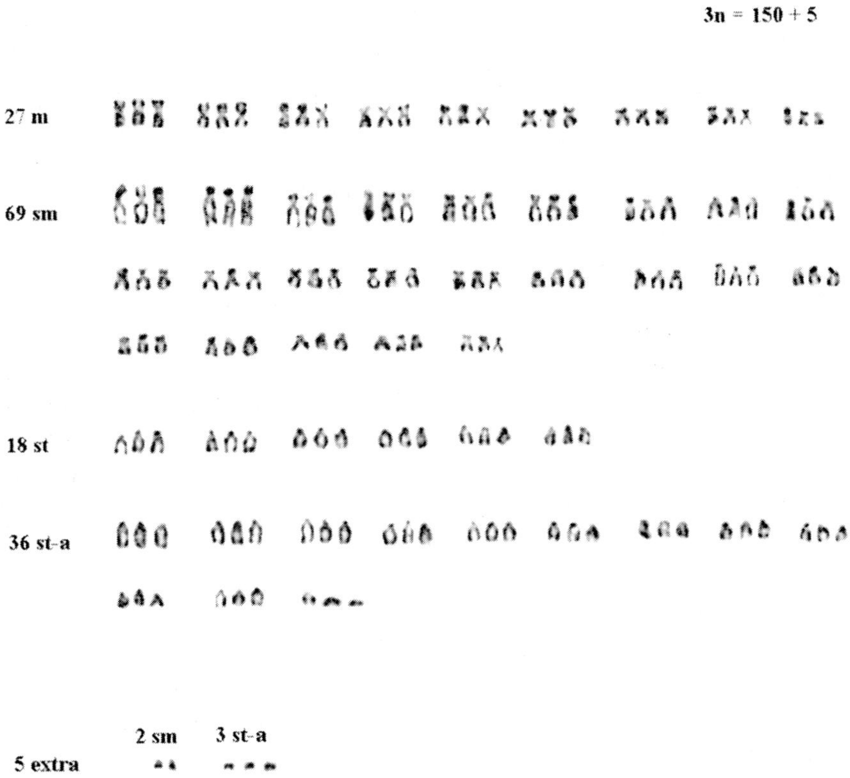

$3n = 150 + 5$

27 m

69 sm

18 st

36 st-a

2 sm   3 st-a

5 extra

Fig. 2.1.7 Karyotype of triploid female of *Carassius gibelio* arranged from conventionally Giemsa-stained chromosomes showing the 150 + 5 chromosome frequently occurring form.

Psephurus) exhibit classical polyploid interrelationships in terms of different evolutionary ploidy levels (Birstein *et al.*, 1997; Ludwig *et al.*, 2001 and references therein). A review of cytogenetic and molecular approaches to the study of polyploidy in Acipenseriformes is provided by Fontana *et al.* (Chapter 4.2 present volume). The species of these genera can be separated into three different ploidy classes: evolutionary tetraploid taxa (4n) with approximately 120 chromosomes (representatives of all extant genera); evolutionary octaploid taxa (8n) (*Acipenser baerii*, *A. gueldenstaedti*, *A. dabryanus*, *A. fulvescens*, *A. medirostris*, *A. naccari*, *A. persicus*, *A. schrenckii*, *A. sinensis*, *A. transmontanus*) with approximately 240 chromosomes; and two species (the Pacific sturgeons *A. mikadoi* and *A. brevirostrum*) with nuclear DNA content corresponding to hexadecaploid level (16n) (Birstein *et al.*, 1993; Blacklidge and Bidwell, 1993), and with a hypothetical chromosome number around 500. However, a recent karyotype study of the shortnose sturgeon *A. brevirostrum* (Kim *et al.*, 2005) has revealed that the chromosome number is around 360, i.e., dodecaploid level. Acipenserid fishes are notoriously known for their interspecific hybridizations. Birstein *et al.* (1997) have listed more than 10 various hybrid interspecfic combinations in nature, and even between species with different ploidy levels. Using the process of flow cytometry, Flajšhans and Vajcová (2000) revealed (in a single hatchery stock presumably identifiable as *A. gueldenstaedti*) hybrids of evolutionary tetraploid *A. ruthenus* (4n, ~ 120 chromosomes) and octaploid *A. gueldenstaedti* (8n, ~ 240 chromosomes) with intermediate ploidy corresponding to the hexaploid level (4n × 8n = 6n, ~ 180 chromosomes), but surprisingly also hybrids of such an intermediate hybrid with both parental species with pentaploid (4n × 6n = 5n, ~ 150 chromosomes) and heptaploid (8n × 6n = 7n, ~ 210) ploidy levels. They further hypothesized that such odd-ploidy specimens arose due to hybridization of hybrids with intermediate hexaploid level with either tetraploid or octaploid specimens. This discovery has stimulated large-scale screenings (using FC profiling, combined with direct chromosome examination in expected but unusual ploidy levels) of hatchery sturgeon populations across several European countries, and has resulted in the finding of specimens with ploidy level 9n (~ 270 chromosomes), 10n (~ 300 chromosomes), 12n (~ 360 chromosomes) (Fig. 2.1.8), 13n (~ 390 chromosome). A specimen with an FC profile corresponding to 14n (~ 420 chromosomes) was, unfortunately, lost during the catastrophic flood in summer 2002, so that such a high ploidy level could not be verified by direct chromosome examination (Ráb *et al.*, 2004). To summarize the data, ploidy levels 10n ~ 300, 11n ~ 330, 15n ~ 450 and 16n ~ 480 have not been recorded as yet, but will very likely be discovered in the near future. These recent findings strongly suggest mutual fertility of various hybrids irrespective of ploidy level in sturgeons. Remarkable and apparent multiples of about 30 chromosomes might suggest that these semi-haploid units (n = 60, n/2 = 30) still reflect the original haploid genome of the extinct sturgeon diploid ancestor similar to gar of the genus *Lepisosteus* (Ráb *et al.*, 1999). Moreover, they reinforce the hypothesis of Vasil'ev (1999) about the hybrid origin of sturgeons through the reticulate mode of speciation.

Fig. 2.1.8 Metaphase cell of a sturgeon individual identified as *A. gueldenstaedti* with approximately 360 chromosomes corresponding to dodekaploid (12n) ploidy level.

The numbers of acipenserid fishes are rapidly diminishing and becoming highly endangered, due both to massive habitat destruction, and to legal and illegal overfishing (Birstein, 1993). They are listed as a CITES I category. The long-term keeping and breeding hatchery stocks of acipenserid species, i.e., aquaculture, with subsequent stockings, has been identified as the most effective recent measure in their conservation (Bronzi *et al.*, 1999). Enormous ploidy diversity among sturgeons and their easy hybridization, irrespective of the ploidy level, is a great challenge for large-scale hatchery/stocking practices. The need for cytogenetics methods is obvious.

Cytogenetics in Fish Diversity Research and Fish Conservation

Due to the undoubted informative value of cytogenetics, one could try to understand why cytogenetic methods are infrequently applied to fish diversity research and whether their absence from the toolbox of investigative methods is due to misunderstanding. The methodology of cytogenetics is based on the preparation of chromosomes from living, dividing cells. These are arrested by mitostatic agents in a suitable stage of the cell division, fixed and spread onto microscope slides. The chromosomes are then studied after either conventional staining techniques or molecular cytogenetic procedures. The more distant phylogenetically the vertebrate group is from humans, the more difficult is the cytogenetic approach. The reasons are obviously historically related to different rates of research intensity in the various groups of organisms, and are dependent on a number of factors such as the number of scientists and laboratories involved, the possibilities of raising funds, technical problems in the modification of cytogenetic protocols and the characteristics of the chromosomes in the particular group. Fishes, the most speciose and phylogenetically diverse group of vertebrates, are no exception. Compared to mammals, including humans, they sometimes have very small and numerous chromosomes. They exhibit seasonal cycles in cell activity, which sometimes prevents the discovery of any dividing cells. Chromosomes must be prepared using either direct *in vivo* methods, with dissection of actively dividing tissues on anaesthetized specimens, or appropriate cell culture methods. Some of the banding methods such as G-banding used mainly in mammals usually fail to produce reproducible results in fish species. Moreover, the logistical problems in sampling are sometimes enormous, especially in remote regions such as tropical or polar areas. Fish cytogenetics has, therefore, unjustly received a reputation of being difficult and not too rewarding, with the result that relatively few research laboratories dealing with fish are involved in cytogenetic studies, the chromosomal level of fish genome organization remaining ignored.

Instead, as explained by the examples provided, cytogenetics is a unique tool in collection of information that can rarely or never be collected by other genetic methods. Although cytogenetics was theoretically considered to be a powerful tool in the elucidation of chromosomal speciation, presently fish cytogenetics can provide some suited examples (cytogenetic differentiation within populations and between species found in e.g., *Gobio, Romanogobio, Cobitis, Neogobius* may represent beginning of speciation, sorting of 'forms' according to habitat) to study these predictions in the field. Another important problem deals with chromosomal irregularities because the present lack of knowledge does not allow discriminating between natural cytogenetic variability, natural rate and/or effects of such abnormalities caused by cytotoxic effects of aquatic habitats.

The degree of variability observed in cytogenetically unusual fishes (e.g., reproductive modes in hybrid asexual complexes; triploid, bisexually reproducing males in *Carassius*; extensive hybrid ploidy diversity in sturgeons) exceeds the theoretically predicted degree of complexity. This indicates a much more widespread and surprising source of biological diversity than formerly believed, which must therefore be considered in conservation. We predict that in the near future, cytogenetic studies in European freshwater fishes (as in other continents) will reveal more exciting examples, exceptions and biological discoveries than have been recently understood.

Moreover, at present, many of the technical problems that really hampered the application potential of fish cytogenetics have been overcome. More standardized chromosome preparation and cultivation procedures have enabled the preparation of chromosomes from practically all fish species by means of non-invasive approach. The use of karyotyping software and digital imaging has significantly shortened the time for chromosome analyses, and large population screenings are now possible. The large-scale application of flow cytometry to fishes and the significant recent advent of molecular cytogenetic methods have made cytogenetics a totally competent tool in fish biodiversity research.

Acknowledgements

This study was supported by projects: IRP IAPG No. AV0Z50450515 (J.B., M.R.), USB RIFCH No. MSM6007665809 (P.R., M.F.) and CAU No. MSM6046070901 (L.K.).

References

Alves, M.J., M.M. Coelho and M.J. Collares-Pereira. 2001. Evolution in action through hybridisation and polyploidy in an Iberian freshwater fish: A genetic review. *Genetica* 111: 375–385.

Arai, K., K. Matsubara and R. Suzuki. 1991. Karyotype and erythrocyte size of spontaneous tetraploidy and triploidy in the loach *Misgurnus anguillicaudatus*. *Bulletin of the Japanese Society of Scientific Fisheries* 57: 2167–2172.

Benfey, T.J. 1989. A bibliography of triploid fish, 1943–1988. *Canadian Technical Reports of Fisheries and Aquatic Sciences* 1682: 1–33.

Bianco, P.G., G. Aprea, E. Balletto, T. Capriglione, D. Fulgione and G. Odierna. 2004. The karyology of the cyprinid genera *Scardinius* and *Rutilus* in southern Europe. *Ichthyological Research* 51: 274–278.

Birstein, V.J. 1993. Sturegons and paddlefishes: Threatened fishes in need of conservation. *Conservation Biology* 7: 773–787.

Birstein, V.J., A.I. Poletaev and B.F. Goncharov. 1993. DNA content in Eurasian sturgeon species determined by flow cytometry. *Cytometry* 14: 377–383.

Birstein, V.J., R. Hanner and R. DeSalle. 1997. Phylogeny of the Acipenseriformes: Cytogenetic and molecular approaches. *Environmental Biology of Fishes* 48: 127–156.

Blacklidge, K.H. and C.A. Bidwell. 1993. Three ploidy levels indicated by genome quantification in Acipenseriformes of North America. *Journal of Heredity* 84: 427–430.

Bohlen, J. and P. Ráb. 2001. Species and hybrid richness in spined loaches of genus *Cobitis* (Teleostei: Cobitidae) with a checklist of European forms and suggestions for conservation. *Journal of Fish Biology* 59(A): 75–89.

Bohlen, J., A. Perdices, I. Doadrio and P.S. Economidis. 2006. Vicariance colonisation and fast local speciation in Asia Minor and the Balkans as revealed from the phylogeny of spined loaches (Cobitidae). *Molecular Phylogenetics and Evolution* 39: 552–561.

Boroń, A. 1994. Karyotypes of diploid and triploid silver crucian carp *Carassius auratus gibelio* (Bloch). *Cytobios* 80: 117–124.

Boroń, A. 1999. Banded karyotype of spined loach *Cobitis taenia* and triploid *Cobitis* from Poland. *Genetica* 105: 293–300.

Boroń, A. 2000. Cytogenetic characterization of the loaches of the genera *Sabanejewia*, *Misgurnus* and *Barbatula* (Pisces, Cobitidae). *Folia Zoologica* 49 (Supplement 1): 37–44.

Boroń, A. 2001. Comparative chromosomal studies on two fish, *Phoxinus phoxinus* (Linnaeus, 1758) and *Eupallasella perenurus* (Pallas, 1814): An associated cytogenetic-taxonomic considerations. *Genetica* 111: 387–395.

Boroń, A. 2003. Karyotypes and cytogenetic diversity of the genus *Cobitis* (Pisces, cobitidae) in Poland: A review. Cytogenetic evidence for a hybrid origin of some *Cobitis* triploids. *Folia Biologica-Krakow* 51(Supplement S): 49–54.

Boroń, A., C. Ozouf-Costaz, J.P. Coutanceau and K. Woroniecka. 2004. FISH mapping of 28S and 5S ribosomal genes in pure and diploid-tetraploid Polish populations of the spined loach *Cobitis taenia* (Pisces, Cobitidae). *Cytogenetic and Genome Research* 106: 12.

Bronzi, P., H. Rosenthal, G. Arlati, and P. Williot. 1999. A brief overview on the status and prospects of sturgeon farming in Western and Central Europe. In: *Proceedings of the 3rd International Symposium on sturgeon. Journal of Applied Ichthyology* 15: 224–227.

Collares-Pereira, M. J. and P. Ráb. 1999. NOR polymorphism in the Iberian species *Chondrostoma lusitanicum* (Pisces: Cyprinidae) Re-examination by FISH. *Genetica* 105: 301–303.

Collares-Pereira, M.J., J. Madeira, and P. Ráb. 1995. Spontaneous triploidy in the stone loach *Noemacheilus barbatulus* (Balitoridae). *Copeia* 1995: 483–484.

Collares-Pereira, M.J., I.G. Cowx, J.A. Rodrigues and I. Rogado. 2002. A conservation strategy for *Anaecypris hispanica*: A picture of LIFE for a highly endangered Iberian fish. In: *Conservation of Freshwater Fishes: Options for the Future*, M.J. Collares-Pereira, M.M. Coelho and I.G. Cowx (eds.), Fishing News Books, Blackwell Science, Oxford, pp. 186–197.

Ene, C. and R. Suciu. 2000. Chromosome study of *Misgurnus fossilis* from Danube Delta Biosphere Reserve, Romania. *Folia Zoologica* 49 (Supplement 1): 91–95.

Flajšhans, M. and V. Vajcová. 2000. Odd ploidy levels in sturgeons suggest a backcross of interspecific hexaploid sturgeon hybrids to evolutionarily tetraploid and/or octaploid parental species. *Folia Zoologica* 49 (Supplement 1): 133–138.

Gaffaroglu, M. 2003. *The Karyological Analysis of Some Species from Cyprinidae in Karakaya Dam Lake*. Ph.D. Thesis. Inonu University, Turkey.

Ganthe, H.F., M.J. Collares-Pereira and M.M. Coelho. 2004. Introgressive hybridisation between two Iberian *Chondrostoma* species (Teleostei, Cyprinidae) revisited: New evidence from morphology, mitochondrial DNA, allozymes and NOR-phenotypes. *Folia Zoologica* 53: 423–432.

Gilles, A., G. Lecointre, E. Faure, R. Chappaz and G. Brun. 1998. Mitochondrial phylogeny of the European cyprinids: Implications for their systematics, reticulate evolution and colonization time. *Molecular Phylogenetics and Evolution* 10: 132–143.

Gromicho, M., C. Ozouf-Costaz and M.J. Collares-Pereira. 2005. Lack of correspondence between CMA3-, Ag-positive signals and 28S rDNA loci in two Iberian minnows (Teleostei, Cyprinidae) evidenced by sequential banding. *Cytogenetic and Genome Research* 109: 507–511.

Hänfling, B. and R. Brandl. 2000. Phylogenetics of European cyprinids: insights from allozymes. *Journal of Fish Biology* 57: 265–276.

Hitotsumachi, S. M., M. Sasaki and Y. Ojima. 1969. A comparative study in several species of Japanese loaches (Pisces, Cobitidae). *Japanese Journal of Genetics* 44: 157–161.

Holčík, J. 1980. Possible Reason for the Expansion of *Carassius auratus* (Linnaeuus, 1758) (Teleostei, Cyprinidae) in the Danube River Basin. *Internationale Revue der Gesamten Hydrobiologie* 65: 673–679.

Hosoya, K. 2000. Cyprinidae. In: *Fishes of Japan with Pictorial Keys to the Species*. T. Nakabo, (ed.), 2nd Edition. Tokai University Press, Tokyo.

Howes, G.J. 1991. Sytematics and biogeography: An overview. In: *Cyprinid Fishes. Systematics, Biology and Exploitation*, I. Winfield and J. Nelson (eds.). Chapman and Hall, London, pp. 1–33.

Janko, K., P. Kotlik and P. Ráb. 2003. Evolutionary history of asexual hybrid loaches (*Cobitis*: Teleostei) inferred from phylogenetic analysis of mitochondrial DNA variation. *Journal of Evolutionary Biology* 16: 1280–1287.

Janko, K., P. Kotlík, M.A. Culling and P. Ráb. 2005. Ice age cloning – comparison of Quaternary evolutionary histories of sexual and clonal forms of European loaches (*Cobitis*: Teleostei) using the analysis of mitochondrial DNA variation. *Molecular Ecology* 14: 2991–3004.

Kalous, L. 2005. *Contribution to Revision of Carassius auratus Complex in the Czech Republic*. Ph.D. Thesis. Czech Agricultural University, Prague.

Kalous, L. and V. Šlechtová. 2004. *Carassius gibelio* autochthonous or exotic species in Europe: Molecular phylogenetic evidence. ECI XI, Tallin 2004, Abstract Book: 122.

Kalous, L., J. Bohlen and P. Ráb. 2004. What fish is *Carassius gibelio*: Taxonomic and nomenclatoric notes. ECI XI, Tallin 2004, Abstract Book: 26–27.

Kalous, L., I. Doadrio, M. Rábová and P. Ráb. 2006. Note on the banded karyotype of cyprinid fish *Chrondrostoma arrigonis*. Cybium. (In press).

Kim, D.S., Y.K. Nam, J.K Noh, C.H. Park and F.A. Chapman. 2005. Karyotype of North American shortnose sturgeon *Acipenser brevirostrum* with the highest chromosome number in the Acipenseriformes. *Ichthyological Research* 52: 94–97.

Kottelat, M. 1997. European freshwater fishes. *Biologia (Bratislava)* 52/Supplement Cybium 5: 1–271.

Kottelat, M., 2001. Freshwater Fishes of Northern Vietnam. A preliminary check-list of the fishes known or expected to occur in northern Vietnam with comments on systematics and nomenclature. *Environment and Social Development Sector Unit East Asia and Pacific Region*, The World Bank, 122 pp.

Kraus, F. 1995. The conservation of unisexual vertebrate populations. *Conservation Biology* 9: 956–959.

Lee, G.Y., S.I. Jang, Y.N. Oh, S.J. Lee and J.N. So. 1987. Nucleolus organizer chromosomes in a cobitid fishes of *Misgurnus mizolepis* and *M. anguillicaudatus*. *Korean Journal of Limnology* 20: 171–176.

Li, K., Y. LI and D. Zhou. 1983. A comparative study of the karyotypes in two species of mud loaches. *Zoological Research* 4: 75–81.

Ludwig, A., N.M Belfiore, Ch. Pitra, V. Svirksy and I. Jenneckens. 2001. Genome duplications events and functional reduction of ploidy levels in Sturgeon (*Asipenser, Huso* and *Scaphirhynchus*). *Genetics* 158: 1203–1215.

Morishima, K., S. Horie, E. Yamaha and K. Arai. 2002. A cryptic clonal line of the loach *Misgurnus anguillicaudatus* (Teleostei: Cobitidae) evidenced by induced gynogenesis, interspecific hybridization, microsatellite genotyping and multilocus DNA fingerprinting. *Zoological Science* 19: 565–575.

Murakami, M., C. Matsuba and H. Fujitani. 2001. The maternal origins of the triploid ginbuna (*Carassius auratus langsdorfi*): phylogenetic relationships within the *C. auratus* taxa by partial mitochondrial D-loop sequencing. *Genes Genetic Systematics* 76: 25–32.

Nalbant, T.T., P. Ráb, J. Bohlen and K. Saitoh. 2001. Evolutionary success of the loaches of the genus *Cobitis* (Pisces: Ostariophysi: Cobitidae). *Travaux du Museum National d´Histoire Naturelle Grigore Antipa* 43: 277–289

Ojima, Y. and A. Takai. 1979. The occurrence of polyploidy in the Japanese common loach, *Misgurnus anguillicaudatus*. *Proceedings of the Japan Academy Series B*, 58: 56–59.

Park, L.K. and P. Moran. 1994. Developments in molecular techniques in fisheries. *Reviews in Fish Biology and Fisheries* 4: 272–299.

Perdices, A. and I. Doadrio. 2001. The molecular systematics and biogeography of the European Cobitids based on mitochondrial DNA sequences. *Molecular Phylogenetics and Evolution* 19: 468–478.

Perdices, A., J. Bohlen and I. Doadrio. 2004. Double origin of the European spined loaches of the genus *Cobitis* (Teleostei: Cobitidae) based on molecular data. In: ECI XI, Tallin 2004, Abstract Book: 124.

Post, A. 1965. Vergleichende untersuchungen der chromosomenzahlen bei süßwasser-teleosteem. *Zeitschrift fur Zoologische Systematik und Evolutions-forschung* 3:47–93.

Ráb, P. and M.J. Collares-Pereira. 1995. Chromosomes of European cyprinid fishes (Cyprinidae, Cypriniformes): A review. *Folia Zoologica* 44: 193–214.

Ráb, P. and J. Bohlen. 2000. A new problem in fish conservation: impact of polyploid and/or hybrid lineages. *International Symposium on Freshwater Fish Conservation: Options for Future*. Montechoro, Portugal, Abstract Book: 76.

Ráb, P. and O. Slavík. 1996. Diploid-triploid-tetraploid complex of the spined loach, genus *Cobitis* in Pšovka Creek: The first evidence of the new species of *Cobitis* in the ichthyofauna of Czech Republic. *Acta Universitatis Carolinae - Biologica* 39: 201–214.

Ráb, P., P. Roth and A. Aref´ev.1990. Chromosome studies in European leuciscine fishes (Pisces, Cyprinidae). Karyotype of *Aspius aspius*. *Caryologia* 43: 249–256.

Ráb, P., M. Rábová, K.M. Reed and R.B Phillips. 1999. Chromosomal characteristics of ribosomal DNA in the primitive semionotiform fish, longnose gar *Lepisosteus osseus*. *Chromosome Research* 7: 475–480.

Ráb, P., M. Rábová, J. Böhlen and S. Lusk. 2000. Genetic differentiation of the two hybrid diploid-polyploid complexes of loaches, genus *Cobitis* (Cobitidae) involving C. *taenia*, C. *elongatoides* and C. spp. in the Czech Republic. I. Karyotypes and cytogenetic diversity. *Folia Zoologica* 49 (Supplement 1): 55–66.

Ráb, P., M. Flajšhans, A. Ludwig, D. Lieckfeldt, C. Ene, M. Rábová, V. Piačková and T. Paaver. 2004. The second highest chromosome count among vertebrates is associated with extreme ploidy diversity in hybrid sturgeons. *Cytogenetic and Genome Research* 106: 24

Rábová, M., P. Ráb and C. Ozouf-Costaz. 2001. Extensive polymorphism and chromosomal characteristics of ribosomal DNA in a loach fish, *Cobitis vardarensis* (Ostariophysi, Cobitidae) detected by different banding techniques and fluorescence *in situ* hybridization (FISH). *Genetica* 111: 413–422.

Rábová, M., P. Ráb, C. Ozouf-Costaz, C. Ene and J. Wanzeböck. 2003. Comparative cytogenetics and chromosomal characteristics of ribosomal DNA in the fish genus *Vimba* (Cyprinidae). *Genetica* 118: 83–91.

Rábová, M., P. Ráb, A. Boroň, J. Bohlen, K. Janko, V. Šlechtová and M. Flajšhans. 2004. Cytogenetics of bisexual species and their asexual hybrid clones in European spined loaches, genus *Cobitis*. I. Karyotypes and extensive of major ribosomal sites in four parental species. *Cytogenetic and Genome Research* 106: 16.

Raicu, P. and E. Taisescu. 1972. *Misgurnus fossilis*, a tetraploid fish species. *Journal of Heredity* 63: 2–94.

Richardson, M.J., F.G. Whoriskey and L.H. Roy. 1995. Turbidity generation and biological impacts of an exotic fish *Carassius auratus* introduced into shallow seasonality anoxic ponds. *Journal of Fish Biology* 47: 576–585.

Rodrigues, E. and J. Collares-Pereira. 1999. NOR polymorphism in the Iberian species *Chondrostoma lusitanicum* (Pisces: Cyprinidae). *Genetica* 98: 59–63.

Slavík, O. and L. Bartoš. 2004. What are the reasons for the Prussian carp expansion in the upper Elbe River, Czech Republic? *Journal of Fish Biology* 65 (Supplement A). 240–253.

Šlechtová, V., V. Lusková, V. Šlechta, S. Lusk, K. Halačka and J. Bohlen. 2000. Genetic differentiation of two diploid-polyploid complexes of spined loach, genus *Cobitis* (Cobitidae), in the Czech Republic, involving *C. taenia*, *C. elongatoides* and *C.* spp.: Allozyme interpopulation and interspecific differences. *Folia Zoologica* 49 (Supplement 1): 67–78.

Sofradzija, A. and L. Berberovic. 1978. Diploid-triploid sexual dimorphism in *Cobitis taenia* L. (Cobitidae, Pisces). *Acta Biologica Jugoslavica – Genetika* 10: 389–397.

Ueno, K., H. Senou and I.S. Kim. 1985. A chromosome study of five species of Korean cobitid fish. *Japanese Journal of Genetics* 60: 539–544.

Vasil'ev, V.P. 1985. *Evolutionary Karyology of Fishes*. Nauka Press, Moscow. (In Russian).

Vasil'ev, V.P. 1999. Polyploidization by reticular speciation in Acipenseriform evolution: working hypothesis. *Journal of Applied Ichthyology* 15: 29–31.

Vasil'eva, E.D. 2001. Loaches (genus *Misgurnus*, Cobitidae) of the Asiatic part of Russia. 1. Species composition of the genus in Russian waters (with the description of a new species) and some nomenclatural and taxonomic problems of the related forms from the adjacent countries. *Voprosy Ikhtiologi* 41: 581–592.

Vasil'ev, V.P. and E.D. Vasil'eva. 1982. Novyj diploidno-polyploidnoj kompleks u ryb [*A new diploid/polyploid complex in fishes*]. *Doklady Akademii Nauk SSSR* 266: 250–252. (In Russian).

Vrijenhoek, R.C., R.M. Dawley, C.J. Cole and J.P. Bogart. 1989. A list of the known unisexual vertebrates. In: *Evolution and Ecology of Unisexual Vertebrates*, R.M. Dawley and J.P. Bogart (eds.). Bulletin 466, New York State Museum, Albany, New York, pp. 19–23.

Vujoševič, M., S. Zivkovic, D. Rimsa, S. Jurisic and P. Cakic. 1983. The chromosomes of 9 fish species from Dunav basin in Yugoslavia. *Acta Biologica Jugoslavica—Icthyologia* 15: 29–40.

Yang, L., Y. Shu-Ting, W. Xue-Hong and J.F. Gui. 2001. Genetic diversity among different clones of the gynogenetic Silver Crucian Carp *Carassius auratus gibelio* revealed by transferrin and isozyme markers. *Biochemical Genetics* 39: 213–225.

Yu, X., T. Zhou, Y. Li., K. Li and M. Zhou. 1989. *Chromosomes of Chinese Fresh-Water Fishes*. Science Press, Beijing.

Cytogenetic Studies and Reproductive Strategies of an Invasive Fish Species, the Silver Crucian Carp (*Carassius auratus gibelio* Bloch)

Eszter Patakiné Várkonyi[1*] and Balázs Tóth[2]

INTRODUCTION

The geohistoric evolution and actual climatic zone arrangement of the Earth's surface determine the floral and faunal patterns that have evolved as a result of a natural process, under natural conditions. This process was associated with immensely slow genetic changes measurable on a time scale of several million years, and involved the generation and extinction of species. According to our present knowledge, 100 to 250 million species have evolved on the Earth since the appearance of life, but 90% of these

Address for Correspondence: [1*]Institute for Small Animal Research, Gödöllő, P.O. Box 417, Hungary. E-mail: eszter@katki.hu.
[2]Duna-Ipoly National Park Directorate, Budapest, 1021, Hungary.

species have already disappeared and become extinct. As a result of human effects the rate of extinction has changed substantially from approximately 1 species per 1000 years to 1000 species per year, in the 1970s (Bakonyi, 1995).

Among the human effects, one of the biggest changes was caused by the resettlement of species from one geographical unit to another. During the continental drift, oceans served as an isolation border, causing the inland waters of individual continents to create well-separated water systems, each of which supported the development of local species diversity and peculiar biocenoses, according to the interrelationships among the different species. Within this continuous dynamic change, the appearance and disappearance of species (i.e., the speed of genetic change) was proportional to the speed of environmental changes that occurred in the system as a whole. The long period of time enabled the evolution of a large number of species, which resulted in biological diversity. As the interrelationships among species also developed over a long period of time, the rate of permanent change due to competition remained quite low. The resettlement of species among different biocenoses, i.e., the introduction of foreign species put an end to this situation. Such elements were introduced into our waters which had developed outside and which—on account of their characteristic features—either could not establish themselves in the system or were displaced the indigenous species occupying a similar niche. A good example is provided by the silver crucian carp, which is a competitor of the crucian carp, the tench and the common carp. The silver crucian carp not only displaces the tench and the crucian carp from their habitat but, through its intensive sexual-parasitic mode of reproduction, it also influences the spawning success of species occupying some other feeding niches (Fig. 2.2.1). Generally, this means that the appearance of a new species (such as the silver crucian carp) will have an impact not only on the displaced species but also on the entire system. This stresses the necessity of protecting indigenous species and preventing the expansion of invasive species in order to preserve all the genetic diversity that forms the basis of the dynamic changes of our waters, taking place as a slow but stable process.

In addition to the above considerations, the protection of indigenous species is justified by their geohistoric significance. The following fishes are endemic in the water system of the river Danube and today relict species: the Ukrainian brook lamprey (*Eudontomyzon mariae*), the Danube

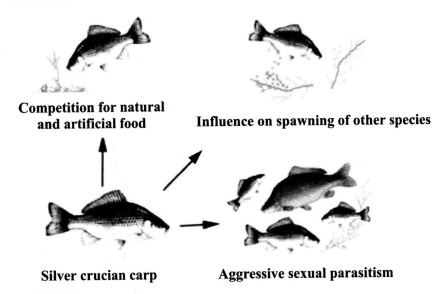

Fig. 2.2.1 The dangers posed by the introduction of silver crucian carp in the waters (from Váradi and Tóth, 1998).

gudgeon (*Gobio uranoscopus*), the striped ruffe (*Gymnocephalus schraetzer*), the huchen (*Hucho hucho*), the asprete (*Romanichthys valsanicola*), the Romanian loach (*Cobitis romanica*), Kessler's goby (*Gobio kessleri*), the zingel (*Zingel zingel*), the Danube streber (*Zingel streber*) and the mudminnow (*Umbra krameri*) (Balon, 1967).

If we look back upon the figures presented in the introduction, we can easily realize the importance of protecting these species. The survival of genetic combinations created during millions of years has become uncertain, and the possible extinction of these combinations cannot be attributed to natural causes. However, because of the transformation of habitats, these species are present in smaller and smaller areas. The majority of these species are specialists and thus the possibility of their transfer, as opposed to that of the generalist species, is rather limited. As a result of deliberate human activity, in the hope of economic or ecological benefits, numerous fish species have been transferred to ecosystems distant from their original habitat. These transfers usually involved more damage than benefit. Fish species thus introduced into Europe include: the brown bullhead (*Ictalurus nebulosus*), the black bullhead (*I. melas*), the pumpkinseed (*Lepomis gibbosus*), the green sunfish (*L. cyanellus*) and the redbreast sunfish (*L. auritus*) from North America, and the stone moroco

(*Pseudorasbora parva* Schlegel) and the silver crucian carp (*Carassius auratus gibelio* Bloch) from the Far East (Terofal, 1996). Among these non-indigenous fish species regarded as harmful from the perspective of nature conservation, the silver crucian carp has attracted the attention of cytogeneticists because of its unique reproductive features and cytogenetic variability. Because of its aggressive expanding strategy, the silver crucian carp endangers the genetic diversity of our natural waters and may act as an ecological competitor for many valuable fish species whose populations have become extremely sparse (Váradi and Tóth, 1998). In the present report we shall review the complicated reproductive mechanism of the silver crucian carp and the composition of the different populations according to reproductive and cytogenetic studies.

The Origin of the Silver Crucian Carp and Its Spread into Central and Eastern Europe

The original home of the silver crucian carp is Southeast Asia, but now this fish species occurs in the overwhelming majority of the European countries (Harka, 1997). The circumstances of its appearance in the water system of the River Danube are not clear. According to Balon (1967), this species was able to move from the water system of Asian rivers into that of the River Danube by a natural route. However, in the 1950s, breeding fish for use in fish farms were imported into some Eastern European countries from Russia and Bulgaria. From then on, the silver crucian carp started its stemless spread into the European countries from the east to the west (Figs. 2.2.2 and 2.2.3).

Due to the wide tolerance to extreme conditions (Pintér, 2002), we cannot exclude the fact that these characteristics facilitate the natural migration of this fish species. In its native land the silver crucian carp is an economically important fish species cultured in pond farms, as it shows great adaptability and good tolerance of poor water quality, rapid reproduction and relatively fast growth. However, it can provide economic benefit only if it is kept under extensive conditions. The silver crucian carp does not enjoy popularity in any part of Europe because of its aggressive expansion strategy resulting in the displacement of indigenous species. Its economic importance is negligible. Solutions to control its spread are being sought by making use of scientific methods.

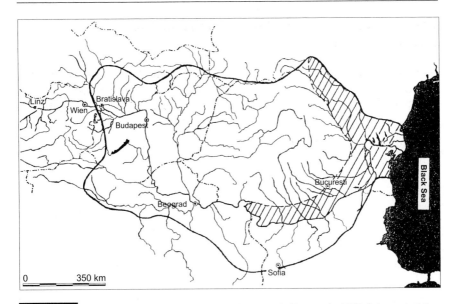

Fig. 2.2.2 Distribution areas of the silver crucian carp in Europe in 1970 (interrupted line and shaded area) and 1980 (heavy line) (from Holcík, 1980).

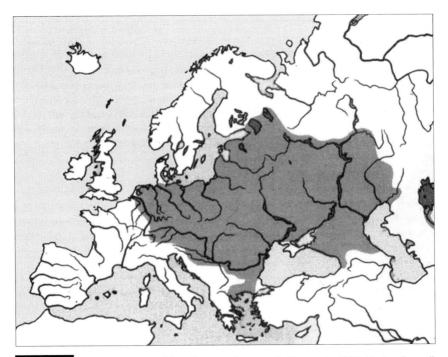

Fig. 2.2.3 Distribution areas of the silver crucian carp in Europe in 1996 (dark coloured) (from Terofal, 1996).

The Unique Reproductive Strategy of the Silver Crucian Carp

Since the first gynogenetic fish, the Amazon molly (*Poecilia formosa*), was found by Hubbs and Hubbs (1932), about 50 unisexual species composed of all-female individuals have been reported in lower vertebrates (Vrijenhoek *et al.*, 1989). These all-female species can propagate by asexual reproduction modes (Stanley and Sneed, 1973), such as gynogenesis, hybridogenesis or parthenogenesis (Gui, 1989; Beukeboom and Vrijenhoek, 1998), where meiosis (Horváth and Orbán, 1995) and syngamy (Maynard Smith, 1989) are absent. Table 2.2.1 provides the glossary of sperm-dependent parthenogenesis, according to Beukeboom and Vrijenhoek (1998). Owing to the absence of genetic recombination and the accumulation of deleterious mutations, unisexual species were generally thought to be short-lived on an evolutionary time scale because

Table 2.2.1 Glossary of sperm-dependent parthenogenesis (from Beukeboom and Vrijenhoek, 1998).

Term	Definition
Androgenesis	Development from a fertilized egg followed by the disintegration of the maternal pronucleus prior to syngamy so that the zygote possessed only paternally derived chromosomes.
Cryptoparthenogenesis	Synonym of pseudogamous parthenogenesis.
Gynogenesis	Pseudogamous parthenogenesis involving sperms from males of a related sexual species.
Hybridogenesis	A hemiclonal form of reproduction in interspecific hybrids that transmit only a maternally derived chromosome set to ova. The paternal genome is excluded during oogenesis and the hybrid condition is restored in each generation by mating with males of a suitable host species.
Parthenogenesis (strict sense)	Virgin birth, or all-female clonal reproduction without the involvement of sperm.
Paternal genome loss	A general term for systems in which paternal chromosomes are eliminated at various developmental stages prior to oogenesis. Pseudogamous parthenogenesis, gynogenesis and hybridogenesis are special cases.
Pseudofertilization	Synonymous with pseudogamous parthenogenesis.
Pseudogamous parthenogenesis	Initiation of parthenogenetic development by the penetration of a sperm into an ovum, without the sperm genome contributing genetic information to the zygote.
Pseudogamy (strict sense)	Synonym of pseudogamous parthenogenesis.
Pseudogamy (ad sense)	A term widely used by plant biologists to refer to the initiation of apomictic (i.e., parthenogenetic) seed production following pollination and fertilization of the primary endosperm.

of Muller's ratchet (Muller, 1964; Crow, 1999). Actually, comparative studies on mitochondrial genome sequences in hybridogenetic *Poeciliopsis* fish (Quattro *et al.*, 1992) suggested that unisexual species were genuinely old in evolutionary terms. However, how do the species evolve and manage their longevity and long history without recombination?

The silver crucian carp, *Carassius auratus gibelio*, shows peculiar genetic background, population structure and reproduction mechanisms. The population distributed throughout Byelorussia exists in the form of unisexual gynogenetic triploids with 150± chromosomes in somatic cells (Cherfas, 1966). Kobayashi *et al.* (1970) and Liu *et al.* (1978) reported three chromosomal polymorphisms characterized by diploidy, triploidy and tetraploidy in a related crucian carp, *Carassius auratus langsdorfii*. The triploids and tetraploids appeared to be unisexual populations propagating by gynogenesis (Kobayashi *et al.*, 1970; Ueda and Ojima, 1978).

In north-eastern China, the silver crucian carp was considered to be triploid, as the chromosomal number in somatic cells is 156 and the population propagates by gynogenesis (Zhou *et al.*, 1983). In Hungary, Tóth *et al.* (2000) assessed the occurrence of both a diploid and triploid populations: the triploid unisexual population has a chromosome number of 150±4 while the diploid population, consisting of both males and females which reproduce sexually, has 100±2 chromosomes.

Moreover, the species also exists as a bisexual triploid population that consists predominantly of females with a variable proportion of males up to 25%. The triploid males were found to produce milt with normal morphology, motility and life-span. The process of spermatogenesis in males did not show any abnormalities at the light and through the electron microscope and the DNA contents of the erythrocytes and spermatozoa gave a ratio of 1.96:1. Thus, these results indicated that the males could carry out the entire meiotic division and that they were effectively diploid derivatives of triploid or hexaploid conditions (Fan and Shen, 1983a,b; Shen *et al.*, 1984).

Fan and Liu (1990), Fan and Shen (1990) and Zhou *et al.* (2000) were able to locate 5–25% males among the gynogenetic progenies of triploid females. Their findings suggested that the sex-determining genes of males of other species have been transmitted to the offspring. According to their hypothesis, female silver crucian carp might produce two kinds of eggs in a normal environment: 'G' (gynogenesis)-type eggs, constituting about 50–90% of all eggs, and 'B' (bisexual)-type eggs constituting about 10–50%. The 'G'-type eggs can carry out only one of the two meiotic divisions and develop gynogenetically. The 'B'-type eggs undergo the whole process of meiosis to form haploid female gametes. The ratio of 'G' to 'B'-type eggs might be influenced by environmental conditions such as temperature, hydrostatic pressure or by factors such as the age, health and nutritional condition of the fish. Different individuals might produce different ratios of 'B' to 'G'-type eggs (Fan and Shen, 1990).

One study conducted by Zhou *et al.* (2000) has already provided direct genetic evidence that triploid silver crucian carp females can be produced by a gonochoristic mode of reproduction with triploid silver crucian carp males. RAPD (Random Amplification of Polymorphic DNA) analysis showed that the paternal DNA appeared in the offspring and the phenotype of the paternal line was also hereditary. Tóth *et al.* (2005) have demonstrated that a triploid female in a crossing with a goldfish male reproduces sexually and that the offspring are triploid interspecific hybrids. RAPD analysis showed that the goldfish-specific DNA appeared in the offspring but the maternal phenotype was strongly expressed. In the case of triploid females, most of the offspring were clones, since they generally reproduce by natural gynogenesis, but some interspecific hybrids could also be found. The appearance of interspecific hybrids among the offspring can be explained by the incorporation of genetic material of the spermatozoon into the female pronucleus.

These results imply that the silver crucian carp is not an exclusively gynogenetic species but also employs the process of sexual reproduction. Therefore, it is possible that the foreign genetic material is transferred to the 'gynogenetic' populations of the species. Since the silver crucian carp spawns with other cyprinid males in natural habitats, this type of hybridization may occur in natural waters (Stranai, 1999). As a result, the natural populations of silver crucian carp can also contain genes from other species, thus probably explaining its exceptional adaptability and the heterozygosity within the gynogenetic populations.

Cytogenetic Investigations

Several authors (Penáz *et al.*, 1979; Fister and Soldatovic, 1989b) have called attention to the great variability of the chromosome numbers of silver crucian carp populations even within a given ploidy level. In the diploid bisexual population(s), the chromosome number varied between $2n=94$–104. The most commonly found chromosome number was

2n=100 (Ojima and Hitotsumachi, 1967; Kobayashi *et al.*, 1970, 1973; Muramoto, 1975; Ruiguang, 1982; Fister, 1989a). Triploid populations, consisting mostly of females, showed even higher variation in the chromosome number with values 3n=141–166. Chromosome numbers between 150 and 160 were found to be the most common. There was no morphological difference between the male and female karyotypes.

Table 2.2.2 provides the cytogenetic data summarized by the authors, grouped according to subspecies. Examples of chromosome metaphases from a diploid silver crucian carp with 2n=100 and from a triploid silver crucian carp clone with 3n=150 are provided in Fig. 2.2.4.

On the basis of the above data, we can state that a species-specific chromosome number cannot be assigned to the silver crucian carp. In addition, the chromosome analyses suggest that certain clonal lines acquire extra chromosomes through reproduction by way of hybridogenesis. It can be seen that triploid gynogenetic females almost never represent a real autotriploid, i.e., it is impossible to obtain their chromosome number by simple multiplication of the basic 'n' value of a diploid bisexual population found at the same or a neighbouring locality (Fister and Soldatovic, 1989b). These results are coherent to data by Kobayashi *et al.* (1970, 1973) who found values of n=50, 3n+6=156 and 4n+6=206. In the latter case, both triploid and tetraploid forms contained an excess of 6 chromosomes, which the real autotriploid or tetraploid should not possess. Moreover, analyzing *Carassius auratus gibelio* from Kunming Lake, Ruiguang (1982) found two extra chromosomes (n=50, 3n+2=152).

In 1983, Vujosevic *et al.*, found an excess of 10 chromosomes (3n+10=160) in *Carassius auratus gibelio*, if the basic 'n' amount was 50. DNA-level and karyotype analyses based on different propagation experiments have also proved (Tóth *et al.*, 2005) that the triploid silver crucian carp does not reproduce exclusively by natural gynogenesis but it can increase its genetic variability by hybridogenesis.

Possible Origin of Different Karyotypes of Silver Crucian Carp Populations

Although fish chromosomes, especially those of ancient tetraploid forms where numerous and tiny elements are found, are very difficult to prepare, in the authors' opinion, the differences in the number of chromosomes found by several other authors reflect real differences and do not result

Table 2.2.2 Selected cytogenetic data in subspecies of *Carassius auratus*. m: metacentric; sm: submetacentric; a: acrocentric chromosomes; NF: Fundamental number.

Chromosome number	Karyotype formula (pairs)	NF	References
	Carassius auratus gibelio Bloch		
2n=104	23m- sm+29a	150	Post, 1965
2n=100±4	31-32m- sm+18-21a	165–166	Ohno and Atkin, 1966; Ohno et al., 1967
2n=94	–	–	Cherfas, 1966
2n=100	24m- sm+26a	148	Ojima and Hitotsumachi, 1967
2n=104	–	–	Chiarelli et al., 1969
2n=100	30m- sm+20a	160	Muramoto, 1975
2n=100	12m+17sm+20a+XY	160	Fister, 1989a
2n=98	24m-sm+25a	146	Raicu et al., 1981
2n=100	26m- sm+24a	152	Ruiguang, 1982
2n=100	–	–	Tóth et al., 2005
3n=141	–	–	Cherfas, 1966
3n=150	–	–	Sofradzija et al., 1978
3n=160	23m+41sm, st+16a	–	Penáz et al., 1979
3n=152	–	–	Ruiguang, 1982
3n=160	–	–	Vujosevic et al., 1983
3n=158	18m+27sm+34a	–	Fister and Soldatovic, 1989b
3n=166	16m+20sm+9st+ 38t	238	Minrong et al., 1996
3n=152±4	–	–	Tóth et al., 2005
	Carassius auratus auratus		
2n=104	10m+42sm, st	–	Chiarelli et al., 1969
2n=104	23m+8st+21a	–	Ohno et al., 1967
2n=100	6m+18sm+26a	–	Ojima and Hitotsumachi, 1967
2n=100	10m+20sm+20a	–	Kobayashi, 1977
	Carassius auratus langsdorfii		
2n=100	10m+20sm+20a	160	Kobayashi et al., 1970, 1973
2n=100	10m+20sm+20a	–	Muramoto, 1975
3n=150	–	–	Muramoto, 1975
3n=156	17m+31sm+30a	–	Kobayashi et al., 1970, 1973
3n=157	32m+59sm+62a+4 microchromosomes	–	Muramoto, 1975
3n=165	32m+59sm+62a+9 microchromosomes	–	Muramoto, 1975
4n=206	22m+41sm, st+40a	–	Kobayashi et al., 1970, 1973

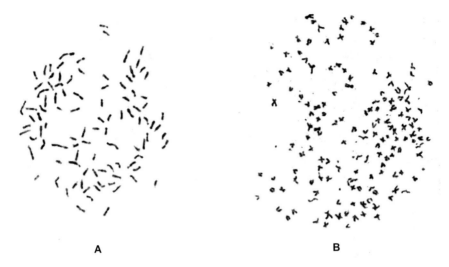

Fig. 2.2.4   A: Metaphase from a diploid silver crucian carp (2n=100), **B:** Metaphase from a triploid silver crucian carp clone (3n=150).

from individual errors. These differences in chromosome number are not easy to explain even in the case of species of tetraploid origin, because no species can lose or gain its genetic material without consequences, i.e., without phenotypic and especially genotypic changes. It is possible that our approaches are not precise enough to make it possible to recognize such an effect.

The appearance of different karyotype clones could originate from changes in old clones or through eventual *de novo* formation. Changes in the old clones could include karyotypic rearrangements, resulting from intrinsic genetic factors (e.g., ploidy level or tetraploid origin) and/or from mutagenic action of various compounds found in the environment. These actions could lead to the appearance of an aberrant viable individual able to reproduce and give a clone with new characteristics (Fister and Soldatovic, 1989b).

De novo genesis of the clones could originate from hybridogenesis. Experimental evidences have been provided by some recent studies (Zhou *et al.*, 2000; Yi *et al.*, 2003; Tóth *et al.*, 2005). For example, Yi *et al.* (2003) found some paternal chromosome fragments of bluntnose black bream incorporated into the silver crucian carp clone by using molecular cytogenetic techniques. Hybridogenesis is a hemiclonal form of reproduction in interspecific hybrids that transmits only a maternally derived chromosome set to the ova. The paternal genome is excluded during oogenesis and the hybrid condition is restored in each generation by mating with males of a suitable host species (Beukeboom and Vrijenhoek, 1998). To further complicate this subject, some triploid all-female fishes superimpose elements of hybridogenesis on top of gynogenesis (reviewed by Dawley, 1989). For example, a gynogenetic triploid might occasionally produce reduced ova that fuse with sperm and generate a new triploid. Over a period of time, these replacements may produce triploid lineages that contain none of the original nuclear genomes (Kraus and Miyamoto, 1990). Such genomic replacements may be responsible for the apparent evolutionary longevity of some gynogenetic species (Hedges *et al.*, 1992; Spolsky *et al.*, 1992).

The evolution and speciation of fishes has also involved mechanisms of chromosome rearrangement and duplication (Ohno, 1970; Chrisman *et al.*, 1990). A natural genome duplication process (Force *et al.*, 1999) in the *Cyprinidae* family is believed to have occurred as far as 25 million years ago (Allendorf and Thoorgard, 1984), which means that the 25 pairs of chromosomes changed into 50 pairs. So, each gene of ancient tetraploids has twice as many alleles as do the other species. Since two alleles are needed for survival, the remaining alleles will possess the possibility to undergo free mutations, possibly enhancing the adaptive plasticity in different natural situations (Allendorf and Thoorgard, 1984). The ancient tetraploidization was followed by a new spontaneous triploidization (Cherfas, 1966; Horváth and Orbán, 1995). If that is the case, the diploid form of *Carassius auratus gibelio* is actually tetraploid and the triploid form is hexaploid in origin. This means that the triploid silver crucian carp has 6 variable alleles in each gene locus and, therefore, it has much higher chance of adaptation to the changing environment than do species having only two alleles, because new functions may evolve following the gene duplications (Force *et al.*, 1999) (Fig. 2.2.5).

During phase I, genes may experience one of three alternative fates. First, one copy may incur a null mutation in the coding region, which subsequently drifts to fixation, leading to gene loss (nonfunctionalization). Nonfunctionalization can also occur if all of the regulatory regions of one duplicate are destroyed. Second, one copy may acquire a mutation conferring a new function, which becomes fixed through positive Darwinian selection (neofunctionalization). Assuming this new function results in the loss of an essential ancestral function, neofunctionalization

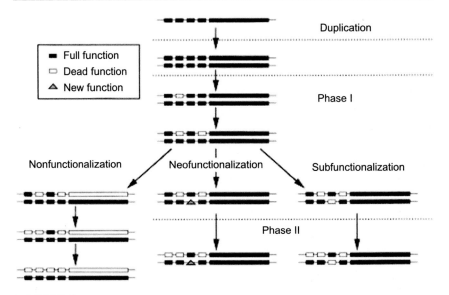

Fig. 2.2.5 Three potential fates of duplicate gene pairs with multiple regulatory regions (from Force *et al.*, 1999). The small boxes denote regulatory elements with unique functions, and the large boxes denote transcribed regions. Solid boxes denote intact regions of a gene, while open boxes denote null mutations, and triangles denote the evolution of a new function.

insures the preservation of the non-mutated copy. Third, each duplicate may experience loss or reduction of expression for different subfunctions by degenerative mutations. In such a case, the combined action of both gene copies is necessary to fulfil the requirements of the ancestral locus (subfunctionalization). If this happens, then complementation of subfunctions between duplicate genes will preserve both partially degenerated copies.

Relevance of Cytogenetic Studies of Invasive Species

Because of its rapid expansion and adaptability, there is no doubt that the silver crucian carp causes problems to natural biocenoses. Cytogenetic studies of recent years have provided evidence that this species can integrate genes of related fish species into their own populations. This can greatly contribute to its adaptability. Another great advantage arises from the fact that in the triploid silver crucian carp, each gene may have 6 alleles, which increases genetic variability and may result in favourable gene interactions, the detrimental effect related to genetic overloading being rapidly eliminated by selection. In view of the above scenario, it can be established that the silver crucian carp is a genetically unstable species. Its populations and different clonal lines exhibit very high variability, which may greatly improve the ecological competitiveness of the species. This can pose a serious danger especially in smaller natural waters bodies still existing in which the crucian carp and the tench occur. This problem could have been easily prevented if at the time of introduction we were aware of the cytogenetic characteristics enhancing the expansion and adaptability of the silver crucian carp. In that case, either the danger could be recognized in time and the introduction implemented more carefully or it could have been altogether avoided. At present, we cannot stop the expansion of this fish but further cytogenetic studies may lead to the development of new methods and strategies suitable for the regulation of silver crucian carp populations.

The primary responsibility of nature conservation is the preservation of biological diversity. To lay down the foundations for the nature conservation treatment of waters, we have to select the approach by which we wish to attain the intended goal. For instance, we can protect the undisturbed course of the natural processes. The continuous alteration of the abiotic elements of the system by humans makes such an approach unrealistic in the European civilization. The continuously ongoing expansion of the silver crucian carp cannot be regarded as a natural process. Today, the crucian carp and the tench can be protected only by artificial means. The ban on fishing and the protection of habitats have not been successful, and therefore the populations of the above two fish species can be increased only by artificial propagation and introduction. The small ponds created in the framework of habitat reconstruction programmes provide an excellent opportunity for realising this effort. There are old, far-gone succesioned lakes and ponds in national park areas. The experts of the national parks usually carry out programmes to rehabilitate those particular water bodies. After rehabilitation, the area can be developed into a valuable wetland habitat for indigenous species. In order to avoid the spontaneous ecesis of non-indigenous species, settlement of indigenous species is necessary as soon as possible after the reconstruction.

The usefulness of some non-indigenous species cannot be doubted. For instance, in farm ponds, the grasscarp (*Ctenopharyngodon idella*) efficiently utilizes the food base not useful for other species, and in other areas it slows down the growth of pondweed in water drainage channels. Therefore, in some cases, the introduction of the grasscarp can be justified. For instance, cytogenetic research has enabled the use of triploid grasscarp populations to clean out pondweed from canals in the United States. Such triploid grasscarp are incapable of reproduction and, thus, their introduction achieves the desired result without causing any harmful effects (Allen and Wattendorf, 1986; Thompson *et al.*, 1987). The main reason for production of triploid fish is to control sex. Sterile fish usually grow better; their stocking excludes unwanted overpopulation of natural waters (Nagy, 1987). The reason of the sterility is not known exactly. The research work of ploidy manipulations of fish has been enabled by cytogenetic studies.

Acknowledgement

We would like to thank Dr László Váradi for doing pioneer work in this subject. We are grateful to Dr András Székely, the editor *of Acta Veterinaria Hungarica* for revision of the English translation of the manuscript.

References

Allen, S.K. and R.J. Wattendorf. 1986. A review of the production and quality control of triploid grass carp and progress in implementing 'sterile' triploids as management tools in the U.S. *Aquaculture* 57: 359–379.

Allendorf, F. W. and G. H. Thoorgard. 1984. Tetraploidy and the evolution of salmonid fishes. In: *Evolutionary Genetics of Fishes*, Turner, B.J. (ed.) Plenum Press, New York pp. 1–53.

Balon, K.E. 1967. History of the fish fauna of the Danube river, and prognosis for changes, caused by constructive works of water management. *Biologické Práce*, Bratislava. (In Slovak).

Bakonyi, G. 1995. *Zoology*. (In Hungarian).

Beukeboom, L.W. and R.C. Vrijenhoek. 1998. Evolutionary genetics and ecology of sperm-dependent parthenogenesis. *Journal of Evolutionary Biology* 11: 755–782.

Cherfas, N.B. 1966. Natural triploidy in the females of the unisexual variety of the silver crucian carp (*Carassius auratus gibelio* Bloch). *Genetika* 2: 16–24. (In Russian).

Chiarelli, B., O. Ferrantelli and C. Cucchi. 1969. The karyotype of some teleostean fish obtained by tissue culture *in vitro*. *Experientia* 25: 426–427.

Chrisman, C.L., K.H. Blacklidge and P.K. Riggs. 1990. Chromosomes of fish. In: *Domestic Animal Cytogenetics. Advances in Veterinary Science and Comparative Medicine*, R. A. McFeely (ed.) Academic Press, San Diego, Vol. 34, pp. 209–227.

Crow, J.F. 1999. The odds of losing at genetic roulette. *Nature (London)* 397: 293–294.

Dawley, R.M. 1989. An introduction to unisexual vertebrates. In: *Evolution and Ecology of Unisexual Vertebrates*, Dawley, R. M. and Bogart, J. P. (eds.). Bulletin 466, pp. 1-18. New York State Museum, Albany, NY.

Fan, Z. and G. Liu. 1990. The ploidy and reproductive mechanism of crucian carp, *Carassius auratus gibelio*. *Journal of Fish Biology* 36: 415–419.

Fan, Z. and J. Shen. 1983a. Studies of spermatozoa biology in *Carassius*. *Science Reports of Heilongjiang Fish Research Institute of Chinese Academy of Fisheries Science* 21: 3–8. (In Chinese).

Fan, Z. and J. Shen. 1983b. Observation of the processes of spermatogenesis in *Carassius* by electron microscope. *Science Reports of Heilongjiang Fish Research Institute of Chinese Academy of Fisheries Science* 21: 16–20. (In Chinese).

Fan, Z. and J. Shen. 1990. Studies on the evolution of bisexual reproduction in crucian carp (*Carassius auratus gibelio* Bloch). *Aquaculture* 84: 235–244.

Fister, S. 1989a. Karyotype analysis of a male *Carassius auratus gibelio* Bloch (*Pisces, Cyprinidae*) caught in the river Tamis. *Acta Veterinaria Beograd* 39: 99–108.

Fister, S. and B. Soldatovic. 1989b. Karyotype analysis of a gynogenetic population of *Carassius auratus gibelio* Bloch (*Cyprinidae*) from Pancevacki rit. *Acta Veterinaria Beograd* 39: 259–268.

Force, A., M. Lynch, F.B. Pickett, A. Amores, Y. Yan and J. Postlethwait. 1999. Preservation of duplicate genes by complementary, degenerative mutations. *Genetics* 151: 1531–1545.

Gui, J.F. 1989. Evolutionary genetic of unisexual vertebrates. *Nature (London)*. 12: 116–122.

Harka, Á. 1997. *Our Fish Species*. Pictorial Identification and Expansion Guide. Harka, Á. (ed.) Foundation of Teachers for Nature and Environmental Protection, Budapest. (In Hungarian).

Hedges, S.B., Bogart, J.P. and L.R. Maxson. 1992. Ancestry of unisexual salamanders. *Nature (London)* 356: 708–710.

Holcík, J. 1980. *Carassius auratus* (Pisces) in the Danube river. *Acta scientiarum naturalium Academiae scientiarum bohemoslovacae Brno* 14: 1–43.

Horváth, L. and L. Orbán. 1995. Genome and gene manipulation in the common carp. *Aquaculture* 129: 157–181.

Hubbs, C.L. and L.C. Hubbs. 1932. Apparent parthenogenesis in nature, in a form of fish of hybrid origin. *Science* 76: 628–630.

Kobayashi, H. 1977. Hybridization in Japanese funa *Carassius auratus*. In: *Proceedings of the Fifth Japan-Soviet Joint Symposium on Aquaculture*, S. Motoda (ed.), Tokai University, Tokyo pp. 193–208.

Kobayashi, H., Y. Kawashima and N. Takeuchi. 1970. Comparative chromosome studies in the genus *Carassius*, especially with a finding of polyploidy in the ginbuna. *Japanese Journal of Ichthyology* 17: 153–160.

Kobayashi, H., H. Ochi and N. Takeuchi. 1973. Chromosome studies of the silver crucian carp (*Carassius auratus gibelio*) from the valley of the Amur river, and their progenies. *Journal of Japan Women's University (Home Economics)* 20: 83–88.

Kraus, P. and M.M. Miyamoto. 1990. Mitochondrial genotype of unisexual salamander of hybrid origin is unrelated to either of its nuclear haplotypes. *Proceedings of the National Academy of Sciences of the United States of America* 87: 2235–2238.

Liu, S., K. Sezaki, K. Hashimoto, H. Kobayashi and M. Nakamura. 1978. Simplified techniques for determination of polyploidy in ginbuna, *Carassius auratus langsdorfii*. *Bulletin of the Japanese Society of Scientific Fisheries* 44: 601–606.

Maynard Smith, J. 1989. *Evolutionary Genetics*. Oxford University Press, Oxford.

Minrong, C., Y. Xingqi, Y. Xiaomu and C. Hongxi. 1996. Karyotype studies on the bisexual natural gynogenetic crucian carp (*Carassius auratus*) of Pengze. *Acta Hydrobiologica Sinica* 20: 25–31.

Muller, H.J. 1964. The relation of recombination to mutational advance. *Mutation Research* 1: 2–9.

Muramoto, J. 1975. A note on triploidy in the funa (*Cyprinidae, Pisces*). *Proceedings of the Japan Academy* 51: 101–103.

Nagy, A. 1987. Genetic manipulations performed on warm water fish. *Proceedings of World Symposium on Selection, Hybridization and Genetic Engineering in Aquaculture*, Vol. 2, Berlin, pp. 163–174.

Ohno, S. 1970. *Evolution by Gene Duplication*. Springer-Verlag, New York.

Ohno, S. and N.B. Atkin. 1966. Comparative DNA values and chromosome complements of eight species of fishes. *Chromosoma* 18: 455–466.

Ohno, S., J. Muramoto and L. Christian. 1967. Diploid-tetraploid relationship among old world members of the fish family Cyprinidae. *Chromosoma* 23: 1–9.

Ojima, Y. and S. Hitotsumachi. 1967. Cytogenetic studies in lower vertebrates. A note on the chromosomes of the carp in comparison with those of the funa and the goldfish. *Japanese Journal of Genetics* 42: 163–167.

Penáz, M., P. Ráb and M. Prokes. 1979. Cytological analysis, gynogenesis and early development of *Carassius auratus gibelio*. *Acta scientiarum naturalium Academiae scientiarum bohemoslovacae Brno* 13: 1–33.

Pintér, K. 2002. *Fish Fauna of Hungary*. Mezõgazdasági Kiadó, Budapest. (In Hungarian).

Post, A. 1965. Vergleichende untersuchungen der Chromosomenzahlen bei Süsswasser-Zeleosteen. *Zeitschrift Für Zoologische Systematik und Evolutionsforschung* 3: 47–93.

Quattro, J.M., J.C. Avise and R.C. Vrijenhoek. 1992. An ancient clonal lineage in the fish genus *Poeciliopsis*. *Proceedings of the National Academy of Sciences of the United States of America* 89: 348–352.

Raicu, P., E. Taisescu and P. Banarescu. 1981. *Carassius carassius* and *C. auratus*, a pair of diploid and tetraploid representative species (*Pisces, Cyprinidae*). *Cytologia* 46: 233–240.

Ruiguang, Z. 1982. Studies of sex chromosomes and C-banding karyotypes of two forms of *Carassius auratus* in Kunming Lake. *Acta Genetica Sinica* 9: 32–39.

Shen, J., Z. Fan, S. Li, R. Sheng and S. Xue. 1984. Comparative studies of the somatic cell and spermatozoon DNA contents and ploidy of Fangzeng crucian carp and Zhalong lake goldfish. *Acta Zoologica Sinica* 30: 7–13.

Sofradzija, A., L. Berberovic and R. Hadziselimovic. 1978. Chromosome analysis of crucian carp (*Carassius auratus*) and silver crucian carp (*Carassius auratus gibelio*). *Ichthyologia* 10: 135–143.

Spolsky, C.M., C.A. Philips and T. Uzzell. 1992. Antiquity of clonal salamander lineages revealed by mitochondrial DNA. *Nature* (*London*) 356: 706–708.

Stanley, J. and E. Sneed. 1973. Artificial gynogenesis and its application in genetics and selective breeding of fish. In: *The Early Life History of Fish*. J.H.S. Baxter (ed.). Proceedings of International Symposium, Oban, Scotland. Springer-Verlag, Berlin, pp. 527–536.

Stranai, I. 1999. The find of a natural hybrid of *Carassius gibelio* (Bloch, 1782) × *Cyprinus carpio* (Linnaeus, 1758). *Czech Journal of Animal Science* 44: 515–522. (In Slovak with English abstract).

Terofal, F. 1996. *Süsswasserfische*. Mosaik Verlag GmbH, München.

Thompson, B.Z., R.J. Wattendorf, R.S. Hestand and J.L. Underwood. 1987. Triploid grass carp production. *The Progressive Fish-Culturists* 49: 213–217.

Tóth, B., L. Váradi, E.Várkonyi and A. Hidas. 2000. Silver crucian carp (*Carassius auratus gibelio* Bloch) in the Danube river basin. *Tiscia monograph series*, 61-65.

Tóth, B., E. Várkonyi, A. Hidas, E. Edviné Meleg and L. Váradi. 2005. Genetic analysis of offspring from intra- and interspecific crosses of the silver crucian carp (*Carassius auratus gibelio* Bloch) by chromosome and RAPD analysis. *Journal of Fish Biology* 66: 784-797.

Ueda, T. and Y. Ojima. 1978. Differential chromosomal characteristics in the Funa subspecies (*Carassius*). *Proceedings of Japan Academy* 54B: 283–288.

Váradi, L. and B. Tóth. 1998. Reproduction biology of silver crucian carp, a new strategy of evolution. I. Csengeri (ed.): *Halászatfejlesztés 21 – Fisheries and Aquaculture Development* 21: 102–107 (In Hungarian with English abstract).

Vrijenhoek, R.C., R.M. Dawley, C.J. Cole and J.P. Bogard. 1989. A list of known unisexual vertebrates. In: *Evolution and Ecology of Unisexual Vertebrates*, R.M. Dawley and J.P. Bogard (eds.) New York State Museum, Albany, pp. 19–23.

Vujosevic, M., S. Zivkovic, D. Rimsa, C. Jurisic, and P. Cakic. 1983. The chromosomes of 9 fish species from Dunav basin in Yugoslavia. *Acta Biologica Yugoslavia Ichthyologia* 15: 29–40.

Yi, M.S., Y.Q. Li, J.D. Liu, L. Zhou, Q.X. Yu and J.F. Gui. 2003. Molecular cytogenetic detection of paternal chromosome fragments in allogynogenetic gibel carp, *Carassius auratus gibelio* Bloch. *Chromosome Research* 11: 665–671.

Zhou, L., J. Shen and M. Liu. 1983. A cytological study on the gynogenesis of Fangzheng crucian carp of Heilongjiang province. *Acta Zoologica Sinica* 29: 11–16.

Zhou, L., Y. Wang and J. Gui. 2000. Genetic evidence for gonochoristic reproduction in gynogenetic silver crucian carp (*Carassius auratus gibelio* Bloch) as revealed by RAPD assays. *Journal of Molecular Evolution* 51: 498–506.

SECTION

3

Fish Cytogenetics in Stock Assessment and Aquaculture

Chromosomal Analysis in Population Structuring and Stock Identification: Robertsonian Polymorphism in the White Sea Herring (*Clupea pallasi marisalbi*)

Dmitry L. Lajus

INTRODUCTION

The chromosomal analysis is usually applied to fish in order to analyze relationships between subspecies, species or taxons of a higher range. Associations between populations and stock identification are usually studied with more sensitive methods such as protein electrophoresis or nucleic acids analysis. For instance, chromosomal studies of populations of the stripped mullet *Mugil cephalus* from different geographical areas did not reveal any differences even when molecular cytogenetic techniques

Address for Correspondence: St. Petersburg State University, Faculty of Biology and Soil Sciences, Department of Ichthyology and Hydrobiology, 16 Linia V.O., 29. 199178. St. Petersburg, Russia. E-mail: dlajus@yahoo.com

were used (Rossi *et al.*, 1996) but they could be clearly differentiated by allozyme or mtDNA analysis (Crosetti *et al.*, 1994; Rossi *et al.*, 1998). According to Galetti and co-authors (2000), chromosomal rearrangements are difficult to be fixed in populations of marine fishes due to the high mobility of the species (on eggs, larvae or adults stage) and also as a result of the absence of geographical barriers in marine environment. But there is always an exception to the rule. Such exception will be considered in the present chapter, which deals with a study on the White Sea herring.

In the White Sea herring *Clupea pallasi marisalbi,* chromosomal analysis revealed differences between allopatric populations. This is especially noticeable as the herring is an extremely mobile fish with pelagic larvae in its life cycle, and the White Sea is a rather small and young (post-glacial) sea with a very dynamic environment and absence of clear geographical barriers. In addition, allozyme analysis did not show differences between these populations.

The chromosomal studies of the White Sea herring and their population interpretation also have an interesting historical dimension. Since 1960s, population studies of the White Sea herring were accompanied by intense discussions and chromosomal data were in the very centre of the debate. The need for karyological data was essential because no other genetic data were available at that time. Then, techniques of chromosomal analysis were rather imperfect. This combination resulted in premature conclusions, which were readily used to support one or another concept.

The White Sea Herring Population Structure: Patterns and Concepts

The White Sea herring is considered to be a subspecies of the Pacific herring, *C. pallasi*. It has occupied the White Sea in post-glacial time, 10,000 years ago (Derjugin, 1928; Andriashev, 1957) and distributed over all the sea (Fig. 3.1.1). It formed many stocks during, presently representing an excellent example of high intra-species diversity within a comparatively small body of water. Slow- and fast-growing herring are distinguishable; at 4-5 years of age, they differ 3-5 times in weight (Dmitriev, 1946; Tambovtsev, 1957). In the north-western part of the sea (Kandalaksha Bay), herring have two main spawning periods. The slow-growing herring spawn in April, usually under the ice. The fast-growing

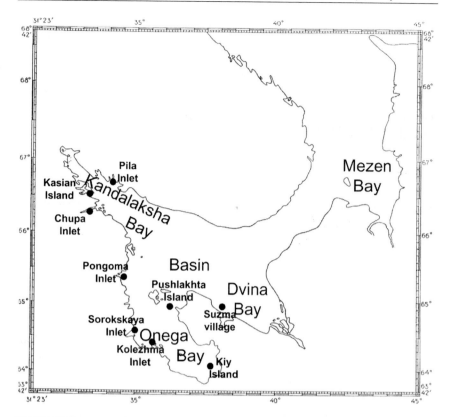

Fig. 3.1.1 Sampling locations of herring in the White Sea: 1. Chavanga Village/Basin (66°03'N, 37°46'E); 2. Pila Inlet/Kandalaksha Bay (66°44'N, 34°10'E); 3. Kasian Island/ Kandalaksha Bay (66°34'N, 33°28'E); 4. Chupa Inlet/Kandalaksha Bay (66°28'N, 33°30'E); 5. Pongoma Inlet/Basin-Karelsky coast (65°22'N, 34°28'E); 6. Sorokskaya Inlet/ Onega Bay (64°32'N, 34°49'E); 7. Kolezhma Inlet/Onega Bay (64°25'N, 36°00'E); 8. Kiy Island/Onega Bay (64°15'N, 37°55'E); 9. Pushlakhta Inlet/Basin-Liamitzky coast (64°49'N, 36°34'E); 10. Suzma Village/Dvina Bay (64°45'N, 36°02'E).

herring spawn in June, at temperature of 8-12°C. In southern parts of the sea (Onega and Dvina bays), the spawning period lasts from the middle of May up to the end of June. Slow- and fast-growing herring usually spawn here together. In some locations, the fast-growing or slow-growing herring may spawn separately in June.

Lepekhin (1805) conducted first descriptions of the White Sea herring in the eighteenth century. Problems and even heated debates have surrounded the research on White Sea herring population structure during the entire period (Lajus, 2002). The main question of these debates was '...whether or not the White Sea herring is a single stock' (Danilevsky, 1862, p. 16). Since 1920s, the White Sea herring was considered in a number of isolated stocks of Pacific origin (Rabinerson, 1925; Averintsev, 1927, 1928). Most researchers believed that the growth rate was a main feature of a stock, and the fast- and slow-growing stocks received Latin names at sub-subspecies level. In all, more than ten self-dependent stocks were described (Dmitriev, 1946; Altukhov, 1975).

The view of self-dependence of the White Sea herring stocks has been disputed in 1960s by a concept of a single population, 'a single biological entity' (Lapin, 1966). It was argued that herring inhabit different parts of the sea at different stages of their life cycle, which are most appropriate for a particular ontogenetic stage (Lapin, 1966, 1978; Lapin and Pokhiliuk, 1993). This concept of a single biological entity quickly became popular, probably because of its convenience for the fishery. The main gear for fishing herring in the White Sea is the bottom trawl, but its use is limited due to the rocky bottom. Thus, acceptance of the concept of a single population allowed fishermen to take the whole herring quota in the few locations that were most convenient for trawling without danger of over-fishing of entire stock. In a short period, some researchers developed an opposite concept of polyphyletic origin of the White Sea herring, whose consequence would be reproductive isolation (Dushkina, 1975).

Chromosomal Analysis for Population Structuring: First Attempts

Polyphyletic origin suggests profound genetic differentiation of the forms and, hence, quite probable differences of their karyotypes. Moreover, in 1960-70s, some good examples of chromosomal differentiation of related forms were reported (Ohno, 1970; Schell, 1972; Hartley and Horney, 1982). This explains the reason why in 1970s, chromosomal analysis has been applied to study the population structure of the White Sea herring. The first studies reported clear differentiation between the stocks. The spring-spawning slow-growing herring of the Kandalaksha Bay were found to have $2n = 52$, while sympatric summer-spawning fast-growing herring and slow-growing herring from the Onega and Dvina bays have $2n = 54$ (Skvortsova, 1975). The paper appeared in a book edited by Dushkina and was considered as a very strong support to the polyphyletic origin of the White Sea herring. However, a book edited by Lapin data on intra-population polymorphism in the stock from the Onega Bay was published indicating individuals with 2n = 52 and 2n = 54 (Krysanov, 1978). As a consequence, previous results by Skvortsova could hardly be considered as indisputable proof of the reproductive isolation of the herring populations.

Therefore, chromosomal data obtained by 1980s did not answer the question whether the White Sea herring is a single stock or not. However, these studies demonstrated some potential of chromosomal analysis for resolving problems of its population structure. Moreover, they definitely were very interesting from a more general karyological point of view. In reviewing the karyology of the family Clupeidae, Doucette and Fitzsimons (1988) reported that most of the Clupeidae possess diploid chromosome number (2n) of 44-54 with a number of chromosomal arms (NF) of 46-60. The karyotypes are close to the presumed ancestral karyotype of Teleostei, made of 48 acrocentrics (Ohno, 1970). This provided evidence of a possible low rate of chromosomal evolution of Clupeidae in comparison with other fishes. Thus, the finding of differences between the White Sea herring stocks served a good stimulus towards continued cytogenetic research.

The main difficulty that the researchers faced while karyotyping specimens from the supposed different populations was a low number of dividing cells in adult fishes. Data which would allow reliable interpretation of the population structure must be obtained on spawning individuals, in a period of the life cycle when mitotic activity is very low because of inhibition of growth processes in spawning season and low temperatures. In fact, the spawning of most stocks takes place at slightly positive or even negative temperatures. Moreover, the analysis can be done only once a year in the field, as the key spawning grounds are difficult to access because of remoteness and presence of ice.

Attempts to use the techniques of Skvortsova (1975) and Krysanov (1978) failed. The squash technique on spawning fish yielded only a few poorly distributed metaphases, and the use of blastula embryos resulted in a large number of overlays of chromosomes. Much better results were obtained by using cell suspensions from kidney tissue with consequent sedimentation by hand centrifuge (Gold, 1974), with modification of Shelenkova (1986) and adaptation to field conditions by Lajus (1987, 1989a).

Analyses of specimens from the Kandalaksha and Onega bays carried out according to this technique did not reveal differences between the stocks. No individuals with 2n = 54 were found (Lajus, 1987). Attention has also been paid to embryonic metaphases because in addition to differences in the number of chromosomes, certain differences in the number of secondary constrictions and satellite chromosomes were also reported. These data were obtained using squash technique on blastula (Skvortsova, 1975). Fast-growing herring from the Kandalaksha Bay and slow-growing herring from the Onega and Dvina bays had 2-4 secondary constrictions and the satellite chromosomes each, whereas in slow-growing herring from the Kandalaksha Bay these structures were totally absent. These structures are well visible on the metaphase plates obtained from blastula cells due to low condensation of chromatin, but it is more difficult to examine them on plates obtained from adults. The secondary constrictions and satellite chromosomes can also be considered as a population marker. Moreover, the number of secondary constrictions should also be taken into account.

Checking, whether the number of secondary constrictions is different among populations, is a difficult task while counting chromosomes in blastula cells. To cope with this problem, only blastula embryos with known number of chromosomes were used in the analysis (Lajus, 1989a, 1996a). The karyotypes were determined on parents and on embryos of latter stages of development from the same series. Absence of mortality was a guarantee that number of chromosomes in blastula embryos is known. In such an experimental design, differences between populations, if present, indeed deal with number of secondary constrictions. The analysis included three stocks: slow- and fast-growing stocks of the Kandalaksha Bay and slow-growing herring of the Onega Bay. In total, 171 embryos from nine series (three from each stock) were karyotyped. The average number of secondary constrictions per metaphase varied from 2,56 to 3,00 and the number of satellite chromosomes from 2,55 to 2,95. Differences between populations were not found (Lajus, 1996a).

These studies did not confirm all the earlier conclusions about chromosomal differences between the White Sea herring stocks showing, instead, their chromosomal homogeneity. Therefore, the results questioned any further application of karyological analyses for population studies of the White Sea herring, although the question could arise whether these conclusions reflected the will of the researchers or the biological reality.

Robertsonian Polymorphism: Intra-population Patterns

The first analyses showed that the most individuals have 2n = 52, NF = 60 (32 specimen from the Kandalaksha Bay and 27 specimens from the Onega Bay). But one specimen from the Onega Bay had the karyotype 2n = 51, NF = 60. In this karyotype, there was one well-visible large metacentric instead of two acrocentrics in karyotype 2n = 52, indicating that the difference may be caused by the Robertsonian translocation. Potentially, it can serve a convenient population marker, but its appearance only in one individual and very low frequency indicated a high probability that it was mutation but not polymorphism. The frequency of translocation was even less than 1% (taking into account that 60 specimens of the White Sea herring were analyzed by that time). As it is commonly accepted that the arbitrary cut-off point between a mutation and a polymorphism is a frequency of 1%, a probability that it was due to a mutation was very high. A mutation, as a unique phenomenon, would be useless in population analysis.

Further, intensive analyses proved that the Robertsonian translocation in the White Sea herring is caused by polymorphism. The translocation was found to be widely distributed in different stocks. In all locations, however, individuals with 2n = 52 predominated. Almost in all the samples, individuals with 2n = 51 were also found, and in some samples individuals with 2n = 50 with two large metacentrics were also observed (Fig. 3.1.2).

Karyotypes 2n = 52, 2n = 51 and 2n = 50 were compared with each other as also between four populations (from slow- and fast-growing forms of the Kandalaksha, Onega and Dvina bays) by way of measuring the chromosomes. In total, the relative length (RL) and centromeric index (CI) of 148 metaphase plates from 115 individuals have been measured (Lajus, 1989a). This analysis indicated no differences between karyotypes with the same number of chromosomes from different locations.

Chromosomal formula of the karyotype 2n = 52, according to classification by Levan and co-authors (1964), is 6m + 2sm + 44a. The absolute length of the chromosomes varies from 1.5 to 6.0 μm. Three pairs of metacentric and one pair of submetacentric chromosomes may be identified by way of length and centromeric index (Fig. 3.1.3). Among the acrocentric elements, it was possible to identify three pairs of the smallest ones, the remaining chromosomes forming a row gradually decreasing in

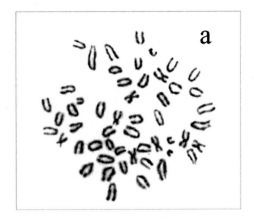

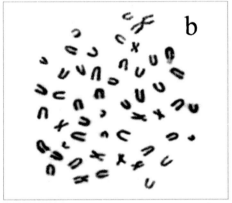

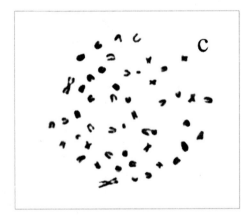

Fig. 3.1.2 Metaphase chromosomes from the White Sea herring: (a) 2n = 52, NF = 60 (slow-growing, Chupa/Kandalaksha Bay); (b) 2n = 51, NF = 60 (slow-growing, Kolezhma/Onega Bay); (c) 2n = 50, NF = 60 (slow-growing, Pushlakhta/Onega Bay).

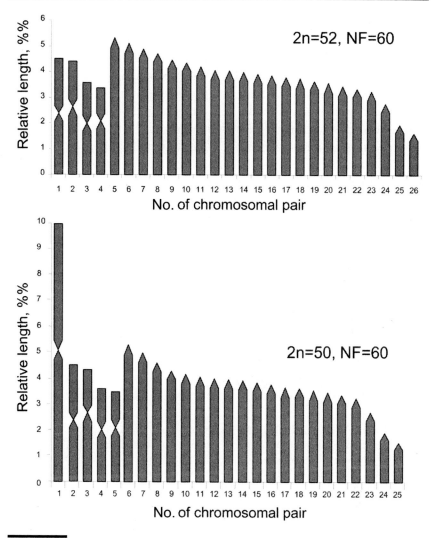

Fig. 3.1.3 Morphometry of the White Sea herring karyotype: 2n=52 (based on 26 metaphase plates from 26 specimens from Chupa/Kandalaksha Bay); 2n=50 (based on 10 metaphase plates from 2 specimens from Pushlakhta/Onega Bay).

length and indistinguishable. In some specimens, 1-2 subtelocentric chromosomes instead of middle-sized acrocentrics were found. They are similar to the subtelocentrics found in Atlantic herring (Lajus, 1994). In the metaphase plates with low condensation of chromatin, satellite chromosomes and chromosomes with secondary constrictions were sometimes visible.

Comparison of morphometry of chromosomes from different karyotypes revealed that the Robertsonian translocation is caused by the fusion of two large acrocentric in a metacentric chromosome with almost equal arms (CI = 49%). The length of these metacentric chromosomes is twice as much as length of other greatest elements in the complement and they are very well distinguishable. Very rarely, an intra-individual variation in the Robertsonian translocation was observed. In the localities where the proportion of fishes with 2n = 51 and 2n = 50 was relatively high (Pushlakhta and Suzma), individuals with modal chromosome number 2n = 52 sometimes possessed cells with 2n = 51.

My data on structure of the White Sea herring karyotype are in good agreement with the results published by Klinkhardt (1993a, b) who studied 21 slow-growing and 18 fast-growing herring from the Kandalaksha Bay. All specimens had karyotype 2n = 52 and formula 6m + 2st + 44a, NF = 58, without any polymorphism. The fourth chromosome pair was classified as subtelocentric, whereas I classified it as submetacentric according to centromeric index (38-40%, Fig. 3.1.3; Table 3.1.1). Probably Klinkhardt used metaphase plates for analysis with more condensed chromatin and, consequently, obtained a smaller value of arm ratio of bi-armed chromosomes. Identical to my observations (Lajus, 1989a), Klinkhardt noted some heteromorphism of large acrocentric chromosomes which is not due to the presence of a nucleolar organizer

Table 3.1.1   Relative length (RL, %%) (M±m) and centromeric index (CI, %%) (M ± m) of chromosomes in karyotypes of the White Sea herring with 2n = 52 (based on 26 metaphase plates from 26 slow-growing individuals from the Chupa Inlet) and with 2n = 50 (based on 10 metaphase plates from 2 individuals from the Pushlakhta Inlet) (from Lajus, 1996a).

Chromosome pairs	2n = 52		Chromosome pairs	2n = 50		Type
	RL	CI		RL	CI	
–	–	–	1	9.98 ± 0.12	49.0 ± 0.4	m
1	4.47 ± 0.03	47.2 ± 0.7	2	4.52 ± 0.10	47.1 ± 0.9	m
2	4.38 ± 0.02	40.0 ± 0.5	3	4.33 ± 0.07	39.2 ± 0.5	m
3	3.58 ± 0.05	44.3 ± 0.7	4	3.60 ± 0.07	43.8 ± 1.0	m
4	3.37 ± 0.05	38.2 ± 0.8	5	3.47 ± 0.11	39.8 ± 1.0	sm
5-23	from 5.32 to 3.24	–	6-22	from 5.32 to 3.25	–	a
24	2.76 ± 0.04	–	23	2.71 ± 0.08	–	a
25	1.94 ± 0.03	–	24	1.91 ± 0.08	–	a
26	1.62 ± 0.05	–	25	1.52 ± 0.12	–	a

region, as revealed by Silver staining (Klinkhardt, 1993a; Fig. 9a) but may be associated with one of the pairs of satellite chromosomes, which are well visible in blastula metaphase plates (Lajus, 1989a). Klinkhardt, however, did not observe the Robertsonian translocation because he has been studying only Kandalaksha herring where the frequency of such rearrangement is too low.

Robertsonian Polymorphism: Relationship between the Forms from the Same Location

The Robertsonian translocation 2n = 52, 51, 50, NF=60 is the only chromosomal marker of the White Sea herring found till date. This translocation is not related to sex (Lajus 1989a). Also, there were no differences between samples collected in different years from the same localities, which allowed using pooled samples. In some cases, herrings during spawning, and also during feeding periods were analyzed. There were no differences between them in frequency of the translocation (Lajus, 1996a). It should be stressed, however, that specimens collected on spawning grounds were sufficient to arrive at conclusions on significant differences between stocks in frequency of the translocation.

Particular attention has been paid to comparing slow- and fast-growing herring from the same localities, because growth rate was always considered the main feature of the White Sea herring stocks. If this is the case, one might expect that the slow-growing herring from different parts of the White Sea are genetically closer to each other than slow- and fast-growing herring from the same location. But this was not true. In the Kandalaksha Bay, both slow- and fast-growing forms were characterized by very low frequencies of the translocation: only one of 144 (0.7%) slow-growing and one of 84 (1.2%) fast-growing herrings had 2n = 51. All others had 2n = 52. Such low frequencies make the comparison difficult, but results show similarity rather than difference between the forms. Samples from the Onega Bay, where frequencies were higher, provided more data for such comparison. In the Sorokskaya Inlet, the ratio of herring with 2n = 51 to those with 2n = 52 was 8:44 (15.4%) in slow growing and 2:12 (14.3%) in fast-growing herrings. In the Kolezhma Inlet, this ratio was 7:86 (7.5%) and 2:18 (10.0%), respectively (Table 3.1.2). These analyses showed no differences between the fast- and slow-growing herrings from the same location. Thus, in further analyses, all the samples collected in one location were pooled together.

Table 3.1.2 Summary of the White Sea herring karyotyping.

Location and form	Technique	Number of studied specimens	Results (karyotypes and number of specimens in parentheses)	Reference
Kandalaksha Bay (Chupa Inlet), slow-growing	Squash on embryos at blastula stage	–	2n = 52, no secondary constrictions, no satellite chromosomes	Skvortsova, 1974a, 1975
Kandalaksha Bay (Chupa Inlet), fast-growing	"	–	2n = 54, 2-4 secondary constrictions and 2-4 satellite chromosomes per plate	Skvortsova, 1974a, 1975
Onega Bay (Kii Island), slow-growing	"	–		Skvortsova, 1974a, 1975
Dvina Bay (Yagry Island), slow-growing	"	–		Skvortsova, 1974a, 1975
Kandalaksha Bay (Chupa Inlet), slow-growing	Squash on larvae	–	2n=52, NF=60	Skvortsova, 1974a, 1975
Onega Bay (Kolezhma Inlet), slow-growing	Squash on adults	18	2n=52, NF=58 (16), 2n=54, NF=58 (2)	Krysanov, 1978
Basin (Chavanga), slow-growing	Cell suspension on adults	61	2n = 52, NF=60 (61)	Lajus, 1989a, 1996a
Kandalaksha Bay (Pila Inlet), slow-growing	"	29	2n = 52, NF=60 (29)	Lajus, 1989a, 1996a
Kandalaksha Bay (Chupa Inlet), slow-growing	"	144	2n = 52, NF=60 (143), 2n=51, NF=60 (1)	Lajus, 1989a, 1996a
Kandalaksha Bay (Chupa Inlet), slow-growing	Cell suspension on embryos at eye-pigmentation stage	759	2n = 52, NF=60 (755), 2n=51, NF=60 (4)	Lajus, 1989a, 1996a
Kandalaksha Bay (Chupa Inlet), fast-growing	Cell suspension on adults	84	2n = 52, NF=60 (83), 2n=51, NF=60 (1)	Lajus, 1989a, 1996a
Kandalaksha Bay (Chupa Inlet), fast-growing	Cell suspension on larvae	85	2n = 52, NF=60 (85)	Lajus, 1989a, 1996a

(Table 3.1.2 contd.)

(Table 3.1.2 contd.)

Kandalaksha Bay (Chupa Inlet), slow-growing	Squash on embryos at blastula stage	60 (obtained from 9 parents with 2n=52, NF=60)	In an average, 2.7 secondary constrictions and 2.6 satellite chromosomes per plate	Lajus, 1989a, 1996a
Kandalaksha Bay (Chupa Inlet), fast-growing	"	49 (obtained from 9 parents with 2n=52, NF=60)	In average, 2.9 secondary constrictions and 2.8 satellite chromosomes per plate	Lajus, 1989a, 1996a
Kandalaksha Bay (Kassian Island), slow-growing	Cell suspension on embryos at eye-pigmentation stage	321	2n = 52, NF=60 (320), 2n=51, NF=60 (1)	Lajus, 1989a, 1996a
Basin (Pongoma Inlet), slow-growing	Cell suspension on adults	32	2n = 52, NF=60 (32)	Lajus, 1989a, 1996a
Onega Bay (Sorokskaya Inlet), slow-growing	"	52	2n = 52, NF=60 (44), 2n=51, NF=60 (8)	Lajus, 1989a, 1996a
Onega Bay (Sorokskaya Inlet), fast-growing	"	14	2n = 52, NF=60 (12), 2n=51, NF=60 (2)	Lajus, 1989a, 1996a
Onega Bay (Kolezhma Inlet), slow-growing	"	93	2n = 52, NF = 60 (86), 2n = 51, NF = 60 (7)	Lajus, 1989a, 1996a
Onega Bay (Kolezhma Inlet), fast-growing	"	20	2n = 52, NF=60 (18), 2n = 51, NF=60 (2)	Lajus, 1989a, 1996a
Onega Bay (Kolezhma Inlet), fast- and slow-growing	Cell suspension on embryos at eye-pigmentation stage	713	2n = 52, NF=60 (649), 2n=51, NF=60 (61), 2n=50, NF=60 (1)	Lajus, 1989a, 1996a
Onega Bay (Kolezhma Inlet), slow-growing	Squash on embryos at blastula stage	62 (obtained from 15 parents with 2n=52, NF=60)	In average, 2.8 secondary constrictions 2.8 satellite chromosomes per plate	Lajus, 1989a, 1996a
Onega Bay (Kiy Island), slow-growing	Cell suspension on adults	71	2n = 52, NF=60 (61), 2n=51, NF=60 (9), 2n = 50, NF=60 (1)	Lajus, 1989a, 1996a

(Table 3.1.2 contd.)

(Table 3.1.2 contd.)

Onega Bay (Pushlakhta Inlet), slow-growing	"	97	2n = 52, NF=60 (80); 2n=51, NF=60 (14), 2n = 50, NF=60 (3)	Lajus, 1989a, 1996a
Dvina Bay (Suzma village), slow-growing	"	45	2n = 52, NF=60 (34), 2n=51, NF=60 (10), 2n = 50, NF=60 (1)	Lajus, 1989a, 1996a
Kandalaksha Bay (Chupa Inlet), slow-growing	Squash on adults	21	2n =52, NF=58 (21), NOR is located at 4th-6th pair of acrocentrics	Klinkhardt, 1993a,b
Kandalaksha Bay (Chupa Inlet), fast-growing	"	18	2n = 52, NF=58 (18)	Klinkhardt, 1993a,b

Table 3.1.3 Frequency of the Robertsonian translocation (2n = 52, 51, 50) in the White Sea herring (from Lajus, 1996b). ad – adults, em – embryos at the stage of eye pigmentation.

Location	Material used	Number of specimens observed	Number of specimens with 2n =			Frequency	significance NS - $p>0.05$; * - $p<0.05$; ** - $p<0.01$; *** - $p<0.001$								
			52	51	50		1	2	3	4	5	6	7	8	9
1. Chavanga	ad	61	61	0	0	0.000	#								
2. Pila	ad	29	29	0	0	0.000	NS	#							
3. Kasian	em	321	320	1	0	0.002	NS	NS	#						
4. Chupa	ad, em	1130	1124	6	0	0.003	NS	NS	NS	#					
5. Pongoma	ad	32	32	0	0	0.000	NS	NS	NS	NS	#				
6. Sorokskaya	ad	66	56	10	0	0.076	**	*	***	***	*	#			
7. Kolezhma	ad, em	826	756	69	1	0.043	**	NS	***	***	NS	NS	#		
8. Kiy	ad	71	61	9	1	0.078	***	*	***	***	*	NS	NS	#	
9. Pushlakhta	ad	97	80	14	3	0.103	***	**	***	***	**	NS	**	NS	#
10. Suzma	ad	45	34	10	1	0.133	***	**	***	***	***	NS	**	NS	NS

Robertsonian Polymorphism: Frequency in Embryos

Low frequencies of the translocation required analyses of a large number of individuals, thus seriously limiting the utility of chromosomal investigation for population studies. In order to overcome the problem, a technique for assessing the frequency of Robertsonian translocation on early stages of development has been attempted. Developing eggs of the White Sea herring are being attached to littoral and sublittoral vegetation (mostly to brown macroalgae *Fucus* and *Ascophyllum*) and are easily available for sampling at all stages without special equipment. Importantly, the observations of the herring's spawning behavior indicated a wide dispersion of eggs and sperm in the spawning grounds (Blaxter and Hunters, 1982; Lajus, 1989b). So, even in small areas of a spawning ground, there existed offspring of many parental fishes. Therefore, the sample of eggs from a small part of a spawning ground may represent an entire population.

Taking these circumstances into account, an original modification of chromosomal analysis of embryos has been elaborated (Lajus 1989a, 1996a). Capsules of eggs were separated by passing them through suitable gauze in Petri dishes with physiological solution. The tissues were consequently treated with colchicine, hypotonic solution and then fixed. Next, the tissues were suspended and used to prepare slides. The quality and quantity of metaphases were greatly dependent on the stage of embryogenesis, with the best results at the beginning of the eye-pigmentation stage. Under optimal conditions, it was possible to obtain 200-300 well-spread metaphase plates per slide; 40-150 plates in each slide were usually analyzed. For this technique, it was crucial that karyotypes of the White Sea herring can be easily differentiated by the presence or absence of one or two well-visible large metacentric chromosomes.

The above technique has been applied on three spawning grounds in the Kandalaksha and Onega Bay. Quantity of metaphases analyzed per spawning ground was several hundred, i.e., one or even two orders more than it was possible on adults during the same time. The results were basically similar to those obtained on the adults but allowed a more exact estimation of frequencies of the translocation and obtaining results of higher significance (Table. 3.1.2).

Robertsonian Polymorphism: Geographical Patterns

An overall picture of frequencies of the Robertsonian translocation when the samples obtained from the same location are pooled (adults and embryos, slow- and fast-growing, collected in different years) shows that the frequencies differ greater for more distant locations (Table 3.1.3). The frequency increases from the north-west to the south-east of the White Sea. Herring with low frequency of the rearrangement (0.000-0.005) occur in the Kandalaksha Bay and the adjacent regions of the Basin. In the central part of the Onega Bay, herring have a higher frequency (0.04-0.08). The highest frequency (about 0.13) has been observed in the Dvina Bay. Herring from locations between the Onega and Dvina bays (Pushlakhta) have an intermediate frequency of translocation (0.10) (Fig. 3.1.4).

This finding represents the first evidence of chromosomal polymorphism in the genus *Clupea* and in clupeids. In particular, no translocation has been reported for the Pacific herring (Ohno *et al.*, 1969), although further analysis of a larger number of specimens would be useful in supporting this conclusion. Also, no translocation has been found in

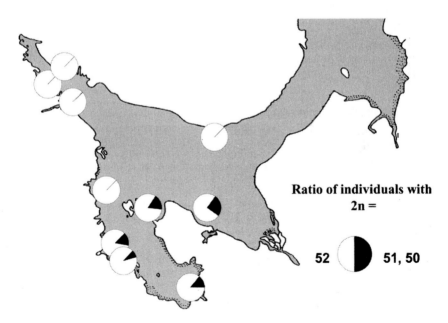

Fig. 3.1.4 Ratio of White Sea herring individuals with 2n=52 and with 2n=51 and 50 in different parts of the sea.

the Atlantic herring *C. harengus* (Klinkhardt, 1993a; Lajus, 1994). The Atlantic herring differ from all stocks of the White Sea herring by way of the structure of one chromosome pair (Lajus, 1994). Unfortunately, no cytogenetic information is available on the samples of interesting herring from the south-eastern part of the Barents Sea, which also represents a subspecies of the Pacific herring, and is very closely related to the White Sea herring.

The presence of the same chromosomal rearrangement in herring from most locations of the White Sea reveals that the White Sea herring stocks are closely related and, evidently, there is gene exchange between them. It is known that herring usually return for spawning to their parental spawning grounds (Blaxter and Hunters, 1982). This prevents the mixing of populations from different locations, maintaining the difference in the frequency of chromosomal rearrangement between them.

Rate of Chromosomal Evolution of the White Sea Herring

Bikham and Baker (1979) in their 'canalization model' predict that chromosomal evolution occurs in three successive stages. The first stage starts immediately after colonization of a new adaptive zone and it is characterized by the highest rate of chromosomal rearrangement owning to extensive selection exerted on newly emergent chromosomal variants. The evolution rate diminishes on subsequent stages. The data on the White Sea herring seem to fit well with this hypothesis. Probably, the translocation appeared anew just after colonization of the sea by herring and we are observing initial steps of their chromosomal evolution.

There are several hypotheses concerning the fate of a chromosomal mutation. Some of them are discussed here, although it is not easy to find evidence to verify them. Wilson and co-authors (1975) consider that conditions under which a chromosomal mutation can be fixed are severely limited and it is extremely unlikely that such a mutation will be fixed in a large outbreeding population. Ten individuals seem to be a maximal size of the population for which chromosomal rearrangement can be fixed. But herring populations are usually large, and individuals have a high mobility (Blaxter and Hunter, 1982). Moreover, herring are adapted to spawn in large intermixing groups. This limits the role of genetic drive and inbreeding, considered most important in fixation of chromosomal rearrangement. According to White (1978) and Hedrick (1981), other factors such as meiotic drive and selection also influence the probability of fixation of a new chromosomal rearrangement. Meiotic drive, even if rather slow, may potentially have serious populational and evolutional consequences, possibly increasing the rate of fixation of chromosomal rearrangement not only in small demesne, but also in large populations (theoretical analysis by Walsch, 1982). However, the actual role of meiotic drive in chromosomal evolution remains quite unclear at present (Gileva, 1990).

Selection seems to act in a more explanatory manner. In the case under study, the frequency of the translocation could have increased due to selective advantages of individuals with 2n = 51 and 50 in comparison with those with 2n = 52 in southern parts of the sea. Selective advantage of reducing chromosome number may be due to new chromosomes, which contain a complex of associated genes (Scheel, 1972; Birshtein, 1987; Kirpichnikov, 1987). Such chromosomal rearrangements limit the recombination level and can be favorable for organisms occupying a narrow ecological niche (Quimsiyeh, 1994).

Another explanation, based on selection, is the enhanced fitness of heterozygotes recorded in some animals (Christensen and Pedersen, 1982; Baker et al., 1983). When the frequencies are low, similar to the White Sea herring, this factor can be an important one in increasing the frequency of chromosomal rearrangement, because almost all individuals bearing the translocation are heterozygotes. In total, among 2553 specimens analyzed in this study, translocation was found in 125 individuals and 119 of them were heterozygotes. While considering factors causing fixation of the translocation, one has to also keep in mind the negative effect caused by disturbance of meiosis, which occur in heterozygous individuals. Fixation of the chromosomal rearrangement may occur only if such a negative effect is week or absent (Baker et al., 1987; King, 1987).

The known age of the White Sea, which does not exceed 10,000 years (Derjugin, 1928; Berg, 1934; Andriashev, 1957), allows to estimate the rate of chromosomal evolution in herring. The ratio between the frequency of the rearrangement (maximal 0.13 for the Dvina Bay population) and the time of rearrangement yields 1.3×10^{-5} rearrangements per year. The mean rate of fixation in Teleostei may be assessed as the ratio between the average number of rearrangements per year per species (about 2.9×10^{-8} according to Bush et al., 1977) and the proportion of polymorphic species (0.01 according to Vasil'ev, 1985). The result, 2.9×10^{-6}, shows that in the White Sea herring (population of the

Dvina Bay), the rate of fixation is about five-fold higher than the average rate in Teleostei. The comparatively high rate may be explained by a possible initial stage of chromosomal evolution of the White Sea herring, according to the model of chromosomal evolutionary change suggested by Bikham and Baker (1979).

Chromosomal Differentiation and Population Structuring

What do the above data mean for the population studies of the White Sea herring? First, and the most important, stocks from different localities differ in the frequency of chromosomal rearrangement, which reveals their reproductive isolation. Hence, the White Sea herring is not a single population, as Lapin and his followers argued (Lapin, 1966; Lapin and Pokhiluk, 1993). Second, the presence of the same translocation in populations from different locations indicates their close genetic similarity and, probably, the exchange of genes between them. This evidence rejects the hypothesis of a different origin of the White Sea herring stocks (Dushkina, 1975). Third, the finding of chromosomal differences between stocks from different parts of the White Sea, but not between fast- and slow-growing herring from the same location, is contradictory to the concept of Rabinerson (1925) and Averintzev (1927, 1928), who divided the White Sea herring into the two large groups of slow- and fast-growing. Therefore, these data allowed the formulation of a new concept of the White Sea herring population structure: the White Sea herring is a system of geographically distinct populations of Pacific origin characterized by poor gene flow (Lajus, 1996b).

Usually chromosomal analysis has lower resolution in population structuring analysis than conventional methods of population genetics such as protein electrophoresis. This is, however, not the case for the White Sea herring. An analysis of variation of polymorphous enzymes which included numerous samples from different populations collected in different years did not show differences between populations from geographically distinct locations (Semenova et al., 2004), although the same study revealed differences between the White Sea herring and other relative forms of Pacific herring occurring in the Chesha and Pechora inlets in the south-eastern part of the Barents Sea.

The White Sea herring is characterized by a complicated hierarchical population structure in which chromosomal translocation marks the first level, constituting the difference between geographically distinct stocks.

In addition to the cytogenetic difference (frequency of chromosomal rearrangement), these stocks are characterized by other distinctive features including morphometric and meristic characters (Lajus, 1996b). The second level of population structure refers to stocks from the same location, differing in spawning time. Such stocks occur in the Kandalaksha Bay (Chupa Inlet) and are different in salinity resistance of larvae (Ivanchenko and Lajus, 1985), morphological characters (Lajus, 1996b, 2001), and also in the frequency of allozymes (Semenova et al., 2004). The third level of structure is represented by stocks that spawn at the same time in the same location, but differ in the growth rate. They have some morphological differences which are, however, smaller than that between populations spawning in different seasons (Lajus, 1996b). Evolution of populations on the two latter levels is unaccompanied by chromosomal rearrangement, or the exchange between them is too large to maintain difference in the frequencies of possible rearrangements.

Previous surveys of both Pacific and Atlantic herring have revealed little or no genetic differentiation of putative stocks (Anderson et al., 1981; Kornfield et al., 1982; Grant, 1984; Grant and Utter, 1984; Ryman et al., 1984; Kornfield and Bogdanowicz, 1987; Schweigert and Withler,1990). Kirpichnikov (1987) considered herring to be a fish with only slightly differentiated population structure. The apparent lack of genetic structuring in coastal and oceanic herring stocks has even lead to a reappraisal of the genetic stock concept (Smith and Jamieson, 1986; Smith et al., 1990). These authors have discussed the validity of the traditional view of a discrete stock model and argue strongly for the acceptance of a more dynamic population concept. Support for the discrete population concept in herring is, however, provided in some studies (Stephenson and Kornfield, 1990; McPherson et al., 2001). Furthermore, in some cases, genetic differences between neighbouring herring stocks may be very large. The electrophoretic data indicate a considerable degree of reproductive isolation between the stocks of herring from some fjords of the Norwegian coast and adjacent oceanic and coastal populations, with genetic distance values suggesting a taxonomic status of subspecies or species (Jorstad et al., 1991). These stocks show high similarity to the Pacific herring (Jorstad, 2004) probably sharing the same origin. This means that the ability of herring to form genetically isolated stocks may take place easily in suitable conditions. The results of the study described in the present chapter show that the White Sea herring is one of such examples.

CONCLUSIVE REMARKS

The White Sea herring provides an example of development of karyological studies in a context of population structuring analysis directed to answer questions raised by population concepts. At the beginning, the replies in some cases were probably determined by *a priori* ideas but not by scientific objectivity, as none of reported differences were confirmed by latter analyses. In part, this was also caused by methodical imperfections of the available cytogenetic techniques. Nevertheless, at the end, chromosomal analysis occupied an important place in population studies of the White Sea herring. A chance to find a Robertsonian polymorphism in the White Sea herring was a first (necessary but not sufficient) condition for successful application of karyotyping in the population analysis. The second condition was the existence of differences in the frequency of the Robertsonian rearrangement between the stocks. We know several examples when chromosomal analysis revealed differences between populations within a polymorphous species, especially among Salmonids (e.g., Hartley and Horne, 1982), but in most cases, cytogenetic approaches were used to study relationships between populations which are known to be isolated. The peculiar distinction of the White Sea herring is that cytogenetic analysis of a chromosomal polymorphism has been used as a tool to analyse population structuring.

Acknowledgments

I thank Eva Pisano and an anonymous referee for useful suggestions improving the manuscript.

References

Altukhov, K.A. 1975. Sel'd' Solovetskikh ostrovov [Herring of Solovetsky Islands]. *Issledovaniia Fauny Morei* (*Exploration of Fauna of Seas*) 16: 53–91. (In Russian).

Anderson, L., N. Ryman, R. Rosenberg and G. Stahl. 1981. Genetic variability in Atlantic herring (*Clupea harengus harengus*): Description of protein loci and population data. *Hereditas* 95: 69–78.

Andriashev, A.P. 1957. Nekotorye voprosy amfiboreal'nogo rasprostraneniia s zamechaniiami o vozmozhnom proiskhozhdenii navagi i malopozvonkovoi sel'di [Same questions of amphiboreal distribution with notes on possible origin of navaga and low-vertebrate herring]. *Materialy po Kompleksnomu Izucheniiu Belogo Moria* [*Materials on Complex Investigation of the White Sea*]. Moskow-Leningrad, Izd. AN SSSR, 1: 117–125. (In Russian).

Averintsev, S.V. 1927. Sel'di Belogo moria ch.1. [Herring of the White Sea, part 1]. *Trudy Nauchnogo instituta rybnogo khoziaistva* [*Proceedings of the Research Institute for Fisheries*] 2: 41–77. (In Russian).

Averintsev, S.V. 1928. Sel'di Belogo moria ch. 2, 3 [Herring of the White Sea, parts 2 and 3]. *Trudy Nauchnogo instituta rybnogo khoziaistva* [*Proceedings of the Research Institute for Fisheries*] 3-4: 73–142. (In Russian).

Baker, R.J., B.F. Koop and M.W. Haiduk. 1983. Resolving systematic relationships with G-bands: A study of five genera of South American cricetine rodents. *Systematic Zoology* 32: 403–416.

Berg, L.S. 1934. Ob amfiboreal'nom (preryvistom) rasprostranenii morskoi fauny v severnom polusharii [On amphiboreal (discontinuous) distribution of marine fauna in North Hemisphere]. *Izvestia Geograficheskogo Obtschestva* [*Proceedings of the Geographical Society*]. 66: 69–78. (In Russian).

Bikham, J.W. and R.J. Backer. 1979. Canalization model of chromosomal evolution. *Bulletin of Carnegie Museum of Natural History, Pittsburg* 13: 70–84.

Birshtein, V.J. 1987. *Thsytogeneticheskie i Molekuliarnye Aspekty Evoliutsii Pozvonochnykh* [*Cytogenetic and Molecular Aspects of Vertebrate Evolution*]. Nauka Press, Moscow. (In Russian).

Blaxter, J.H.S. and H. Hunters. 1982. The biology of the clupeoid fishes. *Advances in Marine Biology* 20: 1–223.

Bush G.L., S.W. Case, A.C. Wilson and J.L. Patton. 1977. Rapid speciation and chromosomal evolution in mammals. *Proceedings of the National Academy of Sciences of the United States of America* 74: 3942–3946.

Christensen, K. and H. Pedersen, 1982. Variation in chromosome number in the blue fox (*Alopex lagopus*) and its effect on fertility. *Hereditas* 97(2): 211–215.

Crosetti, D., W.S. Nelsson and J.C. Avise. 1994. Pronounced genetic structure of mitochondrial DNA among populations of the circumglobally distributed grey mullet (*Mugil sephalus*). *Journal of Fish Biology* 44: 47–58.

Danilevsky, N.J. 1862. Rybnye i zverinye promysly v Belom i Ledovitom Moriakh [Fisheries and marine hunting in the White and Arctic Seas]. *Issledovania o Sostoianii Rybolovstva v Rossii* [*Investigations on the Status of Fisheries in Russia*] 6, St. Petersburg. (In Russian).

Derjugin, K.M. 1928. Fauna Belogo moria i usloviia ee suschestvovaniia [Fauna of the White Sea and environmental conditions of its existence]. *Issledovania Morei SSSR* [*Exploration of the USSR seas*] 7/8, Leningrad. (In Russian).

Dmitriev, N.A. 1946. *Biologia i Promysel Seldi v Belom More* [*Biology and Fishery of Herring in the White Sea*]. Pitshepromizdat Press, Moscow.

Doucette, A.J. and J.M. Fitzsimins. 1988. Karyology of Elopiform and Clupeiform fishes. *Copeia* 1988: 124–130.

Dushkina, L.A. 1975. Pigmentatsia sel'dei roda *Clupea* kak odin iz vozmozhnykh pokazatelei ikh proiskhozhdenia [Pigmentation of herring of genus *Clupea* as one of possible indicators of its origin]. *Issledovaniia Fauny Morei (Exploration of the Fauna of Seas)* 16: 227–254. (In Russian).

Galetti. P.M. Jr., C.T. Anguilar and W.F. Molina. 2000. An overview of marine fish cytogenetics. *Hydrobiologia* 420: 55–62.

Gileva E.A. 1990. *Khromosomnaia Izmenchivost' i Evoliutsiia [Chromosomal Variation and Evolution]*. Nauka Press, Moscow. (In Russian).

Gold J.R.A. 1974. A fast and easy method for chromosome karyotyping in adult teleosts. *Progressive Fish-Culturist* 36: 169–171.

Grant, W.S. 1984. Biochemical population genetics of Atlantic herring, *Clupea harengus*. *Copeia* 1984: 357–364.

Grant, W.S. and F.M. Utter. 1984. Biochemical population genetics of Pacific herring (*Clupea pallasi*). *Canadian Journal of Fisheries and Aquatic Sciences* 41: 856–864.

Hartley, S.E. and M.T. Horney. 1982. Chromosomal polymorphism in the rainbow trout (*Salmo gairdneri* Richardson). *Chromosoma* 87: 461–468.

Hedrick, G.H. 1981. The establishment of chromosomal variants. *Evolution* 35: 322–332.

Ivanchenko, O.F. and D.L. Lajus. 1985. Ustoichivost' lichinok i mal'kov belomorskoi sel'di k nizkim solenostiam [Resistance of larvae and fry of the White Sea herring to low salinity]. *Trudy Zoologicheskogo Instituta [Proceedings of Zoological Institute]* 179: 70–79. (In Russian).

Jorstad, K. 2004. Evidence for two highly differentiated herring groups at Goose Bank in the Barents Sea and the genetic relationship to Pacific herring, *Clupea pallasi*. *Environmental Biology of Fishes* 69: 211–221.

Jorstad, K.E., D.P.F. King and G. Naevdal. 1991. Population structure of Atlantic herring, *Clupea harengus* L. *Journal of Fish Biology* 39 (Supplement A): 43–52.

King, M. 1993. *Species Evolution: The Role of Chromosomal Change*. Cambridge University Press. Cambridge.

Kirpichnikov, V.S. 1987. *Genetika i Selektsiia Ryb [Genetics and Selection of Fishes]*. Nauka Press, Leningrad. (In Russian).

Klinkhardt, M.B., 1993a. Cytogenetics of the herring (*Clupea harengus* L.) 1. Karyotypes of North European stocks. *Cytobios* 75: 113–128.

Klinkhardt, M.B., 1993b. Cytogenetics of the herring (*Clupea harengus* L.) 2. Karyotypes of White Sea herring groups. *Cytobios* 75: 149–156.

Kornfield, I. and S.M. Bogdanowicz. 1987. Differentiation of mitochondrial DNA in Atlantic herring, *Clupea harengus*. *Fisheries Bulletin* 85: 561–568.

Kornfield, I., B.D. Sidell and P.C. Gagnon. 1982. Stock definition in Atlantic herring (*Clupea harengus harengus*): Genetic evidence for discrete fall and spring spawning populations. *Canadian Journal of Fisheries and Aquatic Sciences* 39: 1610–1621.

Krysanov, E.J. 1978. Ob izmenchivosti chisla khromosom u sel'di [On variation of chromosomal number in herring]. In: *Ekologia Ryb Belogo Moria [Ecology of the White Sea Fishes]*, J. Lapin (ed.) Nauka Press, Moscow. pp. 94–98. (In Russian).

Lajus, D.L. 1987. Novye dannye o kariologicheskikh osobennostiakh sel'dei, nerestiaschikhsia v Kandalakshskom zalive Belogo moria. [New data on karyological characteristics of herring spawning in the Kandalaksha Bay of the White Sea]. *Doklady Akademii Nauk SSSR* 297: 254–256. (In Russian) (Translation into English: Laius, D. L. 1987, *Doklady Biological Sciences* 292–297: 698–701).

Lajus, D.L. 1989a. Preobrazovaniia khromosomnykh naborov belomorskoi sel'di (*Clupea pallasi marisalbi* Berg) (Teleostei, Clupeidae) [Chromosomal transformations in the

White Sea herring (*Clupea pallasi marisalbi* Berg) (Teleostei, Clupeidae)]. *Trudy Zoologicheskogo Instituta* (*Proceedings of Zoological Institute*) 192: 113–125. (In Russian).

Lajus, D.L. 1989b. *Osobennosti ekologii i populiatsionnoi struktury seldei belogo moria.* [*Ecology and population structure of herring in the White Sea*]. Ph.D. Thesis. Zoological Institute of Russian Academy of Sciences. Leningrad. (In Russian).

Lajus, D.L. 1994. Opisanie khromosomnogo nabora atlanticheskoi sel'di iz Belogo moria [Description of the chromosome set of *Clupea harengus* from the White Sea]. *Voprosy ikhtiologii.* 1994 34: 572–574. (In Russian). Translation into English: Lajus, D.L. 1995. *Journal of Ichthyology* 35: 132–136.

Lajus, D.L. 1996a. White Sea herring (*Clupea pallasi marisalbi*, Berg) population structure: Interpopulation variation in frequency of chromosomal rearrangement. *Cybium* 20: 279–294.

Lajus, D.L. 1996b. What is the White Sea herring *Clupea pallasi marisalbi* Berg, 1923?: A new concept of the population structure. *Publicaciones Especiales Instituto Espanol de Oceanografia* 21: 221–230.

Lajus, D.L. 2001. Variation patterns of bilateral characters: variation among characters and among populations in the White Sea herring (*Clupea pallasi marisalbi*). *Biological Journal of the Linnean Society* 74: 237–253.

Lajus, D.L. 2002. Long-term discussion on the stocks of the White Sea herring: Historical perspective and present state. *ICES Marine Science Symposia* 215: 315–322.

Lapin, J.E. 1966. Belomorskaia sel'd' kak biologicheskoe tseloe [The White Sea herring as a biological entity]. In: *Zakonomernosti Dinamiki Chislennosti Ryb Belogo Moria i ego Basseina* [*Patterns of Dynamics of Fish Abundance of the White Sea and its Basin*]. Nauka Press, Moscow, pp. 5–28. (In Russian).

Lapin, J.E. 1978. Obschie zakonomernosti dinamiki populiatsii i raznoglasia v ikh traktovke (General patterns of population dynamics and disagreements in their interpretation). *Ekologia Ryb Belogo Moria* [*Ecology of the White Sea Fishes*], J. Lapin (ed.). Nauka Press, Moscow, pp. 98–112. (In Russian).

Lapin, J.E. and V.V. Pokhiliuk. 1993. Prostranstvennaia dinamika pokolenii belomorskoi sel'di *Clupea pallasi marisalbi* [Spatial dynamics of generations of the White Sea herring *Clupea pallasi marisalbi*]. *Voprosy ikhtiologii,* 33: 367–371. (In Russian).

Lepekhin, I.I. 1805. *Dnevnye zapisi puteshestviia doktora i Akademii nauk ad'iunkta Ivana Lepekhina po raznym provintsiiam Rossiiskogo gosudarstva v 1768 i 1769 gg.* [Daily notes of voyages of doctor Ivan Lepekhin to different provinces of the Russian State in 1768 and 1769]. St. Petersburg, pt. 4. (In Russian).

Levan, A., K. Fredga and A.A. Sandberg. 1964. Nomenclature for centromeric position on chromosomes. *Hereditas* 52: 201–220.

McPherson, A.A., R.L. Stephenson, P.T. O'Reilly, M.W. Jones and C.T. Taggart. 2001. Genetic diversity of coastal Northwest Atlantic herring populations: Implications for management. *Journal of Fish Biology* 59 (Supplement A): 356–370.

Ohno, S. 1970. *Evolution by Gene Duplication.* Springer-Verlag, New York.

Ohno, S., J. Muramoto, J. Klein and N. Atkin. 1969. Diploid-tetraploid relationships in clupeoid and salmonoid fish. *Chromosomes Today, II, Oxford Chromosome Conference,* 2: 139–148.

Quimsiyeh, M.B. 1994. Evolution of number and morphology of mammalian chromosomes. *Journal of Heredity* 85: 455–465.

Rabinerson, A.I. 1925. Materialy po issledovaniu belomorskoi sel'di (po dannym 1923–1924 gg.) [Materials for investigation of the White Sea herring (on data from 1923–1924]. *Trudy Instituta po Izucheniu Severa [Proceedings of the Institute for the Exploration of the North]* 25. 146 pp. (In Russian).

Rossi, A.M., M. Capula and D.E. Campton. 1998. Allozyme variation in global populations of stripped mullet, *Mugil cephalus* (Pisces: Mugilidae). *Marine Biology* 131: 203–212.

Rossi, A.M., D. Crosetti, E. Cornung and L. Sola. 1996. Cytogenetic analysis of global populations of *Mugil cephalus* (stripped mullet) by different staining techniques and fluorescent *in situ* hybridization. *Heredity* 76: 77–82.

Ryman, N., U. Lagercrantz, L. Anderson, R. Chakraborty and R. Rosenberg. 1984. Lack of correspondence between genetic and morphologic variability patterns in Atlantic herring (*Clupea harengus*). *Heredity* 53: 687–704.

Scheel, J.J. 1972. Rivuline karyotypes and their evolution (*Rivuline*, Cyprinodontidae, Pisces). *Zeitschrift für Zoologische Systematik und Evolutionsforschung* 10: 180–229.

Schweigert, J.F. and R.E. Withler. 1990. Genetic differentiation of Pacific herring based on enzyme electrophoresis and mitochondrial DNA analysis. *American Fishery Society Symposium* 7: 459–469.

Semenova, A.V., A.P. Andreeva, A.K. Karpov, S.B. Frolov, E.I. Feoktistov, G.G. Novikov. 2004. Geneticheskaia izmenchivost' sel'dei roda *Clupea* Belogo moria [Genetic variation of herring of genera *Clupea* of the White Sea]. *Voprosy ikhtiologii* 2: 207–217. (In Russian).

Shelenkova, N.J. 1986. Kariologiia nerki (*Oncorhynchus nerka* Walb.) dvukh izolirovannykh populiatsii na Kamchatke [Karyology of the sockey (*Oncorhynchus nerka* Walb.) of two isolated populations of Kamchatka]. *Thsytologia* 28: 735–739. (In Russian).

Skvortsova, T.A. 1975. Khromosomnye kompleksy belomorskoi sel'di (*Clupea harengus pallasi n. maris-albi* Berg) i salaki (*Clupea harengus harengus n. membras* L.) [Chromosomal complexes of the White Sea herring (*Clupea harengus pallasi n. maris-albi* Berg) and Baltic herring (*Clupea harengus harengus n. membras* L.)]. *Issledovania Fauny Morei (Exploration of Fauna of Seas)* 16: 104–108. (In Russian).

Smith, P. and A. Jamieson. 1986. Allozyme data and stock discreetness in herring: a conceptual revolution. *Fisheries Research* 4: 223–234.

Smith, P., A. Jamieson and A.J. Birley. 1990. Electrophoretic studies and the stock concept in marine teleosts. *Journal du Counseil* 47: 231–245.

Stephenson, R.L. and I. Kornfield. 1990. Reappearance of spawning Atlantic herring (*Clupea harengus harengus*) on Georges Bank: population resurgence not recolonization. *Canadian Journal of Fisheries and Aquatic Sciences* 47: 1060–1064.

Tambovtzev, B.M. 1957. Malopozvonkovye sel'di Barentseva I Belogo morei [Low-vertebrate herring of the Barents and White Seas]. *Trudy Murmanskoi Biologicheskoi Stantsii [Proceedings of the Murman Biological Station]* 3: 169–174. (In Russian).

Vasil'ev, V.P. 1985. *Evoliutsionnaia Kariologiia Ryb [Evolutionary Karyology of Fishes]*. Nauka Press, Moscow. (In Russian).

Walsch, J.B., 1982. Rate of accumulation of reproductive isolation by chromosome rearrangements. *American Naturalist* 120: 510–522.

White, M.J. 1978. *Models of Speciation*. W.H. Freeman and Co., San Francisco.

Wilson, A.C., G.L. Bush, S.M. Case and M.C. King. 1975. Social structuring of mammalian populations and rate of chromosomal evolution. *Proceedings of the National Academy of Sciences of the United States of America* 72: 5601–5065.

Cytogenetic Characteristics of Interspecific Hybrids and Chromosome Set Manipulated Finfish

Konrad Ocalewicz[1*], Malgorzata Jankun[1] and Miraslau Luczynski[2]

INTRODUCTION

The importance of finfish aquaculture continues to increase due to over-fishing of natural fish stocks in oceans, as reported in recent decades. Understanding and controlling the process of fish reproduction is one of the crucial factors enabling aquaculture development. Modern approaches, including interspecific hybridization and application of chromosome set manipulations, had passed through their experimental phase and have been successfully established in aquaculture (Ihssen *et al.*,

Address for Correspondence: [1*]Department of Ichthyology, University of Warmia and Mazury in Olsztyn, ul. Oczapowskiego 5, 10-957 Olsztyn, Poland. E-mail: con@uwm.edu.pl

[2]Department of Environmental Biotechnology, University of Warmia and Mazury in Olsztyn, ul. Oczapowskiego 5, 10-957 Olsztyn, Poland

1990; Donaldson, 1996; Pandian and Koteeswaran, 1998; Arai, 2001; Beardmore *et al.*, 2001; Hulata, 2001; Babiak *et al.*, 2002b; Gomelsky, 2003).

Successful incorporation of methods establishing fast-growing and sterile fish hybrids, monosex strains or polyploid organisms, strongly depends on the availability of tests for checking out the efficiency of such techniques. Moreover, it is necessary to study all kinds of alterations in genome structure caused by hybridization and chromosomes set manipulations such as: genome asymmetry in hybrids, homozygosity in gynogenetic and androgenetic organisms, chromosomal aberrations in genome manipulated fish, development of polyploids, functional disturbances in sex reversed fish, etc.

The aim of the present chapter is to emphasize the application of cytogenetic techniques in modern aquaculture. Rapid development of fish cytogenetics has focused on constructing standard karyotypes of fish species, chromosomal localization of particular DNA sequences and revealing chromosome anomalies. Applications of modern cytogenetics combining basic staining techniques with fluorescence *in situ* hybridization (FISH) and its modifications, together with flow cytometry and electron microscopy have also been reviewed in this chapter.

Interspecific, Intergeneric and Intraspecific Hybridization of Fishes

Both natural and artificial interspecific hybridization are widespread in fish species (Schwartz, 1981; Bartley *et al.*, 2000; Scribner *et al.*, 2000; Allendorf *et al.*, 2001). Among numerous examples, natural interspecific hybridization takes place between Atlantic salmon (*Salmo salar*) (2n= 58, NF= 74) and brown trout (*Salmo trutta*) (2n= 80, NF= 74) (Garcia de Laniz and Verspoor, 1989; Jansson *et al.*, 1991) and between some cyprinid fish species (Kapusta and Ciesielski, unpublished). On the other hand, interspecific (or even intergeneric) hybridization is often used in aquaculture as a way for obtaining desired changes in commercial fish strains (Chevassus, 1983). Hybrids can show improved growth rate (Song, 1987), higher disease resistance (Kerby and Havell, 1980) or bigger harvest yield (Zhang, 1979). In certain cases, interspecific hybridization can be applied in order to establish monosex or sterile fish stocks (Lahav and Lahav, 1990).

During hybridization, nuclear genomes of different species are placed in the cytoplasm derived from maternal parent. In a variety of cases, hybrid embryos develop without any visible perturbation. In other cases, however, incompatibility of parental genomes (differences in chromosome size, number and morphology) and asynchronization in successive cell divisions (Ye *et al.*, 1989) can cause re-organizations and modifications of genomes. Due to this phenomenon, hybridization can involve partial chromosome elimination or entire genome alterations, like gynogenesis, androgenesis or development of polyploids (Chevassus, 1983). The extreme response to the parental genome incompatibility is the inviability of hybrids, which die before hatching (Arai, 1984; Gray *et al.*, 1993).

Genetic studies of hybrids encompass, among others, methods, which directly or indirectly identify parental chromosomes and enable estimation of the efficiency of hybridization process (determination of the amount of DNA per cell, molecular, biochemical and karyological studies). Identification of parental chromosomes in hybrid fish requires application of techniques of visualization of chromosome-banding patterns (structural and replication banding techniques) towards species taking part in the hybridization process. This, in turn, enables description of the so-called marker chromosomes such as NORs (Nucleolar Organizer Regions) bearing chromosomes, chromosomes with peculiar morphological characteristics or those possessing clusters of AT- or GC-rich chromatin or heterochromatin. Identification of species-specific DNA sequences and their use as probes in fluorescence *in situ* hybridization can make it possible to identify some (or all) parental chromosomes (Phillips and Reed, 1996).

Chromosome elimination in hybrids – Evidence from conventional and molecular cytogenetics

One explanation for the interspecific hybrids' inviability is their chromosome loss during embryonic development. This process has been observed in natural and artificially induced hybrids in a variety of animals and plants (Bennett *et al.*, 1976; Finch, 1983; Breeuwer and Werren, 1990; Reed and Werren, 1995). The cytogenetic analysis of fish hybrids performed with basic cytogenetic techniques provided data about chromosome fragmentation and loss (Goodier *et al.*, 1987; Yamano *et al.*, 1989; Yamazaki *et al.*, 1989).

Silver-banded karyotypes (in which the NORs are identified by silver nitrate staining) were used for cytological analysis of diploid and triploid rainbow trout/brook trout (*Oncorhynchus mykiss/Salvelinus fontinalis*) hybrids. In all observed triploid cells stained with silver nitrate, only rainbow trout NORs were identified (Ueda *et al.*, 1988).

Development of molecular cytogenetics led to a more distinctive analysis of hybrid genomes. For instance, in the process called Genomic *In Situ* Hybridization (GISH) (Snowdon *et al.*, 1997; Barre *et al.*, 1998) or Whole Chromosome Painting (WCP) (Fujiwara *et al.*, 1997), the total genomic DNA coming from one parent is labelled and hybridized with unlabelled chromosome DNA of another parental fish. In the case of inviable, artificially induced hybrids between masu salmon (*Oncorhynchus masou*) female and rainbow trout male, rainbow trout chromosomes were eliminated while masu salmon chromosomes remained in the embryo cells. However, molecular cytogenetic analysis indicated the existence of rainbow trout chromosome fragments and micronuclei in the studied embryos (Fujiwara *et al.*, 1997). The micronuclei had been formed in the course of 'trapping' chromosomes in the mid zone from ana- to telophase. Moreover, FISH with telomeric probe revealed that most of the chromosome fragments missed telomeric caps at their ends. All those anomalies in rainbow trout haploid genome resulted in the death of the hybrid embryos. It has also been shown that paternal chromosomes were preferentially eliminated due to the mitotic anomalies during embryonic development (Fujiwara *et al.*, 1997). This may be caused by a factor (or factors) in maternal cytoplasm and its incompatibility with rainbow trout genome. That would further promote retention of maternal genome and elimination of paternal one. On the other hand, it is still not known why other hybrids can successfully overcome such a 'cytoplasmic barrier'.

Crossbreeding

Crossing individuals derived from the same species but from different lines, populations or strains has to be followed by more precise studies concerning parental genome identification. The multilocus DNA fingerprints have been used successfully in studies of genetic divergence among closely related fish populations (Palti *et al.*, 1997) as well as of their hybrids (Bosworth *et al.*, 1999). Also, analysis of minisatellite and microsatellite DNA, or of mitochondrial DNA proves valid data about spontaneous or intentional hybridization process in fish (Brzuzan, 1998;

Brzuzan et al., 1998). In such cases, karyological studies are useful whenever differences between chromosomes of fish from different populations have been recognized. Different numbers of NORs-bearing chromosomes or different NORs locations on the chromosomes of fish from different strains (Woznicki et al., 2000; Castro et al., 2001), the presence of supernumerary chromosomes or the occurrence of Robertsonian polymorphism observed within some fish species, can make cytogenetic approach useful in such studies. For instance, identification of differentially located NORs on chromosomes of rainbow trout from different strains has been applied in the analysis of interstrain hybrids (reviewed by Porto-Foresti et al., chapter 3.3, present volume).

Interspecific Hybridization—Natural Occurrence

Wild hybrids between Atlantic salmon and brown trout are common and frequent in Europe (Garcia de Laniz and Verspoor, 1989; Jansson et al., 1991; Jordan and Verspoor, 1993; Beall et al., 1997) and North America (Beland et al., 1981; Verspoor, 1988; McGowan and Davidson, 1992). The hybrids show salmon-like or trout-like phenotype and the discrimination of hybrids from salmons or trouts requires application of molecular or cytogenetic techniques. The chromosome number of wild salmon/brown trout equals the mean from diploid number of salmon and brown trout (2n= 69). Moreover, the chromosomal distribution of histone genes clusters (hDNA) indicates that studied individuals are hybrids (Pendas et al., 1994). After FISH with a hDNA probe the fluorescent signals derived from medium-sized acrocentric chromosome correspond to Atlantic salmon, and that of large metacentric chromosome corresponds to brown trout (Perez et al., 1999).

Intentional interspecific and intergeneric hybrids

Interspecific and intergeneric hybridization has been proposed to establish sterile or monosex fish stocks but only a few hybrids were cytogenetically studied (Zhang and Tiersch, 1997). Identification of haploid sets of the parental species in chromosome complement of hybrids between bighead carp (*Aristichthys nobilis*) and silver carp (*Hypophthalmichthys molitrix*) was performed applying conventional cytogenetics. Fish of both species and their hybrids had 2n= 48 chromosomes, and C banding enabled identification of parental chromosomes (Foresti de Almeida Toledo et al., 1995). Hybrids between yellowtail flounder (*Pleuronectes ferrugineus*) and winter flounder (*Pleuronectes americanus*) were identified by their cellular DNA content and identification of the satellites (regions corresponding to NORs) on their chromosomes. In spite of similar morphology of parental karyotypes, fishes differ from each other by one pair of chromosomes with satellites: only winter flounder possess a pair of acrocentrics with satellites, and only one such chromosome was identified in hybrid fish (Park *et al.*, 2003).

Karyotypes of intergeneric hybrids of ictalurids were analyzed with respect to chromosome number and morphology and NORs location (Zhang and Tiersch, 1997). The 2n chromosome number in channel catfish (*Ictalurus punctatus*) and in black bullhead (*Ameiurus melas*) hybrids was 59 and for hybrids between channel catfish and flathead catfish (*Pylodictis divaris*) the 2n equalled the mean (57) between those of parental species (Zhang and Tiersch, 1997). DNA content of diploid cells from hybrids were exactly intermediate to DNA content in cells of parental species (Tiersch and Goudie, 1993). Detailed analysis of chromosomes (length of the chromosomes and centromere location) confirmed stable pattern of chromosome segregation in ictalurid hybrids. Different location of NORs in channel catfish/black bullhead cross causes the hybrids to possess two unpaired NORs.

The karyotypes of hybrids were asymmetric (sets of parental chromosomes differed to each other by chromosome length and morphology) in a process that could be responsible for production of aneuploid gametes or even for adult fish sterility. The disturbances in sex ratios noticed in channel catfish/flathead catfish and channel catfish/ white catfish (*Ameiurus catus*) hybrids seem to be connected with differences in genomes which had been brought together to unusual nuclear environment (Goudie *et al.*, 1993).

Chromosome Set Manipulations in Fishes

It is known that some fishes such as crucian carp (*Carassius auratus gibelio*) (Purdom, 1993), loach (*Misgurnus fosilis*) (Arai and Mukaino, 1997; Momotani *et al.*, 2002; Janko *et al.*, 2003) and several species of poecilids can propagate by gynogenesis, one of the various modes of asexual reproduction (Hubbs and Hubbs, 1932; Schartl *et al.*, 1995). In all female populations of those species, development of diploid embryos is induced by sperm coming from males belonging to another species, but the male DNA is not transferred to the oocyte. Exclusively paternal inheritance—

androgenesis, happens spontaneously in case of crosses between different fish species. Occasionally, natural androgenesis was observed when the female common carp (*Cyprinus carpio*) was crossed with male grass carp (*Ctenopharyngodon idella*) (Chevassus *et al.*, 1983).

Both gynogenesis and androgenesis can be induced intentionally (Donaldson and Hunter, 1982; Thorgaard, 1986; Ihssen *et al.*, 1990; Donaldson, 1996; Pandian and Koteeswaran, 1998; Gomelsky 2003). These manipulations include irradiation of oocyte (androgenesis) or spermatozoa (gynogenesis). The irradiation with UV, gamma rays or X-rays is supposed to inactivate the genome of irradiated gametes. This phenomenon, followed by fertilization and diploidization of a haploid zygote with temperature, hydrostatic pressure (Thorgaard, 1983) or chemical shocks as well (colchicine, colcemide, cytochalasin B and other chemicals affecting chromosomal spindle (Ihssen *et al.*, 1990) results in the retention of the second polar body (early shock, heterozygous or meiotic gynogenesis) or blockage of the first mitotic division (late shock, homozygous or mitotic gynogenesis/androgenesis) of the zygote (Thorgaard, 1983b, 1986). Instead of using shocks for re-diploidization and in order to avoid embryonic mortality after application of the shock, several authors employed insemination with diploid spermatozoa of tetraploid males (Thorgaard *et al.*, 1990) or used diploid ova of tetraploid females (Arai *et al.*, 1991).

When diploidization fails (or is not applied), haploid embryos start developing. Usually, haploids are unviable and their embryonic development is disrupted at some early stage of ontogeny. However, cytogenetic analysis of haploid embryos can provide some interesting information concerning, for instance, the inheritance of Robertsonian polymorphism in fishes. Contrary to haploid individuals, diploid gynogenetic, androgenetic, tri-, tetra-, penta- and even hexaploid fish are viable. The potential application of such genome manipulations includes establishing of inbred lines, recovery of strains from cryopreserved semen or production of sterile fishes (Purdom, 1983; Thorgaard, 1983b, 1986, 1992; Yamazaki, 1983; Ihssen *et al.*, 1990; Thorgaard *et al.*, 1990; Arai *et al.*, 1992; Myers *et al.*, 1995; Pandian and Koteeswaran, 1998; Benfey, 1999; Gomelsky, 2003). But first of all, chromosome set manipulations provide an opportunity to elucidate sex determination systems in fish and also establish monosex fish lines (Pandian and Koteeswaran, 1998; Devlin and Nagahama, 2002). The key problem in the analysis of efficiency of such manipulations is the evaluation of cell ploidy level in genome-manipulated progenies and the examination of sex ratios among progenies with respect to genetic and functional (gonadal) sex of individuals.

Cytogenetic methods for the determination of ploidy

Although autopolyploid fish individuals are usually morphologically indistinguishable from diploids (Thorgaard and Gall, 1979; Pandian and Koteeswaran, 1998), there are studies describing morphological differences between diploids and autotriploids (Tiwary et al., 1999).

Additionally, various cytogenetic methods have been developed for ploidy identification, including direct chromosome counting, measurement of DNA amount per cell by flow cytometry, measuring of nuclei or cells, nucleoli counting (Ihssen et al., 1990; Tambets et al., 1991; Foresti et al., 1994). These may comprise other assays measuring the DNA content of specifically stained cell nuclei than flow cytometry such as spectrofluorometry, yet based upon quantification of fluorescence (Schwarzbaum et al., 1992) or microdensitometry (e.g., Gold and Price, 1985; Gold and Amemiya, 1987; Gold et al., 1990) based upon the measurement of stain absorbency. The latter approach was recently reviewed in detail by Hardie et al. (2002).

Karyotyping is a common, direct method for ploidy determination (Fig. 3.2.1). Chromosomes can be prepared either from fish tissues (head kidney, gill epithelium, spleen) after *in vivo* treatment of the specimens with colchicine or from cell cultures. Mitotic chromosome preparations can also be obtained from fish embryos and larvae (Phillips and Hartley, 1984; Inokuchi et al., 1994). This is an important advantage because of the possibility to observe chromosomes also in those organisms, which will be unviable after hatching.

Many authors proposed different methodologies for cell cultures of lymphocytes (Sanchez et al., 1990; Martinez et al., 1991), fibroblasts (Alvarez et al., 1991) and embryos (Shimizu et al., 2003). Fujiwara et al. (2001) elaborated optimal conditions for salmonid leukocyte cultures. However, this method seems to be useful in case of the detailed cytogenetic studies such as physical mapping of fish chromosomes, and represents a powerful way to trace genetic markers useful for the improvement of the production in aquaculture (Martins et al., 2004).

Flow cytometry, a procedure that uses a laser-powered instrument, allows precise determination of the DNA amount in a cell. The cells are

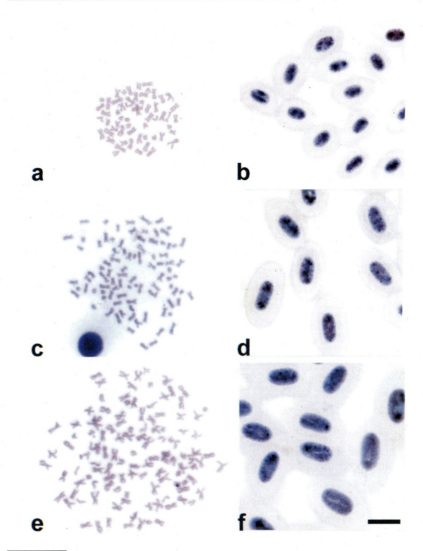

Fig. 3.2.1 Metaphase chromosomes and erythrocytes of rainbow trout; a, b – diploid, 2n=58; c,d – triploid, 3n=91; e,f – tetraploid, 4n=116. Scale bar = 10 μm.

stained with a light-sensitive dye, suspended in a fluid, and passed in a stream through either a laser or another type of light. This method enables researchers to determine ploidy in cells of tissues of fish embryos (Lecommandeur *et al.*, 1994) and larvae (Ewing and Scalet, 1991), and in blood or tissue cells of juvenile and adult specimens (Thorgaard *et al.*,

1983b). Apart from native cells, use of cryopreserved isolated blood cells (Gold *et al.*, 1991) or paraformaldehyde-fixed salmonid embryos in eye-bud stage (Johnstone and Lincoln, 1986; Lecommandeur *et al.*, 1994) was also successful. Ethanol fixation of the latter did not appear to be suitable due to production of high background noise signal, while ethanol-fixed blood cells, e.g., of sturgeons (Birstein *et al.*, 1993) brought slightly more variable but still reasonable results. This is a widespread method which has been applied to evaluate the success of chromosome manipulations in aquaculture species (Thorgaard *et al.*, 1982; Allen, 1983; Allen and Stanley, 1983; Johnson *et al.*, 1987; Tiersch and Chandler, 1989; Cherfas *et al.*, 1993; Arai, 2001; Flajshans *et al.*, 2004), and to estimate the ploidy level and genetic heterogeneity among natural populations of gynogenetic gibel carp clones (Wei *et al.*, 2003). Flow cytometry enabled detection of mosaicism in genetically manipulated kokanee salmon (*Oncorhynchus nerka*) (Tanaka *et al.*, 2003), northern pike (*Esox lucius*) (Lin *et al.*, 2001), and also in natural populations of Amazon molly (*Poecilia formosa*) (Lamatsch *et al.*, 2002). Lin and Dabrowski (1998) have proved the usefulness of flow cytometry technique when studying the effects of ultraviolet (UV) irradiation on denucleation of eggs and induction of androgenesis in muskellunge (*Esox masquinongy*). Applying flow cytometry, Wei *et al.* (2003) confirmed the hybrid origin of the multiple tetraploid gibel carp studying DNA contents in their red blood cells and spermatozoa obtained from different clones of the polyploids.

Coulter counter (measuring volume of cells) is a quick and reliable method to determine fish ploidy based on the analysis of blood samples. It enables rapid identification of triploid grass carp as it has been shown by Watterdorf (1986). This method is unsuitable for analyzing samples obtained from solid tissues. Apart from the original coulter counter and channelizer, several types of automated haematological analyzers are available nowadays, allowing the counting of nucleated erythrocytes (fish, amphibian, reptile, bird) and also measure their volume upon changing impedance when the cells pass between electrodes. Parallel to these automated measurements, there has been the classical conventional haematological approach in computing the cell volume (erythrocyte volume; mean cell volume, MCV) from haematocrit (packed cell volume, PCV) and number of erythrocytes as MCV = (PCV * 1000)/No. of erythrocytes.

Measuring the erythrocyte dimensions is a precise and quick method of the fish ploidy determination (Fig. 3.2.1). This approach has been first used by Swarup (1959) to distinguish diploid and triploid fish. Length of erythrocyte, mostly reported as erythrocyte major axis (longitudinal axis of this elliptic cell) and that of erythrocyte nucleus (erythrocyte nuclear major axis) are generally acknowledged to be the best dimensional predictors of ploidy level in fish (Purdom, 1972; Thorgaard and Gall, 1979; Wolters *et al.*, 1982; Benfey and Sutterlin, 1984; Benfey *et al.*, 1984; Ihssen *et al.*, 1990; Boron, 1994) as the increase in erythrocyte cell and nucleus volume associated with triploidy mainly results from an increase in their major axes. This approach appears to be applicable to also distinguish tetraploidy (Arai *et al.*, 1991) or higher ploidy levels, as shown in the case of polyploid sturgeons and their hybrids, e.g., by Arefjev and Nikolaev (1991) or Flajshans and Vajcova (2000). Nevertheless, some other directly measured cell dimensions such as erythrocyte nuclear area and erythrocyte nuclear perimeter are also used as ploidy level predictors (Flajshans, 1997; Linhart *et al.*, 2001 and others) and they too appear to be reliable.

It is a good practice to combine two methods, for example: (1) determination of ploidy of embryos or larvae on the basis of the karyotype; and (2) specimens from the fish groups containing the highest percentage of polyploid individuals can then be analyzed *intra vitam* once they reach an appropriate size. Based on the differences in the length of erythrocyte nuclei, the polyploid specimens can be distinguished from the diploid ones.

Inspection of blood smears showed that some of the brook trout erythrocytes exhibited segmentation of their nucleus (Wlasow *et al.*, 2004). Nuclei were divided into two parts of similar size (Fig. 3.2.2). The number of erythrocytes with nuclear segmentation was significantly higher in triploids (20%) than in diploids (less than 1%). Triploids also had a significantly higher number of granulocytes with multi-lobed nucleus. Wlasow *et al.* (2004) showed that in brook trout erythrocyte nucleus cleavages, granulocyte nucleus hypersegmentation are correlated with ploidy level, and that these features can be informative in ploidy level determination.

Analyzing the number of nucleoli per nucleus is an inexpensive technique in the determination of fish ploidy (Cherfas and Ilyasova, 1980) but it is useful in case of species possessing only one pair of chromosomes bearing Nucleolar Organizer Regions (NORs) (Phillips *et al.*, 1986; Flajshans *et al.*, 1992). NORs are chromosome regions containing genes

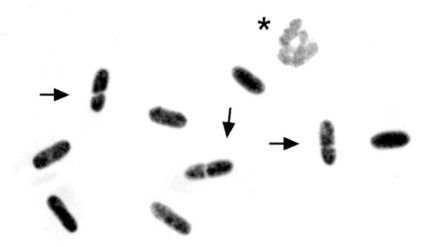

Fig. 3.2.2 Blood cells alteration in triploid brook trout. Arrows indicate erythrocytes with amitotically divided nuclei, asterisk indicates an hypersegmented granulocyte (from Wlasow *et al.*, 2004).

coding ribosomal molecules (rRNA). Since nucleolar proteins conjugated to rRNA can be visualized by silver staining (Howell and Black, 1980), it is possible to visualize nucleoli and NORs in interphase nuclei (Fig. 3.2.3). In case of diploid organisms, one or maximum two nucleoli per cell can be observed. In triploids there are one, two or maximum three nucleoli per cell. Kucharczyk *et al.* (1997) showed that analysis of not more then 40 interphase cells per individual enables positive identification of haploid, diploid and triploid specimens of bream (*Abramis brama*).

Analysis of the number of nucleoli per nucleus was used successfully in case of triploid bream (Kucharczyk *et al.*, 1996), northern pike (Kucharczyk *et al.*, 1999), tench (*Tinca tinca*) (Flajshans *et al.*, 1992,

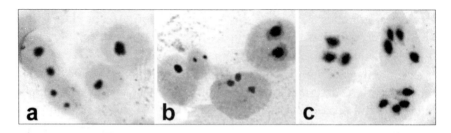

Fig. 3.2.3 Silver-stained nucleoli (black dots) in fin epithelial cells of rainbow trout; a – diploid; b – triploid; c – tetraploid.

1993), rainbow trout (Phillips *et al.*, 1986; Flajshans *et al.*, 1992) and chinook salmon (*Oncorhynchus tshawytscha*) (Phillips *et al.*, 1986). Babiak *et al.* (1998) applied this method to examine rainbow trout subjected to tetraploidization manipulation, and the authors revealed upto 18.7% of tetraploids among examined fishes. However, the technique does not allow reliable identification of fish of higher degree of ploidy (such as 4n or 6n). In such cases, it is only possible to distinguish diploid individuals from polyploid ones.

Transmission of Robertsonian variations into gametes— Haploid fishes as a model for studying inheritance of unpaired chromosomes

In most cases, haploid fish specimens do not survive post-embryonic stages. However, studies of haploid fishes can contribute data about inheritance of chromosomal variations caused by Robertsonian rearrangements, such as fusion of two acrocentrics into one metacentric chromosome. Such chromosome polymorphism occurs quite often in fishes, such as in salmonids (Thorgaard 1983a; Phillips and Kapuscinski, 1988; Castro *et al.*, 1994; Frolov and Frolova, 2000), gobiids (Canapa *et al.*, 2002), cobitids (Zhang and Arai, 2003), in Pleuronectidae and Gadidae (Fan and Fox, 1991) among others. As salmonids are of autotetraploid origin (Allendorf and Thorgaard, 1984), Robertsonian fusions have played important role in evolution (re-diploidization) of their karyotypes (Phillips and Rab, 2001).

Some populations of rainbow trout are characterized by different diploid chromosome numbers, varying from 58 to 64 but showing the same chromosome arm number (NF= 104) (Thorgaard, 1983a). Such variation is explained as the effect of post-zygotic fusions of acrocentrics (Ohno *et al.*, 1965). The pattern of parent-offspring transmission of chromosome variants was provided by the analysis of progeny from crosses between individuals with different chromosome numbers (Ueda and Ojima, 1984). Data that are more precise could be brought by cytogenetic studies of meiosis II or of haploid gynogenetic embryos derived from parents whose chromosome number was known. The advantage of such an approach comes from the fact that all chromosomes of a haploid gynogenetic embryo come from the oocyte. This, in turn, gives the information about meiotic divisions in case of individuals with odd (2n= 59, 61 or 63) chromosome numbers. Karyological studies of females carrying Robertsonian polymorphism (2n= 59, 60 or 61) and their haploid gynogenetic progenies provided information on the inheritance of chromosome variations (Nakayama and Chourrout, 1993). In spite of some exceptions, the inheritance pattern in studied fish did not differ from the theoretical segregation models, which assumed three possible fusions of six acrocentrics (Nakayama and Chourrout, 1993). Similar approach during inheritance studies of Robertsonian polymorphism involved NOR region in pink salmon. Cytogenetic analysis of gynogenetic progenies showed that the mother of the studied offspring was heterozygous for a Robertsonian translocation (Phillips and Kapuscinski, 1988).

Investigating chromosomal sex determination systems

Information about sex determination systems in fish has been obtained by direct observations like cytogenetic and molecular studies (Thorgaard 1977; Devlin et al., 2001) or by indirect approaches such as analysis of sex ratios among families, obtaining functionally sex-reversed individuals and development of monosex fish lines by chromosome set manipulations (Dabrowski et al., 2000; Lin et al., 2001). Contrary to mammals and birds where sex chromosome dimorphism is well developed, only a minority of fish species possess morphologically differentiated sex chromosomes. Among several sex determination systems in fishes, two are most common: male heterogamety (XY system existing in Salmoniformes, Cypriniformes, Cyprinodontiformes) and female heterogamety (WZ system in Cypriniformes, Anguilliformes, Perciformes) (Solari, 1993). In case of a lack of heteromorphic sex chromosomes, the information about sex determination systems is usually provided by indirect methods, among which chromosome set manipulations play the main role (Devlin and Nagahama, 2002).

Artificially induced gynogenesis and androgenesis helped determine the sex chromosome systems in salmonids (Chourrout and Quillet, 1982; Parsons and Thorgaard, 1985; Thorgaard et al., 2002), cyprinids (Komen et al., 1990, 1992; Horvath and Orban, 1995), flatfish (Kakimoto et al., 1994), loach (*Misgurnus anguillicaudatus*) (Suzuki et al., 1985) and other fishes (Pandian and Koteeswaran, 1998). On the other hand, the same approaches pointed out the existence of XY and WZ systems in different but closely related species of catfish *Hypostomus* (Artoni et al., 1998), esocids (Dabrowski et al., 2000; Lin et al., 2001) or tilapia (Campos-Ramos et al., 2001, 2003).

The multidisciplinary approach merging chromosome set manipulations and cytogenetic analysis have been proposed to study the sex determination system in *Oreochromis mossambicus*. The study encompassed meiotic gynogenesis in eggs obtained from both normal females and neofemales, synaptonemal complex analysis in males, females and hybrids between Mozambique tilapia and Nile tilapia (*Oreochromis niloticus*) as well as fluorescence *in situ* hybridization with a probe derived from the *O. niloticus* sex chromosome. Such a complex investigation confirmed a XY sex determination system in the Mozambique tilapia. Moreover, FISH and the analysis of the synaptonemal complex supported the view that the Mozambique tilapia chromosomes from pair 1 are related to sex determination, although they are morphologically less differentiated than those of Nile tilapia and blue tilapia (*Oreochromis aureus*) (Carasco *et al.*, 1999; Campos-Ramos *et al.*, 2003).

Cytogenetic sex markers—Sex composition among gyno-, androgenetic and sex reversed individuals

Phenotypic markers such as sex-specific pigmentation (Angus, 1989; Fernando and Pang, 1989, 1990; Morizot *et al.*, 1991; Khoo *et al.*, 1999), sex-linked isozymes (Nyman *et al.*, 1984; Allendorf *et al.*, 1994) or DNA markers help determine genetic sex in fishes (Devlin and Nagahama, 2002). Markers of female and male sex seem to be limiting factors in studying genetic sex composition among chromosome set manipulated fish stocks, because only some 10% of cytogenetically studied fish species exhibit morphologically differentiated sex chromosomes (Solari, 1993). Histological studies of gonads need to be supported by cytogenetic studies, which enable diagnosing of the genetic sex of fish. This, in consequence, leads to evaluation of the sex composition of genome manipulated fish.

Number and chromosome location of NORs are sex related in some fish species (Li and Gold, 1991; Ren *et al.*, 1993; Khuda Bukhsh and Datta, 1997); 5S rDNA was found to be chromosome X related in rainbow trout (Moran *et al.*, 1996), in chinook salmon (Stein *et al.*, 2001), and Y related in the Antarctic fish *Chionodraco hamatus* (Pisano *et al.*, 2003). Different repetitive DNA sequences as well as pseudogenes and transposable elements have been found and localized on sex chromosomes, mostly in salmonid fish species (Devlin *et al.*, 1991, 1998; Du *et al.*, 1993; Forbes *et al.*, 1994; Iturra *et al.*, 1998, 2001; Nakayama *et al.*, 1998; Zhang *et al.*, 2001), in *Leporinus elongatus* (Nakayama *et al.*,

1994) and in *Chionodraco hamatus* (Capriglione *et al.*, 1994; Ozouf-Costaz *et al.*, 2004). In salmonid fish species, heteromorphic sex chromosomes have only been described in lake trout (*Salvelinus namaycush*) (Phillips and Ihssen, 1984, 1985), brook trout (Phillips *et al.*, 2002; Ocalewicz *et al.*, 2004b), kokanee salmon (*Oncorhynchus nerka*) (Thorgaard, 1978) and in rainbow trout (Thorgaard, 1977). Like other salmonid fish species, rainbow trout possesses XY sex determination system. The early studies on rainbow trout chromosomes using basic staining techniques let for identification of X- and Y-chromosomes, which differ from each other both by the length (Thorgaard, 1977) and by the heterochromatin distribution (Lloyd and Thorgaard, 1988; Ocalewicz, 2002b).

Diploid androgenetic rainbow trout were examined by way of cytogenetic analyses by Parsons and Thorgaard (1985). This karyological study has confirmed fish diploidy and revealed the sex-chromosome makeup of the androgenetic individuals. Based on the length of the pair of subtelocentric sex chromosomes, the authors proved that four out of eight examined specimens were YY 'super males'. Additionally, exclusively paternal inheritance was confirmed by the transmission of albinism from father to progenies, and androgenetic progenies showed only paternal allele coding for LDH-n (lactate dehydrogenase).

The homogametic sex genotypes in androgenetic individuals (XX - female and YY - super male) were confirmed using PRINS (PRimed *IN Situ* labelling) method, enabling *in situ* amplification of 5S rDNA sequences, visualizing the repetitive DNA on chromosome X and identifying NORs bearing chromosomes in rainbow trout (Moran *et al.*, 1996). Androgenetic individuals showed two patterns of distribution of fluorescent spots on chromosomes: androgenetic females showed four fluorescent signals (two X and two NORs bearing chromosomes) and androgenetic males had only two signals, derived from NORs-bearing chromosome (YY genotype) (Fig. 3.2.4). The advantage of this method is its sensitivity, as compared to traditional FISH technique. The existence of 5S rDNA clusters on autosomes and heterochromosomes seems to be extremely useful in case of YY individuals, as the fluorescent spots derived from NORs-bearing chromosomes prove that the method has been performed successfully (Ocalewicz and Babiak, 2003).

Chromosome Y, which was unrecognizable from other acrocentric autosomes in the studied rainbow trout population (Ocalewicz, 2002b), has been identified by FISH using the OmyP9 probe, a DNA sequence

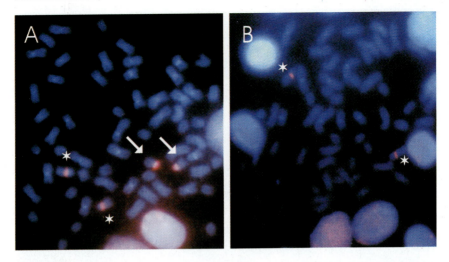

Fig. 3.2.4 5S rDNA sequences localized by PRINS on androgenetic rainbow trout female XX (A) and male YY (B). Arrows point X chromosomes, asterisks indicate NORs-bearing chromosomes, DAPI counterstaining (from Ocalewicz and Babiak, 2003).

isolated from the Y chromosome of rainbow trout from Chile (Iturra *et al.*, 2001). Male individuals belonging to androgenetic line are characterized by the presence of two Y chromosomes in their karyotypes (Fig. 3.2.5). The OmyP9 sex-Y probe hybridized to the interstitial region of the long arm of two uniarmed chromosomes, namely those that correspond to the Y chromosomes (Lam *et al.*, 2003). These results confirmed the YY condition of the examined specimens and reinforced the utility of this marker, which can be used in cytogenetic studies on the efficiency of androgenesis and on the sex ratios among androgenetic offspring.

Polymorphism of Y chromosome in a rainbow trout

Sex chromosomes in fishes have been found to be polymorphic within a species, indicating the dynamic evolutionary nature of this system. In platyfish (*Xiphophorus maculatus*), repetitive Y-linked DNA sequence was reported only in Rio Jamapa strain (Nanda *et al.*, 2000), whereas in *Blennius tentacularis*, Y-autosome translocation exists only in certain males (Carbone *et al.*, 1987). Rainbow trout sex chromosomes exhibit variable levels of differentiation: isolated populations near the edge of the species range maintain undifferentiated sex chromosomes (Thorgaard, 1983a). Moreover, intraspecific polymorphism is observed in those populations

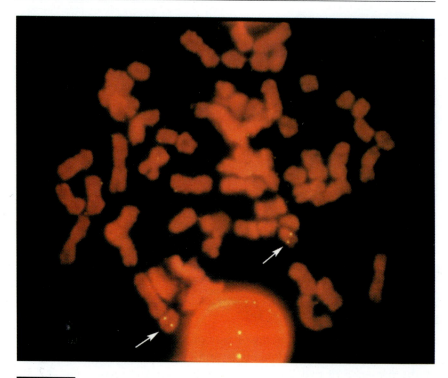

Fig. 3.2.5 Localization of OmyP9 sequences on two Y chromosomes (arrows) in androgenetic rainbow trout male by FISH, Propidium iodide counterstaining (from Lam *et al.*, 2003).

which exhibit morphologically recognizable sex chromosomes (Thorgaard, 1977), what seems to be the key problem in genetic sex diagnosis in this species, including genome-manipulated individuals.

Morphological distinguishing of rainbow trout X from Y chromosome was replaced by more advanced cytogenetic techniques, enabling identification of heterochromosomes. Fluorescent (DAPI) banding showing AT-rich chromatin in centromeric region of rainbow trout X chromosome was used in studying chromosomes of androgenetic progenies of four fathers randomly selected from a strain in Poland where approximately 15% of males possessed Y chromosomes morphologically identical to X chromosomes (Ocalewicz, 2002a). In all cases, androgenetic offspring showed 1:1 female : male sex composition. Cytogenetic analysis showed that females and males from two fathers had similar sex chromosomes. Progenies of two other males differed: only females had two X chromosomes as identified by DAPI (Fig. 3.2.6). Results of similar

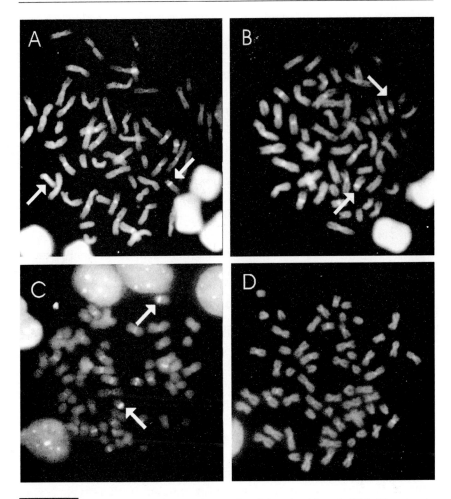

Fig. 3.2.6 Metaphase plates from androgenetic rainbow trout female (XX) (A) and male (YY) (B) with undifferentiated sex (X and Y) chromosomes (arrows) and from androgenetic female (C) and male (D) with morphologically different sex chromosomes (arrows indicate X chromosomes on female metaphase, male does not have any subtelomeric chromosomes).

investigations became confusing when phenotypic males possessed similar X and Y sex chromosomes. The question arose whether those fishes were genetic females with two X chromosomes and male phenotypic sex, or whether they were genetic males with morphologically undifferentiated sex chromosomes. This question became important in the light of spontaneous sex conversion, which had been well documented in fish species (Baroiller et al., 1999). On the other hand, studied rainbow trout strain was established on the basis of individuals originating from five different populations (Dobosz et al., 1999), whose sex chromosomes could be differently advanced in their morphological differentiation (Thorgaard, 1983a). The results did not support the assumption that males with X-like sex chromosomes are genetic females. Instead, they confirmed the existence of males with differentiated Y-chromosomes, ranging in their morphology from the ones looking like X chromosome to the shorter subtelo-acrocentric Y chromosome, lacking almost whole short arm and AT-rich centromeric chromatin (Ocalewicz et al. unpublished).

Such studies indicate the existence of different forms of rainbow trout Y-chromosomes in a strain originating from several ancestral rainbow trout populations. Additionally, this approach proved that merging chromosome set manipulations and modern cytogenetics becomes a useful tool in studying intraspecific sex chromosome polymorphism.

Sex-reversed Individuals—Following Genetic Sex

In most fish species, sex is genetically determined, but the process of sex differentiation can be disturbed by external factors (Francis, 1984; Rubin, 1985; Schultz, 1993; Hostache et al., 1995; Strusmann et al., 1996, 1997; Baroiller et al., 1999) or by hormonal stimulation (Ihssen et al., 1990; Pandian and Koteeswaran, 1998; Demska-Zakes et al., 2000; Gomelsky, 2003). Gonadal sex differentiation in fishes occurs at a defined stage of development and is regulated by gonadal sex steroid hormones. Prior to this stage, the phenotypic sex can be altered through temperature or endocrine manipulations (reviewed by Devlin and Nagahama, 2002). Steroid (androgen or estrogen) treatment can affect phenotypic sex, in producing neo-males (genetic females in which testes are developed) or neo-females (genetic males in which ovaries are developed) (Purdom, 1983; Thorgaard, 1983, 1986; Yamazaki, 1983; Ihssen et al., 1990; Pandian and Koteeswaran, 1998; Benfey, 1999; Gomelsky, 2003; Hendry et al., 2003). Neo-male development (masculinization) can also be induced using non-steroidal agents such as those that block the activity of P450 cytochrome aromatase (Piferrer et al., 1994; Kwon et al., 2000).

Molecular and cytogenetic identification of sex-reversed fish can be applied in order to estimate the efficiency of a given approach. Moreover, such an evaluation is often needed in cases when, for instance, some chemicals present in sewage can mimic estrogens and disrupt process of sex differentiation in fish. Identification of genetic sex of fishes using sex-related cytogenetic or molecular markers together with histological analysis of gonads has great potential in examination of the influence of the so-called 'endocrine disruptors' on fish sex differentiation (Gimeno *et al.*, 1996, 1997; Keith, 1997; Gimeno *et al.*, 1998; Tyler *et al.*, 1998; Sumpter, 1998).

Molecular and cytogenetic sex markers have been used to investigate the process of sex differentiation in several fish species (Madigou *et al.*, 2001; Nagler *et al.*, 2001; Afonso *et al.*, 2002), providing new information regarding the course of this process in organisms inhabiting polluted environments and on the impact of pollution on vertebrate health.

Genetic Sex and Gonad Development in Triploid Individuals

Natural triploids are frequently observed in poeciliids (Schultz, 1969), cyprinids (Ojima and Takai, 1979; Arai *et al.*, 1991; Flajshans *et al.*, 1993; Boron, 1994b, 2003; Collares-Pereira *et al.*, 1995; Zhang and Arai, 1999; Gromicho and Collares-Pereira, present volume), attherinids (Echelle *et al.*, 1988), salmonids (Benfey, 1989; Aegerter and Jalabert, 2004), wels (Varkonyi *et al.*, 1998). In aquaculture, triploidization can be experimentally induced by blocking of the second polar body extrusion with heat, hydrostatic pressure shock or as a result of chemical treatment with cytochalasin, colcemide or colchicine. Additional set of chromosomes in triploids causes cytogenetic incompatibility, which makes most of the triploids functionally sterile. In triploids, sex chromosomes have smaller impact on sex differentiation than in diploids. In case of salmonids with XY system, XXX individuals develop into females (even if they have reduced gonads) and XXY become males (which can be semi-sterile) (Devlin and Nagahama, 2002). Faster growth of female triploids and their functional sterility (Arai, 2001) makes them more advantageous in aquaculture than males.

In the XY system fertilization of oocytes with X-carrying spermatozoa followed by the triploidization results in offspring which are all 3n females. In case of XY species, XX neomales are widely used to produce all-triploid female stocks (Arai, 2001). Here identification of sex chromosomes could be useful in establishing triploid females particularly due to identification of sex genotypes in sex-reversed specimens. Analyses of sex-reversed individuals can confirm XX sex genotype of neomales. In such cases, cytogenetic analysis should be done on material sampled without fish sacrification. The efficiency of establishing brook trout monosex female diploid and triploid stocks and production of sex-reversed neomales has been studied analyzing gonad development and also by way of identification of X chromosomes (Kuzminski, unpublished). In brook trout, karyotype only submetacentric X chromosome possesses discrete cluster of AT-rich chromatin (positively stained with DAPI) in subtelomeric region, what makes it recognizable from the Y chromosome (Fig. 3.2.7) (Phillips *et al.*, 2002; Ocalewicz *et al.*, 2004b).

IDENTIFICATION OF ANEUPLOIDY AND CHROMOSOMAL DISTURBANCES IN ANDROGENETIC, GYNOGENETIC AND POLYPLOID FISHES

High rate of mortality and significant percentage of malformed larvae among chromosome set manipulated fishes are considered to be the consequence of drastic manipulations performed on gametes. Radiation can induce damage of mitochondrial DNA and mRNA in oocytes that play essential roles during the primary phases of embryonic development. Moreover, radiation used for inactivation of nuclear DNA in oocytes and spermatozoa can produce chromosome fragments, and these maternal or paternal residues are able to disturb cell divisions and may interfere with embryonic development of androgenetic or gynogenetic organisms. Gamma irradiation, which easily penetrates the cell and acts by inducing chromosome breaks, usually appears to be the main cause of fragmentation (Chourrout and Quillet, 1982; Parsons and Thorgaard, 1985). Ultraviolet irradiation destroys genome by inducing thymidine dimers, which in turn can be repaired during the photo reactivation process occurring in a visible light (Ijiri and Egami, 1980). It is considered that UV irradiation does not produce chromosome fragments, although maternal residues were reported in haploid embryos of androgenetic loach and muskullenge after inactivation of egg genome by UV light (Arai *et al.*, 1992; Lin and Dabrowski, 1998). This could be an effect of incomplete destruction of the maternal nuclear DNA either because of the imperfect penetration of UV rays or due to photoreactivation process.

Chromosomal Anomalies in Chromosome Set Manipulated Fish-mosaicism

When polyploidy is induced by a shock at the early meiotic phase, no chromosomal disturbances occur because the undamaged polar body is

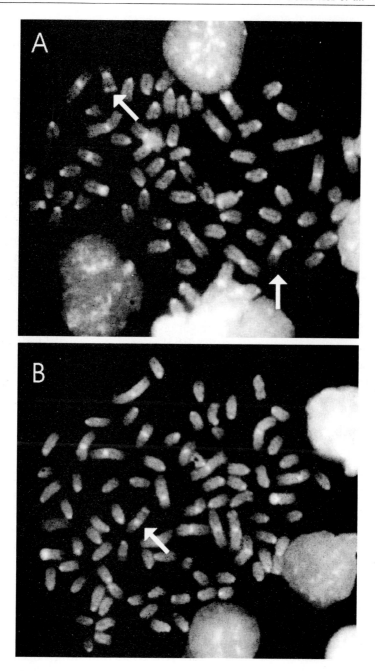

Fig. 3.2.7 Metaphase plates of brook trout female (XX) (A) and male (XY) (B). Arrows point X chromosomes (from Ocalewicz *et al.*, 2004b).

either discarded (resulting in a diploid fish individual) or retained (resulting in a triploid fish). In those embryos whose first mitotic cell division is brought to a halt, mosaicism can occur as described by various researchers (Lin et al., 2001; Tanaka et al., 2003, among others). Mosaicism is most probably caused by a lack of synchronization of cell divisions in treated eggs. Varied rate of development of eggs results from individual variability and cannot be reduced in any other way but by using inbred lines or clones as a source of gametes. Flow cytometry has been found to be the best method for identification of mosaic individuals. For instance, Tanaka et al. (2003) applied flow cytometry and demonstrated different amounts of DNA in cells of *Oncorhynchus nerka* larvae, which indicated that mosaic organisms were built of cells of different ploidy.

Lin and Dabrowski (1998) revealed differences in DNA content between cells of androgenetic and control fish. Nuclear DNA content of the abnormal androgenetic larvae, which were morphologically identical to haploids, was significantly lower than in diploids, and its variation was much bigger than in the control. This suggests that a too low dose of UV irradiation applied in this study might result in gene mutations, alterations in chromosomal conformation and their fragmentation, but did not prevent maternal DNA from participating in mitotic division. All mosaic larvae were abnormal. So far, no mosaic individuals have been detected in fishes appearing to be morphologically well developed.

Remnants of irradiated genetic material in resulting manipulated finfish

The existence of male subgenomic DNA in the form of microchromosomes was proved in Amazon molly (*Poecilia formosa*), a species reproducing through sperm-induced parthenogenesis. The microchromosomes were unstable components of Amazon molly genome and only some of them were transmitted to the offspring (Schartl et al., 1995). Such sub-genome components have been found in approximately 5% of fish coming from natural populations (Schartl et al., 1995; Lamatsch et al., 2000), and the genetic information located in microchromosomes may be normally expressed, as it was observed in *P. formosa* (Lamatsch et al., 2000).

The identification of fragments of the paternal (in gynogenetic) or maternal (in androgenetic) genome is much more difficult than determination of fish ploidy. In fact, all indirect methods are either useless or hardly useful. Gynogenetic and androgenetic individuals, whose genome contains such remnants, do not differ much from their pure (i.e., lacking microchromosomes) counterparts in the amount of DNA, size of cells or the cell nucleus or in the number of nucleoli per nucleus. In tilapia, *Oreochromis aureus*, paternal genome 'contamination' was revealed in gynogenetic specimens using DNA fingerprinting (Carter *et al.*, 1991). The presence of stable maternal DNA residues, detected by microsatellite DNA analysis, has also been found in interspecific androgenetic Atlantic salmon embryos developing in gamma ray-enucleated oocytes of brown trout (Babiak *et al.*, 2002b).

The possibility that unwanted genome fragments of parental origin will contain marker gene loci is relatively small. Although it is possible to identify contaminated gyno- and androgenetic individuals, one can never be certain that the genome of the remaining fish is pure. Although difficult and time consuming, the most reliable method for identification of any chromosomal disturbances in genetically manipulated fish remains karyotyping, which can provide an actual picture of what is happening to the chromosomes of a given individual.

Additional DNA residues (whole chromosomes and/or chromosome fragments) are easy to recognize by way of basic cytogenetic techniques (such as Giemsa staining). Such chromosome fragments have been revealed in gynogenetic haploid and diploid rainbow trout embryos (Chourrout and Quillet, 1982). The authors postulated that those fragments were remnants of paternal chromatin. Chourrout (1984) identified gynogenetic haploid embryos with 1 to 6 chromosome fragments. The embryos were mosaics so that not every cell of an individual possessed the same number of fragments. In chromosome-mediated transgenic rainbow trout, fragments of brook trout chromosomes showed such a high stability in adults that they were transmitted to the next generations (Disney *et al.*, 1987, 1988; Peek *et al.*, 1997). Using conventional Giemsa staining and light microscopy, the chromosome fragments appeared spherical and were closely associated with intact chromosomes. In a few cases, dicentric chromosomes were noticed in what suggests a fusion of biarmed intact chromosomes with chromosome fragments. Analysis of the chromosome fragments showed that some of them did not have sister chromatids. However, fragments examined with scanning electron microscopy showed centromere-like constrictions, suggesting their normal movements during cell divisions and efficient inheritance to daughter cells.

Gibel carp has two modes of reproduction (gonochoristic and gynogenetic triploids) and was found to form clonal all-female stocks differing in chromosome number (Zou *et al.*, 2001). The paternal effect of allogynogenesis is outstanding in an artificial clone produced by a cold shock treatment. The non-activated milt of blunt nose black bream (*Megalobrama amblycephala*) was used to induce gynogenesis in one of the cloned gibel carp (Zou *et al.*, 2001; Yi *et al.*, 2003). Except for a basic set of chromosomes, which equals 156, 1 to 15 microchromosomes were observed. Using microdissected probes and chromosome painting, it has been demonstrated that those microchromosomes originated from the paternal genome and have been incorporated into the genome of the clone (Yi *et al.*, 2003).

Gamma irradiation of oocytes during the androgenesis of rainbow trout-generated chromosome fragments (1-3 per cell) were reported in adult, healthy and morphologically normal 3-year-old YY males (Ocalewicz *et al.*, 2004a). This observation shows that chromosome fragments segregated into daughter cells and did not disturb the cell divisions. Some of the chromosome fragments exhibited bright signals after DAPI staining, in a process that suggested that these structures were composed of AT-rich chromatin similar to the paracentromeric regions of intact chromosomes (Fig. 3.2.8). As the chromosome fragments possess centromere-like structures, they are able to segregate in the same way as unbroken host chromosomes. The fragments negatively stained by DAPI are considered to be centromeric remnants of acrocentric chromosomes which do not possess AT-rich clusters in centromeric regions. The chromosome fragments that do not have functional centromeres might have to undergo a process of end-joining (McClintock, 1941) or their DNA incorporation into the intact chromosomes (Ocalewicz *et al.*, 2004a). Identification of Interstitial Telomeric Sites (ITS) on four metacentric chromosomes seems to confirm that assumption (Fig. 3.2.9A).

Telomeric sequences have been found on both ends of all intact chromosomes and only on few chromosome fragments (Fig. 3.2.9B). Lack of telomeric sequences at the end of the chromosome is supposed to be deleterious for the chromosome function (Lima de Faria, 1983). In the absence of functional telomeres, eukaryotic chromosomes undergo end fusions and a variety of degradation events and become unstable (Blackburn, 2001; de Lange, 2002). Although it does seem that in studied

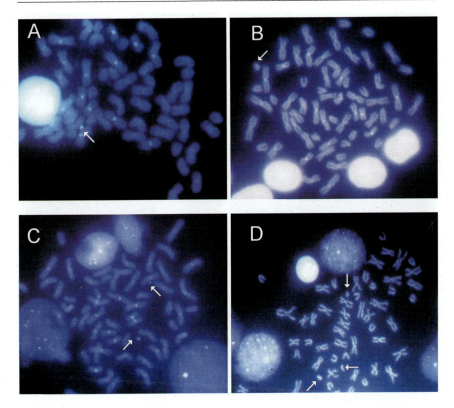

Fig. 3.2.8 Chromosomes of androgenetic rainbow trout stained with DAPI. Arrow indicates one DAPI positive chromosome fragment (A), one DAPI negative (B), two fragments with DAPI signals (C) and three DAPI negative chromosomes fragments (D) (Ocalewicz *et al.*, 2004a).

individuals, telomereless chromosome fragments segregated normally (Ocalewicz *et al.*, 2004a).

Transcriptional activity of genes located on chromosome fragments is as interesting as their stability. The chromosome-mediated transgenic fish having chromosome fragments were mosaics for some pigment genes. Also, androgenetic brook trout had chromosome fragment bearing genes coding for allozyme markers (Disney *et al.*, 1987, 1988), what proved that the genes were transcriptionally active. Additionally, AgNO$_3$ banding indicated NORs activity on those fragments (Disney *et al.*, 1987). The indirect proof of the chromosome fragment activity was odd sex ratios in androgenetic rainbow trout progenies (Scheerer *et al.*, 1991). If such structures carry genes involved in the process of sex determination, their

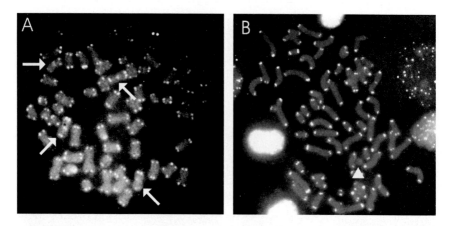

Fig. 3.2.9 Identification of Interstitial Telomeric Sites (ITS) (arrows) on androgenetic rainbow trout chromosomes (A). Telomereless chromosome fragment (arrowhead) detected in cells of androgenetic rainbow trout (from Ocalewicz et al., 2004a).

presence can affect sex determination and lead to the development of testis in XX individuals, or ovaries in YY specimens.

CONCLUSION

The progress in cytogenetic techniques make them an integral part of biotechnological experiments in finfish breeding. As discussed in the present chapter, cytogenetic techniques can largely contribute to studies concerning hybrid and chromosome set manipulated fish. Karyological analysis enables discriminating hybrids from parental species. In hybrids that were viable, cytogenetic approach pointed out that genomes incompatibility and asynchronic nuclear divisions may cause drastic chromosome rearrangements such as chromosome fragmentation or elimination, which results in the death of hybrids during embryogenesis. In chromosome set manipulated fish, cytogenetic analyses help evaluate the efficiency of such processes by checking the ploidy level and sex ratios among manipulated progenies. Additionally, cytogenetic analysis provides original data about residues of parental-irradiated genomes in cells of genome-manipulated fish.

Moreover, the progress in molecular biology techniques, establishing gene libraries of fish species and gene mapping programmes can be the source of DNA sequences, which chromosomal localization can improve usefulness of cytogenetic tools in fish biotechnology. Further development of applied fish cytogenetics and new perspectives of using it in aquaculture are also related to the availability of new chromosome markers of genetic sex, and centromeric DNA sequences, that will improve the analysis of chromosome disturbances in hybrids and in chromosome set manipulated fishes. Moreover, it will be important to implement modifications of procedures that will make cytogenetic protocols easier and faster, such as the PRINS method when repetitive DNA sequences have to be localized.

In a more general context, it is worth noting that some fish species are considered to be model organisms in studies focused on embryonic development (zebrafish) (Allende *et al.*, 1996; Schier *et al.*, 1996), melanoma formation (platyfish) (Witbrodt *et al.*, 1989; Mallitschek *et al.*, 1995; Gutbrod and Schartl, 1999; Nanda *et al.*, 2000), genome evolution (pufferfish, salmonidae, zebrafish) (Aparicio *et al.*, 1997; Amores *et al.*, 1998; Thorgaard *et al.*, 2002), including the evolution of vertebrate sex chromosomes. Therefore, cytogenetic studies in fishes can provide important contribution to comparative genomics, with implications largely encompassing the disciplinary limits of ichthyology.

Acknowledgment

We are grateful to Patricia Iturra and Natalia Lam (Facultad de Medicina, Programa de Genetica Humana, ICBM, Universidad de Chile, Santiago, Chile) for providing figures and unpublished information.

References

Aegerter, S. and B. Jalabert. 2004. Effects of post-ovulatory oocyte ageing and temperature on egg quality and on the occurrence of triploid fry in rainbow trout, *Oncorhynchus mykiss*. *Aquaculture* 231: 59–71.

Afonso, L.O.B., J.L. Smith, M.G. Ikonomou and R.H. Devlin. 2002. Y-chromosomal DNA markers for discrimination of chemical substance and effluent effects on sexual differentiation in *Salmon*. *Environmental Health Perspectives* 110: 881–886.

Allen, S.K. 1983. Flow cytometry: assaying experimental polyploid fish and shelfish. *Aquaculture* 33: 317–328.

Allen, S.K. and J.G. Stanley. 1978. Reproductive sterility in polyploid brook trout, *Salvelinus fontinalis*. *Transaction of the American Fisheries Society* 107: 473–478.

Allende, M.L., A. Amsterdam, T. Becker, K. Kawakami, N. Gaiano and N. Hopkins. 1996. Insertional mutagenesis in zebrafish identifies two novel genes, pescadillo and dead eye, essential for embryonic development. *Genes and Development* 10: 3141–3155.

Allendorf, F.W. and G.H. Thorgaard. 1984. Tetraploidy and the evolution of salmonid fishes. In: *Evolutionary Genetics of Fishes*, B.J. Turner (ed.), Plenum Press, New York, pp. 1–53.

Allendorf, F.W., W.A. Gellman and G.H. Thorgaard. 1994. Sex-linkage of two enzyme loci in *Oncorhynchus mykiss* (rainbow trout). *Heredity* 72: 498–507.

Allendorf, F.W., R.F. Leary, P. Spruell and J.K. Wenburg. 2001. The problems with hybrids: Setting conservative guidelines. *Trends in Ecology and Evolution* 16: 613–622.

Alvarez M.C., J. Otis, A. Amores and K. Guise. 1991. Short-term culture technique for obtaining chromosomes in marine and freshwater fish. *Journal of Fish Biology* 39: 817–824.

Amores, A., A. Force, Y.L. Yan, L. Joly, C. Amemiya, A. Fritz, R.K. Ho, J. Langeland, V. Prince, Y.L. Wang, M. Westerfield, M. Ekker and J.H. Postlethwait. 1998. Zebrafish *hox* clusters and vertebrate genome evolution. *Science* 282: 1711–1714.

Angus, R.A. 1989. Inheritance of melanistic pigmentation in the eastern mosquitofish. *Journal of Heredity* 80: 387–392.

Aparicio, S., K. Hawker, A. Cottage, Y. Mikawa, L. Zou, B. Vankatesh, E. Chen, R. Krumlauf and S. Brenner. 1997. Organization of the *Fugu rubripes* Hox clusters: Evidence for continuing evolution of vertebrate *Hox* complexes. *Nature Genetics* 16: 79–83.

Arai, K. 1984. Developmental genetic studies on salmonids: morphogenesis, isozyme phenotype and chromosomes in hybrid embryos. *Memoirs of Faculty of Fisheries, Hokkaido University* 31: 1–91.

Arai, K. 2001. Genetic improvement of aquaculture finfish species by chromosome manipulation techniques in Japan. *Aquaculture* 197: 205–228.

Arai, A. and M. Mukaino. 1997. Clonal nature of gynogenetically reproduced progeny of triploid loach, *Misgurnus anguillicaudatus. Journal of Experimental Zoology* 278: 412–421.

Arai, K., K. Matsubara and R. Suzuki. 1991. Karyotype and erythrocyte size of spontaneous tetraploidy and triploidy in the loach, *Misgurnus anguillicaudatus. Nippon Suisan Gakkaishi* 57: 2167–2172.

Arai, K., T. Masaoka and R. Suzuki. 1992. Optimum conditions of UV-ray irradiation for genetictic inactivation of loach eggs. *Nippon Suisan Gakkaishi* 58: 1197–1201.

Arefjev, V.A. and A.I. Nikolaev. 1991. Cytological analysis of the reciprocal hybrids between low- and high-chromosome Acipenserids, the Great Sturgeon, *Huso huso* (L.), and the Russion Sturgeon, *Acipenser gueldenstaedtii* Brandt. *Cytologia* 56: 495–502.

Artoni R.F., P.C. Venere and L.A.C. Bertollo. 1998. A heteromorphic ZZ/ZW sex chromosome system in fish, genus *Hypostomus* (Loricariidae). *Cytologia* 63: 421–425.

Babiak, I., S. Dobosz, K. Goryczko, H. Kuzminski and P. Woznicki. 1998. Androgenesis in rainbow trout, *Oncorhynchus mykiss*, using gamma irradiation and heat shock. Aquaculture Europe' 98. *Aquaculture and Water: Fish Culture, Shellfish Culture and Water Usage*. Bordeaux, France. October 7–10, 1998.

Babiak, I., S. Dobosz, K. Goryczko, H. Kuzminski, P. Brzuzan and S. Ciesielski. 2002a. Androgenesis in rainbow trout using cryopreserved spermatozoa: The effect of processing and biological factors. *Theriogenology* 57: 1229–1249.

Babiak I., S. Dobosz, H. Kuzminski, K. Goryczko, S. Ciesielski, P. Brzuzan, B. Urbanyi, A. Horvath, F. Lahnsteiner and J. Piironen. 2002b. Failure of interspecies androgenesis in salmonids. *Journal of Fish Biology* 16: 432–447.

Baroiller, J.F., Y. Guguen and A. Fostier. 1999. Endocrine and environmental aspects of sex differentiation fish. *Cell Molecular Life Science* 55: 910–931.

Barre, P., M. Laysacc, A. D'Hont, J. Louaran, A. Charrier, S. Hamon and M. Noirot. 1998. Relationship between parental chromosomic contribution and nuclear DNA content, in the coffee interspecific hybrid C. *pseudozanguebariae* × C. *liberica vardewevrei. Theoretical and Applied Genetics* 96: 301–305.

Bartley, D.M., K. Rana and A.J. Immink. 2002. The use of inter-specific hybrids in aquaculture and fisheries. *Reviews in Fish Biology and Fisheries* 10: 325–337.

Beall, E., P. Moran, A.M. Pendas, J.I. Izguieredo and E. Garcia-Vazquez. 1997. Hybridization in natural populations of salmonids in south-west Europe and in an Experimental channel. *Bulletin Francais de la Peche et la Pisciculture* 344/345: 271–285.

Beardmore, J.A., G.C. Mair and R.I. Lewis. 2001. Monosex male production in finfish as exemplified by tilapia: Application, problems, and prospects. *Aquaculture* 197: 283–301.

Beland, K.F., F.L. Roberts and R.L. Saunders. 1981. Evidence of *Salmo salar* × *Salmo trutta* hybridization in a North American river. *Canadian Journal of Fisheries and Aquatic Sciences* 38: 552–554.

Benfey, T. 1989. *A bibliography of triploid fish, 1943 to 1988. Canadian Technical Report of Fisheries and Aquatic Sciences* 1682: 33 p.

Benfey, T. and A.M. Sutterlin. 1984a. The haematology of triploid landlocked Atlantic salmon *Salmo salar* L. *Journal of Fish Biology* 24: 333–338.

Benfey, T.J., A.M. Sutterlin and R.J. Thompson. 1984b. Use of erythrocyte measurement to identify triploid salmonids. *Canadian Journal of Fisheries and Aquatic Sciences* 41: 980–984.

Bennett, M.D., R.L. Finch and I.R. Barclay 1976. The time rate and mechanism of chromosome elimination in Hordeum hybrids. *Chromosoma* 54: 175–200.

Birstein, V.J., A.I. Poletaev and B.F. Goncharov. 1993. The DNA content in Eurasian sturgeon species determined by flow cytometry. *Cytometry* 14: 377–383.

Blackburn, E.H. 2001. Switching and signaling at the telomere. *Cell* 106: 661–673.

Boron, A. 1994a. Use of erythrocyte measurements to detect natutal triploids of spined loach *Cobitis taenia* (L.). *Cytobios* 78: 197–202.

Boron, A. 1994b. Karyotypes of diploid and triploid silver crucian carp *Carassius auratus gibelio* (Bloch). *Cytobios* 80: 117–124.

Boron, A. 2003. Karyotypes and cytogenetic diversity of the genus *Cobitis* (Pisces, Cobitidae) in Poland: A review. Cytogenetic evidence for a hybrid origin of some *Cobitis* triploids. *Folia Biologica (Krakow)* 51: 49–54.

Bosworth, B.G., E.A. Dunnington, G.S. Libey and L.C. Stallard. 1994. Restriction enzyme/multi-locus probe combinations useful for DNA fingerprinting of the striped bass, white bass and their F1 hybrid. *Aquaculture* 123: 205–215.

Breeuwer, J.A.J. and J.H. Werren. 1990. Microorganisms associated with chromosome destruction and reproductive isolation between two insect species. *Nature (London)* 346: 558–560.

Brzuzan, P. 1998. DNA length variation and RFLP of the mitochondrial control region in two samples of whitefish, *Coregonus lavaretus*, from Lake Baikal (Russia) and Lake Maroz (Poland). *Archiv fuer Hydrobiologie Special Issues Advanced Limnology* 50: 349–356.

Brzuzan, P., M. Luczynski and P.A. Kuzniar. 1998. Mitochondrial DNA variation in two samples of northern pike, *Esox lucius* L. *Aquaculture Research* 29: 521–526.

Campos-Ramos, R., S.C. Harvey, J.S. Masabanda, L.A.P. Carasco, D.K. Griffin, B.J. McAndrew, N.R. Bromage and D.J. Penman. 2001. Identification of putative sex chromosomes in the blue tilapia, *Oreochromis aureus*, through synaptonemal complex and FISH analysis. *Genetica* 111: 143–153.

Campos-Ramos, R., S.C. Harvey, B.J. McAndrew and D.J. Penman. 2003. An investigation of sex determination in the Mozambique tilapia, *Oreochromis mossambicus*, using synaptonemal complex analysis, FISH, sex reversal and gynogenesis. *Aquaculture* 221: 125–140.

Canapa, A., P.N. Cerioni, M. Barucca, E. Olmo and V. Caputo. 2002. A centromeric satellite DNA may be involved in heterochromatin compactness in gobiid fishes. *Chromosome Research* 10: 297–304.

Capriglione, T., A. Morescalchi, E. Olmo, L. Rocco, V. Stingo and S. Manzo. 1994. Satellite DNAs, heterochromatin and sex chromosomes in *Chiondraco hamatus* (Channichthyidae, Perciformes). *Polar Biology* 14: 285–290.

Carasco, L.A.P., D.J. Penman and N. Bromage. 1999. Evidence for the presence of sex chromosomes in the Nile tilapia *Oreochromis niloticus* from synaptonemal complex analysis of XX, XY and YY genotypes. *Aquaculture* 173: 207–218.

Carbone, P., R. Vitturi, E. Catalano and M. Macaluso. 1987. Chromosome sex determination and Y-autosome fusion in *Blennius tentacularis* Brunnich, 1765 (Pisces, Blennidae). *Journal of Fish Biology* 31: 597–602

Carter, R.E., G.C. Mair, D.O.F. Skibinski, D.T. Parkin and J.A. Beardmore. 1991. The application of DNA fingerprinting in the analysis of gynogenesis in tilapia. *Aquaculture* 95: 41–52.

Castro, J., S. Rodriguez, J. Arias, L. Sanchez and P. Martinez. 1994. A population analysis of Robertsonian and Ag-NOR polymorphisms in brown trout *Salmo trutta*. *Theoretical and Applied Genetics* 89: 105–111.

Castro, J., S. Rodríguez, B.G. Pardo, L. Sanchez and P. Martinez. 2001. Population analysis of an unusual NOR-site polymorphism in brown trout *Salmo trutta* L. *Heredity* 86: 291–302.

Cherfas, N., B. and V.A. Ilyasova,. 1980. Induced gynogenesis in hybrids between crucian carp and common carp. *Genetics (Moscow)* 16: 1260–1269.

Cherfas, N.B., B. Gomelsky, Y. Peretz, N. Ben-Dom and G. Hulata. 1993. Induced gynogenesis and polyploidy in the Israeli common carp line dor-70. *Israel Journal of Aquaculture-Bamidgeh* 45: 59–72.

Chevassus, B. 1983. Hybridization in fish. *Aquaculture* 33: 245–262.

Chourrout, D. 1984. Pressure-induced retention of second polar and suppression of first cleavage in rainbow trout. Production of all triploids, all tetraploids, and heterozygous and homozygous diploid gynogenesis. *Aquaculture* 36: 111–126.

Chourrout, D. and E. Quillet. 1982. Induced gynogenesis in rainbow trout: Sex and survival of progenies. Production of of all triploid populations. *Theoretical and Applied Genetics* 63: 201–205.

Collares-Pereira, M. J., J. M. Madeira and P. Rab. 1995. Spontaneous triploidy in the stone loach *Noemacheilus barbatulus* (Balitoridae). *Copeia* 1995: 483–484.

Dabrowski, K., J. Rinchard, F. Lin, M.A. Garcia-Abiado and D. Schmidt. 2000. Induction of gynogenesis in muskellunge with irradiated sperm of yellow perch proves diploid muskellunge male homogamety. *Journal of Experimental Zoology* 287: 96–105.

de Lange, T. 2002. Protection of mammalian telomeres. *Oncogene* 21: 532–540.

Demska-Zakes, K., M.J. Luczynski, K. Dabrowski, M. Luczynski and J. Krol. 2000. Masculinization of northern pike fry using the steroid, 11-hydroxyandrostedione. *North American Journal of Aquaculture* 62: 294–299.

Devlin, R.H. and Y. Nagahama. 2002. Sex determination and sex differentiation in fish: An overview of genetic, physiological, and environmental influences. *Aquaculture* 208: 191–364.

Devlin, R.H., B.K. NcNeil, T.D.D. Groves. and E.M. Donaldson,. 1991. Isolation of a Y-chromosomal DNA probe capable of determining genetic sex in chinook salmon *Oncorhynchus tshawytscha*. *Canadian Journal of Fisheries and Aquatic Sciences* 48: 1606–1612.

Devlin, R.H., G.W. Stone and D.E. Smailus. 1998. Extensive direct-tandem organization of a long repeat DNA sequence on the Y chromosome of chinook salmon *Oncorhynchus tshawytscha*. *Journal of Molecular Evolution* 46: 277–287.

Devlin, R.H., C.A. Biagi and D.E. Smailus. 2001. Genetic mapping of Y-chromosomal DNA markers in Pacific salmon. *Genetica* 111: 43–58.

Disney, J.E., K.R. Johnson and G.H. Thorgaard. 1987. Intergeneric gene transfer of six isozyme loci in rainbow trout by sperm chromosome fragmentation and gynogenesis. *Journal of Experimental Zoology* 244: 151–158.

Disney, J.E., K.R. Johnson, D.K. Banks and G.H. Thorgaard. 1988. Maintenance of foreign gene expression and independent chromosome fragments in adult transgenic rainbow trout and their offspring. *Journal of Experimental Zoology* 248: 335–344.

Dobosz, S., K. Goryczko, W. Olech, A. Zyczynski, F. Friedrich and K. Kohlman. 1992. Incomplete diallele cross of rainbow trout strains imported to Poland. *Fortschritte der Fischerewissenschaft* 10: 75–82.

Donaldson, E.M. 1996. Manipulation of reproduction in farmed fish. *Animal Reproduction Science* 42: 381–392.

Donaldson, E.M. and G.A. Hunter. 1983. Induced final maturation, ovulation and spermiation in cultured fish. In: *Fish Physiology: Reproduction*, W.S. Hoar, D.J. Randall and E.M. Donaldson (eds.), Academic Press, New York, Vol. 9B, pp. 351–403.

Du, S. J., R. H. Devlin and C. L. Hew. 1993. Genomic structure of growth hormone genes in chinook salmon *Oncorhynchus tshawytscha*: presence of two functional genes,

GH-I and GH-II, and a male-specific pseudogene, GH-psi. *DNA Cell Biology* 12: 739–751.

Echelle, A.A., A.D. Echelle, L.E. Debault and D.W. Durham. 1988. Ploidy levels in the silver side fishes (Atherniidae, *Menidia*) on the Texas coast: Flow cytometric analysis of the occurrence of allotriploidy. *Journal of Fish Biology* 32: 835–844.

Ewing, R.R. and C.G. Scalet. 1991. Flow cytometric identification of larval triploid walleyes. *The Progressive Fish-Culturist* 53: 177–180.

Fan, Z. and D.P. Fox. 1991. Robertsonian polymorphism in plaice *Pleuronectes platessa* L. and cod *Gadus morhua* L. (Pisces Pleuronectiformes and Gadiformes). *Journal of Fish Biology* 38: 635–640.

Fernando, A.A. and V.P.E. Phang. 1989. Inheritance of the tuxedo and blond tuxedo color pattern phenotypes of the guppy, *Poecilia reticulata*. In: *Proceedings of the Second Asian Fisheries Forum*, Tokyo, Japan, The Second Asian Fisheries Forum, Tokyo, Japan, pp. 487–490

Fernando, A.A. and V.P.E. Phang. 1990. Colour pattern inheritance in three domesticated varieties of guppy, *Poecilia reticulata*. Genetic in Aquaculture III 85, p. 320.

Finch, R., A. 1983. Tissue-specific elimination of alternative whole parental genomes in one barley hybrid. *Chromosoma* 88: 386–393.

Flajshans, M. 1997. A model approach to distinguish diploid and triploid fish by means of computer-assisted image analysis. *Acta Veterinaria (Brno)* 66: 101–110.

Flajshans, M. and V. Vajcova. 2000. Odd ploidy levels in sturgeons suggest a backcross of interspecific hexaploid sturgeon hybrids to evolutionarily tetraploid and/or octaploid parental species. *Folia Zoologica* 49: 133–138.

Flajshans, M., P. Rab and S. Dobosz. 1992. Frequencey analyses of active NORs in nuclei of artificially induced triploid fishes. *Theoretical and Applied Genetics* 85: 68–72.

Flajshans, M., O. Linhart and P. Kvasnicka. 1993. Genetic studies of tench *Tinca tinca* L.: Induced triploidy and tetraploidy and first performance data. *Aquaculture* 113: 301–312.

Flajshans, M., M. Kocour, D. Gela and V. Piackova. 2004. The first results on relationships among amphimictic diploid, diploid gynogenic and triploid tench, *Tinca tinca* L. under communal testing. *Aquaculture International* 12: 103–118.

Forbes, S.H., K.L. Knudsen, T.W. North and F.W. Allendorf. 1994. One of two growth hormone genes in coho salmon is sex-linked. *Proceedings of the National Academy of Sciences of the United States of America* 91: 1628–1631.

Foresti, F., C. Oliveira and E. D. Carvalho. 1994. Ploidy evaluation in the pacu fish, *Piaractus mesopotamicus* (Pisces, Characiformes): techniques and comments. *Revista Brasileira de Biologia* 54: 31–37.

Foresti de Almeida-Toledo, L., A. P. Verri Bigoni, G. Benardino and Silvio de Almeida Toledo Filho. 1995. Chromosomal location of NORs and C bands in F_1 hybrids of bighead carp and silver carp reared in Brazil. *Aquaculture* 135: 277–284.

Francis, R. C. 1984. The effects of bidirectional selection for social dominance on agonistic behaviour on sex ratios in the paradise fish *Macropodus opercularis*. *Behaviour* 90: 25–45.

Frolov, S.V. and V.N. Frolova. 2000. Polymorphism and karyotype divergence in taimens of the *Hucho* genus. *Russian Journal of Genetics* 36: 175–178.

Fujiwara, A., S, Abe, E. Yamaha, F. Yamazaki and M.C. Yoshida, M. 1997. Uniparental chromosome elimination in the early embryogenesis of the inviable salmonid hybrids between masu salmon female and rainbow trout male. *Chromosoma* 106: 44–52.

Fujiwara, A., C. Nishida-Umehara, T. Sakamoto, N. Okamoto, I. Nakayama and S. Abe. 2001. Improved fish lymphocyte culture for chromosome preparation. *Genetica* 111: 77–89.

Garcia de Laniz, C. and E. Varspoor. 1989. Natural hybridization between Atlantic salmon, *Salmo salar* and brown trout, *Salmo trutta* in northern Spain. *Journal of Fish Biology* 34: 41–46.

Gimeno, S., A. Gerritsen, T. Bowmer and H. Komen. 1996. Feminization of male carp. *Nature (Lond.)* 384: 221–222.

Gimeno, S., H. Komen, P. W.M. Venderbosch and T. Bowmer. 1997. Disruption of sexual differentiation in genetic male common carp *Cyprinus carpio* exposed to an alkylphenol during different life stages. *Environmental Science and Technology* 31: 2884–2890.

Gimeno, S., H. Komen, A.G.M. Gerritsen and T. Bowmer. 1998. Feminisation of young males of the common carp, *Cyprinus carpio*, exposed to 4-tert-penylphenol during sexual differentiation. *Aquatic Toxicology* 43: 77–92.

Gold, J.R. and C.T. Amemiya. 1987. Genome size variation in North American minnows (Cyprinidae). II. Variation among twenty species. *Genome* 29: 481–489.

Gold, J.R. and H.J. Price. 1985. Genome size variation in North American minnows (Cyprinidae). I. Distribution of the variation in five species. *Heredity* 54: 297–305.

Gold, J.R., C.J. Ragland and L.J. Schliesing. 1990. Genome size variation and evolution in North American cyprinid fishes. *Genetics, Selection, Evolution* 22: 11–29.

Gold, J.R., C.J. Ragland, M.C. Birkner and G.P. Garrett. 1991. A simple procedure for long-term storage and preparation of fish cells for DNA content analysis using flow cytometry. *The Progressive Fish-Culturist* 53: 108–110.

Gomelsky, B. 2003. Chromosome set manipulation and sex control in common carp: A review. *Aquatic Living Resources* 16: 408–415.

Goodier, J., H-F. Ma and F. Yamazaki. 1987. Chromosome fragmentation in and loss in two salmonid hybrids. *Bulletin of the Faculty of Fisheries Hokkaido University* 38: 181–184.

Goudie, C.A., T.R. Tiersch, B.A. Simco, K.B. Davis and Liu, Q. 1993. Early growth and morphology among hybrids of ictalurid catfishes. *Journal of Applied Aquaculture* 3: 235–256.

Gray, A.K., M.A. Evans and G.H. Thorgaard. 1993. Viability and development of diploid and triploid salmonid hybrids. *Aquaculture* 112: 125–142.

Gutbrod, H. and M. Schartl. 1999. Intragenic sex-chromosomal crossovers of X*mrk* oncogene alleles affect pigment pattern formation and the severity of melanoma in *Xiphophorus*. *Genome Research* 6: 102–113.

Hardie, D.C., T.R. Gregory and P.D.N. Hebert. 2002. From pixels to picograms: A beginners' guide to genome quantification by feulgen image analysis densitometry. *Journal of Histochemistry and Cytochemistry* 50: 735–750.

Hendry C.I., D.J. Martin-Robichaud and T. J. Benfey. 2003. Hormonal sex reversal of Atlantic halibut *Hippoglossus hippoglossus* L. *Aquaculture* 219: 769–781.

Horvath, L. and L. Orban. 1995. Genome and gene manipulation in the common carp. *Aquaculture* 129: 157–181.

Howell, W.M. and D.A. Black. 1980. Controlled, silver staining of nucleolus organizer regions with protective colloidal developer: A 1-step method. *Experientia* 36: 1014–1015.

Hostache, G., M. Pascal and C.Tessier. 1995. Influence de la temperature d'incubation sur le rapport male: Female chez l'atipa, *Hoplosternum littorale* Hancock (1828). *Canadian Journal of Zoology* 73: 1239–1246.

Hubbs, C.L. and L.C. Hubbs. 1932. Apparent parthenogenesis in nature, in a form of fish of hybrid origin. *Science* 76: 628–630.

Hulata, G. 2001. Genetic manipulations in aquaculture: A review of stock improvement by classical and modern technologies. *Genetica* 111: 155–173.

Ihssen, P.E., L.R. McKay, I. MCMillan and R.B. Phillips. 1990. Ploidy manipulation and gynogenesis in fishes: Cytogenetic and fisheries applications. *American Fisheries Society* 119: 698–717.

Ijiri, K. and N. Egami. 1980. Hertwig effect causes by UV-irradiation of sperm of *Oryzias latipes* (Teleost) and its photoreactivation. *Mutation Research* 69: 241–248.

Inokuchi T., S. Abe, E. Yamaha, F. Yamazaki and M.C. Yoshida. 1994. BrdU replication banding studies on the chromosomes in early embryos of salmonid fishes. *Hereditas* 121: 255–265.

Iturra, P., J.F. Medrano, M. Bagley, N. Lam, N. Vergara and J.C. Marin. 1998. Identification of sex chromosome molecular markers using RAPDs and fluorescent in situ hybridization in rainbow trout. *Genetica* 101: 209–213.

Iturra, P., M. Bagley, N. Vergara, P. Inbert and J.F. Medrano. 2001. Development and characterization of DNA sequence OmyP9 associated with the sex chromosomes in rainbow trout. *Heredity* 86: 412–419.

Janko, K.P., P. Kotlik and P. Rab. 2003. Evolutionary history of asexual hybrid loaches (*Cobitis*, Teleostei) inferred from phylogenetic analysis of mitochondrial DNA variation. *Journal of Evolutionary Biology* 16: 1280–1287.

Jansson, H., I. Holmgren, K. Wedin and T. Andersson. 1991. High frequency of natural hybrids between Atlantic salmon, *Salmo salar* L. and brown trout, *Salmo trutta* L. in a Swedish river. *Journal of Fish Biology* 39A: 343–348.

Johnson, O.W., F.M. Utter and P.S. Rabinovitch. 1987. Interspecies differences in salmonid cellular DNA identified by flow cytometry. *Copeia* 1987: 1001–1009.

Johnstone, R. and R.F. Lincoln. 1986. Ploidy estimation using erythrocytes from formalin-fixed salmonid fry. *Aquaculture* 55: 145–148.

Jordan, W.C. and E. Verspoor. 1993. Incidence of natural hybrids produce gynogens and triploids when backcrossed to male *Atlantic salmon*. *Aquaculture* 57: 245–258.

Kakimoto, Y., S. Aida, K. Arai and R. Suzuki. 1994. Production of gynogenetic diploids by temperature and pressure treatments and sex reversal by immersion in methyltestosterone in Marbled sole, *Limanda yokohamae*. *Applied Biological Science/Seibutsu Seisangaku Kenkyu* 33: 113–124.

Kapusta, A. and S. Ciesielski. 2004. Identification of cyprinid hybrids based on isozyme electrophoresis and mitochondrial DNA analysis. (In preparation).

Keith, L.H. 1997. *Environmental Endocrine Disruptors*. John Wiley and Sons, New York.

Kerby, J.H. and R.M. Havell. 1980. Hybridization, genetic manipulation, and gene pool conservation of striped bass. In: *Culture and Propagation of Striped Bass and Its Hybrids*. R.M. Harell, J.H. Kerby and R.V. Minton (eds.). Bethesda, Maryland: American Fishery Society, pp. 159–190.

Khoo, G., T.M. Lim, W.K. Chan and V.P.E. Phang. 1999. Sex-linkage of the black caudal-peduncle and red tail genes in the tuxedo strain of the guppy, *Poecilia reticulata*. *Zoological Science* 16: 629–638.

Khuda Bukhsh, A.R. and S. Datta. 1997. Sex-specific difference in NOR-location on metaphase chromosomes of mosquito fish, *Aplocheilus panchax* (Cyprinodontidae). *Indian Journal of Experimental Biology* 35: 1111–1114.

Komen, J. and C.J.J. Richter. 1990. Sex control in carp *Cyprinus carpio*. *Recent Advances in Aquaculture*. 4: 76–86.

Komen, J., P. de Boer and C.J.J. Richter. 1992. Male sex reversal in gynogenetic XX females of common carp *Cyprinus carpio* L. by a recessive mutation in a sex-determining gene. *Journal of Heredity* 83: 431–434.

Kucharczyk, D., M.J. Luczynski, M. Jankun and M. Luczynski. 1996. Preliminary observations on the induction of triploidy in *Abramis brama* by cold shock. *Cytobios* 88: 153–160.

Kucharczyk, D., M. Janku and M. Luczynski. 1997. Ploidy level determination in genetically manipulated bream, *Abramis brama* L., based on the number of active nucleoli per cell. *Journal of Applied Aquaculture* 7: 13–21.

Kucharczyk, D., P. Woznicki, M.J. Luczynski and M. Klinger. 1999. Ploidy level determination in genetically manipulated northern pike based on the number of active nucleoli per cell. *North American Journal of Aquaculture* 61: 38–42.

Kwon J.Y., V. Haghpanah, L.M. Kogson-Hurtado, B.J. McAndrew and D.J. Penman. 2000. Masculinization of genetic female Nile tilapia *Oreochromis niloticus* by dietary administration of an aromatase inhibitor during sexual differentiation. *Journal of Experimental Zoology* 287: 46–53.

Lahav, M. and E. Lahav. 1990. The development of all-male tilapia hybrids in Nir David. *Israel Journal of Aquaculture, Bamidgeh*, 42: 58–61.

Lam, N., K. Ocalewicz and P. Iturra. 2005. OMY-P9 sex probe application to identity the Y-chromosomes of androgenetic rainbow trout from Rutki strain. *Aquaculture* 247: 22.

Lamatsch, D.K., I. Nanda, J.T. Epplen, M. Schmid and M. Schartl. 2000. Unusual triploid males in a microchromosome-carrying clone of the Amazon molly, *Poecilia formosa*. *Cytogenetics and Cell Genetics* 91: 148–156.

Lamatsch, D.K., M. Schmid and M. Schartl. 2002. A somatic mosaic of the gynogenetic Amazon molly. *Journal of Fish Biology* 60: 1471–1422.

Lecommandeur, D., P. Haffray and L. Philippe. 1994. Rapid flow cytometry method for ploidy determination in salmonid eggs. *Fishery Management* 25: 345–350.

Li, Y. and J.R. Gold. 1991. Cytogenetic studies in North American minnows (Cyprinidae): 22. chromosomal nucleolar organizer regions in the genus *Pimephales*. *Canadian Journal of Zoology* 69: 2826–2830.

Lima De Faria, A. 1983. *Molecular Evolution and Organization of the Chromosome*. Elsevier, New York.

Lin, F. and K. Dabrowski. 1998. Androgenesis and homozygous gynogenesis in muskellunge *Esox masquinongy*: Evaluation using flow cytometry. *Molecular Reproduction and Development* 49: 10–18.

Lin, F., K. Dabrowski, M.J. and M. Luczynski. 2001. Mosaic individuals found in genetically manipulated northern pike *Esox lucius* using flow cytometry. *Journal of Applied Ichthyology* 17: 85–88.

Linhart, O., P. Haffray, C. Ozouf-Costaz, M. Flajšhan and M. Vandeputte. 2001. Comparaison of methods for hatchery-scale triploidization of European catfish, *Silurus glanis*. *Journal of Applied Ichthyology* 17: 247–255.

Lloyd, M.A. and G.H. Thorgaard. 1988. Restriction endonuclease banding of rainbow trout chromosomes. *Chromosoma* 96: 171–177.

Madigou, T., P. Le Goff, G. Salbert, J.P. Cravedi, H.F. Segner, F. Pakdel and Y. Valotaire. 2001. Effects of nonylphenol on estrogen receptor conformation, transcriptional activity and sexual reversion in rainbow trout, *Oncorhynchus mykiss*. *Aquatic Toxicology* 53: 173–186.

Mallitschek, B., D. Fornzler and M. Schartl. 1995. Melanoma formation in *Xiphophorus*: A model system for the role of receptor tyrosine kinases in tumorigenesis. *BioEssays* 17: 1017–1023.

Mamotani, S., K. Morisihima, Q. Zhang and K. Arai. 2002. Genetic analysis of the progeny of triploid gynogens induced from unreduced eggs of triploid (diploid female × tetraploid male) loach. *Aquaculture* 2004: 311–322.

Martinez, P., A. Vinas, C. Bouza, J. Arias, R. Amaro and L. Sanchez. 1991. Cytogenetical characterization of hatchery stocks and natural populations of sea and brown trout from Northwestern Spain. *Heredity* 66: 9–17.

Martins, C., C. Oliveira, A.P. Wasko and J.M. Wright. 2004. Physical mapping of the Nile tilapia *Oreochromis niloticus* genome by fluorescent in situ hybridization of repetitive DNAs to metaphase chromosomes: A review. *Aquaculture* 231: 37–49.

McClintock, B. 1941. The stability of broken ends of chromosomes in *Zea mays*. *Genetics* 41: 234–283.

McGowan, C. and W.S. Davidson. 1992. Unidirectional natural hybridization between brown trout *Salmo trutta* and Atlantic salmon *Salmo salar* in Newfoundland. *Canadian Journal of Fisheries and Aquatic Sciences* 49: 1953–1958.

Moran, P., J.L. Martinez, E. Garcia-Vazquez and A.M. Pendas. 1996. Sex chromsomes likage of 5S rDNA in rainbow trout *Oncorhynchus mykiss*. *Cytgenetics and Cell Genetics* 75: 145–150.

Morizot, D.C., S.A. Slaugenhaupt, K. D. Kallman and A. Chakravarti. 1991. Genetic linkage map of fishes of the genus *Xiphophorus* (Teleostei: Poeciliidae). *Genetics* 127: 399–410.

Myers, J.M., D.J. Penman, J.K. Rana, S.F. Bromage, N. Powell, S.F. Powell and B.J. McAndrew. 1995. Application of induced androgenesis with tilapia. *Aquaculture* 135: 150.

Nagler, J.J., J. Bouma, G.H. Thorgaard and D.D. Dauble. 2001. High incidence of a male-specific genetic marker in phenotypic female chinook salmon from the Columbia River. *Environment Health Perspectives* 109: 67–69.

Nakayama, I. and D. Chourrout. 1993. Uniparental reproduction demonstrates the inheritance of Robertsonian variations in the rainbow trout, *Oncorhynchus mykiss*. *Copeia* 1993: 553–557.

Nakayama, I., F. Foresti, R. Tewari, M. Schartl and D. Chourrout. 1994. Sex chromosome polymorphism and heterogametic males revealed by two cloned DNA probes in the ZW/ZZ fish *Leporinus elongatus*. *Chromosoma* 103: 31–39.

Nakayama, I., C.A. Biagi, N. Koide and R.H. Devlin. 1998. Identification of a sex-linked GH pseudogene in one of two species of Japanese salmon *Oncorhynchus masou* and *O. rhodurus*. *Aquaculture* 173: 65–72.

Nanda, I., V. Volff, S. Weis, C. Koerting, A. Froschauer, M. Schmid and M. Schartl. 2000. Amplification of a long terminal repeat-like element on the Y chromosome of the platyfish, *Xiphophorus maculatus*. *Chromosoma* 109: 173–180.

Nyman, L., J. Hammar and R. Gydemo. 1984. Lethal, sex-linked genes associated with homozygosity at the esterase-2 locus in Arctic char. In: *Proceedings of the Third ISACF Workshop on Arctic Char*, ISACF Inf. Ser. **vol. 3**, ISACF Inf. Ser., Tromso, Norway, pp. 125–130.

Ocalewicz, K. 2002a. *Essentials of sex determination process in fish*. Ph.D. Thesis. Department of Evolutionary Genetics, Faculty of Environmental Protection and Fisheries, University of Warmia and Mazyry in Olsztyn, Olsztyn, Poland.

Ocalewicz, K. 2002b. Cytogenetic markers for X chromosome in karyotype of rainbow trout from Rutki strain. *Folia biologica (Krakow)* 50: 10–14.

Ocalewicz, K. and I. Babiak. 2003. Primed in situ labeling (PRINS) detection of 5S rDNA sequences proves absence of X chromosome in supermale androgenetic rainbow trout. *Journal of Fish Biology* 62: 1462–1466.

Ocalewicz K., I. Babiak, S. Dobosz, J. Nowaczyk and K. Goryczko. 2004a. The stability of telomereless chromosome fragments in adult androgenetic rainbow trout. *Journal of Experimental Biology* 207: 2229–2236.

Ocalewicz, K., A. Sliwinska and M. Jankun. 2004b. Autosomal localization of Interstitial Telomeric Site (ITS) in brook trout, *Salvelinus fontinalis* (Pisces, Salmonidae). *Cytogenetic and Genome Research* 105: 79–82.

Ohno, S., C. Stenius, E. Faissit and M.T. Zenzes. 1965. Post-zygotic chromosomal rearrangements in rainbow trout, *Salmo irideus* Gibbons. *Cytogenetics* 4: 117–129.

Ojima, Y. and A. Takai. 1979. The occurrence of spontaneous polyploidy in the Japanese common loach, *Misgurnus angullicaudatus*. *Proceedings of the Japan Academy* 55B: 487–491.

Ozouf-Costaz, C., J. Brandt, C. Korting, E. Pisano, C. Bonillo, J. P. Coutanceau and J.N. Volff. 2004. Genome dynamics and chromosomal localization of the non-LTR retrotransposons *Rex1* and *Rex3* in Antarctic fish. *Antarctic Science* 16: 51–57.

Palti, Y., J.E. Parsons and G.H. Thorgaard. 1997. Assessment of genetic variability among strains of rainbow trout and cutthroat trout using multilocus DNA fingerprints. *Aquaculture* 149: 47–56.

Pandian, T.J. and R. Koteeswaran. 1998. Ploidy induction and sex control in fish. *Hydrobiologia* 384: 167–243.

Park, I., S. Janku, K. Nam, S.E. Douglas, S.C. Johnson and D.S. Kim. 2003. Genetic characterization, morphometrics and gonad development of induced interspecific hybrids between yellowtail flounder, *Pleuronectes ferrugineus* (Storer) and winter flounder, *Pleuronectes americanus* (Walbaum). *Aquaculture Research* 34: 389–396.

Parsons, J.E. and G.H. Thorgaard. 1985. Production of androgenetic diploid rainbow trout. *Journal of Heredity* 76: 177–181.

Peek, A.S., P.A. Wheeler, C.O. Ostberg and G.H. Thorgaard. 1997. A minichromosomes carrying a pigmentation gene and brook trout DNA sequence in transgenic rainbow trout. *Genome* 40: 594–599.

Pendas, A.M., P. Moran and E. Garcia-Vazquez. 1994. Organization and chromosomal location of the major histone DNA cluster in brown trout, Atlantic salmon and rainbow trout. *Chromosoma* 103: 152–157.

Perez, J., J.L. Martinez, P. Moran, P. Beall and E. Garcia-Vasquez. 1999. Identification of Atlantic salmon × brown trout hybrids with a nuclear marker useful for evolutionary studies. *Journal of Fish Biology* 54: 460–464.

Phillips, R.B. and S.E. Hartley. 1984. Fluorescent banding pattern of the chromosomes of the genus *Salmo. Genome* 30: 193–197.

Phillips, R.B. and P.E. Ihssen. 1984. Chromosome banding difference between the X and Y chromosomes of lake trout. *Genetics* 107: 82–83.

Phillips, R.B. and P.E. Ihssen. 1985 Identification of sex chromosomes in lake trout *Salvelinus namaycush. Cytogenetics and Cell Genetics* 39: 14–18.

Phillips, R.B. and A.R. Kapuscinski. 1988. High frequency of translocation heterozygotes in odd year populations of pink salmon *Oncorhynchus gorbuscha. Cytogenetics and Cell Genetics* 48: 178–182.

Phillips, R.B. and P. Rab. 2000. Chromosome evolution in the Salmonidae (Pisces): An update. *Biological Reviews* 76: 1–25.

Phillips, R.B. and K.M. Reed. 1996. Application of fluorescence in situ hubridization (FISH) techniques to fish genetics: A review. *Aquaculture* 140: 197–216.

Phillips, R.B., K.D. Zajicek, P.E. Ihssen and O. Johnson. 1986. Application of silver staining to the identification of triploid fish cells. *Aquaculture* 54: 313–319.

Phillips, R.B., N.R. Konkol, K.M. Reed and J.D. Stein. 2001. Chromosome painting supports lack of homology among sex chromosomes in *Oncorhynchus, Salmo,* and *Salvelinus* (Salmonidae). *Genetica* 111: 119–123.

Phillips, R.B, M.P. Matsuoka and K.M. Reed. 2002. Characterization of charr chromosomes using fluorescence *in situ* hybridization. *Environmental Biology of Fishes* 64: 223–228.

Piferrer, F., S. Zanuy, M. Carrillo, I. Solar, R.H. Devlin and E.M. Donaldson. 1994. Brief treatment with an aromatase inhibitor during sex differentiation causes chromosomally female salmon to develop as normal, functional males. *Journal of Experimental Zoology* 270: 255–262.

Pisano, E., L. Ghigliotti, F. Mazzei, C. Ozouf-Costaz, C. Bonillo and J.P. Coutanceau. 2003. Mapping of 5S ribosomal genes revealed independent evolution of sex-linked heterochromosomes in two closely related teleostean species, *Pagetopsis macropterus* and *Chionodraco hamatus* (suborder Notothenioidei, family Channichthyidae). *ECA Fourth European Cytogenetic Conference, Bologna, September 2003 Annales de Génétique* 46: 98.

Porto-Foresti, F., C. Oliveira, Y.A. Tabata, M.G. Rigolino and F. Foresti. 2002. NORs inheritance analysis in crossing including individuals from two stocks of rainbow trout, *Oncorhynchus mykiss. Hereditas* 136: 227–230.

Purdom, C.E. 1972. Induced polyploidy in plaice, *Pleuronectes platessa* and its hybrid with the flounder, *Platichthys flesus. Heredity* 29: 11–24.

Purdom, C.E. 1983. Genetic engineering by the manipulation of chromosomes. *Aquaculture*, 33: 287–300.

Purdom, C.E. 1993. *Genetics and Fish Breeding*. Chapman and Hall, London.

Reed, K.M. and J.H. Werren. 1995. Induction of paternal genome loss by the paternal-sex-ratio chromosome and cytoplasmic incompatibility bacteria (Wolbachia): A comparative study of early embryonic events. *Molecular Reproduction and Development* 40: 408–418.

Ren, X., O. Yu, J. Cui and Z. Chang. 1993. Fluorescence banding studies in fish chromosomes. *Acta Genetica Sinica* 20: 116–121.

Rubin, D.D. 1985. Effect of pH on sex ratio in cichlids and poeciliids (Teleostei). *Copeia* 1985: 233–235.

Sanchez, L., P. Martinez, C. Bouza and A. Vinos. 1991. Chromosomal heterochromatin differentiation in *Salmo trutta* with restriction enzymes. *Heredity* 66: 421–249.

Schartl, M., I. Nanda, I. Schlupp, B. Wilde, J.T. Epplen, M. Schmid and J. Parzefall. 1995. Incorporation of subgenomic amounts of DNA as compensation for mutational load in a gynogenetic fish. *Nature (London)* 373: 68–71.

Scheerer, P.D., G.H. Thorgaard and F.W. Allendorf. 1991. Genetic analysis of androgenetic rainbow trout. *Journal of Experimental Zoology* 260: 382–290.

Schier, A.F., S.C. Neuhauss, M. Harvey, J. Malicki, L. Solnica-Krezel, D.Y. Stainier, F. Zwartkruis, S. Abdelilah, D.L. Stemple, Z. Rangini, H. Yang and W. Driever. 1996. Mutations affecting the development of the embryonic zebrafish brain. *Development* 123: 165–178

Schultz, R.J. 1969. Hybridization, unisexuality, and polyploidy in the teleost *Poeciliopsis* (Poeciliidae) and other vertebrates. *American Naturalist* 103: 605–619.

Schultz, R.J. 1993. Genetic regulation of temperature-mediated sex ratios in the livebearing fish *Poeciliopsis lucida. Copeia* 1993: 1148–1151.

Schwartz, F.J. 1981. World literature to fish hybrids with an analysis by family, species, and hybrids. Supplement 1. *NOAA Technical Report NMFS SSRF 750*: 507.

Schwartzbaum, P., A. Valcarcel and M.C. Maggese. 1992. Ploidy of South American Catfish *Rhamdia sapo* (Pisces, Pimelodidae) determined by spectrofluorometry. *Journal of Aquaculture in the Tropics* 7: 151–155.

Scribner, K.T., K.S. Page and M. Bartron. 2000. Hybridization in freshwater fishes: A review of case studies and cytonuclear methods of biological inference. *Reviews in Fish Biology and Fisheries* 10: 293–323.

Shimizu, C., H. Shike, D.M. Malicki, E. Breisch, M. Westerman, J. Buchanan, H.R. Ligman, R.B. Phillips, J.M. Carlberg, J. Van Olst and J.C. Burns. 2003. Characterization of a white bass *Morone chrysops* embryonic cell line with epithelial features. *In Vitro Cellular and Developmental Biology – Animal* 39: 29–35.

Snowdon, R.J., W. Köhler, W. Friedt and A. Köhler. 1997. Genomic in situ hybridization in *Brassica* amphidiploids and interspecific hybrids. *Theoretical and Applied Genetics* 95: 1320–1324.

Solari, A.J. 1993. *Sex Chromosomes and Sex Determination in Vertebrates.* CRC Press. Boca Raton.

Song, S.X. 1987. Success in production a new *Epineelus* hybrid, *Epineelus awoara* × *E. fario. Marine Fisheries* 6: 271–272.

Stein, J.D., R.B. Phillips and R.H. Devlin. 2001. Identification of sex chromosomes in chinook salmon *Oncorhynchus tshawytscha. Cytogenetics and Cell Genetics* 98: 108–110.

Strusmann, C.A., J.C. Calsina Costa, G. Phonlor, H. Higuchi and F. Takashima. 1996. Temperature effects on sex differentiation of two South American atherinid, *Odontesthes argentinensis* and Pataagonina hatchery. *Environmental Biology of Fishes* 47: 143–154.

Strusmann, C.A., T. Saito, M. Usui, H. Yamada and F. Takashima. 1997. Thermal threshold and critical period of thermolabile sex determination in two Atherinid fishes, *Odontesthes bonariensis* and Pataagonina hatchery. *Journal of Experimental Zoology* 278: 167–177.

Sumpter, J.P. 1998. Xenoendocrine disrupters—Environmental impacts. *Toxicological Letters* 102-103: 337–342.

Suzuki, R., T. Oshiro and T. Nakanishi. 1985. Survival, growth and fertility of gynogenetic diploids induced in the cyprinid loach, *Misgurnus anguillicaudatus. Aquaculture* 48: 45–55.

Swarup, H. 1959. Effect of triploidy on the body size. General organization and cellular structure in *Gasterosteus aculeatus* (L.). *Journal of Genetics* 56: 141–156.

Tambets, J., T. Paaver, A. Palm, A. Pihlak and R. Gross. 1991. Variability of some cell parameters in di and triploid rainbow trout *Oncorhynchus mykiss. Proceedings of the Estonian Academy of Science of Biology* 40: 129–135.

Tanaka, M., S. Kimura, T. Fujimoto, S. Sakao, E. Yamaha and K. Arai. 2003. Spontaneous mosaicism occurred in normally fertilized and gynogenetically induced progeny of the kokanee salmon *Oncorhynchus nerka. Fisheries Science* 69: 176–180.

Thorgaard, G.H. 1977. Heteromorphic sex chromosomes in male rainbow trout. *Science* 196: 900–902.

Thorgaard, G.H. 1978. Sex chromosomes in the sockeye salmon: A Y-autosome fusion. *Canadian Journal of Genetics and Cytology* 20: 349–354.

Thorgaard, G.H. 1983a. Chromosomal differences among rainbow trout populations. *Copeia* 1983: 650–662.

Thorgaard, G. H. 1983b. Chromosome set manipulation and sex control in fish. In: *Fish Physiology,* W.S. Hoar, D.J. Randall and E.M. Donaldson (eds.), Academic Press, New York, Vol. 9B. pp. 405–434.

Thorgaard, G.H. 1986. Ploidy manipulations and performance. *Aquaculture* 57: 57–64.

Thorgaard, G.H. 1992. Application of genetic technologies to rainbow trout. *Aquaculture* 100: 85–97.

Thorgaard, G.H. and G.A.E. Gall. 1979. Adult triploids in a rainbow trout family. *Genetics* 93: 961–973

Thorgaard, G.H., P.S. Rabinovitch, M.W. Shen, G.A. Gall, J. Propp and F. M. Utter. 1982. Triploid rainbow trout identified by flow cytometry. *Aquaculture* 29: 305–309.

Thorgaard, G.H., P.D. Schereer, W.K. Hershberger and J.M. Myers. 1990. Androgenetic rainbow trout produced using sperm from tetraploid males show improved survival. *Aquaculture* 85: 215–221.

Thorgaard, G.H., G.S. Bailey, D. Williams, D.R. Buhler, S.L. Kkaattari, S.S.Ristow, J.D. Hansen, J.R. Winton, J.L. Bartholomew, J.J. Nagler, P.J. Walsh, M.M. Vijayan, R.H. Devlin, R.W. Hardy, K.E. Overturf, W.P. Young, B.D. Robinson, C. Rexroad and Y. Palt. 2002. Status and opportunities for genomics research with rainbow trout. *Comparative Biochemistry and Physiology B* 133: 78–86.

Tiersch, T.R. and R.W. Chandler. 1989. Chicken erythrocytes as an internal reference for analysis of DNA content by flow cytometry in grass carp. *Transaction of the American Fisheries Society* 118: 713–717.

Tiersch, T.R. and C.A. Goudie. 1993. Inheritance and variation of genome size in half-sib families of hybrid catfishes. *Journal of Heredity* 84: 122–125.

Tiwary, B.K., R. Kirubagaran and A.K. Ray. 1999. Altered body shape as a morphometric indicator of triploidy in Indian catfish, *Heteropneustes fossilis* (Bloch). *Aquaculture Research* 30: 907–910.

Tyler, C.R., S. Jobling and J.P. Sumpter. 1998. Endocrine disruption in wildlife: A critical review of the evidence. *Critical Reviews in Toxicology* 28: 319–361.

Ueda, T. and Y. Ojima. 1984. Cytogenetical characteristics of the progeny from the heteroploidy in the rainbow trout. *Proceedings of the Japan Academy* 60: 183–186.

Ueda, T., R. Sato and J. Kobayashi. 1988. Silver-banded karyotypes of the rainbow trout and the brook trout and their hybrids: Disappearance in allotriploids of Ag-NOR originated from the brook trout. *Japanese Journal of Genetics* 63: 219–226.

Varkonyi, E., M. Bercsenyi, C. Ozouf-Costaz and R. Billard. 1998. Chromosomal and morphological abnormalities caused by oocyte ageing in *Silurus glanis*. *Journal of Fish Biology* 52: 899–906.

Verspoor, E. 1988. Widespread hybridisation between Atlantic salmon, *Salmo salar*, and introduced brown trout, *S. trutta*, in eastern Newfoundland. *Journal of Fish Biology* 32: 327–334.

Wattendorf, R.J. 1986. Rapid identification of triploid grass carp with a coulter counter and channelyzer. *Progressive Fish-Culturist* 48: 125–132.

Wei, W.H., J. Zhang, Y.B. Zhang, L. Zhou and J.F. Gui. 2003. Genetic heterogeneity and ploidy level analysis among different gynogenetic clones of the polyploid gibel carp. *Cytometry* 56A: 46–52.

Wittbrodt, J., D. Adam, B., Malitschek, F. Raulf, A. Telling, S.M. Robertson and M. Schartl. 1989. Novel putative receptor tyrosine kinase encoded by the melanoma inducing *Tu* locus in *Xiphophorus*. *Nature* (London) 341: 415–421.

Wlasow, T., H. KuŸminski, P. Woznicki and E. Ziomek. 2004. Blood cells alteration in triploid brook trout *Salvelinus fontinalis* (Mitchill). *Acta Veterinaria (Brno)* 73: 25–29.

Wolters, W.R., C.L. Chrisman and G.S. Libey. 1982. Erythrocyte nuclear measurement of diploid and triploid channel catfish, *Ictalurus punctatus* (Rafinesque). *Journal of Fish Biology* 20: 253–258.

Woznicki, P., L. Sanchez, P. Martinez, B.G. Pardo and M. Jankun. 2000. A population analysis of the structure and variability of NOR in *Salmo trutta* by Ag, CMA$_3$ and ISH. *Genetica* 108: 113–118.

Yamano, K., E. Yamaha and F. Ymazaki. 1989. Increased viability of allotriploid pink salmon × Japanese charr hybrids. *Nippon Suisan Gakkaishi* 54: 1477–1481.

Yamazaki, F. 1983. Sex control and manipulation in fish. *Aquaculture* 33: 329–354.

Yamazaki, F., J. Goodier and K. Yamano 1989. Chromosomal aberrations caused by aging and hybridization in charr, masu salmon and related salmons. *Physiology and Ecology Japan Special Issue* 1: 529–542.

Ye, Y., Q. Wu and R. Chen. 1989. Studies on cytology of crosses between grass carp and carp—asynchronization between nucleus and cytoplasm in distant hybridization of fishes. *Acta Hydrobiologia Sinica* 13: 234–239.

Yi, M.S., Y.Q. Li, J.D. Liu, L. Zhou, Q.X. Yu and J.F. Gui. 2003. Molecular cytogenetic detection of paternal chromosome fragments in allogynogenetic gibel carp, *Carassius auratus gibelio* Bloch. *Chromosome Research* 11: 665–671.

Zhang, J.S. 1979. Studies and applications of the F$_1$ hybrid of *Cyprinus carpio* × *C. pellegrin*. *Freshwater Fisheries* 2: 14–18.

Zhang, Q. and K. Arai. 1999. Distribution and reproductive capacity of natural triploid individuals and occurrence of unreduced eggs as a cause of polyploidization in the loach, *Misgurnus anguillicaudatus*. *Ichthyological Research* 46: 153–161.

Zhang, Q. and K. Arai. 2003. Extensive karyotype variation in somatic and meiotic cells of the loach *Misgurnus anguillicaudatus* (Pisces: Cobitidae). *Folia Zoologica* 52: 423–429.

Zhang, Q. and T.R. Tiersch. 1997. Chromosomal inheritance patterns of intergeneric ictalurid catfishes: Odd diploid numbers with equal parental contributions. *Journal of Fish Biology* 51: 1073–1084.

Zhang, Q., I. Nakayama, A. Fujiwara, T. Kobayashi, I. Oohara, T. Masaoka, S. Kitamura and R.H. Devlin. 2001. Sex identification by male-specific growth hormone pseudogene (GH-Y) in *Oncorhynchus masou* complex and a related hybrid. *Genetica* 111: 111–118.

Zou, Z., Y. Cui, J. Gui and Y. Yang. 2001. Growth and feed utilization in two clones of gibel carp, *Carassius auratus gibelio*: Paternal effects in a gynogenetic fish. *Journal of Applied Ichthyology* 17: 54–58.

NOR Markers in the Identification and Management of Cultured Fish Species: The Case of Rainbow Trout Stocks Reared in Brazil

Fábio Porto-Foresti[1*], Claudio Oliveira[2], Yara Aiko Tabata[3], Marcos Guilherme Rigolino[3] and Fausto Foresti[4]

INTRODUCTION

Rainbow trout (*Oncorhynchus mykiss*) is the most extensively cultured salmonid species and artificial populations can be found worldwide, except

---

Address for Correspondence: [1*]Laboratório de Genética de Peixes, Departamento de Ciências Biológicas, Faculdade de Ciências de Bauru, Universidade Estadual Paulista, Brazil. E-mail: fpforesti@fc.unesp.br

[2]Laboratório de Biologia e Genética de Peixes, Departamento de Morfologia, Instituto de Biociências, Universidade Estadual Paulista, Brazil.

[3]Estação Experimental de Salmonicultura de Campos do Jordão, Secretaria de Agricultura e Abastecimento do Estado de São Paulo, Instituto de Pesca, São Paulo, Brazil.

[4]Departamento de Morfologia, Instituto de Biociências, Universidade Estadual Paulista, Botucatu, SP, Brazil.

for Antarctica. Easily adapted to captivity conditions, rainbow trout is one of the oldest cultured fish for human food supply, just behind the common carp. Besides presenting important economic features, this species also constitutes a useful biological experimental model for research in ecology, genetics, nutrition, pathology, and toxicity.

Over the last decades, new technologies have been introduced into aquaculture programmes, particularly in fish culture, determining a revolution in species management strategies by the adoption of methodologies of genome control and manipulation (Foresti, 2000). Such methods, based on precise identification of natural and commercial stocks, involve the utilization of different genetic markers for supporting the application of modern biotechnological methodologies. After specific adjustments, the in-depth knowledge in this area has been extended to different fish species, especially rainbow trout, bringing out relevant contributions to the improvement, management and production of this species, reflected in the increase of productivity rates and in higher quality products (Dunham, 1990). Initially reared close to its origin localities, the rainbow trout was then widespread to different parts of the world. Nowadays, it represents the basis of fish culture programmes in many countries. In this sense, the process of identification of the stocks used in such programmes, in association with the application of modern and refined available methodologies, can constitute a powerful tool for a proper management of rainbow trout at captivity conditions.

Introduction of the Rainbow Trout to Brazil

In the beginning of the previous century, two pioneer surveys on fertilized trout eggs imported from Europe were performed in Brazil. The main goal of these experiments was the introduction of this species to cold and clear-water streams at high altitude regions of the State of Rio de Janeiro, considered adequate for the culture of salmonid fishes. In the first introduction, 10,000 fertilized eggs of the European trout (*Salmo fario*) were imported from England, which arrived at Rio de Janeiro in January 1913 (Sociedade de Amigos dos Aquários Públicos, 1996). Such trial was unsuccessful, since all embryos died within 10 days of development at local conditions.

The second attempt was performed later in same year, using rainbow trout (*Salmo irideus*, presently *Oncorhynchus mykiss*) eggs, a species that seemed to present better possibilities for acclimatization. This second stock arrived at Rio de Janeiro in May 1913; due to hard trip conditions, there was a great loss of embryos. However, 150 fry were obtained from this experiment, demonstrating the viability of trout egg hatchery in that region, representing the production of the first trouts raised in Brazil (Sociedade de Amigos dos Aquários Públicos, 1996).

In April, 1949, the salmonid introduction process was restarted by the Agriculture Ministery, and nearly 5.000 eyed eggs of rainbow trouts (identified as *Salmo gairdneri irideus*) were imported from Esbjerg, Denmark. Prior to this procedure, limnological studies in Mountain Rivers on Southeastern Brazil were carried out by Ascânio de Faria (Faria, 1953). The objective of such actions was to promote an alternative fishery for local populations by the introduction of the species in that specific environment. Fertilized eggs were then transported to Bocaina Hills at Bananal municipality (SP) in the month of May the same year, which hatched in artisanal incubators and the offspring was released into Jacu Pintado and Bonito rivers (Faria, 1953).

In May 1950, in coincidence with the construction of a small laboratory for development of trout eggs at the Jacu Pintado river margins, the Hunting and Fishery Division of the Agriculture Ministery received 50,000 eyed-embryos from Denmark, which were incubated in the new facilities. In that same year, the offspring obtained was released into Jacu Pintado River as well as into other regional rivers. In July 1, 1951, a migratory reproduction behavior of individuals of this species was observed in a large extension of Bonito River until the headwaters of the Jacu Pintado River, with the spawning of several mature individuals in many river sites (Faria, 1953).

Despite further trout introduction trials, the activities carried out in 1949 are considered the starting point of the salmonid introduction in Brazil, since the stocks introduced by that time gave rise to the naturalized populations found today in the Brazilian rivers (Porto-Foresti *et al.*, 2002a). Moreover, most of the stocks reared at present on the southeastern and southern Brazil are also related to the individuals introduced by that time.

From the stock generated by the Division of Hunting and Fishery of the Agriculture Ministery at the State of Rio de Janeiro, 1,500 eyed eggs were taken to the region of Campos do Jordão (SP), where they were released in local small rivers (Azevedo *et al.*, 1961). In 1955, local fishermen reported, as expected, the occurrence of an unknown small fish species, lately identified as rainbow trout, living in streams at the region of Campos do Jordão (SP), confirming the acclimatization of the species to the peculiarities of the hydrographic system in that region (Azevedo *et al.*, 1961). Despite the initial difficulties, the program for the acclimatization of the rainbow trout in Campos do Jordão was considered a success and this species was able to be introduced in other hydrographic river basins of similar characteristics, distributed along the Bocaina Hills, in Cunha municipality (SP) (Azevedo *et al.*, 1961).

New and successive rainbow trout introductions were carried out right from 1962, when 200,000 eyed-eggs were imported from California (USA). The fries originated from this stock were distributed to breeders over the region by the Division of Hunting and Fishery of the Agriculture Ministry in order to initiate a new campaign to encourage and boost rainbow trout culture.

Following these introductions, a remarkable degree of interest for the culture and production of rainbow trout occurred in Brazil, with the improvement of processes for rearing associated with the high adaptation ability of this species to both local wild and captivity conditions. The Experimental Center of Salmon Culture of the Fishery Institute at Campos do Jordão (SP), inaugurated in 1964, was initially created to perform the diffusion and technical assistance to particular farmers. In 1974, the first works of artificial reproduction were also initiated in Brazil (Stempniewski, 1997) and the Center of Salmon Culture began to develop research in distinct areas as Nutrition, Reproduction, Biotechnology, and Genetics (Tabata, 2004). Currently, it represents the most important center for diffusion, guiding information and distribution of rainbow trout specimens for fish culture projects all over the country. This institute has received fertilized eggs belonging to stocks of different origins, among them, Chile, Argentina, Japan (Golden line), United States (Shasta line), Canada (Kamloops line) and Denmark.

Rearing facilities in the culture of rainbow trout became popular during the last decade, although a local private breeder at Campos do Jordão (SP) dates from nearly four decades ago. Some reasons considered for the delaying in the development of commercial activity with this species point to the lacking of rearing technology diffusion and the scarcity of adequate brood stocks in Brazil. Thus, only a reduced number of fries are still frequently available at the beginning of culture seasons, besides occasional difficulties in the acquisition of balanced trout food in the

Brazilian market (Tsukamoto, 1988). However, this situation is changing fast and nowadays, with the efforts of the private sector in the establishment of a commercial culture directed to production and regional tourism activities, the trout culture has grown quickly, providing new job opportunities and motivating business related initiatives (Porto-Foresti, 2002a).

Cytogenetic Markers Applied to Fish Culture

Over the last decades, the advances in the knowledge of general biology of the species and the improvement of methods for the management and rearing of fishes, particularly involving genetic aspects can be considered remarkable. The identification of genetic markers in species of zootechnical interest and their utilization in breeding programmes played a significant role in the development of specific lineages, accomplished by individual genome manipulation and, mainly, by the application of modern techniques of monitoring traits related to production over generations (Foresti, 2000).

In genetic studies of fish species, phenotypic markers related to external appearance and meristic characters, like coloration, body shape, scale pattern, number of fin rays among others, as well as genotypic markers, involving biochemical, molecular and chromosomal features, intrinsically related to the cell genetic material and functioning, should be considered. While phenotypic markers are more useful for the morphological identification of species, genetic markers directly reflect the particularities of the hereditary material or represent gene products, thus providing tools for the identification of individuals. The chromosomal features are inherent to cell nuclei and reveal the structural components of the genetic material. Each organism presents a unique chromosomal set, which can differ from that of other organisms in number, shape, size and microstructure. In particular cases, such divergences can involve structural and functional aspects of some chromosomal segments.

Until recently, the karyotypical surveys in fish were mainly regarded as a tool for basic and evolutionary studies, but useless in the stock management and monitoring. Recent advances in the methodology of identification and analysis of chromosomes have showed that the cytogenetic patterns of fish species can play a significant role also in fish culture (Toledo-Filho *et al.*, 1992). When conventional analyses are not sufficient to characterize precisely the individuals, the identification can be based on structural chromosomal markers. These markers indicate the chromosomal location of specific chromosomal segments as, for instance, the position of Nucleolar Organizer Regions (NORs) and the pattern of heterochromatin distribution (C-bands). Techniques for the identification of more specific chromosomal regions, involving the utilization of fluorochromes (chromomycin and actinomycin) and base-substitutes (5-bromo-2-deoxyuridine) can further provide refined information about the structure of the genetic material in chromosomes. Moreover repeated sequences and single genes can be also localized by Fluorescence *In Situ* Hybridization (FISH).

Chromosomal analyses are also useful in the characterization of cryptic species and chromosomal races, avoiding the occurrence of unviable crosses related to chromosomal incompatibilities between selected individuals (Toledo-Filho *et al.*, 1994). Otherwise, cytogenetics can also be helpful in the identification of genetic stocks of commercial species in fish culture and in hybridization projects, in which the identification of ploidy levels of inter-crossed offspring is required. Basically, such crosses could give rise to gynogenetic, androgenetic, triploid and tetraploid individuals, and/or diploid hybrids (Toledo-Filho *et al.*, 1994, 1996).

Detection of NORs by Silver Nitrate Staining

Nucleolar organizer regions were formerly characterized in cell nuclei by Heitz (1931) and McClintock (1934). Currently, they are identified as chromosomal regions involved in the transcription of ribosomal genes. If these regions were active during the interphase prior to mitosis, they can be detected by silver nitrate staining (Goodspasture and Bloom, 1975; Howell and Black, 1980; Hudbell, 1985) since silver nitrate specifically stains a set of acidic proteins related to ribosomal synthesis process; this technique actually reveals active NORs (Ag-NORs) and not the rDNA associated to NORs (Howell, 1977; Jordan, 1987).

Several reports in different fish species have demonstrated that most of species present single NORs, i. e., positive Ag-NOR marks on a single pair of homologous chromosomes (Almeida-Toledo *et al.*, 2000). Some fish species can also present multiple NORs, showing more than one chromosome pair bearing the Ag-NOR marks (Almeida-Toledo *et al.*, 1985, Galetti Jr. *et al.*, 1985; Foresti *et al.*, 1989; Sola *et al.*, 1990; Wasko and Galetti Jr., 1999; Jankun *et al.*, 2001; among others). Variability of

NORs patterns in fish were frequently reported, involving number (Miyazawa and Galetti Jr., 1994), location (Gold *et al.*, 1990; Feldberg *et al.*, 1992) and size polymorphism (Foresti *et al.*, 1981; Moreira-Filho *et al.*, 1984; Martínez *et al.*, 1991; Vitturi *et al.*, 2005).

In several groups, the number and position of NORs are specific to each species or each population (Vênere and Galetti Jr., 1989; Moreira Filho and Bertollo, 1991; Maistro *et al.*, 1998). This character has been used with relative success in the establishment of interrelationships hypothesis within some fish groups (Amemiya and Gold, 1988; Phillips *et al.*, 1989). Studies on NORs size and position have showed that these regions can be extensively polymorphic (Foresti *et al.*, 1981; Martínez *et al.*, 1993). The application of silver nitrate staining for the NORs detection in salmonids revealed that nearly all species present a single active NOR-bearing chromosome pair (Phillips and Ihssen, 1985; Phillips *et al.*, 1986). However, in some individuals, other minor active NORs can be occasionally detected (Phillips *et al.*, 1989; Pendás *et al.*, 1993b; Castro *et al.*, 1997).

Analysis of chromosomes from rainbow trout firstly revealed the NORs located at subterminal position on the short arms in a pair of submetacentric chromosomes in individuals of both wild and cultured stocks (Schmid *et al.*, 1982; Phillips and Ihssen, 1985; Mayr *et al.*, 1986; Ueda and Kobayashi, 1988). Nevertheless, studies carried out by Oliveira *et al.* (1996) in rainbow trouts individuals from a stock reared in Brazil, at the Experimental Center in Campos de Jordão, demonstrated that NORs were located at subterminal position on the long arms of a submetacentric chromosomal pair. According to these authors, such difference in chromosomal shape and NORs location could be related to the condensation level of the metaphasic chromosomes analyzed. Furthermore, they observed that 13 out of 19 specimens analyzed presented two kinds of NOR-bearing chromosomes. In one type the NOR was displayed as a single block and in the other one two separated blocks could be visualized when chromosome preparations were stained by silver nitrate (Fig. 3.3.1a, b). This same situation was found among the individuals analyzed from the Gavião river, as it will be discussed later (Fig. 3.3.1c).

The occurrence of two chromosome types in the rainbow trout population from Campos do Jordão detected by silver nitrate staining, was initially thought to be due to a differential activation of usually inactive

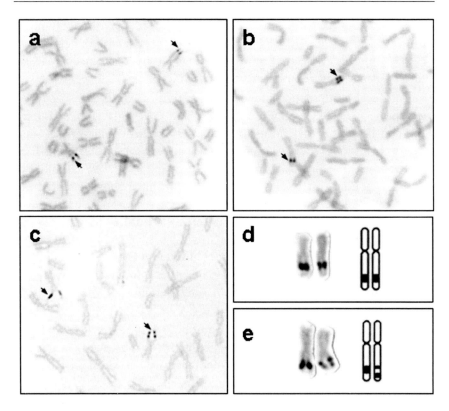

Fig. 3.3.1 Chromosomal location of the nucleolar organizing regions (NORs) in samples of rainbow trout (*Oncorhynchus mykiss*), after silver staining (arrows). (a) Chromosomes from a homozygous individual for NOR phenotype (N1N1), b-c) chromosomes from individuals for the NOR heterozygote phenotype (N1N2) found in the stock reared in the Núcleo Experimental de Salmonicultura de Campos do Jordão and Gavião River, respectively; and d-e) details of the NOR-bearing chromosome pairs in homozygous (d) and heterozygous (e) individuals.

rDNA cistrons, unequal exchanges leading to amplification of a segment bearing rDNA, or by the occurrence of a paracentric inversion involving the chromosomal segment carrying portions of rDNA cistrons. Analysis of chromosomal pairing during meiosis, through the visualization of the synaptonemal complex, clarified that the morphological difference observed in the NORs-bearing chromosomes was determined by a paracentric inversion, a chromosomal rearrangement never described before in salmonids (Oliveira *et al.*, 1996).

Conventional Giemsa staining analysis of chromosome preparations revealed that the paracentric inversion observed in some individuals of this specific trout stock does not determine any other recognizable alteration in chromosomal morphology.

Later cytogenetic analysis performed in 66 specimens of rainbow trout of this same stock by Porto-Foresti *et al.* (2004b), confirmed the NORs location on the long arms of a submetacentric chromosomal pair, as described by Oliveira *et al.* (1996), stating that the difference in NORs position was rather related to the occurrence of a paracentric inversion in NOR-bearing chromosomes, instead of reflecting a particular chromosomal condensation stage. In the same work the preliminary observation by Oliveira *et al.* (1996) that the NOR sites could be present as a single block (N1 condition) or as double blocks (N2 condition), both located at the subterminal position on the chromosomes, was confirmed. The study revealed the occurrence of two NOR phenotypes, N1N1 and N1N2 (Porto-Foresti *et al.*, 2004b); individuals with the phenotype N2N2, putative homozygotes for the inversion were absent in the sample.

Based on the identification of individuals from this stock, Porto-Foresti *et al.* (2004b) carried out controlled crossings involving animals bearing distinct NOR phenotypes in order to evaluate the heritage pattern for such markers. No significant differences between observed and expected frequencies were detected in the offspring resulted from crosses involving specimens with the phenotypes N1N1 × N1N2 and N1N2 × N1N1, which resulted in the expected proportion (1:1). However, significant differences were detected between the expected and observed frequencies in the results from crosses involving individuals with the phenotypes N1N2 × N1N2, which resulted in offspring presenting only the phenotypes N1N1 and N1N2; individuals with the N2N2 phenotype were not observed. The absence of N2N2 individuals (homozygotes for the inversion) in the broodstock and particularly in the offspring from crosses involving both parents presenting the heterozygous condition (N1N2 × N1N2) suggests that this condition might be lethal in this specific stock of rainbow trout. The union of gametes bearing inverted segments for the NORs could result in a structural condition that should express a deleterious effect, thus determining the inviability of the individuals during some stage of the embryonic development. Nevertheless, the possibility that such inverted condition might exhibit earlier effects, leading to determinant fails during oocyte formation, can not be discarded since the affected region is associated to the process of ribosomal subunits production and therefore, to protein synthesis, with implications for the correct yolk production in the eggs. The absence of differences in the number of oocytes produced by females presenting N1N1 and N1N2 phenotypes, as reported by Porto-Foresti et al. (2004b) seems to support this hypothesis.

Analysis of NORs Location in Different Samples of Rainbow Trout (*Oncorhynchus mykiss*) Introduced to Brazil

The studies of the NORs identification in individuals belonging to cultivated or natural populations show that this chromosomal region can be considered as an important cytogenetical marker for fish culture research (Ferguson et al., 1995; Porto-Foresti et al., 2002a), especially considering that several works demonstrate that NOR polymorphisms are inherited (Markovic et al., 1978; Arruda and Monteagudo, 1989). The analysis by Porto-Foresti et al. (2002b) on specimens belonging to different stocks of rainbow trout reared in Brazil and originated from different countries (Japan, USA, Denmark), and localities (such as Man Island, Shasta Hills, Kamloops, Teresópolis, Nova Friburgo) and from particular stocks cultivated by breeders such as AQUA Ltda. and São João Hotel, revealed that most of them present NORs located at subterminal position on the short arms of the chromosomes in a submetacentric pair (Fig. 3.3.2a), as previously reported by Schmid et al. (1982), Phillips and Ihssen (1985), and Phillips et al. (1989), among others.

An interesting situation was reported for specimens from stocks identified as AQUA Ltda., Teresópolis and from Japanese lineages by Porto-Foresti et al. (2002a). In these samples, most of the individuals analyzed presented the NORs at subterminal position on the short arms of a submetacentric chromosomal pair (Fig. 3.3.2a), as previously described by various authors (Schmid et al., 1982; Phillips and Ihssen, 1985, among others). However, 20% of individuals from both localities and 30% of specimens descendant from the Japanese lineage presented one NOR located at subterminal position on the short arms of a submetacentric chromosome and the other one at subterminal position on the long arms also of a submetacentric chromosome (Fig. 3.3.2b). Such results were expected since the presence of NORs at subterminal position on the long arms of a submetacentric chromosome is typical in individuals from the Experimental Center of Salmon Culture at Campos do Jordão stock (Oliveira et al., 1996; Porto-Foresti et al., 2004a, b) and individuals

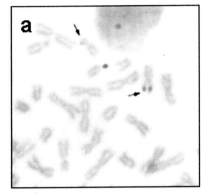 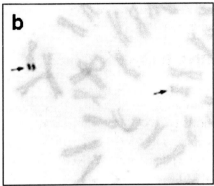

Fig. 3.3.2 Silver-stained chromosomes from samples of rainbow trout (*Oncorhynchus mykiss*) recently introduced to Brazil. Arrows indicate the NOR-bearing chromosomes with marks on the short arms (a) and on the short and long arm in individuals resulting from inter-population crosses (b).

from this locality have been widely distributed during the last decades to several trout culture farms located at mountain areas in the State of São Paulo, Rio de Janeiro and Minas Gerais (Porto-Foresti *et al.*, 2002a, b). Therefore, the specimens bearing both NOR types should represent hybrid products, originated by inadvertent mixing of distinct stocks during reproductive practices in the commercial rearing of this species, as pointed out by Porto-Foresti *et al.* (2002b).

Individuals collected at Gavião river showed NORs located interstitially on the long arms of a submetacentric chromosomal pair, as reported by Oliveira *et al.* (1996) (Fig. 3.3.1c). Porto-Foresti *et al.* (2002a) also observed that, in this population, one submetacentric chromosome could present double marks at terminal portion of long arms, a characteristic found in the captive stock from the Experimental Center of Salmon Culture at Campos de Jordão (Oliveira *et al.*, 1996; Porto-Foresti *et al.*, 2002b, 2004b). As individuals found in Gavião river seems to belong to the same primitive stock used to compose the broodstock from the Experimental Center of Salmon Culture at Campos de Jordão (MacCrimmon, 1971), the cytogenetic data besides, by confirming the stock identity can also suggest that the presence of NORs on the long arms of a submetacentric chromosomal pair in both populations were probably due to the occurrence of a paracentric inversion in the original stock imported from California (Porto-Foresti *et al.*, 2002a). And possibly, the origin of N1N1 and N1N2 NOR phenotypes, as the result of the inversion, may have occurred specifically only in this stock (Porto-Foresti *et al.*, 2004b).

RNA Ribosomal Genes (45S and 5S rDNA) and NORs

Ribosomal RNA genes (rRNA) are organized in two distinct multigenic families in higher eukaryotes, represented by 45S rDNA and 5S rDNA, both composed by tandem repeat units from hundred to thousand of copies. The 45S ribosomal DNA comprises the genes coding for 18S, 5,8S, and 28S ribosomal rRNAs, separated by transcribed internal spacers (ITS1 and ITS2) and flanked by external non-transcribed spacers (ETS1 and ETS2) (Long and David, 1980; Pardo, 2001). Multiple copies of these units correspond to the nucleolar organizer regions.

The application of fluorescence *in situ* hybridization technique using rDNA probes for NORs identification in fish chromosomes has become quite common and the results have usually confirmed and complemented previous data obtained by conventional staining. In the Atlantic salmon (*Salmo salar*), FISH showed that probes for the major rDNA (18S, 5,8S, and 28S) hybridized with the entire heterochromatic arms of a chromosomal pair bearing secondary constrictions, while previous silver nitrate staining suggested that NORs were restricted to the region of secondary constriction (Pendás *et al.*, 1993a).

Cytogenetic mapping of the rRNA genes in the brown trout by FISH revealed a major single NOR-bearing chromosome pair, as previously detected by silver nitrate staining, but also showed additional minute NORs on eight chromosomal pairs (Pendás *et al.*, 1993b), thus indicating the limits of conventional staining in detecting minor chromosomal rDNA sites. However several works using Ag-NOR banding and FISH with rDNA probes provided coincident results (Reed and Phillips, 1995; Castro *et al.*, 1997; Rossi *et al.*, 1997). Dual-color FISH with probes of 45S and 5S rDNA in Atlantic salmon revealed that both genes families were located at a single chromosomal pair, in adjacent clusters, without overlapping (Pendás *et al.*, 1994). In *Anguilla anguilla*, the simultaneous mapping of 45S and 5S ribosomal genes showed that they are distributed in distinct chromosomes (Martínez *et al.*, 1996). Fujiwara *et al.* (1998) demonstrated that, in *Oncorhynchus masou*, one chromosomal pair carries the 45S rRNA genes and three chromosomal pairs bear 5S rRNA genes while in *Oncorhynchus mykiss*, 45S ribosomal genes are located on a single chromosomal pair and two pairs present 5S rRNA genes. In both species

5S rDNA and 45S rDNA segments are located in different chromosomes, except for one chromosome pair in rainbow trout. Such chromosomes display both ribosomal gene clusters in adjacent arrays (Fujiwara et al., 1998). Similar results about the location of these genes were reported in five Neotropical fish species of the genus Astyanax (Almeida-Toledo et al., 2002).

Chromosomal Location of 45S and 5S Ribosomal Genes in Different Samples of Rainbow Trout (*Oncorhynchus mykiss*) Introduced to Brazil

The application of FISH using 18S rDNA probes in chromosomal preparations of individuals of the rainbow trout stock from the Experimental Center at Campos do Jordão revealed hybridization signals at subterminal position on the long arms of a single chromosomal pair of submetacentric chromosomes arranged as a single (N1 condition) or double (N2 condition) mark. Therefore, specimens bearing the phenotypes N1N1 and N1N2 were identified (Fig. 3.3.3a), as previously reported by Oliveira et al. (1996) and Porto-Foresti et al. (2004b) using conventional Ag-NOR staining. In the chromosomal preparations from the specimens from Gavião river, positive hybridization marks were obtained in the same segments of the submetacentric chromosomal pair, with one of these chromosomes bearing double signals at the terminal portion of the long arms (Fig. 3.3.3b) (Porto-Foresti et al., 2004a), displaying a similar pattern found in individuals from the Experimental Center of Salmon Culture at Campos de Jordão (Oliveira et al., 1996; Porto-Foresti et al., 2004b).

The application of FISH using a 18S rDNA probe in specimens from the Shasta and Teresópolis stocks revealed signals on the short arms on the chromosomes of a submetacentric pair (Fig. 3.3.3c), as reported by Porto-Foresti et al. (2002a). However, in some individuals of the Teresópolis stock, one mark was located on the short arms of a submetacentric chromosome and the other one on the long arms of a submetacentric chromosome (Fig. 3.3.3d), confirming the mixing of individuals with distinct karyotypical features in the rearing program of this species.

Fujiwara et al. (1998) described the results of FISH with 5S rDNA probes in four populations, showing two chromosomal pairs positively marked. In this species, the marks were identified on the chromosomes of

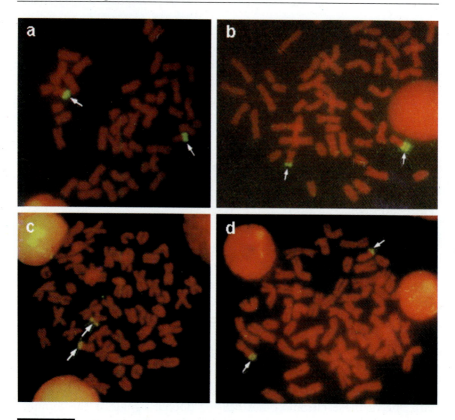

Fig. 3.3.3 Mitotic chromosomes in samples of rainbow trout (*Oncorhynchus mykiss*) reared in Brazil, after FISH of a 18S rDNA probe. Arrows indicate in a, the location of major ribosomal genes on two chromosomes in the homozygote phenotype (N1N1) and in b, the heterozygote phenotype (N1N2). In c, the fluorescent marks can be visualized on the short arms of the chromosomes in the sample of Mount Shasta. In the hybrid forms (d) one hybridization signal is located on the long arm in one chromosome and on the short arm in the other (from Porto-Foresti *et al.*, 2004a).

both subtelocentric and submetacentric pairs. Moreover in the sub metacentric pair, the two 45S rRNA and 5S rRNA genes families are adjacent to each other (Fujiwara *et al.*, 1998; Porto-Foresti *et al.*, 2004a).

No differences were observed among the four analyzed populations in relation to the marks at the pericentromeric region in the subtelocentric chromosomal pair. Nevertheless, in individuals from the Experimental Center at Campos do Jordão and from Gavião river, the marks were located at the interstitial region on the long arms of the chromosomes in a submetacentric pair (Fig. 3.3.4a, b); while in specimens from Shasta

Mountain and Teresópolis, the marks were located at subterminal position on the short arms of a pair of submetacentric chromosomes (Fig. 3.3.4c). In some individuals from Teresópolis, 5S rDNA sequences were located at different positions in the submetacentric chromosomal pair, being one hybridization signal at sub terminal position on the short arms and another at subterminal position on the long arms (Fig. 3.3.4d) (Porto-Foresti *et al.*, 2004a).

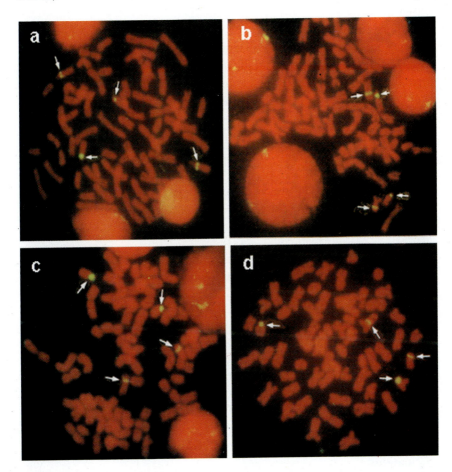

Fig. 3.3.4 Mitotic chromosomes in samples of the rainbow trout (*Oncorhynchus mykiss*) reared in Brazil after FISH of a 5S rDNA probe. Arrows indicate the position of 5S ribosomal genes on two chromosome pairs in individuals from the Núcleo Experimental de Salmonicultura de Campos do Jordão (a), Gavião River (b) and Teresópolis (c, d) (from Porto-Foresti *et al.*, 2004a).

The differences related to the nucleolar organizer regions location in the studied samples suggest the occurrence of a pericentric inversion, involving the chromosomal portion bearing 45S rRNA and 5S rRNA genes. Otherwise, the polymorphism observed in some specimens of Teresópolis confirms the occurrence of the hybridization process involving individuals from recent imported stocks and those from the original stock cultivated in the Experimental Center of Salmon Culture at Campos de Jordão.

The results of the hybridization of 45S and 5S rDNA probes in the chromosomes of individuals collected in the Gavião river reinforce the hypothesis that the sample initially imported from California in 1962 and introduced in this river shares a common origin with the stock reared at the Experimental Center of Salmon Culture at Campos de Jordão (MacCrimmon, 1971; Porto-Foresti et al., 2002a)

According to the polymorphism in the chromosomal location of NORs found in the analyzed samples, also confirmed by FISH, the Nucleolar Organizing Regions seem to constitute an important genomic marker for rainbow trout stocks reared in Brazil, which can be used in the identification of actual cultured and 'naturalized' populations and also to control the possible future introductions.

Implications of the cytogenetic studies for rainbow trout rearing programmes in Brazil

The introduction of biotechnology practices, mainly over the last two decades, has induced significant changes and fast development in the exploitation process of aquatic organisms. However, basic concepts of animal breeding are still poorly applied in fish culture and little is known about the genetics of important economical traits for most of captive fish species (Foresti, 2000). Therefore, it is essential to improve the available methodologies and increase the number of studies approaching the application of these technologies for the development of more adequate systems of zootechnical, genetic, and conservation management of fish populations.

Conservation programmes require monitoring of the genetic variation in fish stocks in order to avoid genetic variability losses (Ward and Grewe, 1994). According to Fergunson et al. (1995), the genetic characterization of individuals is important in studies of management and conservation of wild or domesticated biological stocks, since they are frequently differentiated by genomic parameters. In such context, the utilization of cytogenetic and molecular markers can be of great value.

Cytogenetic studies have revealed their importance as structural markers used in genetic and zootechnique management of fish populations. The use of silver nitrate staining technique for identification of nucleolar organizer regions showed that all salmonids present a single pair bearing active NORs commonly located at subterminal position on the short arms of a submetacentric chromosomal pair in all lines until studied (Schmid *et al.*, 1982; Phillips and Ihssen, 1985). However, in specimens reared at the Experimental Center of Salmon Culture at Campos do Jordão, the Ag-NORs were found at subterminal position on the long arms in a pair of submetacentric chromosomes, displayed as single or double blocks (Oliveira *et al.*, 1996; Porto-Foresti *et al.*, 2002a, b, 2004b). Such disposition give rise to phenotypes identified as normal when single NORs are present in homozygosis or heterozygote for the inversion in a situation when the normal type is coupled with individuals presenting the inverted arrangement. Homozygotes for the inverted condition were found to be always absent in this stock (Porto-Foresti *et al.*, 2002a, 2004b).

The use of silver nitrate staining in cytogenetic studies performed by Porto-Foresti *et al.* (2004b) in this stock showed that crosses involving breeders presenting the heterozygote condition for the analyzed character should result in 25% of N2N2 individuals (therefore, lethal) in the F_1 generation. Such crosses would lead to a loss of 25% of individuals in the offspring, which can be considered an expressive value at production condition for this fish species. Crosses involving N1N1 and N1N2 breeders, which result in 50% of N1N1 and 50% of N1N2 offspring, seemed to present a higher survival rate in the F_1 generation, which could be considered an advantage under the focus of productivity, considering the absence of N2N2 individuals. Under this meaning, N1N2 individual should be preserved in the actual stocks but, as long they were identified and tagged, these animals could be used as breeders in crossings involving partners exclusively homozygotes (N1N1).

In a complementary, study NOR analysis performed in different reared stocks of rainbow trout recently introduced to Brazil, showed that they frequently present NORs in a subterminal position on the short arms of the chromosomes in a submetacentric pair, as commonly described all over the world. Cytogenetic analysis performed in such stocks revealed that some individuals presented one mark located at subterminal position on the short arms of a submetacentric chromosome and the other one located also at subterminal position but on the long arms of another submetacentric chromosome of similar size (Porto-Foresti et al., 2002b). These data reveal the occurrence of a mixture of different morphotypes due to crosses involving individuals from recent introductions and those belonging to the ancient stocks originated at the Experimental Center of Salmon Culture at Campos de Jordão.

Controlled crosses performed among rainbow trout specimens from these stocks resulted, without exception, in positive marks of silver nitrate in two submetacentric chromosomes repeating the distinct position found respectively on the short and long arms, and thus evidencing the effective hybridization process. However, as the chromosomes involved present similar size and morphology, the real identity of these chromosomes remains unknown and only the application of longitudinal banding techniques or crossings involving individuals of the F_1 generation could provide better understanding. Such information could be considered important for the management of rainbow trout stocks. In the case that the NORs were located in distinct chromosomes, a new condition for generating not viable zygotes could give rise, once gametes lacking nucleolar chromosomes could be originated as the result of a single segregation process and be combined by chance during fertilization.

Analysis of differential growth performance amongst individuals in the stock reared in the Center of Salmon Culture at Campos do Jordão, according to the different NOR phenotypes, were performed in the offspring resultant from crossings between homozygote and heterozygote breeders for their NOR characteristics (Porto-Foresti et al., 2006). The results failed in reveal growth differences inferred by weight gain and total length values of analyzed specimens, since the mean values of these parameters were not significantly different in all sampled periods regarding to the presence or absence of the inversion involving the NOR bearing chromosomal segment. Such results seem to confirm the neutral condition of the N2 NOR phenotype when in heterozygous condition and it can be supposed that when in single dose both viability and either growth characteristics of individuals seems to be not affected by the presence of the chromosome with inverted segment. Therefore, the effect of such chromosomal modification would be apparently restricted to the possibility of a not viable condition that could appear in individuals presenting the N2N2 NOR phenotype, as demonstrated by Porto-Foresti *et al.* (2004a).

CONCLUDING REMARKS

The main results of cytogenetic analysis of nucleolar organizer regions carried out in different stocks of rainbow trout currently reared in Brazil can be summarized here with focus on its management:

1. In the rainbow trout stock reared at the Experimental Center of Salmon Culture at Campos de Jordão, SP, Brazil, the NORs were located at a subterminal position on the long arms of a submetacentric chromosome pair. Due to the occurrence of an inversion involving the NOR segment, two different types of chromosomes were identified when chromosome preparations were stained by silver nitrate. In one type, the NOR was displayed as a single block (N1) and in the other one two separated blocks could be visualized (N2). Analysis of sampled specimens in this stock revealed the occurrence of two NOR phenotypes (N1N1 and N1N2); individuals with the phenotype N2N2, putative homozygotes for the inversion, were not observed in the surveyed sample.

2. Individuals collected at Gavião river from "naturalized" stocks revealed the same pattern of NORs as the individuals belonging to stock reared in the Experimental Center of Salmon Culture at Campos de Jordão. As the cultivated and "naturalized" stocks share the same origin, the cytogenetic data besides confirming stock identity also suggest that such chromosomal rearrangement may have been fixed in the original stock imported from California (USA) before the introduction in this country.

3. In a similar way, two structural conditions regarding to the NOR location characterize the specimens from eleven analyzed stocks of rainbow trout actually reared in Brazil. The presence of nucleolar organizer regions on the short arms of a submetacentric chromosomal pair is a frequent condition found in this species all over the world and characterizes all the recently introduced stocks. Some cultivated stocks showed individuals with the NORs located in different positions on two morphologically similar chromosomes. In one, the NOR was located at subterminal position on the short

arms and in the other one, at subterminal position on the long arms. It can be considered that an intra-populational hybridization process took place by inadvertent mixing of distinct stocks during reproductive practices in the commercial rearing process of this species. It involved individuals proceeding from the old established stock reared and distributed by the Experimental Center of Salmon Culture at Campos do Jordão and of the new lines recently introduced. However, the effective differences or homeology involving the NOR-bearing chromosomes in the lineages still need to be clarified either by banding techniques or by the analysis of crosses among individuals from F_1 generations.

4. FISH analysis of individuals obtained from intraspecific crosses involving different lineages of rainbow trout, with the NORs located on the short arm and the long arm of the submetacentric bearing chromosome, revealed that the entire chromosomal portion bearing adjacent 18S and 5S ribosomal genes, migrated as a block during the chromosomal modification process.

5. No statistically significant differences in the growth performance, as inferred by weight gain and total length, were detected in the individuals cytogenetically characterized by NOR phenotypes identified as N1N1 and N1N2 in the stock reared in the Center of Salmon Culture at Campos do Jordão. Such results seem to confirm the neutral role of the N2 condition when in heterozygous and apparently showing no effect on the viability and growth of individuals bearing such characteristic.

6. Controlled crosses among specimens from the old stock and recently introduced lineages repeated the production of individuals with the NORs in different chromosomal positions, as occasionally found in the cultivated stocks of private producers. The silver nitrate reaction resulted without exception in positive marks located on the short arms in one of the submetacentric chromosomes and on the long arm in the other. This occurrence could be identified as the result of an effective intraspecific hybridization process. In the case that the NORs bearing chromosomes are homologues, the phenotypic characteristics will reflect the effects of the inversion process. Otherwise, in case that the NORs were located in non-homologous chromosomes, problems affecting zygote viability could be detected in the next

generations, possibility due to the combination of deficient gametes lacking nucleolar chromosomes originated during the segregation process of meiosis and combined by chance during fertilization. The economical importance of such information may be considered for the management of these rainbow trout stocks.

References

Almeida-Toledo, L.F., F. Foresti, M.F.Z. Daniel and S.A. Toledo. 2000. Sex chromosome evolution in fish: The formation of the neo-Y chromosome in *Eigenmannia* (Gymnotiformes). *Chromosoma* 109: 197–200.

Almeida-Toledo, L.F. and F. Foresti. 1985. As regiões organizadoras de nucléolos em peixes. *Ciência e Cultura* 37: 448–453.

Almeida-Toledo, L.F., C. Ozouf-Costaz, F. Foresti, F. Porto-Foresti, C.E. Lopes, M.F.Z. Daniel and S.A. Toledo Filho. 2002. Conservation of 5S bearing pair and co-localization with major rDNA clusters in five species of *Astyanax* (Pisces, Characidae). *Cytogenetic and Genome Research* 97: 229–233.

Amemiya, C.T. and J.R. Gold. 1988. Chromosomal NORs as taxonomic and systematic characters in North American cyprinid fishes. *Genetica* 76: 81–90.

Arruda, M.V. and L.V. Monteagudo. 1989. Evidence of Mendelian inheritance of the nucleolar organizer regions in the Spanish common rabbit. *The Journal of Heredity* 80: 85–86.

Azevedo, P., J.O. Vaz and W.B. Parreira. 1961. Aclimatação da truta arco-íris em algumas águas de São Paulo. *Boletim de Industria Animal* 19: 75–105.

Castro, J., S. De Lucchini, I. Nardi, L. Sánchez and P. Martínez. 1997. Molecular analysis of a NOR site polymorphism in brown trout (*Salmo trutta*): Organization of rDNA intergenic spacers. *Genome* 40: 916–922.

Dunham, R.A. 1990. Production and use of monosex or sterile fishes in aquaculture. *Review in Aquatic Science* 2: 1–17.

Faria, A. 1953. *Dados sobre a biologia da truta arco-íris*. Ministério da Agricultura - Departamento Nacional de Produção Animal Divisão de Caça e Pesca Press: Rio de Janeiro.

Feldberg, E., J.I.R. Porto and L.A.C. Bertollo. 1992. Karyotype evolution in Curimatidae (Teleostei, Characiformes) on the Amazon region. I. Studies on the genera *Curimata, Psectrogaster, Steindachnerina* and *Curimatella*. *Brazilian Journal of Genetics* 15: 369–383.

Fergunson, A., J.B Taggart, P.A Prodöhl, O. Mcmeel, C. Thompson, C. Stone, P. Mcginnity and R.A. Hynes. 1995. The applications of molecular markers to the study and conservation of fish populations with special reference to *Salmo*. *Journal of Fish Biology* 47: 103–126.

Foresti, F. 2000. Biotechnology and fish culture. *Hydrobiologia* 420: 45–47.

Foresti, F., L.F. Almeida-Toledo and S.A. Toledo-Filho. 1981. Polymorfic nature of nucleous organizer regions in fishes. *Cytogenetics and Cell Genetics* 31: 137–144.

Foresti, F., L.F. Almeida-Toledo and S.A. Toledo. 1989. Supernumerary chromosome system, C-banding pattern characterization and multiple nucleolus organizer regions in *Moenkhausia sanctafilomenae* (Pisces, Characidae). *Genetica* 79: 107–114.

Fujiwara, A., S. Abe, E. Yamaha, F. Yamazaki and M.C. Yoshida. 1998. Chromosomal localization and heterochromatin association of ribosomal RNA gene loci and silver-stained nucleolar organizer regions in salmonid fishes. *Chromosome Research* 6: 463–471.

Galetti Jr., P.M., E.B. Silva and R.T. Germinaro. 1985. A multiple NOR system in the fish *Serrasalmus spilopleura* (Serrasalminae, Characidae). *Brazilian Journal of Genetics* 8: 479–484.

Gold, J.R., Y.C. Li, N.S. Shipley and P.K. Powers. 1990. Improved methods for working with fish chromosomes with a review of metaphase chromosome banding. *Journal of Fish Biology* 37: 563–575.

Goodspature, C. and S.E. Bloom. 1975. Visualization of nucleolar organizer region in mammalian chromosomes using silver nitrate staining. *Chromosoma* 53: 37–50.

Heitz, E. 1931. Die ursache der gesetzmässignen zahl, lage, form und grösse pflanglicher nucleolen. *Planta* 12: 775–844.

Howell, W.M. 1977. Visualization of ribosomal gene activity: silver stains proteins associated with rRNA transcribed from oocyte chromosomes. *Chromosoma* 62: 361–367.

Howell, W.M. and D.A. Black. 1980. Controlled silver-staining of nucleolus organizer regions with a protective colloidal developer: A 1-step method. *Experientia* 36: 1014–1015.

Hudbell, H.R. 1985. Silver staining as an indicator of active ribosomal genes. *Stain Technology* 60: 285–294.

Jankun, M., P. Martinez, B.G. Pardo., L. Kirtikilis, P. Rab, M. Rabova and L. Sanchez. 2001. Ribosomal genes in coregonid fishes (*Coregonus lavaretus, C. albula* and *C. peled*) (Salmonidae): Single and multiple nucleolus organizer regions. *Heredity* 87: 672–679.

Jordan, G. 1987. At the heart of the nucleolus. *Nature* (Lond.) 329: 489–490.

Long, E.O. and I.B. David. 1980. Repeated genes in eukaryotes. *Annals Biochemistry* 49: 727–764.

MacCrimmon, H.R. 1971. World distribution of rainbow trout (*Salmo gairdneri*). *Journal of the Fisheries Research Board of Canada* 28: 663–704.

Maistro, E.L., C. Oliveira and F. Foresti. 1998. Comparative cytogenetic and morphological analysis of *Astyanax scabripinnis paranae* (Pisces, Characidae, Tetragonopterinae). *Genetics and Molecular Biology* 21: 201–206.

Markovic, V.D., R.G. Worton and J.M. Berg. 1978. Evidence for the inheritance of silver-stained nucleolus organizer regions. *Human Genetics* 41: 181–187.

Martínez, P., A.Vinãs, C. Bouza, J. Arias, R. Amaro and L. Sánchez. 1991. Cytogenetical characterization of hatchery stocks and natural populations of sea and brown trout from northwestern Spain. *Heredity* 66: 9–17.

Martínez, P., A. Vinãs, C. Bouza, J. Castro and L. Sanchez. 1993. Quantitative analysis of the variability of nucleolar organizer regions in *Salmo trutta*. *Genome* 36: 1119–1123.

Martínez, J.L., P. Morán, E. García-Vázquez and A.M. Pendás. 1996. Chromosomal localization of the major and 5S rRNA genes in the European eel (*Anguilla anguilla*). *Cytogenetics and Cell Genetics* 73: 149–152.

Mayr, B., P. Rab and M. Kalat. 1986. Localization of NORs and counterstain-enhanced fluorescence studies in *Salmo gairdneri* and *Salmo trutta* (Pisces, Salmonidae). *Theoretical and Applied Genetics* 71: 703–707.

McClintock, B. 1934. The relation of a particular chromosomal element to the development of the nucleoli in *Zea mays*. *Zeitschrift für Zellforschung* 21: 294–328.

Miyazawa, C.S. and P.M. Galetti Jr. 1994. First cytogenetical studies in *Characidium* species (Pisces: Characiformes, Characidiinae). *Cytologia* 59: 73–79.

Moreira-Filho, O., L.A.C. Bertollo, and P.M. Galleti Jr. 1984. Structure and variability of nucleolar organizer region in parondotidae fish. *Canadian Journal of Genetic and Cytology* 26: 564–568.

Moreira-Filho, O. and L.A.C. Bertollo. 1991. *Astyanax scabripinnis* (Pisces, Characidae): An especies complex. *Brazilian Journal of Genetics* 14: 331–357.

Oliveira, C., F. Foresti, M.G. Rigolino and Y.A. Tabata. 1996. Paracentric inversion involving a NOR-bearing chromosome of rainbow trout (*Oncorhynchus mykiss*): Electron microscopy studies of the synaptonemal complex. *Caryologia* 49: 335–342.

Pardo, B.G. 2001. *Estudio citogenético y molecular de los genes ARNr en el Orden Pleuronectiformes*. Lugo. Universidad de Santiago de Compostela, Departamento de Biologia Fundamental, Doctoral Thesis. Departamento de Biologia Fundamental. Universidade de Santiago de Compostela, Lugo, Spain.

Pendás, A.M., P. Moran and E. Garcia-Vazquez. 1993a. Ribosomal RNA genes are interspersed throughout a heterochromatic chromosome arm in Atlantic salmon. *Cytogenetics and Cell Genetics* 63: 128–130.

Pendás, A.M., P. Morán and E. García-Vazquez. 1993b. Multi-chromosomal location of ribosomal RNA genes and heterochromatin association in brown trout. *Chromosome Research* 1: 63–67.

Pendás, A.M., P. Moran, J.P. Freije and E. Garcia-Vazquez. 1994. Chromosomal mapping and nucleotide sequence of two tandem repeats of Atlantic salmon 5S rDNA. *Cytogenetics and Cell Genetics* 67: 31–36.

Phillips, R. and P.E. Ihssen. 1985. Chromosome banding in salmonid fish: nucleolar organizer regions in *Salmo* and *Salvelinus*. *Canadian Journal of Genetics and Cytology* 27: 433–440.

Phillips, R.B., K.D. Zajicek and F.M. Utter. 1986. Chromosome banding in salmonid fishes: nucleolar organizer regions in *Oncorhynchus*. *Canadian Journal of Genetics and Cytology* 28: 502–510.

Phillips, R.B., K.A. Pleyte and P.E. Ihssen. 1989. Patterns of chromosomal nucleolar organizer regions (NOR) variation in fishes of the genus *Salvelinus*. *Copeia* 1989: 47–53.

Porto-Foresti, F., C. Oliveira, Y.A. Tabata, M.G. Rigolino and F. Foresti. 2002a. Analysis of NOR distribution in cultivated and naturalized stocks of rainbow trout (*Oncorhynchus mykiss*). *Genetics and Molecular Biology* 25: 409–412.

Porto-Foresti, F., C. Oliveira, Y.A. Tabata, M.G. Rigolino and F. Foresti. 2002b. NORs inheritance analysis in crossings including individuals from two stocks of rainbow trout (*Oncorhynchus mykiss*). *Hereditas* 136: 227–230.

Porto-Foresti, F., C. Oliveira, Y.A. Tabata, M.G. Rigolino and F. Foresti. 2004a. Chromosome location of the ribosomal genes 18S and 5S in four stocks of rainbow trout (Oncorhynchus mykiss). Cytology 69: 175–179.

Porto-Foresti, F., C. Oliveira, E.A. Gomes, Y.A. Tabata, M.G. Rigolino and F. Foresti. 2004b. A lethal effect associated with polymorphism of the NOR-bearing chromosomes in rainbow trout (Oncorhynchus mykiss). Genetics and Molecular Biology 27: 51–54.

Porto-Foresti, F., C. Oliveira, Y.A. Tabata, M.G. Rigolino and F. Foresti. 2006. Relationships among growth and different nor phenotypes in a specific stock of rainbow trout (Oncorhynchus mykiss). Brazilian Journal of Biology. (In press).

Reed, K.M. and R.B. Phillips. 1995. Molecular cytogenetic analysis of the double-CMA$_3$ chromosome of lake trout, Salvelinus namaycush. Cytogenetics and Cell Genetics 70: 104–107.

Rossi, A.R., E. Gornung and D. Crosetti. 1997. Cytogenetic analysis of Liza ramada (Pisces, Perciformes) by different staining techniques and fluorescent in situ hybridization. Heredity 79: 83–87.

Schmid, M., C. Loser, J. Schmidtke and W. Engel. 1982. Evolutionary conservation of a common pattern of activity of nucleolous organizer during spermatogenesis in vertebrates. Chromosoma 86: 149–179.

Sociedade De Amigos Dos Aquários Públicos. 1996. Revista Aquarium 2: 16–17.

Sola, L., P.J. Monaco and E.M. Rasch. 1990. Cytogenetics of bisexual/unisexual species of Poecilia. I. C-bands, Ag-NORs polymorphisms and sex-chromosomes in three populations of Poecila latipinna. Cytogenetics and Cell Genetics 53: 148–154.

Stempniewski, H.L. 1997. Retrospectiva dos serviços de pesca da Secretaria de Agricultura e Abastecimento e o jubileu de prata do Instituto de Pesca. São Paulo, Instituto de Pesca, Coordenadoria da Pesquisa Agropecuária, Secretaria de Agricultura e Abastecimento.

Tabata, Y.A. 2004. Obtenção e desemprenho de progênies ginogenéticas meioticas e mitóticas de truta arco-íris (Oncorhynchus mykiss). Doctoral Thesis. Instituto de Biociências, Botucatu, Brazil.

Tsukamoto, R.Y. 1988. Efeitos fisiológicos da indução da triploidia associada à hibridação em peixes salmonóides: uma análise comparativa de viabilidade e crescimento em híbridos de distintos níveis taxonômicos. Doctoral Thesis. Instituto de Biociências, São Paulo, Brazil.

Toledo-FIlho, S.A., L.F. Almeida-Toledo, F. Foresti, E. Galhardo and E. Donola. 1992. Conservação genética de peixes em projetos de repovoamento de reservatórios. Cadernos de Ictiogenética 1: 1–39.

Toledo-Filho, S.A., L.F. Almeida-Toledo, F. Foresti, G. Bernardino and D. Calcagnotto. 1994. Monitoramento e conservação genética em projeto de hibridação entre pacu e tambaqui. Cadernos de Ictiogenética 2: 1–49.

Toledo-Filho, S.A., F. Foresti and L.F. Almeida-Toledo. 1996. Biotecnologia genética aplicada à piscicultura. Cadernos de Ictiogenética 3: 1–60.

Ueda, T. and J. Kobayashi. 1988. Disappearance of Ag-NORs originated from brown trout in the allotriploid female rainbow trout and male brown trout. Proceedings of the Japan Academy 63B: 51–55.

Vênere, P.C. and P.M. Galetti Jr. 1989. Chromosome evolution and phylogenetic relationships of some Neotropical Characiformees of the family Curimatidae. *Brazilian Journal of Genetics* 12: 17–25.

Vitturi, R., M. Colomba, S. Vizzini, A. Libertini, R. Barbieri and A. Mazıola. 2005. Chromosomal location polymorphism of major rDNA sites in two Mediterranean populations of the killifish *Aphanius fasciatus* (Pisces: Cyprinodontidae). *Micron* 36: 243–246.

Ward, R.D. and P. Grewe. 1994. Appraisal of molecular genetic techniques in fisheries. *Reviews in Fish Biology and Fisheries* 4: 300–325.

Wasko, A.P. and P.M. Galetti Jr. 1999. Extensive NOR variability in fishes of the genus *Bryconamericus* (Characidae, Tetragonopterinae). *Cytologia* 64: 63–67.

SECTION

4

Fish Cytogenetics and Genomics

FISH Analysis of Fish Transposable Elements: Tracking Down Mobile DNA in Teleost Genomes

Agnès Dettai[1,4], Laurence Bouneau[2], Cécile Fischer[2], Christina Schultheis[1], Cornelia Schmidt[1], Qingchun Zhou[1], Jean-Pierre Coutanceau[3], Catherine Ozouf-Costaz[3] and Jean-Nicolas Volff[1]*

INTRODUCTION

In the previous decade, a wave of new scientific discoveries has considerably challenged the view that transposable elements (TEs) are only selfish sequences parasiting the genome of living organisms (Doolittle

Address for Correspondence: *E-mail: volff@ens-lyon.fr

[1]Biofuture Research Group, Physiologische Chemie I, Biozentrum, University of Würzburg, Am Hubland, D-97074 Würzburg, Germany.

[2]Genoscope/Centre National de Séquençage and CNRS-UMR 8030, 2 rue Gaston Crémieux, CP5706, F-91057 Evry Cedex 06, France.

[3]CNRS UMR 7138 "Systématique, Adaptation, Evolution" MNHN, Département Systématique et Evolution, C.P. 26, 57 rue Cuvier, 75231 Paris Cedex 05, France.

[4]UMR 7138, Muséum National d'Histoire Naturelle, Département de Systématique et Evolution, 43, rue Cuvier, F-75231 Paris Cedex 05, France.

and Sapienza, 1980; Orgel and Crick, 1980; Brosius and Gould, 1992; Brosius, 1999, 2003; Kazazian, 2004). Even if most TEs are certainly merely 'genome hitchhikers' (Burke et al., 1998) with neutral and occasionally deleterious effects on their hosts, their role in and during evolution has been considerably re-evaluated based on new genetic, genomic and functional data. There is, for example, substantial evidence that TEs can serve as a dynamic reservoir for genetic innovation, and provide the raw material for novel regulatory and coding sequences (Nekrutenko and Li, 2001; Kidwell and Lisch, 2002; Brosius, 2003; Jordan et al., 2003; van de Lagemaat et al., 2003; Brandt et al., 2005). Very important cellular functions, including that of telomerase, which catalyzes the replication of eukaryotic chromosome ends, as well as the Rag1/2 recombination-activating proteins, involved in V (D) J recombination during lymphocyte development, are probably derived from TEs (Eickbush, 1997; Agrawal et al., 1998; Hiom et al., 1998; Kapitonov and Jurka, 2005).

Transposable elements are major constituents of most eukaryotic genomes. For example, over 40% of the human genome consists of TEs, most of them being retrotransposable elements (International Human Genome Sequencing Consortium, 2001). The proportion of the genome constituted by such sequences is even higher in some plants (SanMiguel et al., 1996). TEs significantly participate in the modulation of the regional structure of chromosomes and chromatin, and can influence the expression of neighboring genes (Conte et al., 2002). Particularly, transposable elements are important factors for the formation and evolution of heterochromatin and for the epigenetic regulation of host genes (Karpen and Spradling, 1992; Lippman et al., 2004) and are target for cytosine methylation (Yoder et al., 1997).

TEs are generally not distributed randomly within genomes. The localization of a mobile element within particular genomic regions or sequences can be due, for instance, to a more or less stringent preference for certain target sites (Burke et al., 1987; Cantrell et al., 2001). Insertion preference or specificity can be mediated either directly through interaction between recombinase/transposase/integrase and target DNA or through accessory proteins that link target DNA and recombinase (Craig, 1997). Uneven distribution of TEs within genomes can also be related to selection pressures linked to the fitness of the host (Eickbush and Furano, 2002). TEs can inactivate genes through insertional mutagenesis and ectopic homologous recombination between non-allelic copies, which can induce a variety of DNA rearrangements (Gray, 2000).

Hence, negative selection might act against accumulation of TEs in gene-rich euchromatic regions (Eickbush and Furano, 2002; Volff *et al.*, 2003a). In some cases, preferential accumulation of TEs on particular chromosomes or within particular chromosomal regions might be linked to positively selected advantageous functions. For example, telomere-specific retrotransposons are necessary for the maintenance of chromosomes in *Drosophila*, and preferential accumulation of LINE1 non-LTR retrotransposons on the X chromosome in mammals might reflect a role in X chromosome inactivation (Lyon, 2000; Pardue and DeBaryshe, 2003; Dobigny *et al.*, 2004).

A panoramic view of the genomic distribution of TEs is necessary to understand the interactions and co-evolution of mobile elements with their host genome. Unfortunately, such information is unlikely to be provided using a whole genome shotgun sequencing approach, due to the difficulty in correctly assembling TEs and other repetitive sequences, particularly if they have been recently active. Cytogenetic analyses are, therefore, essential to get a complete picture of the localization of TEs in genomes.

Within vertebrates, transposable elements have been well characterized in human and other mammals. Particularly for retrotransposons, the number of ancient phylogenetic groups present in human and mouse genomes was surprisingly low compared to invertebrates (Volff *et al.*, 2003a; Kazazian, 2004). Therefore, a detailed analysis of another major group of vertebrates was necessary for a better understanding of the evolution of TEs in the vertebrate lineage. The existence of genome-sequencing projects for several teleost fish species, including the pufferfishes *Takifugu rubripes* (Aparicio *et al.*, 2002) and *Tetraodon nigroviridis* (Jaillon *et al.*, 2004) as well as the zebrafish *Danio rerio* and the medaka *Oryzias latipes* provided an excellent opportunity to characterize *in silico* the different families of TEs present in fish genomes, and to analyze their genomic distribution through cytogenetic analysis in different fish species. Several new groups of vertebrate TEs have been also identified through the partial sequencing of the sex chromosomes of the platyfish *Xiphophorus maculatus* (Froschauer *et al.*, 2002).

Diversity of Transposable Elements in Fish Genomes

According to their mode of transposition, TEs are divided into retrotransposable elements (aka retroelements, RTEs), which duplicate through mRNA reverse transcription and cDNA integration (retrotransposition), and DNA transposable elements (DTEs), which generally transpose through a transposase-mediated cut-and-paste mechanism (Curcio and Derbyshire, 2003). Both groups contain active non-autonomous elements, which require *in trans* the enzymatic machinery of coding elements for their transposition. Non-autonomous TEs include SINEs (short interspersed nuclear elements) for the RTEs and MITEs (miniature inverted transposable elements) for the DTEs. Processed intronless (pseudo) genes can also be generated from cellular mRNA molecules through *in trans* retrotransposition.

Autonomous and non-autonomous RTEs and DTEs have all been detected in fish genomes. Different major groups of transposase-encoding autonomous DTEs have been identified in various fish species, including members from the (super) families hAT (Koga and Hori, 1999; Aparicio *et al.*, 2002; Fischer *et al.*, 2004), Tc1/mariner (Radice *et al.*, 1994; Izsvak *et al.*, 1995; Liu *et al.*, 1999), Harbinger (Kapitonov and Jurka, 2004), PiggyBac (Sarkar *et al.*, 2003), P (Hammer *et al.*, 2005) and En/Spm (Kapitonov and Jurka, 2003). Some autonomous DTEs like Sleeping Beauty and Tol2 are promising mutagenesis and gene delivery tools in both biotechnology and medicine (for review, Miskey *et al.*, 2005). Non-autonomous MITEs with mutagenic activity have been identified in fish genomes (Izsvàk et al., 1999; C. Schultheis *et al.*, 2006). Finally, Helitrons, a particular class of transposable elements encoding a 'rolling circle' replication protein and a helicase, have been detected in the genome of several fish species (Poulter *et al.*, 2003; Zhou *et al.*, 2006).

Based on both common phylogenetic origin and shared structural features, autonomous reverse transcriptase-encoding RTEs are classified into several major categories: long terminal repeat (LTR) retrotransposons (without envelope gene) and retroviruses (with envelope gene), non-LTR retrotransposons (aka long interspersed nuclear elements, LINEs) and Penelope-like retrotransposons. LTRs are repeated sequences flanking the coding sequences, which are necessary for proper transcription and integration of the retroelement.

All major groups of LTR retroelements are represented in fish genomes, including Ty3/Gypsy, Ty1/Copia, DIRS1 and BEL retrotransposons (Poulter and Butler, 1998; Frame *et al.*, 2001; Goodwin and Poulter, 2001; Volff *et al.*, 2001b, 2003a,b). Endogenous 'vertebrate' retroviruses, which correspond to remnants of ancient infections, have been identified in zebrafish and other fish species as well (Volff *et al.*, 2003b; Shen and Steiner, 2004). Non-coding LTR retrotransposons are present in the genome of the platyfish, where they are able to inactivate genes through insertional mutagenesis (Schartl *et al.*, 1999).

Numerous families of non-LTR retrotransposons have been identified in fish, including elements encoding an apurinic/apyrimidinic endonuclease (Duvernell and Turner, 1998; Oliveira *et al.*, 1999; Poulter *et al.*, 1999; Volff *et al.*, 1999, 2000, 2001d, 2003a,b) and a restriction enzyme-like endonuclease (Volff *et al.*, 2001c; Bouneau *et al.*, 2003), as well as Penelope-related retrotransposons (Lyozin *et al.*, 2001; Volff *et al.*, 2001a). Fish genomes also contain different families of non-autonomous SINEs, which use the enzymatic insertion machinery of non-LTR retrotransposons for their retrotransposition. One widespread group of such elements is the V-SINEs superfamily. V-SINEs, which display an astonishing level of conservation between different groups of fishes, are not only present in most Actinopterygians (ray-finned fishes) but also in non-amniote sarcopterygians (amphibians, lungfishes) as well as in more basal vertebrates (cartilaginous and jawless fishes) (Ogiwara *et al.*, 2002; Dettai *et al.*, unpublished). These sequences have been very successful in some teleost fish species. For example, DANA elements comprise about 10% of the zebrafish genome (Izsvàk *et al.*, 1996). SINE elements are used as phylogenetic markers for the resolution of fish phylogenies (Terai *et al.*, 2004).

Interestingly, fish transposable elements present a level of diversity not observed in mammals and birds, indicating that numerous ancient phylogenetic groups of TEs have been lost during the evolution of the tetrapod lineage (Aparicio *et al.*, 2002; Volff *et al.*, 2003a; Jaillon *et al.*, 2004). This is particularly true for several ancient groups of RTEs, including LTR retrotransposons (Ty3/Gypsy, Ty1/Copia, DIRS1 and BEL), non-LTR retrotransposons with restriction enzyme-like endonuclease and Penelope-like elements, which are present in invertebrates and teleost fish, but absent in mammals and birds. The success of a particular group of mobile element is host-dependent. For example, DIRS1 elements have been particularly successful in the zebrafish *Danio rerio*, but not in the pufferfishes *Takifugu rubripes* and *Tetraodon nigroviridis*. Even between these two related species, clear differences in copy number have been observed for many different families of TEs (Volff *et al.*, 2003a). Transposable elements have frequently formed ancient paralogous groups of sequences, which have invaded fish genomes with varying success (Volff *et al.*, 2000; Duvernell *et al.*, 2004; Furano *et al.*, 2004). Future analyses of the multitude of fish species, which have not been investigated so far, will provide new information on TE distribution and evolutionary dynamics and might also uncover new groups of transposable elements not characterized until now.

Distribution of Transposable Elements in *Tetraodon nigroviridis* and Other Fish Genomes

The distribution of TEs in fish genomes is frequently not on a random basis. Such a phenomenon might be explained in some cases by a more or less pronounced target sequence preference, as observed for some MITEs in fish (Izsvàk *et al.*, 1999), or might result from selection, for example, against insertions in gene-rich regions (Volff *et al.*, 2003a).

Fluorescence *in situ* hybridization (FISH) and other methods of cytogenetic analysis revealed that some TEs accumulate in particular, generally heterochromatic regions of fish genomes. This is the case for the non-LTR retrotransposon Zebulon as well as for other types of TEs, satellite repeats and pseudogenes in the genome of the pufferfish *Tetraodon nigroviridis* (Dasilva *et al.*, 2002; Bouneau *et al.*, 2003; Fischer *et al.*, 2004; Fig. 4.1.1). These repeated elements are located in heterochromatic short arms of subtelocentric chromosomes and pericentromeric regions, which are brightly stained by DAPI after denaturation treatment (Fischer *et al.*, 2000). The multicopy non-LTR retrotransposon Zebulon accumulates preferentially in such regions on five chromosome pairs, with additional weaker signals at the end of the arms of subtelocentric chromosomes and in pericentromeric regions (Bouneau *et al.*, 2003). Rex3, another non-LTR retrotransposon with approx. 500 copies per haploid genome, shows multiple and intense hybridization signals in short arms and pericentromeric regions of small submeta- or subtelocentric pairs. In addition, faint hybridization spots scattered on the chromosomes were detected. This observation was consistent with the presence of short truncated versions of non-LTR retrotransposons detected in gene-rich euchromatic region through genomic DNA sequencing (Fischer *et al.*, 2002). The Tol2 DTE (seven copies) is located on at least six small metacentric, submetacentric or subtelocentric chromosomes. Particularly, Tol2 elements are located in the small arms of three chromosome pairs, and at the end of the long arm in a submetacentric pair (Fischer *et al.*,

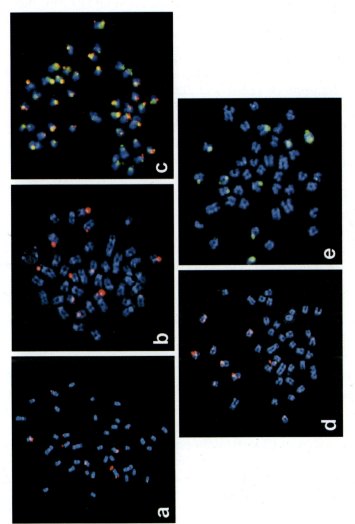

Fig. 4.1.1 FISH analysis of the distribution of transposable elements in the genome of the pufferfish *Tetraodon nigroviridis*. (a): Babar (non-LTR retrotransposon); (b): Zebulon (non-LTR retrotransposon); (c): double FISH of the non-LTR retrotransposons Zebulon (red) and Babar (green) showing their partial co-localization; (d): Rex1 (non-LTR retrotransposon); (e): Hopper (DNA transposable element). Detection was performed using either anti-digoxigenin-rhodamine (red) or avidin-fluorescein (green). All retroelements accumulate in pericentromeric regions and/or short arms of submeta- and subtelocentric chromosomes.

2004). Buffy1, another DTE, is mainly found on small arms of four to five small submetacentric or subtelocentric chromosomes pairs.

Interestingly, a series of double-color FISH analyses with probes specific for different groups of RTEs, DTEs, minisatellites and amplified pseudogenes revealed the preferential co-localization of TEs with other types of repeats within certain common heterochromatin regions of the genome of the pufferfish *T. nigroviridis* (Dasilva *et al.*, 2002; Bouneau *et al.*, 2003; Fischer *et al.*, 2004; Fig. 4.1.1). Superposition of major FISH signals was observed between TEs from a same type, for example between the non-LTR retrotransposons Babar and Zebulon (Fig. 4.1.1), Rex3 and Zebulon (Bouneau *et al.*, 2003), Rex3 and Babar (Fischer *et al.*, 2004), Rex3 and Maui, Zebulon and Maui, and Babar and TX1-1_Tet (Fischer *et al.*, 2005), as well as between the DTEs Tol2 and Buffy1 (Fischer *et al.*, 2004). Co-localization was also observed between different types of TEs, for example between the DIRS1 LTR retrotransposon DIRS1_Tet and the Ty3/Gypsy LTR retrotransposon Jule, the Ty3/Gypsy LTR retrotransposon Barthez and the non-LTR retrotransposons TX1-1_Tet and Maui (Fischer *et al.*, 2005), the non-LTR retrotransposon Dm-Line and the DTE TC1-like (Dasilva *et al.*, 2002) and the non-LTR retrotransposon Rex3 and the DTE Tol2 (Fisher *et al.*, 2004). TEs even co-localize with other categories of repeats in some heterochromatic regions of the genome of *T. nigroviridis*, including minisatellites (Fischer *et al.*, 2004) and pseudogenes like *iSET* and *Trapeze*, which have been amplified in such regions by retrotransposition and segmental duplication, respectively (Dasilva *et al.*, 2002).

The co-localization of different types of repeats within common genomic regions was confirmed through Southern blot hybridization of several specific probes on a bacterial artificial chromosome (BAC) genomic library of *T. nigroviridis* (Dasilva *et al.*, 2002). The probes used were specific for two non-LTR retrotransposons (Dm-Line, and Dr-Line), a LTR retrotransposon (Copia-like), a DTE (TC1-like) as well as for a 10-bp tandem repeat specifically detected in short heterochromatic arms of approximately 10 pairs of subtelocentric chromosomes (Fischer *et al.*, 2000). Strikingly, the four probes hybridized to an unexpected high number of BAC clones in common, indicating a preferential accumulation of repeats in regions representing about 5% of the pufferfish genome (Dasilva *et al.*, 2002). Such repeat-rich BAC clones hybridize on DAPI-positive chromosomal regions in FISH experiments (Fischer *et al.*, 2005).

Taken together, this series of results revealed the extreme degree of compartmentalization of the compact genome of the pufferfish *T. nigroviridis*, with clear-cut separation between heterochromatic repeat-rich, gene-poor regions and euchromatic repeat-poor, gene-rich segments. Such a situation, which was also found in several very compact genomes from other phyla, is not observed in the human genome, indicating that distinct vertebrate species can have a completely different genome organization (Volff *et al.*, 2003a). Genome sequence analysis has suggested the presence of repeat-rich regions in the compact genome of the Fugu *Takifugu rubripes* also (Poulter *et al.*, 1999), but this has not been confirmed by FISH analysis so far.

In order to assess whether the situation observed in *T. nigroviridis* is representative of the general structure of fish genomes, or if this compartmentalization is only associated with the extreme degree of compactness of its genome, the genomic distribution of TEs needs to be analyzed in other teleost fish species. No information is available at the moment concerning the distribution of TEs in the genomes of the zebrafish *Danio rerio* and the medaka *Oryzias latipes*, which are both objects of almost completed whole genome sequencing projects (for review see Volff, 2005). In zebrafish, only several non-TE repeats have been mapped to centromeric and paracentromeric heterochromatic regions (Sola and Gornung, 2001). The genome of another fish model, the Nile Tilapia *Oreochromis niloticus*, contains about 5000 copies of CiLINE2, a non-LTR retrotransposon from the LINE2/Maui family with apurinic/apyrimidinic endonuclease (Oliveira *et al.*, 1999). FISH analysis showed that CiLINE2 sequences are organized in small clusters dispersed over all chromosomes of *O. niloticus*, with a particular high concentration near the chromosome ends.

Antarctic teleosts from the suborder Notothenioidei are a group of species in which the genomic localization of TEs has been investigated through cytogenetic analysis. Due to their phenotypic and karyotypic variability as well as their highly specialized adaptations to extreme conditions, notothenioids represent a very interesting model to study the evolution of genomes in response to environmental challenges. The localization of a Tc1-like TDE has been studied in the genome of *Chionodraco hamatus* (Capriglione *et al.*, 2002). Accumulation of this element was observed in heterochromatic regions at pericentric or telomeric positions and more rarely interstitially along the chromosome arms. Furthermore, several RTEs have been analyzed in 13 species from 5 of the 8 families described in notothenioids, Bovichtidae, Nototheniidae, Artedidraconidae, Bathydraconidae and Channichthyidae (Ozouf-Costaz *et al.*, 2004). The non-LTR retrotransposon with restriction enzyme-like endonuclease Rex6 could not be detected by polymerase chain reaction in any of the species tested. In contrast, both non-LTR retrotransposons with apurinic/apyrimidinic endonuclease Rex1 and Rex3 were identified in all the species tested. Partial sequences were isolated from four species: *Notothenia coriiceps*, *Trematomus newnesi*, *Dissostichus mawsoni* (Nototheniidae) and *Gymnodraco acuticeps* (Bathydraconidae). These sequences were analyzed at the phylogenetic level and used as probes in FISH experiments (Fig. 4.1.2). The high degree of sequence similarity between certain copies indicated that Rex1 and Rex3 were probably relatively recently active in the species studied. Species-specific phylogenetic clades were generally not observed, except for Rex3 in *N. coriiceps*, indicating that this element actively retrotransposed in the genome of this species after its divergence from the other notothenioids analyzed. In comparison with other species, *N. coriiceps* showed a more

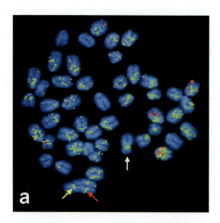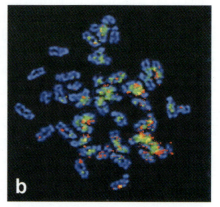

Fig. 4.1.2 FISH analysis of the genomic distribution of the non-LTR retrotransposons Rex1, detected with anti-digoxygenin-rhodamine (red) and Rex3, detected with avidin-fluorescein (greenish) in the Antarctic notothenioids *Chionodraco hamatus* (a) and *Neopagetopsis ionah* (b). *Chionodraco hamatus* shows a more dispersed pattern with hybridization spots scattered throughout the chromosomes and only a few accumulation zones as, for instance, in the centromere (yellow arrow) and on the long arm (red arrow) of the neo-Y sex chromosome (see text). There are two X chromosomes in *C. hamatus*; only the X1 chromosome (white arrow) is easily recognizable on the basis of its morphology. *Neopagetopsis ionah* presents a massive accumulation of both non-LTR retrotransposons.

pronounced compartmentalized distribution pattern for Rex3, often with pericentromeric regions of accumulation. Interestingly, *N. coriiceps* has a much-derived karyotype probably formed by Robertsonian fusions having produced a reduced number of large chromosomes. Hence, a high activity of Rex3 might have been associated with drastic karyotype rearrangements in this species (Ozouf-Costaz *et al.*, 2004).

The FISH signals of Rex1 on notothenioid chromosomes were more variable but often superposed with those of Rex3. Rex3 was also hybridized on chromosomes of *Chionodraco hamatus* (Channichthyidae; Fig. 4.1.2). Strikingly, the FISH pattern observed was very similar to that observed for a Tc1-like DNA transposon (Capriglione *et al.*, 2002). Hence, although the distribution of TEs within notothenioid genomes is highly variable and species-dependent, a clustering of different mobile elements in preferential regions of the genome has been observed in some notothenioid species.

Cytogenetic studies on Neoselachians (sharks and rays) are still scanty, although an increasing number of karyotypes have been determined (Stingo and Rocco, 2001; Rocco, present volume). A few transposable elements have been characterized in this group, including SINEs from the Mermaid and HE1 families, as well as non-LTR retrotransposons from the HER1 family (Shimoda *et al.*, 1996; Ogiwara *et al.*, 1999). The SINEs *HpaI* from salmonids have been hybridized using FISH on *Torpedo ocellata* (common torpedo) chromosomes. This revealed a centromeric and paracentromeric localization on biarmed chromosomes and a centromeric and/or telomeric localization on uniarmed chromosomes (Stingo and Rocco, 2001; Rocco, present volume). Even less is known about the cytogenetics of dipnoans and other 'fishy' sarcopterygians (Morescalchi *et al.*, 2002) and, to our knowledge, no study on the localization of mobile elements has been done for these genomes.

TE-containing regions can be highly polymorphic, probably due to frequent unequal crossovers between non-allelic repetitive elements. In the chromosome pairs of *T. nigroviris*, which could be precisely identified either because of their particular size or morphology or through the mapping of various single copy sequences, intra- and inter-individual size polymorphism of DAPI-positive blocks, corresponding to AT-rich regions where most repetitive elements accumulate, was frequently observed. Size polymorphism of heterochromatic blocks has been also reported in various other fish species (see, for example, Mantovani *et al.*, 2000). In

T. nigroviridis, heteromorphism within a given chromosome pair was detected not only using DAPI staining but also through FISH analysis using different TEs and repetitive sequences as probes (Fischer *et al.*, 2004). The size difference pattern obtained ranged from homeomorphic to strongly heteromorphic, suggesting a dynamic variation of the number of copies of mobile elements in these regions. Mandrioli and coworkers (2000) showed the existence of similar heteromorphisms in the nuclear organizer regions (NORs) of the black goby (*Gobius niger*). The preferential localization of mariner-like DTEs in NOR-associated heterochromatin suggested a relationship between NOR heteromorphism and the presence of TEs in these regions (Mandrioli *et al.*, 2000). Similar observations have been done for the short arms of chromosomes 5 and 12 in *Tetraodon fluviatilis* (Mandrioli and Manicardi, 2001), and NOR heteromorphism has been also observed in *T. nigroviridis* (Fisher *et al.*, 2000).

Transposable Elements on B and Sex Chromosomes in Fish

In fish genomes, TEs can accumulate not only within special regions of 'normal' chromosomes, but also within particular types of chromosomes, like the sex chromosomes and the B chromosomes. B chromosomes are supernumerary elements of the genome that rarely carry active genes and might, therefore, accumulate TEs without disturbing important genomic functions. The B chromosome of the cyprinid *Alburnus alburnus* is the largest found so far in vertebrates and is detected at different frequencies (from 11% to more than 50% of individuals), depending on the population. FISH analysis showed that a Gypsy/Ty3 LTR retrotransposon is strongly amplified on the B chromosome of *A. alburnus* (Ziegler *et al.*, 2003). In contrast, this element is apparently absent or present at a very low copy number in the 'normal' A chromosomes and was not detected on the B chromosome of the related species *Rutilus rutilus*, suggesting either an independent origin for both B chromosomes or the amplification of the Gypsy/Ty3 LTR retrotransposon in *A. alburnus* after the divergence of both species.

All major types of sex determination systems have been described in fish, including hermaphroditism (rarely synchronous or protandrous, more often protogynous) as well as environmental and genetic sex determination, with an occasional association between different systems. Particularly, all known forms of genetic sex determination such as polygenic sex determination and sex determination controlled by dominant factors with or without autosomal influences have been reported in teleosts (Volff and Schartl, 2001; Devlin and Nagahama, 2002). At least 10% of fish species analyzed thus far have cytogenetically distinct sex chromosomes (Devlin and Nagahama, 2002). Sex chromosome systems with male heterogameity (XX/XY, the system at work in mammals) and female heterogameity (WZ/ZZ, the sex determination mechanism in birds) have both been observed in fish, in some cases within a same genus or even a same species. To date, the master sex determining gene has been identified only in the medaka *Oryzias latipes* and several related species (Matsuda *et al.*, 2002; Nanda *et al.*, 2002). This gene, which corresponds to a relatively recent Y chromosome-specific duplicate of the autosomal gene *dmrt1*, is present only in some species of the genus *Oryzias*, and is absent from more divergent species (Kondo *et al.*, 2004).

As a result of the independent origin and the young age of sex chromosomes in different teleost lineages (Nanda *et al.*, 2002; Kondo *et al.*, 2004; Peichel *et al.*, 2004), fish are particularly interesting models for the characterization of the initial steps driving the divergent evolution of different types of gonosomes after their formation from a common autosomal ancestor. Such studies are not possible in most mammals and birds, because of the old age and relative stability of the systems determining sex in these animals.

TEs are involved in the evolution of sex chromosomes in different organisms and might play an active role in the molecular differentiation and suppression of homologous recombination between gonosomes (Steinemann and Steinemann, 2000; Liu *et al.*, 2004). Accordingly, RTEs and other types of repeats accumulate on the sex chromosomes of the platyfish *Xiphophorus maculatus* (Volff and Schartl, 2001; Froschauer *et al.*, 2002; Volff *et al.*, 2003b) and the density of TEs is higher on the Y than on the X chromosome in the sex-determining region of the threespine stickleback *Gasterosteus aculeatus* (Peichel *et al.*, 2004). The Y-specific region carrying the master sex-determining gene of the medaka *O. latipes* is also infested by TEs (Nanda *et al.*, 2002).

Cytogenetic analysis confirmed the preferential accumulation of certain types of TEs on fish sex chromosomes. In the Nile Tilapia *Oreochromis niloticus*, a species with monomorphic X and Y sex chromosomes, numerous copies of the non-LTR retrotransposon CiLINE2 are located on the long arms of the putative pair of gonosomes (Oliveira et al., 1999). About 20% of the sequences obtained by microdissection of these chromosomes are from TE origin (Harvey et al., 2003). TEs accumulate on both types of gonosomes in this fish, but the FISH signals differ between the X and Y chromosomes. Such differences in the amount and distribution of repeated sequences could, through the delaying of the pairing of sex chromosomes during meiosis, reduce the level of recombination and accelerate the process of differentiation between gonosomes. In the platyfish X. maculatus, another species without any well-differentiated sex chromosomes (Traut and Winking, 2001), high-copy number amplification of the long terminal repeat-like element XIR has been observed in the sex-determining region on the Y but not on the X chromosome (Nanda et al., 2000). This event might also correspond to an initial step toward gonosome differentiation. In Synodontidae, a fish family with a well-conserved ZW system involving a small W chromosome, a W-specific marker containing a non-LTR retrotransposon from the L1 clade has been isolated from Aulopus japonicus (Ota et al., 2003), suggesting a possible involvement of this transposable element in the degeneration of the W chromosome. Other types of repetitive sequences are preferentially found on one particular type of gonosome as observed, for example, on the Y chromosome of the chinook salmon (Oncorhynchus tshawytscha) (Devlin et al., 1998).

In some cases, TEs or TE-containing regions might have been involved in the formation of sex chromosomes. In the Antarctic teleost Chionodraco hamatus (Channichthyidae), both a Tc1-like DNA transposon and the Rex3 non-LTR retrotransposon have been localized by FISH within an intercalary band in the long arm of the male neo-Y chromosome (Capriglione et al., 2002; Ozouf-Costaz et al., 2004; Fig. 4.1.2). Interestingly, this intercalary band might correspond to the short arm of one of the autosomes involved in the tandem fusion which generated the new Y chromosome. Hence, these results point towards an involvement of a TE-rich region in the fusion event that led to the formation of a new male-inducing chromosome in C. hamatus.

Finally, transposition itself might induce the formation of neo-sex chromosomes in fish. In salmonids, the major sex-determining locus is located on varying linkage groups in different species. One of the hypotheses explaining this phenomenon is the frequent transposition of the sex-determining locus between chromosomes in this fish family (Woram et al., 2003).

CONCLUSIONS

Due to their amazing genetic, genomic and phenotypic variability, fish are particularly attractive models in the study of the molecular mechanisms involved in evolution and speciation (for review: Volff, 2005). Recent analyses have revealed an unexpected diversity of transposable elements in teleost genomes. These sequences probably play an important role in the structure and evolution of genes, chromosomes and genomes in fish. Of particular interest is the evolutionary impact of TEs on sex chromosomes and sex determination, both of which present a level of diversity not observed in mammals and birds. Why fishes have so many different families of TEs as compared to mammals and birds remains an unsolved question. Maybe, tetrapods have developed new defense mechanisms having inactivated numerous groups of mobile sequences being still active in fish. Alternatively, hyperactive families of TEs might have eliminated through competition other groups of elements.

Completed as well as ongoing entire genome-sequencing projects have already considerably contributed to our understanding of the structure and evolution of fish genomes. For example, they have confirmed that an event of genome duplication has taken place early during the evolution of the fish lineage (Jaillon *et al.*, 2004; for review: Volff, 2005). However, the 'shotgun' sequencing strategy adopted by such genome projects is generally unsuitable for the analysis of TE-rich regions, due to the difficulties in accurately assembling repeat-rich segments. Hence, such genome projects, when providing an excellent overview of the structure of gene-rich euchromatin, will fail to resolve the structure of heterochromatic regions and subsequently display an incomplete picture of the general organization of genomes.

The different chapters of this book demonstrate without any ambiguity that cytogenetics, and particularly FISH, represent a complementary approach able to integrate sequencing data into a broader genomic context. This complementarity between genomics and cytogenetics was perfectly illustrated in the 'cytogenomic' strategy used in the genome project of the pufferfish *Tetraodon nigroviridis*, where sequenced genomic clones were anchored by FISH on chromosomes (Jaillon *et al.*, 2004).

Cytogenetic studies, particularly on the pufferfish *T. nigroviridis*, on Antarctic notothenioids as well as on cichlids have already provided important new information concerning the distribution of TEs in fish genomes. TEs generally accumulate in heterochromatic regions in fish. If these sequences are actively involved in the formation of heterochromatin (Lippman *et al.*, 2004) or are only better tolerated in such gene-poor regions remains to be determined. Particularly, the compact genome of *T. nigroviridis* presents an extreme degree of compartmentalization, with clear-cut separation between heterochromatic repeat-rich, gene-poor regions and euchromatic repeat-poor, gene-rich segments. Such a situation is not observed in humans, but reminiscent of the structure of the genome of the fruit fly *Drosophila melanogaster* (Volff *et al.*, 2003a). A strong selection against insertions in euchromatin due, for example, to frequent deleterious rearrangements through ectopic homologous recombination between non-allelic TE sequences, might explain the compartmentalization observed in *T. nigroviridis*.

Due to the restricted number of species and TE families analyzed, our knowledge of the evolution and interactions of the multiple groups of mobile sequences in fish genomes is still extremely fragmentary. Comparative cytogenetic studies on both closely and distantly related species are now required to grasp the impact of TEs on the amazing level of biological diversity observed in fish. As molecular markers for comparative genomics, low copy number TE insertions will certainly help in analyzing karyotype evolution in fish and contribute to resolving controversial phylogenies in teleosts.

Acknowledgements

Our work is supported by the BioFuture program of the German Ministry for Research and Education (BMBF) (J.-N.V.), by the French Ministère de la Recherche et de la Technologie and the Centre National de la Recherche Scientifique (CNRS) (C.F. and L.B.), and by the French Muséum National d'Histoire Naturelle and the Centre National de la Recherche Scientifique (CNRS) (C.O. and J.-P.C.). A.D. is recipient of a postdoctoral fellowship from the Alexander von Humboldt foundation (Germany). Many thanks to Manfred Schartl (Würzburg) for fruitful discussions and encouragement.

References

Agrawal, A., Q.M. Eastman and D.G. Schatz. 1998. Transposition mediated by RAG1 and RAG2 and its implications for the evolution of the immune system. *Nature (London)* 394: 744–751.

Aparicio, S., J. Chapman, E. Stupka, N. Putnam, J.M. Chia, P. Dehal, A. Christoffels, S. Rash, S. Hoon, A. Smit, M.D. Gelpke, J. Roach, T. Oh, I.Y. Ho, M. Wong, C. Detter, F. Verhoef, P. Predki, A. Tay, S. Lucas, P. Richardson, S.F. Smith, M.S. Clark, Y.J. Edwards, N. Doggett, A. Zharkikh, S.V. Tavtigian, D. Pruss, M. Barnstead, C. Evans, H. Baden, J. Powell, G. Glusman, L. Rowen, L. Hood, Y.H. Tan, G. Elgar, T. Hawkins, B. Venkatesh, D. Rokhsar and S. Brenner. 2002. Whole-genome shotgun assembly and analysis of the genome of *Fugu rubripes*. *Science* 297: 1301–1310.

Bouneau, L., C. Fischer, C. Ozouf-Costaz, A. Froschauer, O. Jaillon, J.-P. Coutanceau, C. Körting, J. Weissenbach, A. Bernot and J.-N. Volff. 2003. An active non-LTR retrotransposon with tandem structure in the compact genome of the pufferfish *Tetraodon nigroviridis*. *Genome Research* 13: 1686–1695.

Brandt J., S. Schrauth, A.-M. Veith, A. Froschauer, T. Haneke, C. Schultheis, M. Gessler, C. Leimeister and J.-N. Volff. 2005. Transposable elements as a source of genetic innovation: expression and evolution of a family of retrotransposon-derived neogenes in mammals. *Gene* 345: 101–111.

Brosius, J. 1999. Genomes were forged by massive bombardments with retroelements and retrosequences. *Genetica* 107: 209–238.

Brosius, J. 2003. The contribution of RNAs and retroposition to evolutionary novelties. *Genetica* 118: 99–116

Brosius, J. and S.J. Gould. 1992. On 'genomenclature': A comprehensive (and respectful) taxonomy for pseudogenes and other 'junk DNA'. *Proceedings of the National Academy of Sciences of the United States of America* 89: 10706–10710.

Burke, W.D., C.C. Calalang and T.H. Eickbush. 1987. The site-specific ribosomal insertion element type II of *Bombyx mori* (R2Bm) contains the coding sequence for a reverse transcriptase-like enzyme. *Molecular and Cellular Biology* 7: 2221–2230.

Burke, W.D., H.S. Malik, W.C. Lathe 3rd and T.H. Eickbush. 1998. Are retrotransposons long-term hitchhikers? *Nature* (*Lond.*)392: 141–142.

Cantrell, M.A., B.J. Filanoski, A.R. Ingermann, K. Olsson, N. DiLuglio, Z. Lister and H.A. Wichman. 2001. An ancient retrovirus-like element contains hot spots for SINE insertion. *Genetics* 158: 769–777.

Capriglione, T., G. Odierna, V. Caputo, A. Canapa and E. Olmo. 2002. Characterization of a Tc1-like transposon in the Antarctic ice-fish, *Chionodraco hamatus*. *Gene* 295: 193–198.

Conte, C., B. Dastugue and C. Vaury. 2002. Coupling of enhancer and insulator properties identified in two retrotransposons modulates their mutagenic impact on nearby genes. *Molecular and Cellular Biology* 22: 1767–1777.

Craig, N.L. 1997. Target site selection in transposition. *Annual Review of Biochemistry* 66: 437–474.

Curcio, M.J. and K.M. Derbyshire. 2003. The outs and ins of transposition: from mu to kangaroo. *Nature Reviews Molecular Cell Biology* 4: 865–877.

Dasilva, C., H. Hadji, C. Ozouf-Costaz, S. Nicaud, O. Jaillon, J. Weissenbach and H. Roest Crollius. 2002. Remarkable compartmentalization of transposable elements and pseudogenes in the heterochromatin of the *Tetraodon nigroviridis* genome. *Proceedings of the National Academy of Sciences of the United States of America* 99: 13636–13641.

Devlin, R.H. and Y. Nagahama. 2002. Sex determination and sex differentiation in fish: An overview of genetic, physiological, and environmental influences. *Aquaculture* 208: 191–364.

Devlin, R.H., G.W. Stone and D.E. Smailus. 1998. Extensive direct-tandem organization of a long repeat DNA sequence on the Y chromosome of chinook salmon *Oncorhynchus tshawytscha*. *Journal of Molecular Evolution* 46: 277–287.

Dobigny, G., C. Ozouf-Costaz, C. Bonillo and V. Volobouev. 2004. Viability of X-autosome translocations in mammals: An epigenomic hypothesis from a rodent case-study. *Chromosoma* 113: 34–41.

Doolittle, W.F. and C. Sapienza. 1980. Selfish genes, the phenotype paradigm and genome evolution. *Nature* (*London*) 284: 601–603.

Duvernell, D. D. and B. J. Turner. 1998. Swimmer 1, a new low-copy-number LINE family in teleost genomes with sequence similarity to mammalian L1. *Molecular Biology and Evolution* 15: 1791–1793.

Duvernell, D.D, S.R. Pryor and S.M. Adams. 2004. Teleost fish genomes contain a diverse array of L1 retrotransposon lineages that exhibit a low copy number and high rate of turnover. *Journal of Molecular Evolution* 59: 298–308.

Eickbush, T.H. 1997. Telomerase and retrotransposons: Which came first? *Science* 277: 911–912.

Eickbush, T.H. and A.V. Furano. 2002. Fruit flies and humans respond differently to retrotransposons. *Current Opinion in Genetics and Development* 12: 669–674.

Fischer, C., C. Ozouf-Costaz, H. Roest Crollius, C. Dasilva, O. Jaillon, L. Bouneau, C. Bonillo, J. Weissenbach and A. Bernot. 2000. Karyotype and chromosome location of characteristic tandem repeats in the pufferfish *Tetraodon nigroviridis*. *Cytogenetics and Cell Genetics* 88: 50–55.

Fischer, C., L. Bouneau, C. Ozouf-Costaz, T. Crnogorac-Jurcevic, J. Weissenbach and A. Bernot. 2002. Conservation of the T-cell receptor alpha/delta linkage in the teleost fish *Tetraodon nigroviridis*. *Genomics* 79: 241–248.

Fischer, C., L. Bouneau, J.-P. Coutanceau, J. Weissenbach, J.-N. Volff, C. Ozouf-Costaz. 2004. Global heterochromatic colocalization of transposable elements with minisatellites in the compact genome of the pufferfish *Tetraodon nigroviridis*. *Gene* 336: 175–183.

Fischer C., L. Bouneau, J.-P. Coutanceau, J. Weissenbach, C. Ozouf-Costaz and J.-N. Volff. 2005. Diversity and clustered distribution of retrotransposable elements in the compact genome of the pufferfish *Tetraodon nigroviridis*. *Cytogenetic and Genome Research* 110: 522–536.

Frame, I.G., J.F. Cutfield and R.T. Poulter. 2001. New BEL-like LTR-retrotransposons in *Fugu rubripes*, *Caenorhabditis elegans*, and *Drosophila melanogaster*. *Gene* 263: 219–230.

Froschauer, A., C. Körting, T. Katagiri, T. Aoki, S. Asakawa, N. Shimizu, M. Schartl and J.-N. Volff. 2002. Construction and initial analysis of bacterial artificial chromosome (BAC) contigs from the sex-determining region of the platyfish *Xiphophorus maculatus*. *Gene* 295: 247–254.

Furano, A.V., D.D. Duvernell and S. Boissinot S. 2004. L1 (LINE-1) retrotransposon diversity differs dramatically between mammals and fish. *Trends in Genetics* 20: 9–14.

Goodwin, T.J. and R.T. Poulter. 2001. The DIRS1 group of retrotransposons. *Molecular Biology and Evolution* 18: 2067–2082.

Gray, Y.H. 2000. It takes two transposons to tango: transposable element-mediated chromosomal rearrangements. *Trends in Genetics* 16: 461–468.

Hammer, S.E., S. Strehl and S. Hagemann. 2005. Homologs of *Drosophila* P transposons were mobile in zebrafish but have been domesticated in a common ancestor of chicken and human. *Molecular Biology and Evolution* 22: 833–844.

Harvey S.C., C. Boonphakdee, R. Campos-Ramos, M.T. Ezaz, D.K. Griffin, N.R. Bromage and P. Penman. 2003. Analysis of repetitive DNA sequences in the sex chromosomes of *Oreochromis niloticus*. *Cytogenetic and Genome Research* 101: 314–319.

Hiom, K., M. Melek and M. Gellert. 1998. DNA transposition by the RAG1 and RAG2 proteins: a possible source of oncogenic translocations. *Cell* 94: 463–470.

International Human Genome Sequencing Consortium. 2001. Initial sequencing and analysis of the human genome. *Nature (London)* 409: 860–921.

Izsvàk, Z., Z. Ivics and P.B. Hackett. 1995. Characterization of a Tc1-like transposable element in zebrafish (*Danio rerio*). *Molecular and General Genetics* 247: 312–322.

Izsvàk, Z., Z. Ivics, D. Garcia-Estefania, S.C. Fahrenkrug and P.B. Hackett. 1996. DANA elements: a family of composite, tRNA-derived short interspersed DNA elements associated with mutational activities in zebrafish (*Danio rerio*). *Proceedings of the National Academy of Sciences of the United States of America* 93: 1077–1081.

Izsvàk, Z., Z. Ivics, N. Shimoda, D. Mohn, H. Okamoto and B.P. Hackett. 1999. Short inverted-repeat transposable elements in teleost fish and implications for a mechanism of their amplification. *Journal of Molecular Evolution* 48: 13–21.

Jaillon, O., J.-M. Aury, J.-L. Petit, N. Stange-Thomann, E. Mauceli, L. Bouneau, F. Brunet, C. Fischer, C. Ozouf-Costaz, A. Bernot, S. Nicaud, D. Jaffe, S. Fisher, G. Lutfalla, C. Dossat, B. Segurens, C. Dasilva, M. Salanoubat, M. Levy, N. Boudet, S. Castellano, V. Anthouard, C. Jubin, V. Castelli, M. Katinka, B. Vacherie, C. Biémont, Z. Skalli, L. Catolico, J. Poulain, S. Duprat, P. Brottier, J.-P. Coutanceau, J. Gouzy, G. Parra, G. Lardier, C. Chapple, K.J. McKernan, P. McEwan, S. Bosak, J.-N. Volff, R. Guigó, M. Zody, J. Mesirov, K. Lindblad-Toh, B. Birren, C. Nusbaum, D. Kahn, M. Robinson-Rechavi, V. Laudet, V. Schaechter, F. Quetier, W. Saurin, C. Scarpelli, P. Wincker, E.S. Lander, J. Weissenbach and H. Roest Crollius. 2004. Genome duplication in the teleost fish *Tetraodon nigroviridis* reveals the early vertebrate proto-karyotype. *Nature (London)* 431: 946–957.

Jordan, I.K., I.B. Rogozin, G.V. Glazko and E.V. Koonin. 2003. Origin of a substantial fraction of human regulatory sequences from transposable elements. *Trends in Genetics* 19: 68–72.

Kapitonov, V.V. and J. Jurka. 2003. EnSpm1_DR, an autonomous En/Spm DNA transposon from zebrafish. *Repbase Reports* 3–152.

Kapitonov, V.V. and J. Jurka. 2004. Harbinger transposons and an ancient *HARBI1* gene derived from a transposase. *DNA and Cell Biology* 23: 311–324.

Kapitonov, V.V. and J. Jurka. 2005. RAG1 core and V(D)J recombination signal sequences were derived from Transib transposons. *PLoS Biology* 3:e181 [Epub ahead of print].

Karpen, G.H. and A.C. Spradling. 1992. Analysis of subtelomeric heterochromatin in the *Drosophila* minichromosome Dp1187 by single P element insertional mutagenesis. *Genetics* 132: 737–753.

Kazazian H.H. Jr. 2004. Mobile elements: Drivers of genome evolution. *Science* 303: 1626–1632.

Kidwell, M.G. and K. Lisch. 2002. Transposable elements as sources of genomic variation. In: *Mobile DNA II*, N.L. Craig, R. Craigie, M. Gellert and A.M. Lambowitz (eds.). ASM Press, Washington pp. 59–90.

Koga, A. and H. Hori. 1999. Homogeneity in the structure of the medaka fish transposable element Tol2. *Genetic Research* 73: 7–14.

Kondo, M., I. Nanda, U. Hornung, M. Schmid and M. Schartl. 2004. Evolutionary origin of the medaka Y chromosome. *Current Biology* 14: 1664–1669.

Lippman, Z., A.V. Gendrel, M. Black, M.W. Vaughn, N. Dedhia, W.R. McCombie, K. Lavine, V. Mittal, B. May, K.D. Kasschau, J.C. Carrington, R.W. Doerge, V. Colot and R. Martienssen. 2004. Role of transposable elements in heterochromatin and epigenetic control. *Nature* (*London*) 430: 471–476.

Liu, Z., P. Li, H. Kucuktas and R. Dunham. 1999. Characterization of nonautonomous Tc1-like transposable elements of channel catfish (*Ictalurus punctatus*). *Fish Physiology and Biochemistry* 21: 65–72.

Liu, Z., P.H. Moore, H. Ma, C.M. Ackerman, M. Ragiba, Q. Yu, H.M. Pearl, M.S. Kim, J.W. Charlton, J.I. Stiles, F.T. Zee, A.H. Paterson and R. Ming. 2004. A primitive Y chromosome in papaya marks incipient sex chromosome evolution. *Nature* (*London*) 427: 348–352.

Lyon M.F. 2000. LINE-1 elements and X chromosome inactivation: A function for 'junk' DNA? *Proceedings of the National Academy of Sciences of the United States of America* 97: 6248–6249.

Lyozin, G.T., K.S. Makarova, V.V. Velikodvorskaja, H.S. Zelentsova, R.R. Khechumian, M.G. Kidwell, E.V. Koonin and M.B. Evgen'ev. 2001. The structure and evolution of Penelope in the *virilis* species group of *Drosophila*: an ancient lineage of retroelements. *Journal of Molecular Evolution* 52: 445–456.

Mandrioli, M. and G.C. Manicardi. 2001. Cytogenetic and molecular analysis of the pufferfish *Tetraodon fluviatilis* (Osteichthyes). *Genetica* 111: 433–438.

Mandrioli, M., G.C. Manicardi, N. Machella and V. Caputo. 2000. Molecular and cytogenetic analysis of the goby *Gobius niger* (Teleostei, Gobiidae). *Genetica* 110: 73–78.

Mantovani, M., L.D. dos Santos Abel, C.A. Mestriner and O. Moreira-Filho. 2000. Accentuated polymorphism of heterochromatin and nucleolar organizer regions in *Astyanax scabripinnis* (Pisces, Characidae): Tools for understanding karyotypic evolution. *Genetica* 109: 161–168.

Matsuda, M., Y. Nagahama, A. Shinomiya, T. Sato, C. Matsuda, T. Kobayashi, C.E. Morrey, N. Shibata, S. Asakawa, N. Shimizu, H. Hori, S. Hamaguchi and M. Sakaizumi. 2002. DMY is a Y-specific DM-domain gene required for male development in the medaka fish. *Nature* (*London*) 417: 559–563.

Miskey, C., Z. Izsvak, K. Kawakami and Z. Ivics. 2005. DNA transposons in vertebrate functional genomics. *Cellular and Molecular Life Sciences* 62: 629–641.

Morescalchi, M.A., L. Rocco and V. Stingo. 2002. Cytogenetic and molecular studies in a lungfish, *Protopterus annectens* (Osteichthyes, Dipnoi). *Gene* 295: 279–287.

Nanda, I., J.-N. Volff, S. Weis, C. Körting, A. Froschauer, M. Schmid and M. Schartl. 2000. Amplification of a long terminal repeat-like element on the Y chromosome of the platyfish, *Xiphophorus maculatus*. *Chromosoma* 109: 173–180.

Nanda, I., M. Kondo, U. Hornung, S. Asakawa, C. Winkler, A. Shimizu, Z. Shan, T. Haaf, N. Shimizu, A. Shima, M. Schmid and M. Schartl. 2002. A duplicated copy of *DMRT1* in the sex-determining region of the Y chromosome of the medaka, *Oryzias latipes*. *Proceedings of the National Academy of Sciences of the United States of America* 99: 11778–11783.

Nekrutenko, A. and W.H. Li. 2001. Transposable elements are found in a large number of human protein-coding genes. *Trends in Genetics* 17: 619–621.

Ogiwara, I., M. Miya, K. Ohshima and N. Okada. 1999. Retropositional parasitism of SINEs on LINEs: Identification of SINEs and LINEs in elasmobranchs. *Molecular Biology and Evolution* 16: 1238–1250.

Oliveira, C., J.S. Chew, F. Porto-Foresti, M.J. Dobson and J.M. Wright. 1999. A LINE2 repetitive DNA sequence from the cichlid fish, *Oreochromis niloticus*: Sequence analysis and chromosomal distribution. *Chromosoma* 108: 457–468.

Orgel, L.E. and F.H. Crick. 1980. Selfish DNA: The ultimate parasite. *Nature (London)* 284: 604–607.

Ota, K., Y. Tateno and T. Gojobori. 2003. Highly differentiated and conserved sex chromosome in fish species (*Aulopus japonicus*: Teleostei, Aulopidae). *Gene* 317: 187–193.

Ozouf-Costaz, C., J. Brandt, C. Körting, E. Pisano, C. Bonillo, J.-P. Coutanceau and J.-N. Volff. 2004. Genome dynamics and chromosomal localization of the non-LTR retrotransposons Rex1 and Rex3 in Antarctic fish. *Antarctic Science* 16: 51–57.

Pardue, M.L. and P.G. DeBaryshe. 2003. Retrotransposons provide an evolutionarily robust non-telomerase mechanism to maintain telomeres. *Annual Review of Genetics* 37: 485–511.

Peichel, C.L., J.A. Ross, C.K. Matson, M. Dickson, J. Grimwood, J. Schmutz, R.M. Myers, S. Mori, D. Schluter and D.M. Kingsley. 2004. The master sex-determination locus in threespine sticklebacks is on a nascent Y chromosome. *Current Biology* 14: 1416–1424.

Poulter, R. and M. Butler. 1998. A retrotransposon family from the pufferfish (fugu) *Fugu rubripes*. *Gene* 215: 241–249.

Poulter, R., M.I. Butler and J. Ormandy. 1999. A LINE element from the pufferfish (fugu) *Fugu rubripes* which shows similarity to the CR1 family of non-LTR retrotransposons. *Gene* 227: 169–179.

Poulter, R.T., T.J. Goodwin and M.I. Butler. 2003. Vertebrate helentrons and other novel helitrons. *Gene* 313: 201–212.

Radice, D.A., B. Bugaj, H.D. Fitch and W.S. Emmons. 1994. Widespread occurrence of the Tc1 transposon family: Tc1-like transposons from teleost fish. *Molecular and General Genetics* 244: 606–612.

SanMiguel, P., A. Tikhonov, Y.K. Jin, N. Motchoulskaia, D. Zakharov, A. Melake-Berhan, P.S. Springer, K.J. Edwards, M. Lee, Z. Avramova and J.L. Bennetzen. 1996. Nested retrotransposons in the intergenic regions of the maize genome. *Science* 274: 765-768.

Sarkar, A., C. Sim, Y.S. Hong, J.R. Hogan, M.J. Fraser, H.M. Robertson and F.H. Collins. 2003. Molecular evolutionary analysis of the widespread piggyBac transposon family and related 'domesticated' sequences. *Molecular Genetics and Genomics* 270: 173–180.

Schartl, M., U. Hornung, H. Gutbrod, J.-N. Volff and J. Wittbrodt. 1999. Melanoma loss-of-function mutants in *Xiphophorus* caused by *Xmrk*-oncogene deletion and gene disruption by a transposable element. *Genetics* 153: 1385–1394.

Schultheis, C., Q. Zhou, A. Froschauer, I. Nanda, Y. Selz, C. Schmidt, S. Matschl, M. Wenning, A.-M. Veith, M. Naciri, R. Hanel, I. Braasch, A. Dettai, A. Böhne, C. Ozouf-Costaz, S. Chilmonczyk, B. Ségurens, A. Couloux, S. Bernard-Samain, M. Schmid, M. Schartl and J.-N. Volff. 2006. Molecular analysis of the sex-determining region of the platyfish *Xiphophorus maculatus*. *Zebrafish* 3(3): 295–305.

Shen, C.H. and L.A. Steiner. 2004. Genome structure and thymic expression of an endogenous retrovirus in zebrafish. *Journal of Virology* 78: 899–911.

Shimoda, N., M. Chevrette, M. Ekker, J. Kikuchi, Y. Hotta and H. Okamoto. 1996. Mermaid: A family of short interspersed repetitive elements widespread in vertebrates. *Biochemical and Biophysical Resarch Communications* 220: 226–232.

Sola, L. and E. Gornung. 2001. Classical and molecular cytogenetics of the zebrafish, *Danio rerio* (Cyprinidae, Cypriniformes): An overview. *Genetica* 111: 397–412.

Steinemann, M. and S. Steinemann. 2000. Common mechanisms of Y chromosome evolution. *Genetica* 109: 105–111.

Stingo, V. and L. Rocco. 2001. Selachian cytogenetics: A review. *Genetica* 111: 329–347.

Traut, W. and H. Winking. 2001. Meiotic chromosomes and stages of sex chromosome evolution in fish: Zebrafish, platyfish and guppy. *Chromosome Research* 9: 659–672.

Terai, Y., N. Takezaki, W.E. Mayer, H. Tichy, N. Takahata, J. Klein and N. Okada. 2004. Phylogenetic relationships among East African haplochromine fish as revealed by short interspersed elements (SINEs). *Journal of Molecular Evolution* 58: 64–78.

van de Lagemaat, L.N., J.R. Landry, D.L. Mager and P. Medstrand. 2003. Transposable elements in mammals promote regulatory variation and diversification of genes with specialized functions. *Trends in Genetics* 19: 530–536.

Volff, J.-N. 2005. Genome evolution and biodiversity in teleost fish. *Heredity* 94: 280–294.

Volff, J.-N. and M. Schartl. 2001. Variability of genetic sex determination in poeciliid fishes. *Genetica* 111: 101–110.

Volff J.-N., C. Körting, K. Sweeney and M. Schartl. 1999. The non-LTR retrotransposon Rex3 from the fish *Xiphophorus* is widespread among teleosts. *Molecular Biology and Evolution* 16: 1427–1438.

Volff, J.-N., C. Körting and M. Schartl. 2000. Multiple lineages of the non-LTR retrotransposon Rex1 with varying success in invading fish genomes. *Molecular Biology and Evolution* 17: 1673–1684.

Volff, J.-N., U. Hornung and M. Schartl. 2001a. Fish retroposons related to the Penelope element of *Drosophila virilis* define a new group of retrotransposable elements. *Molecular Genetics and Genomics* 265: 711–720.

Volff, J.-N., C. Körting, J. Altschmied, J. Duschl, K. Sweeney, K. Wichert, A. Froschauer and M. Schartl. 2001b. Jule from the fish *Xiphophorus* is the first complete vertebrate Ty3/Gypsy retrotransposon from the Mag family. *Molecular Biology and Evolution* 18: 101–111.

Volff, J.-N., C. Körting, A. Froschauer, K. Sweeney and M. Schartl. 2001c. Non-LTR retrotransposons encoding a restriction enzyme-like endonuclease in vertebrates. *Journal of Molecular Evolution* 52: 351–360.

Volff, J.-N., C. Körting, A. Meyer and M. Schartl. 2001d. Evolution and discontinuous distribution of Rex3 retrotransposons in fish. *Molecular Biology and Evolution* 18: 427–431.

Volff, J.-N., L. Bouneau, C. Ozouf-Costaz and C. Fischer. 2003a. Diversity of retrotransposable elements in compact pufferfish genomes. *Trends in Genetics* 19: 674–678.

Volff, J.-N., C. Körting, A. Froschauer, Q. Zhou, B. Wilde, C. Schultheis, Y. Selz, K. Sweeney, J. Duschl, K. Wichert, J. Altschmied, M. Schartl. 2003b. The *Xmrk* oncogene can escape nonfunctionalization in a highly unstable subtelomeric region of the genome of the fish *Xiphophorus*. *Genomics* 82: 470–479.

Woram, R.A., K. Gharbi, T. Sakamoto, B. Hoyheim, L.E. Holm, K. Naish, C. McGowan, M.M. Ferguson, R.B. Phillips, J. Stein, R. Guyomard, M. Cairney, J.B. Taggart, R. Powell, W. Davidson and R.G. Danzmann. 2003. Comparative genome analysis of the primary sex-determining locus in salmonid fishes. *Genome Research*. 13: 272–280.

Yoder, J.A., C.P. Walsh and T.H. Bestor. 1997. Cytosine methylation and the ecology of intragenomic parasites. *Trends in Genetics* 13: 335–340.

Zhou, Q., A. Froschauer, C. Schultheis, C. Schmidt, P. Bienert, M. Wenning, A. Dettai and J.-N. Volff (2006). *Helitron* transposons on the sex chromosomes of the platyfish *Xiphophorus maculatus* and their evolution in animal genomes. *Zebrafish* 3(1): 39–52.

Ziegler, C.G., D.K. Lamatsch, C. Steinlein, W. Engel, M. Schartl and M. Schmid. 2003. The giant B chromosome of the cyprinid fish *Alburnus alburnus* harbours a retrotransposon-derived repetitive DNA sequence. *Chromosome Research* 11: 23–35.

Polyploidy in Acipenseriformes: Cytogenetic and Molecular Approaches

**Francesco Fontana[1*], Lorenzo Zane[2],
Anastasia Pepe[2] and Leonardo Congiu[1]**

INTRODUCTION

The order Acipenseriformes is the only extant one among Chondrostea whose origin dates back as early as Jurassic (about 200 millions years ago). Phylogenetic ancestry and anatomical characteristics make the order especially relevant for the reconstruction of evolutionary history of vertebrates (Bemis *et al.*, 1997).

Acipenseriformes include two families, Acipenseridae (25 species) and Polyodontidae (2 species). The first one, Acipenseridae, is commonly divided into two subfamilies, Scaphyirhynchinae and Acipenserinae (Rochard *et al.*, 1991; Bemis *et al.*, 1997). The subfamily

Address for Correspondence: *E-mail: fon@unife.it

[1]Dipartimento di Biologia, Università di Ferrara, Via L. Borsari 46, 44100 Ferrara, Italy.

[2]Dipartimento di Biologia, Università di Padova, Via Ugo Bassi 58/B, 35121, Padova, Italy.

Scaphyirhynchinae, widespread in North America and Central Asia, includes two genera: *Scaphirhynchus* (with the species *S. platorynchus, S. albus* and *S. sutkusi*, Mississippi basin and Gulf of Mexico) and *Pseudoscaphirhynchus* (with the species *P. kaufmanni, P. hermanni* and *P. fedtschenkoi*, Aral Sea basin). The subfamily Acipenserinae contains the genera *Acipenser* (17 species) and *Huso* (2 species). The family Polyodontidae includes the freshwater species *Psephurus gladius* and *Polyodon spathula*.

According to the data by Artyukhin and Andronov (1990) and Artyukhin (1995), the spreading of Acipenserinae began in four main regions: Ponto-Caspian, China and western North American, Atlantic and North American ones. The Ponto-Caspian region shows the highest diversity among species. The groups in this area include the European-Asian species, *A. gueldenstaedtii, A. persicus, A. stellatus, A. ruthenus, A. nudiventris, A. baerii, A. sturio, Huso huso* (Pirogovsky *et al.*, 1989), the two species endemic to the Amur river, *A. schrenckii, A. mikadoi* and *H. dauricus* (Artyukhin, 1994) and the one endemic to Adriatic Sea, *A. naccarii* (Tortonese, 1989; Rossi *et al.*, 1991). The second group, widespread in China and western North America, includes the species *A. sinensis, A. dabryanus, A. medirostris* and *A. transmontanus* (Findeis, 1993). The third group includes sturgeons of the Atlantic area, on both European and American sides, *A. sturio* and *A. oxyrinchus* (Vladykov and Greeley, 1963). The fourth geographical group includes the freshwater sturgeons of North Eastern America, *A. fulvescens* and *A. brevirostrum* (Artyukhin and Andronov, 1990). Presently, the relationships among these four groups of species remain unknown (Birstein *et al.*, 1997).

The research on systematic relationships among the above species is complicated by the presence in the wild of interspecific and intergeneric hybrids and by the widespread anatomic and ontogenetic polymorphism characterizing the family Acipenseridae (Berg, 1962). The most common taxonomical criteria are based on morphological, ecological and ethological features and on the biogeographical distribution of species (Rochard *et al.*, 1991; Artyukhin, 1995).

Recently, cytogenetic and molecular data gave further insights into species differentiation processes within the order, suggesting new taxonomical and phylogenetic relationships among species groups. The most relevant information concerns a peculiarity of sturgeons, namely polyploidization events. However, not all authors agree on the ploidy degree and events leading to it. In this chapter, we will discuss this issue on the basis of the latest cytogenetic and molecular biology techniques.

Cytogenetics

First data on the nuclear DNA amount were obtained in sturgeons around the 1950s. Mirsky and Ris (1951) and later Vialli (1957) observed through histophotometric methods that A. *sturio* exhibited a nuclear DNA amount of 3.2 pg, a significantly higher value than the modal one previously found in teleosteans. This result supported the hypothesis that in teleosteans, the more primitive species had a higher DNA amount than the more recent and more specialized ones.

The first data on sturgeon chromosome number were obtained around 1965 by Russian authors, especially Serebryakova and co-workers (see review by Serebryakova, 1972). These authors observed 60 chromosomes in both H. *huso* and A. *ruthenus*. However, the metaphase plates on which they scored the results were of low quality, since they were obtained by squash of blastomeres and gill mucosa cells.

At around the same time, by more efficient techniques, Ohno *et al.* (1969) observed that the karyotype of shovelnose sturgeon, *Scaphirhynchus platorynchus*, was composed of 112 chromosomes, 48 of which were microchromosomes. They also observed that the nuclear DNA amount was 3.6 pg. The large difference in chromosome number in comparison to what had been observed by the Russian researchers was probably due to the low resolution of the former techniques, which did not allow one to observe microchromosomes. The newly discovered high chromosome number, together with the high amount of nuclear DNA, led Ohno and co-workers to cautiously advance the hypothesis that S. *platorynchus* was tetraploid. However, they stressed the fact that the hypothesis could have gained support only when 60-chromosome species could be identified.

Further cytogenetic advances were made possible by studies on Italian sturgeons. By the air-drying method in combination with pre-treatment with mitotic inhibitors, Fontana and Colombo (1974) found that H. *huso* and A. *sturio* had a karyotype of about 116 chromosomes and A. *naccarii* of about 240. All karyotypes showed many microchromosomes. Also, the Danube basin species A. *ruthenus* exhibited a similar karyotype to H. *huso* and A. *sturio* (Fontana *et al.*, 1975). A study of nuclear DNA amount of the three species of Italian sturgeons yielded the following result: H. *huso*

3.60 pg, *A. sturio* 3.58 pg and *A. naccarii* 6.26 pg (Fontana, 1976). The data suggested that all sturgeon species studied till that time had a chromosome number of 110-120 and a nuclear DNA amount of 3.2-3.6 pg, except for one (*A. naccarii*) which had a chromosome number and a DNA amount twice as high.

In 1976, Dingerkus and Howell found that the North American paddlefish, *Polyodon spathula*, had a karyotype 2n=120. The authors believed that chromosomes could be gathered into groups of 4 purely on morphological basis. Thus, they stated that the species was tetraploid. The presence of 4 nucleoli in interphase nuclei of *P. spathula* apparently supported their conclusion. However, later one of the authors, on the basis of a report attributing 240 chromosomes to *A. naccarii* (Fontana and Colombo, 1974), stated that the species should have been octoploid (Dingerkus, 1979).

Around 1980, the karyotypic data on sturgeons increased, mostly by works of Russian researchers:

H. dauricus	2n = 120	Burtzev *et al.*, 1976
A. baerii	2n = 250	Vasil'ev *et al.*, 1980
A. schrenckii	2n = 240	Vasil'ev *et al.*, 1980
A. nudiventris	2n = 118	Arefjev, 1983
A. stellatus	2n = 118	Birstein and Vasil'ev, 1987
A. gueldenstaedtii	2n = 250	Birstein and Vasil'ev, 1987
A. sinensis	2n = 264	Yu *et al.*, 1987

All these data supported the presence of two sturgeon groups, one characterized by a karyotype with 116-120 chromosomes (group A) and the other with 240-260 chromosomes (group B). Thus, all authors maintained that the ploidy relationship between the two groups was a tetra-octoploid one. However, even on the basis of questionable cytological markers (such as the the number of nucleoli per nucleus), some data clearly did not agree with the above hypothesis. In this case, the supporters of the tetra-octoploid hypothesis justified the results by rather unclear diploidization processes (Birstein and Vasil'ev, 1987). Only Arefjev (1983) expressed doubts about the fact that the ship sturgeon, *A. nudiventris*, with 118 chromosomes, could be considered tetraploid, since it was impossible to divide the karyotype in groups of 4 similar chromosomes.

The onset of flow cytometry techniques made quantitative determination of nuclear DNA content easier and faster: the large amount of data collected made them more reliable. The genome size of seven species of American sturgeons was found to range from 4.6 to 13.1 pg (Blacklidge and Bidwell, 1993). The authors then divided the 7 species into 3 groups, according to a 1:2:3 ratio. They considered tetraploid to be the species with the lowest value, octaploid the intermediate ones and dodecaploid those with the highest number. Slightly different results were obtained on 10 sturgeon species from American and Eurasian regions by Birstein et al. (1993). They also gathered the 10 species into three ploidy groups, although slightly different (4n, 8n and 16n). According to these data, the 16n North Eastern Asian species, A. mikadoi, for which no cytogenetic data are available, should have about 500 chromosomes. Apart from discussions on the last group, all researchers maintained that group A species were tetraploid and group B ones octoploid. Table 4.2.1 reports the chromosome number and nuclear DNA content in the three ploidy groups.

Chromosome-banding techniques, developed in the early 1970s, greatly contributed to the knowledge of vertebrate karyotype. However, the lack of compartimentalization in fish DNA base composition greatly limited the application of such techniques (Medrano et al., 1988). A significant contribution to the understanding of ploidy relationships between A and B sturgeon groups was provided by chromosome staining with silver nitrate to show ribosomal gene activity and identify nucleolar organizing regions (NORs) (Goodpasture and Bloom, 1975). The NOR number in group A species ranges from 4 to 6, while in group B ranges from 8 to 13 (Fontana et al., 2001). Although it is difficult to identify fish chromosome morphology, because of their small size, it is clear that in group A species, the NOR chromosomes can be gathered in pairs (Fig. 4.2.1A,a), while in group B ones they can be gathered in sets of four similar chromosomes (Fig. 4.2.1A,b,c).

Further insights about the ploidy level of sturgeons can be inferred by fluorescence *in situ* hybridization (FISH). A satellite DNA family, isolated from the genome of A. naccarii by HindIII restriction enzyme treatment, was employed as FISH probe on metaphase plates of 8 species. The hybridization signals were clearly recognizable on the centromeric regions of all species examined, except for one (A. sturio). In group A species, the signals were detected in pericentromeric regions of 8-10 chromosomes,

Table 4.2.1 Chromosome number and DNA content in the Acipenseriformes.

Species	Chromosome numbers	DNA content in pg	References
Group A			
Acipenser nudiventris	2n = 118±3	–	Arefjev, 1983
	–	3.9	Birstein et al., 1993
Acipenser oxyrinchus	2n = 99-112	–	Li et al., 1985
	–	4.55	Blacklidge and Bidwell, 1993
Acipenser ruthenus	2n = 116±4	–	Fontana et al., 1975
	–	3.74	Birstein et al., 1993
Acipenser stellatus	2n = 118±2	–	Birstein and Vasil'ev, 1987
	–	3.74	Birstein et al., 1993
Acipenser sturio	2n = 116±4	–	Fontana and Colombo, 1974
	–	3.2	Mirsky and Ris, 1951
Huso dauricus	2n = 120	–	Burtzev et al., 1976
	–	3.78	Birstein et al., 1993
Huso huso	2n = 116±4	–	Fontana and Colombo, 1974
	–	3.6	Fontana, 1976
Scaphirhynchus	2n = 112	–	Ohno et al., 1969
platyrhynchus	–	3.6	Ohno et al., 1969
Polyodon spathula	2n = 120	–	Dingerkus and Howell, 1976
	–	3.9	Tiersch et al., 1989
Group B			
Acipenser baerii	2n = 249±5	–	Vasil'ev et al., 1980
	–	8.3	Birstein et al., 1993
Acipenser fulvescens	2n = 262±6	–	Fontana et al., 2004
	–	8.9	Blacklidge and Bidwell, 1993
Acipenser gueldenstaedtii	2n = 250±8	–	Birstein and Vasil'ev, 1987
	–	7.87	Birstein et al., 1993
Acipenser medirostris	2n = 249±8	–	Van Eenennaam et al., 1999
	–	8.82	Blacklidge and Bidwell, 1993
Acipenser naccarii	2n = 239±7	–	Fontana and Colombo, 1974
	–	6.26	Fontana, 1976
Acipenser persicus	2n =258±2	–	Nowruzfashkhami et al., 2000
Acipenser schrenckii	2n = 240	–	Vasil'ev et al., 1980
	–	6.07	Zhang et al., 1999
Acipenser sinensis	2n = 264±4	–	Yu et al., 1987
	–	9.07	Zhang et al., 1999
Acipenser transmontanus	2n = 271	–	Van Eenennaam et al., 1998
	–	9.55	Blacklidge and Bidwell, 1993
Group C			
Acipenser mikadoi	–	14.2	Birstein et al., 1993
Acipenser brevirostrum	2n = 372	–	Kim et al., 2005
	–	13.08	Blacklidge and Bidwell, 1993

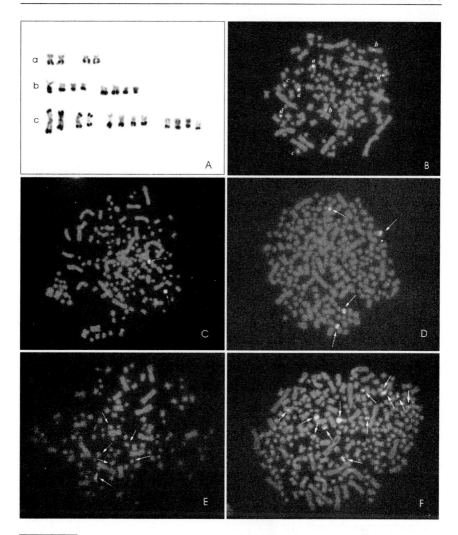

Fig. 4.2.1 (A) Grouping of NOR chromosomes after Ag-staining: (a) *H. huso*, in pairs; (b) and (c) respectively, *A. naccarii* and *A. gueldenstaedtii*, in quadruplets. (B) Location of the HindIII satellite DNA in *H. huso* by FISH: the chromosomes with hybridization signals are grouped in 4 morphologically identifiable pairs (*a, b, c, d*). (C) and (D) Location of 5S rDNA by FISH, respectively, in *H. huso* and *A. naccarii*: arrows indicate hybridization signals. (E) and (F) Location of major ribosomal genes, after FISH with a 28S rDNA probe, respectively, in *H. huso* and *A. naccarii*: arrows indicate hybridization signals.

while in group B species, the signals ranged from 40 to 80 (Lanfredi *et al.*, 2001; Fontana *et al.*, 2004). By this technique, it was possible to identify the morphology of the 4 chromosome pairs showing hybridization signals in *H. huso* (Fig. 4.2.1B).

Moreover, a probe constructed with 5S ribosomal DNA sequences, labelled with digoxygenin and hybridized by FISH on metaphase plates of group A species, showed intense fluorescence signals on telomeric regions, respectively of 2 chromosomes in group A (Fig. 4.2.1C) and 4 chromosomes in group B (Fig. 4.2.1D). The *in situ* hybridization by a probe with 18S and 28S ribosomal DNA sequences showed signals interspersed in telomeric regions of 6-8 chromosomes (Fig. 4.2.1E). In group B species, the signals were found on 10-12 chromosomes (Fig. 4.2.1F).

All these cytogenetic data strongly support the hypothesis that sturgeons with 120-chromosomes belonging to group A are diploid, while those with 240-260 chromosomes belonging to group B are tetraploid.

Recently, the karyotype of *A. brevirostrum* (2n=372) has been described by Kim *et al.* (2005). Blacklidge and Bidwell (1993) previously observed that this species has a nuclear DNA content of 13.08 pg, that is 2.78 times the mean content of group A and 1.44 times the mean content of group B. According to these authors, *A. brevirostrum* is a derived dodecaploid (12C) species whose origin could be either a triploidization event or, more likely, a hybridization between a group A and a group B species. However, Birstein *et al.* (1993) maintain that *A. brevirostrum* is 16-ploid, as (so they state) *A. mikadoi*. The results by Kim *et al.* support the hypothesis of Blacklidge and Bidwell (1993) but they do not choose between a hexaploid or a dodecaploid *A. brevirostrum*. More recently Fontana and coworkers, beside supporting the karyotype data by Kim *et al.*, also observed by *in situ* 5S rDNA hybridization on *A. brevirostrum* that 6 chromosomes exhibited hybridization signals, while the same signals were 2 on group A species and 4 on group B ones (unpublished). These results support the hypothesis that *A. brevirostrum* is a hexaploid species, probably a hybrid between one group A and one group B species, as suggested by Blacklidge and Bidwell (1993).

Figure 4.2.2 summarizes our hypothesis concerning the speciation events occurred in sturgeons by genome duplication, diploidization and hybridization. The first chromosome duplication marks the origin of Actinopterigia: the chromosome number probably increased from 2n=60 to 4n=120. This event is believed to have occurred at least 200 Myr ago, since no Acipenseridae species with 60 chromosomes presently survives. Later, these tetraploid species underwent a diploidization event which originated diploid ones (2n=120). The diploid species evolved into different ones through gene diversification and duplication events.

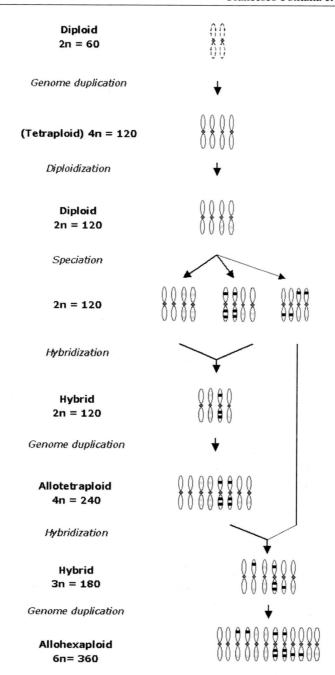

Diploid
2n = 60

Genome duplication

(Tetraploid) 4n = 120

Diploidization

Diploid
2n = 120

Speciation

2n = 120

Hybridization

Hybrid
2n = 120

Genome duplication

Allotetraploid
4n = 240

Hybridization

Hybrid
3n = 180

Genome duplication

Allohexaploid
6n= 360

Fig. 4.2.2 Sturgeon speciation hypothesis through genome duplication, diploidization, divergent evolution and hybridization.

Hybridization among the different diploid species resulted in hybrids which, after genome duplication, became allotetraploid (4n=240-260). This process may have independently occurred several times. In at least one tetraploid species, the white sturgeon (*A. transmontanus*), the allotetraploidy is supported by the study of synaptonemal complexes, which did not show any multivalent in the first meiotic prophase (Van Eenennaam, 1998). The above mentioned recent data on cytogenetic features of *A. brevirostrum* suggest a third polyploidization event caused by hybridization of a 2n=120 species with a 2n=240 one, followed by genome duplication: thus *A. brevirostrum* became allohexaploid (2n=360). Since hybrids are easily produced in the wild among sturgeons, we can not rule out the possibility that other polyploidization events may have occurred, for example between two group B species, originating a species with 2n= about 500.

Molecular Genetics

Apart from the direct visualization of chromosomes that characterizes the cytogenetic approach, an indirect analysis of polyploidy in sturgeons may be attempted by molecular biology techniques. Several phylogenetic reconstructions based on molecular markers are available for sturgeons (Birstein *et al.*, 1997, 1998; Krieger *et al.*, 2000; Zhang *et al.*, 2000; Ludwig *et al.*, 2001). Therefore, the evolution of ploidy can be studied by mapping the chromosome numbers onto the phylogenetic trees, following the approach first introduced by Maddison and Maddison (1992). Recently, this approach has been used by Ludwig and coworkers (2001), producing a molecular phylogeny for 22 out of 27 species of Acipenseriformes based on the entire mitochondrial gene sequence of cytochrome b. Their results indicate that *P. spathula* is a basal genus, reflecting the ancient divergence of the families Polyodontidae and Acipenseridae. Within Acipenseridae, three separate clusters are recognizable. The first one includes three Scaphirhynchinae species, the second is formed by the *A. oxyrinchus/A. sturio* clade, and the third includes all the other species. The third group is composed of at least three well-supported clusters, partially reflecting the geographical distribution of the species (Ludwig *et al.*, 2001).

Based on the superimposition of the chromosome numbers to the molecular-based phylogenetic tree, all three basal clusters (Polyodontidae, Scaphirhynchinae and the *A. oxyrinchus/A. sturio* clade) were found to include only species with 120 chromosomes. Accordingly, the 120 chromosome condition can be considered to be ancestral for Acipenseriformes. Three independent events of genome duplication must be invoked to explain the 120/240 chromosome species distribution across tree topology (Ludwig *et al.*, 2001). A fourth event of genome duplication is required to reach the putative 500 chromosome condition of A. *mikadoi*. This reconstruction shows how polyploidization events may have played a relevant role in the evolution of Acipenseriformes. However, the above approach does not clarify the debated ploidy level associated to each chromosome number.

Molecular markers can be used to directly infer the ploidy level of the genome by analysing the maximum number of alleles found at a given locus in one individual. For example, a maximum of two alleles are expected in a diploid locus, while a tetraploid locus should have a maximum of four alleles. This approach, however, provided contradictory results when applied on DNA coding regions or directly on protein markers. For example, gene duplications have been found with allozymes in 120-chromosome species, up to 6% in *P. spathula* (Carlson *et al.*, 1982) and 31% in A. *stellatus* (Ryabova and Kutergina, 1990). These findings have been considered to support a tetraploid ancestral condition (Birstein *et al.*, 1997). On the contrary, the analysis of protein coding genes revealed only two forms of vitellogenin and growth hormone genes in two 240-chromosome species (Bidwell *et al.*, 1991; Yasuda *et al.*, 1992), suggesting a diploid condition. Moreover, two forms of proopiomelanocortin were found in *P. spathula* and in A. *transmontanus* (Danielson *et al.*, 1999), respectively, a 120- and a 240-chromosome species. Nevertheless, we should consider that the protein-coding genes (and consequently the allozyme markers) can be exposed to high selective pressure and that their application as evolutionary markers can be biased, mainly for what concerns the numbers of variants.

The above limits can be overcome by using noncoding DNA markers. Accordingly, the most widely used markers applied to the analysis of ploidy level are microsatellites. These are tandemly repeated motifs of 1-6 bases found in all prokaryotic and eukaryotic genomes analyzed to date, and mainly present in noncoding regions. They are usually characterized by a high degree of length polymorphism, very likely due to slippage events during DNA replication (Schlötterer and Tautz, 1992): this makes them an extremely valuable tool for the study of genetic variability (Zane *et al.*, 2002).

Microsatellite markers have been first isolated from a 240-chromosome species, *A. fulvescens*, and tested for cross amplification on several species of the *Acipenser* and *Scaphirhynchus* genera (May *et al.*, 1997) (Table 4.2.2). Six of these microsatellites have been used to amplify genomic DNA of eight 120-chromosome species, providing no more than two alleles per individual (Ludwig *et al.*, 2001) (Table 4.2.2), thus strongly supporting the diploid condition of the group. Nine 240-chromosome species were analyzed at the same set of loci. In this case, four of these six loci showed a maximum of four alleles per individual, thus supporting the tetraploid condition of this group of species. The other two loci showed a maximum of eight alleles, apparently supporting a higher ploidy level. However, the same authors reported that the latter two loci were duplicated, therefore introducing a bias in the information.

However, the microsatellite approach can be affected by some problems. First, the presence of no more than two alleles may indicate diploidy at the locus, but that in itself does not contradict an higher ploidy level. Second, a single locus duplication may result in an apparent polyploidy. Third, multilocus analysis can provide contrasting information about ploidy level across loci, because after genome duplication events, relaxed selection allows a functional reduction of ploidy. Consequently, the sequence divergence between duplicated loci may prevent amplification by PCR.

Besides the study of allele number to directly infer the ploidy level, microsatellites have also been applied in segregation analyses to identify disomic, tetrasomic or octasomic transmission patterns in *A. fulvescens* (Pyatskowit *et al.*, 2001; McQuown *et al.*, 2002). Discrimination between disomic and tetrasomic inheritance may represent an indirect way to know whether a tetraploid species originated by allotetraploidy or autotetraploidy. Indeed, disomic inheritance is typical of allotetraploids, in which the homologous chromosomes from the two parental species are so different from one another to prevent a random segregation (Olson, 1997), while tetrasomic inheritance is typical of autotetraploids, in which the four homologous chromosomes are structurally identical. In allotetraploidy, actually, the four chromosomes bearing the same tetraploid locus functionally behave as bearing two diploid loci, thus only one chromosome from each pair segregates into each gamete. This latter

Table 4.2.2 Microsatellite loci isolated and tested by May *et al.* (1997) (**M**). Some of these loci were also used by Ludwig *et al.* (2001) (**L**). **p** = polymorphic; **m** = monomorphic; **n** = no activity

Locus	LS19		LS22		LS23		LS34		LS39		LS54		LS57		LS58		LS62		LS68		LS69	
Species	M	L	M	L	M	L	M	L	M	L	M	L	M	L	M	L	M	L	M	L	M	L
A. baerii	–	p	–	–	–	–	–	p	–	p	–	p	–	n	–	–	–	–	–	p		
A. brevirostrum	p	p	n	–	m	–	p	p	p	p	p	p	p	n	m	–	p	–	p	p	–	–
A. fulvescens	p	p	p	–	m	–	m	n	p	p	m	p	p	p	m	–	p	–	p	p	p	–
A. gueldenstaedtii	–	p	–	–	–	–	–	p	–	p	–	p	–	n	–	–	–	–	–	p	p	–
A. medirostris	p	p	m	–	m	–	m	p	m	n	m	p	p	p	m	–	p	–	p	p	–	–
A. mikadoi	–	p	–	–	–	–	–	p	–	n	–	m	–	p	–	–	–	–	–	p	n	–
A. naccarii	–	p	–	–	–	–	–	p	–	p	–	p	–	n	–	–	–	–	–	p	–	–
A. nudiventris	–	p	n	–	–	–	–	m	–	m	–	p	–	n	m	–	–	–	–	p	–	–
A. oxyrinchus oxyrinchus	p	p	p	–	m	–	m	p	m	p	p	p	m	n	–	–	p	–	p	p	m	–
A. persicus	–	p	–	–	–	–	–	p	–	p	–	p	–	n	–	–	–	–	–	p	–	–
A. ruthenus	–	p	–	–	–	–	–	p	–	p	–	n	–	n	–	–	–	–	–	p	–	1
A. schrenkii	–	p	–	–	–	–	–	p	–	m	–	p	–	n	–	–	–	–	–	p	–	–
A. sinensis	–	p	–	–	–	–	–	p	–	p	–	p	–	n	–	–	–	–	–	p	–	–
A. stellatus	–	p	–	–	–	–	–	p	–	m	–	p	–	n	–	–	–	–	–	p	–	–
A. sturio	–	p	–	–	–	–	–	p	–	p	–	p	–	n	–	–	–	–	–	p	–	–
A. transmontanus	m	n	n	n	n	–	m	p	m	m	m	m	p	n	m	–	p	–	p	p	m	–
H. huso	–	p	n	–	–	–	–	p	–	p	–	p	–	n	–	–	–	–	–	p	–	–
H. dauricus	–	p	–	–	–	–	–	p	–	p	–	p	–	n	m	–	–	–	–	p	–	–
S. albus	p	m	n	m	n	–	p	p	m	p	p	m	p	p	m	–	p	–	p	p	m	–
S. platorynchus	p	p	n	p	n	–	p	p	m	p	p	p	p	p	m	–	p	–	p	p	m	–
A. oxyrinchus desotoi	p	–	n	–	n	–	m	–	p	–	–	–	m	–	m	–	m	–	p	–	p	–

result may be confused with the one yielded by a re-diploidized autopolyploid species, which also shows a disomic pattern; in this case, however, a maximum of two alleles per individual is expected. The presence of more than two alleles at a given locus in some individuals allows to exclude a diploid condition.

In A. *fulvescens*, a 240-chromosome species, a total of 10 microsatellite loci have been investigated for what concerns their inheritance pattern. Of these, 6 have been found to follow a disomic inheritance model, whereas 4 have been considered tetrasomic. No evidence of octoploid inheritance was found (Pyatskowit et al., 2001), but aberrant genotypes with more alleles than expected were found in some crosses (McQuown et al., 2002). Unfortunately, these studies rely on inference of parental genotypes by gene dosage estimation through band intensity (Jenneckens et al., 2001), which is difficult to perform. In addition, a high sample size (hard to obtain in sturgeons) is required to discriminate between allele transmission frequencies expected with different models of inheritance (Rodzen and May, 2002).

A more powerful approach could be first to predict all possible parental genotypes compatible with a given banding pattern (phenotype), for any possible ploidy level, disregarding the gene dosage determination by indirect methods. Second, for all possible combinations of parental genotypes, to determine the banding pattern (or the allele frequencies) expected in the progeny, based on a given inheritance model. Finally, for any parental genotype combination, to compare the observed frequencies of progeny phenotypes with the expected ones by a significance test. Similar approaches are under development for plants, in which details of transmission mechanisms in polyploids have always been considered relevant (Luo et al., 2000).

Six microsatellite loci have been tested for tetrasomic inheritance in A. *naccarii*, a 240-chromosome species, using the above approach implemented in TetraploidMap software (Luo et al., 2000). The study was performed on two progenies from two independent crosses, respectively, consisting of 90 and 65 individuals. All bands observed in the progeny were always present in the parents and no unexpected profiles were found. The results allowed to consistently reject a tetrasomic inheritance model at all loci. The approach is now under development to test disomic inheritance pattern and more complex models of segregation.

In conclusion, all results so far obtained with molecular genetic approaches to segregation analysis are still preliminary. On the contrary, reliable data on the ploidy level have been obtained by analyzing the maximum number of microsatellite alleles per individual. The available results provide strong support for an ancestral diploid condition of 120-chromosome species and suggest a three-fold independent evolution of a tetraploid condition in the 240-chromosome species.

CONCLUDING REMARKS

So far, most of the available data about the sturgeon ploidy are provided by cytogenetic analyses, which strongly suggest that the group of species with 120 and 240 chromosome are, respectively, diploid and tetraploid, at least at the functional level. Although supported by the majority of available molecular data, these results are based on a limited number of markers; thus their extension to the whole genome is not granted. The sturgeon genome should probably be considered as a highly dynamic system, modeled by different evolutionary processes such as interspecific hybridization and functional reduction of ploidy. Different levels of ploidy in different parts of the genome cannot be excluded, as also supported by some molecular results. To constrain these heterogeneous genomes into predefined models may be limiting. For this reason, we think that the above results should be supported by a higher number of cytogenetic markers, with the aim of collecting information from as many chromosomes as possible. Unfortunately, a clear *in situ* detection requires the labelling of several thousands of nucleotides. Thus, only tandemly repeated sequences have been used as markers up to date. The possibility to construct, screen, amplify and label very long fragments (20-30.000 bp) from cosmidic or BAC libraries, to be employed as locus specific probes has never been explored and could represent a rich source of new markers. Again, the now available techniques for chromosome microdissection can help in producing FISH probes, specific for entire chromosomes or for limited regions (Bohlander et al., 1992). By these methods, it would be possible to identify possible synthenic groups with other species and chromosome rearrangements, even in interphase cells. After extending the application of these new techniques to other sturgeon species yet to be investigated, more details on polyploidization events may be expected in the very near future.

References

Arefjev, V.A. 1983. Polykaryogram analysis of ship, *Acipenser nudiventris* Lovetsky (Acipenseridae, Chondrostei). *Voprosy Ichthyology* 23: 209–216. (In Russian).

Artyukhin, E.N. 1994. On the relationships within of the Amur sturgeon, *Acipenser schrencki*. *The Sturgeon Quarterly* 2: 7.

Artyukhin, E.N. 1995. On biogeography and relationships within the genus *Acipenser*. *The Sturgeon Quarterly* 3: 6–8.

Artyukhin, E.N. and A.E. Andronov. 1990. A morphological study of the green sturgeon, *Acipenser medirostris* (Chondrostei, Acipenseridae), from the Tumnin (Datta) River and some aspects of the ecology and zoogeography of Acipenseridae. *Zoologicheskii Zhurnal* 69: 81–91.

Bemis, W.E. and B. Kynard. 1997. Sturgeon rivers: An introduction to acipenseriform biogeography and life history. *Environmental Biology of Fishes* 48: 167–184.

Berg, L.S. 1962. *Freshwater Fishes of the USSR and Adjacent Countries*, Vol. 1. (Translated by Israel Program for Scientific Translation, Jerusalem) Oldbourne Press, London.

Bidwell, C.A., K.J. Kroll, E. Severud, S.I. Doroshov and D.M. Carlson. 1991. Identification and preliminary characterization of white sturgeon (*Acipenser transmontanus*) vitellogenin mRNA. *General and Comparative Endocrinology* 83: 415–24.

Birstein, V.J. and R. DeSalle. 1998. Molecular phylogeny of Acipenserinae. *Molecular Phylogenetics and Evolution* 9: 141–155.

Birstein, V.J. and V.P. Vasil'ev. 1987. Tetraploid-octoploid relationships and karyological evolution in the order Acipenseriformes (Pisces): karyotypes, nucleoli, and nucleolus-organizer regions in four acipenserid species. *Genetica* 73: 3–12.

Birstein, V.J., A.I. Poletaev, and B.F. Goncharov. 1993. The DNA content in Eurasian sturgeon species determined by flow cytometry. *Cytometry* 14: 337–383.

Birstein, V.J., R. Hanner and R. DeSalle. 1997. Phylogeny of the Acipenseriformes: cytogenetic and molecular approaches. *Environmental Biology of Fishes* 48: 127–155.

Blacklidge, K.H. and C.A. Bidwell. 1993. Three ploidy levels indicated by genome quantification in Acipenseriformes of North America. *The Journal of Heredity* 84: 427–430.

Bohlander, S.K., R. Espinosa, M.M. Le Beau, J.D. Rowley and M.O. Diaz. 1992. A method for the rapid sequence-independent amplification of microdissected chromosomal material. *Genomics* 13: 1322–1324.

Burtzev, J.A., J. Nikoljukin and E.V. Serebryakova. 1976. Karyology of the Acipenseridae family in relation to the hybridization and taxonomy problems. *Acta Biologica Jugoslavica. Serija E. Ichthyologia* 8: 27–34.

Carlson, D.M., M.K. Kettler, S.E. Fisher and G.S. Whitt. 1982. Low genetic variability in paddlefish populations. *Copeia* 1982: 721–725.

Danielson, P.B., J. Alrubaian, M. Muller, J.M. Redding and R.M. Dores. 1999. Duplication of the POMC gene in the paddlefish (*Polyodon spathula*): Analysis of gamma-MSH, ACTH, and beta-endorphin regions of ray-finned fish POMC. *General and Comparative Endocrinology* 116: 164–177.

Dingerkus, G. 1979. Chordate cytogenetic studies: An analysis of their phylogenetic implications with particular reference to fishes and the living coelacanth. *Occasional Papers of the California Academy of Sciences* 134: 111–127.

Dingerkus, G. and W.M. Howell. 1976. Karyotypic analysis and evidence of tetraploidy in the North American paddlefish, *Polyodon spathula. Science* 177: 664–669.

Findeis, E.K. 1993. Osteology of the North American shovelnose sturgeon *Scaphirhynchus platorynchus* Rafinesque 1820, with comparisons to other Acipenseridae and Acipenseriformes. Ph.D. Thesis. University of Massachusetts, Amherst.

Fontana, F. 1976. Nuclear DNA content and cytometric of erythrocytes of *Huso huso* L., *Acipenser sturio* L. and *Acipenser naccarii* Bonaparte. *Caryologia* 29: 127–138.

Fontana, F. and G. Colombo. 1974. The chromosomes of Italian sturgeons. *Experientia* 30: 739–742.

Fontana, F., D. Jankovic and S. Zivkovic. 1975. Somatic chromosome of *Acipenser ruthenus* L. *Arhiv Bioloskih Nauka Beograd* 27: 33–35.

Fontana, F., J. Tagliavini and L. Congiu. 2001. Sturgeon genetics and cytogenetics: Recent advancements and perspectives. *Genetica* 111: 359–373.

Fontana, F., R.M. Bruch, F.P. Binkowski, M. Lanfredi, M. Chicca, N. Beltrami and L. Congiu. 2004. Karyotype characterization of the lake sturgeon, *Acipenser fulvescens* (Rafinesque, 1817) by chromosome banding and fluorescent *in situ* hybridization. *Genome* 47: 742–746.

Goodpasture, C. and S.E. Bloom. 1975. Visualization of nucleolar organizer regions in mammalian chromosomes using silver staining. *Chromosoma* 53: 37–50.

Jenneckens, I., J.N. Meyer, G. Hörstgen-Schwark, B. May, L. Debus and A. Ludwig. 2001. A fixed allele at microsatellite LS-39 is characteristic for the black caviar producer *Acipenser stellatus. Journal of Applied Ichthyology* 17: 39–42.

Kim, D.S., Y.K. Nam, J.K. Noh, C.H. Park and F.A. Chapman. 2005. Karyotype of North American shortnose sturgeon *Acipenser brevirostrum* with the highest chromosome number in the Acipenseriformes. *Ichthyological Research* 52: 94–97.

Krieger, J., P.A. Fuerst and T.M. Cavender. 2000. Phylogenetic relationships of the North American sturgeon (Order Acipenseriformes) based on mitochondrial DNA sequences. *Molecular Phylogenetics and Evolution* 16: 64–72.

Lanfredi, M., L. Congiu, M.A. Garrido-Ramos, R. De La Herrán, M. Leis, M. Chicca, R. Rossi, J. Tagliavini, C. Ruiz Rejón, M. Ruiz Rejón, and F. Fontana. 2001. Chromosomal location and evolution of a satellite DNA family in seven sturgeon species. *Chromosome Research* 9: 47–52.

Ludwig, A., N.M. Belfiore, C. Pitra, V. Svirsky and I. Jenneckens. 2001. Genome duplication events and functional reduction of ploidy levels in sturgeon (*Acipenser, Huso* and *Scaphirhynchus*). *Genetics* 158: 1203–15.

Luo, Z.W., C.A. Hackett, J.E. Bradshaw, J.W. McNicol and D. Milbourne. 2000. Predicting parental genotypes and gene segregation for tetrasomic inheritance. *Theoretical and Applied Genetics* 100: 1067–1073.

Maddison, W.P. and D.R. Maddison. 1992. *Macclade. Analysis of Phylogeny and Character Evolution.* Version 3. Sinauer Associates Inc., Sunderland, Massachusetts.

May, B., C.C. Krueger and H.L. Kincaid. 1997. Genetic variation at microsatellite loci in sturgeon: primer sequence homology in *Acipenser* and *Scaphirhynchus*. *Canadian Journal of Fisheries and Aquatic Sciences* 54: 1542–1547.

McQuown, E., G.A.E. Gall and B. May. 2002. Characterization and inheritance of six microsatellite loci in lake sturgeon (*Acipenser fulvescens*). *Transactions of the American Fisheries Society* 131: 299–307

Medrano, L., G. Bernardi, J. Couturier, B. Dutrillaux and G. Bernardi. 1988. Chromosome banding and genome compartmentalization in fishes. *Chromosoma* 96: 178–183.

Mirsky, A.E. and H. Ris. 1951. The DNA content of animal cells and its evolutionary significance. *Journal of General Physiology* 34: 451–462.

Ohno, S., J. Muramoto, C. Stenius, L. Christian and W.A. Kitterell. 1969. Microchromosomes in holocephalian, chondrostean and holostean fishes. *Chromosoma* 226: 35–40.

Olson, M.S. 1997. Bayesian procedures for discriminating among hypotheses with discrete distributions: Inheritance in the tetraploid *Astilbe biternata*. *Genetics* 147: 1933–1942.

Pirogovskii, M.I., L.I. Sokolov and V.P. Vasil'ev. 1989. *Huso huso* (Linnaeus, 1758). In: *The Freshwater Fishes of Europe*, J. Holčik (ed.). AULA-Verlag, Wiesbaden, Vol. 1, Pt. II, *General Introduction to Fishes, Acipenseriformes*, pp. 156–200.

Pyatskowit, J.D., C.C. Krueger, H.L. Kincaid and B. May. 2001. Inheritance of microsatellite loci in the polyploid lake sturgeon (*Acipenser fulvescens*). *Genome* 44: 185–191.

Rochard, E., P. Williot, G. Castelnaud and M. Lepage. 1991. Elements de systematique et de biologie des populations sauvages d'esturgeons. In: *Acipenser*, P. Williot (ed.). Cemagref Publications Bordeaux, pp. 475–507.

Rodzen, J.A. and B. May. 2002. Inheritance of microsatellite loci in the white sturgeon (*Acipenser transmontanus*). *Genome* 45: 1064–1076.

Rossi, R., G. Grandi, R. Trisolini, P. Franzoni, A. Carrieri, B.S. Dezfuli and E. Vecchietti. 1991. Osservazioni sulla biologia e la pesca dello storione cobice *Acipenser naccarii* Bonaparte nella parte terminale del fiume Po. *Atti della Società Italiana di Scienze Naturali (Milano)* 132: 121–142.

Ryabova, G.D. and I.G. Kutergina. 1990. Analyses of allozyme variation in the stellate sturgeon. *Acipenser stellatus* (Pallas) from the northern Caspian Sea. *Genetika* 26: 902–911.

Schlötterer, C. and D. Tautz. 1992. Slippage synthesis of simple sequence DNA. *Nucleic Acids Research* 20: 211–215.

Serebryakova, E.V. 1972. Some data on the chromosome complexes in Acipenseridae. In: *Genetics, Selection, and Hybridization of Fish*, B.I. Cherfas (ed.). Translated from Russian by Israel Program for Scientific Translations. Keter Press Binding: Wiener Bindery Ltd., Jerusalem, pp. 98–106.

Tiersch, T.R., R.W. Chandler, S.S. Wachtel and S. Elias. 1989. Reference standards for flow cytometry and application in comparative studies of nuclear DNA content. *Cytometry* 10: 706–710.

Tortonese, E. 1989. *Acipenser naccarii* (Bonaparte, 1836). In: *The Freshwater Fishes of Europe*, J. Holčík (ed.). AULA-Verlag, Wiesbaden, Vol. 1, Pt. II, *General Introduction to Fishes, Acipenseriformes*, pp. 284–293.

Van Eenennaam, A.L., J.D. Murray and J.F. Medrano. 1998. Synaptonemal complex analysis in spermatocytes of white sturgeon, *Acipenser transmontanus* Richardson (Pisces, Acipenseridae), a fish with a very high chromosome number. *Genome* 41: 51–61.

Vasil'ev, V.P., L.I. Sokolov and E.V. Serebryakova. 1980. Karyotype of the Siberian sturgeon *Acipenser baerii* Brandt from the Lena River and some questions of the acipenserid karyotypic evolution. *Voprosy Ichthyology* 23: 814–822.

Vialli, M. 1957. Volume et contenu en ADN par noyau. *Experimental Cell Resesearch Supplement* 4: 284–293.

Vladykov, V.D. and J.R. Greeley. 1963. Order Acipenseroidei. In: *Fishes of the Western North Atlantic*, H.B. Bigelow, C.M. Breder, D.M. Cohen, G.W. Mead, D. Merriman, Y.H. Olsen, W.C. Schroeder, L.P. Schultz and J. Tee-Van (eds.). *Memoir Sears Foundation for Marine Research* 1: 24–60.

Yasuda, A., K. Yamaguchi, T. Noso, H. Papkoff, A.L. Polenov, C.S. Nicoll and H. Kawauchi. 1992. The complete amino acid sequence of growth hormone from sturgeon (*Acipenser guldenstadtii*). *Biochimica et Biophysica Acta* 1120: 297–304.

Yu, X. 1987. On the karyosystematics of cyprinid fishes and a summary of fish chromosome studies in China. *Genetica* 72: 225–236.

Zane, L., L. Bargelloni and T. Patarnello. 2002. Strategies for microsatellite isolation: A review. *Molecular Ecology* 11: 1–16.

Zhang, S., Y. Yang, H. Deng, Q. Wei and Q. Wu. 1999. Genome size, ploidy characters of several species of sturgeons and paddlefishes with comment on cellular evolution of Acipenseriformes. *Acta Zoologica Sinica* 45: 200–206.

Zhang, S., Y. Zhang, X. Zheng, Y. Chen, H. Deng, D. Wang, Q. Wie, Y. Zhang, L. Nie and Q. Wu. 2000. Molecular phylogenetic systematics of twelve species of Acipenseriformes based on mtDNA ND4L-ND4 gene sequence analysis. *Science in China* (C) 43: 129–137.

Diversity in Isochore Structure and Chromosome Banding in Fish

Gloria G. Fortes, Carmen Bouza, Ana Viñas, Paulino Martínez and Laura Sánchez*

INTRODUCTION

Chromosome Banding

The term chromosome banding refers both to the process and to the pattern of bands that can be found in chromosomes. A great diversity exists in the processes and in the patterns; nevertheless, all chromosome-banding techniques evidence the heterogeneity of chromosome organization. The banding and the study of its mechanisms contribute to understand the complex organization of the genome, behind chromosome structure (Sumner, 1990).

Address for Correspondence: Departamento de Genética, Facultad de Veterinaria, Universidad de Santiago de Compostela, 27002 Lugo, Spain.
*E-mail: lasanche@lugo.usc.es

The first studies of banding began in 1894 (Metzner, revised by Levan, 1946). During more than 50 years, these techniques were limited generally to observations carried out in fixed material.

Modern cytogenetics was born when a series of technical innovations permit accumulation of cells in metaphase by means of colchicine, the use of hypotonic shock and fixation. These methods are considered universal, with certain exceptions, as the technique of squashing used for insect and plants chromosomes (Babu and Verma, 1987). It is important to emphasize that these methods constitute the basis of modern cytogenetics and that the techniques of banding have been designed for material prepared in this way.

The modern age of chromosome banding began in 1968 when Caspersson and collaborators published their article describing the use of the quinacrine to induce a characteristic and reproducible banding pattern. From this moment onwards, the techniques were generalized and extended to different types of organisms, especially warm-blooded vertebrates (birds and mammals). These techniques constitute the basis of chromosome pair identification, and permit the development of clinical cytogenetics and of studies of chromosome evolution and gene mapping.

The modern banding techniques, that intend to understand the underlying mechanisms of chromosome banding, have prompted an extensive investigation in the field of chromosome organization. The universality of these techniques is evidenced by the fact that once developed as a technique for a specific organism, this can be applied to another living being, plant or animal, vertebrate or invertebrate, with minor modifications (Commings, 1978).

The most common banding methods are G, R, Q and T. All of them are based on a common principle: the chromatin denaturation, and/or enzymatic digestion, followed by the incorporation of specific DNA staining. So, G banding results from the proteolytic digestion of chromosomes followed by Giemsa staining. It has been suggested that G bands are produced by the interaction of the components of Giemsa staining with DNA and histones (Craig and Bickmore, 1993). R bands, or reverse to G banding can be obtained by the following two methods: the first one involves a partial denaturation by heat in a salt solution followed by Giemsa staining, with the result of a preferential denaturation of AT rich DNA, followed by the staining of GC rich regions. In the second system, the induction of banding pattern is produced by specific antibiotics recognizing GC regions: chromomycin, olivomycin or mithramycin, indicating the GC richness of R bands. The T bands correspond to the more intense R bands, employing a stronger processing of heat or a colouring combination of fluorochromes (Holmquist, 1992). Q banding is a fluorescent pattern obtained using AT-specific quinacrine for staining. Globally, G and Q banding produce a complementary pattern to R banding (being R bands, the clear bands obtained with Giemsa).

Replication banding explains the observation that R bands replicates before G bands (Dutrillaux *et al.*, 1976). The incorporation of the base analogous 5-bromodeoxyuridire (BrdU) in the chromosomes at the moment of replication traps the fluorescence of Hoechst 33528. The early or late incorporation of BrdU in the S phase will result in R and G bands, respectively. Assuming that single copy genes replicate earlier and the repeated DNA later, a correspondence between genes and repeated DNA, with the R and G bands, respectively, was established (Craig and Bickmore, 1993).

The Isochore Model

In spite of the fact that on an average basis, the genome of vertebrates presents among 35-40% of GC bose pairs, the distribution is far from being uniform. Thus, the genome is a mosaic of long DNA segments (> 300 kb in length) that are homogeneous in base composition but differ in GC content from less than 30% at more than 60% (Bernardi, 1995). The existence of these segments of homogeneous composition of DNA called isochores ('equal regions') has been shown by means of analytical equilibrium centrifugation of DNA preparations in $CsSO_4$ density gradients in the presence of specific ligands of specific bases (Bernardi *et al.*, 1985). Each fraction of a specific gradient contains DNA banded in a specific density as a direct reflection of its base composition. These fractions have been classified in five isochore classes: the AT rich (L1 and L2 classes) and those of growing richness in GC (H1, H2 and H3 classes).

The isochore classes differ not only in their base composition, but also in their non-uniform distribution in the genome. The distribution is not at random within and among chromosomes, and the isochore classes are surprisingly different in gene density (Mouchiroud *et al.*, 1991; Saccone *et al.*, 1992, 1993).

The gene content of the different isochore classes has been determined by the means of three different methods: (1) by means of hybridization of diverse cDNA and genetic probes belonging to the $CsSO_4$ fractions on the genomic DNA, that would indicate the base composition of the corresponding genomic region; (2) by analyzing genomic sequences of genes from databases (Mouchiroud et al., 1987; Ikemura et al., 1990); (3) by calculation of the GC level in the third codon position (GC3) of the DNA sequences, that directly correlates with the genomic GC content in the region in which the gene is inserted (Ikemura and Wada, 1991).

The combination of the results from the various methodological approaches has shown a highly asymmetric gene distribution (Aissani et al., 1991). In spite of the fact that the isochores L1 and L2 represent almost two-thirds of the genome (in mammals), these only contain around a third of the genes. The main gene density is detected in the isochores H3 which contain more than one quarter part of the genes (Clay et al., 1996), while representing approximately the 3% of the genome.

The location of the different isochore classes has been undertaken in different ways. In first place, the isolated DNA from $CsSO_4$ fractions has been used directly as probes in experiments of in situ hybridization onto metaphase chromosomes (Saccone et al., 1992, 1993). The isochores L1 and L2 produced a pattern that recalls G banding but is more diffuse, this result being consistent with the fact that the AT-rich material also appears in some R bands. The hybridizations with isochores H1 and H2 produced an approximate pattern to R banding and the isochores H3 identified a class of R bands, the T bands.

Chromosome Banding in Cold-blooded Vertebrates

The development of the modern cytogenetics of vertebrates has been dominated by spectacular results obtained in mammals. Nevertheless, the level of resolution obtained in the cold-blooded vertebrates (fish, amphibians and reptiles) has not been comparable (Sumner, 1990). The fundamental difference would be summarized in the absence of euchromatic banding (G, Q, R) in cold-blooded vertebrates. Diverse authors have suggested that the technical cytogenetic developments in these groups are extremely limited and that the techniques should be modified in function of the organism (Sumner, 1990). Bernardi (1989) has proposed that the different genome organization of warm- and cold-blooded vertebrates, relating to their isochore structure, would explain the absence of banding in cold-blooded taxa. In other words, the genome of warm-blooded vertebrates would be quite compartmentalized, with GC-rich isochores, type H3, while the genome of cold-blooded taxa would be characterized by a low level of compartmentalization and H3 isochores.

The most extensive group of cold-blooded vertebrates analyzed from the cytogenetic point of view has been fish (Brum and Galetti, 1997). Thus, the data of banding in this group are relatively numerous, fundamentally in the case of C banding (that detects constitutive heterochromatin) and the Ag-staining (that detects the nucleolar organizer region) (e. g. Sánchez *et al.*, 1991; Martínez *et al.*, 1993; Galetti, 1998; Ueda, present volume). Nevertheless, the data of euchromatic banding have been scarce, with the exception of the replication bands (Amaro *et al.*, 1996; Viñas *et al.*, 1996; Amaro and Sánchez, 1997) and the case of the G banding in some particular fish species such as the *Anguilla* (Sola *et al.*, 1984; Medrano *et al.*, 1988).

The development of unconventional cytogenetic techniques such as *in situ* treatment with restriction enzymes, capable of digesting the DNA of fixed chromosomes revealing enriched regions in tagged sites, permitted the induction of different types of banding in mammals (Babu *et al.*, 1988; Sánchez *et al.*, 1991). Miller and collaborators (1983) proposed that this technique could be used in the genomes of fish and amphibians in order to induce euchromatic banding.

The genome of *Anguilla* species seemed to be unique among the genomes of the fish studied until now, because G banding could be induced in their chromosomes (Wiberg, 1983; Sola *et al.*, 1984), as well as in Q and R banding, by means of quinacrine and acridine (Medrano *et al.*, 1988). Medrano *et al.* (1988) also found a correlation between chromosome banding and the compartmentalization of the composition of the *Anguilla* genome after centrifugation in ClCs gradient. The capacity of the restriction enzymes to induce chromosome banding in fish (Lloyd and Thorgaard, 1988; Sánchez *et al.*, 1990, 1991; Lozano *et al.*, 1991) was used for the analysis of the chromosomes of *Anguilla*, revealing high resolutive G patterns. This was the first time that G banding pattern by means of *in situ* digestion with restriction enzymes was described in fish in the literature (Viñas *et al.*, 1994, 1998). The fact that enzymes with different tagged sites, namely *Hae*III (GGCC) and *Mse*I (AATT), produced a similar banding pattern, indicated that other factors, aside from the composition of bases, such as the different organization of the chromatin, could be influencing in this result, as it was previously suggested (Mezzanotte and Ferrucci, 1984; Burkholder, 1989).

In spite of the fact that banding patterns induced by the restriction enzymes depend on the capacity to induce multiple cuts in the DNA, the quantity of DNA extracted will also depend on the extraction of specific associated proteins. This would imply that only an enzyme with a high frequency of tagged sites would give rise to a clear banding pattern. In order to amplify the result of *in situ* digestion, the technique of digestion/ nick translation was applied. This consists of the incorporation of biotinylated precursors by means of 'nick translation' after the processing with restriction enzymes and results in approximately two orders more sensitive in magnitude than the simple *in situ* digestion (Kerem *et al.*, 1984; Adolph and Hameister, 1990). This technique had been earlier applied only to mammal chromosomes (Sumner, 1990) and invertebrates (De la Torre *et al.*, 1993). The first time application of the *in situ* digestion/nick translation on the fish chromosomes (in a cell line of Atlantic salmon) produced a pattern of euchromatic banding depending on the time and the different enzymes used. *Dra*I, *Hae*III and *Hinf*I enzymes, induced clearly euchromatic patterns of G/R type (Abuín *et al.*, 1995).

Historically, the incapacity to show G bands in the genome of the cold-blooded vertebrates has also been attributed to technical factors. A careful revision of the literature reveals that in some fish species, it is possible to induce this banding after treatment with proteinases (trypsin) or by denaturalization by heat (Gold *et al.*, 1990). The following two modifications of the G banding protocol: (1) ASG (acetic acid, salt solution, Giemsa, Sumner *et al.*, 1971) and (2) STG (hot salt solution, trypsin, Giemsa, Gold *et al.*, 1990), permitted to reveal a pattern of G banding in the chromosomes of two species of the *Oncorhynchus* genus (Abuín *et al.*, 1996). The method STG showed a greater resolution and reproducibility, permitting a complete classification of the homologous pairs.

CpG Island and R Banding

The study of the chromosome distribution of the CpG islands has been undertaken by means of diverse techniques (Barbin *et al.*, 1994; Craig and Bickmore, 1994), revealing a certain regional variation.

The CpG islands are sequences of DNA of 500-1000 base pairs identified in all vertebrates, with a high GC content (60% in humans, smaller in fish), and located in the 5´extremes (at times in 3') of the majority of the 'housekeeping' genes as also any tissue specific genes (Cross *et al.*, 1991; Cross and Bird, 1995).

A notable and common result of all these studies is that the location of the CpG islands shows a pattern of R bands, being generally accepted that the distribution of the genes in the chromosomes is correlated with the distribution of the CpG islands and the R bands (Craig and Bickmore, 1994).

Andersen and co-authors (1998) analyzed the distribution of the CpG islands in human chromosomes by means of a technique called Self-Primed *In Situ* Labelling (S-PRINS). S-PRINS is based on the fact that the 3'OH free extremes of DNA *in situ* digested by restriction enzymes, can act as DNA substrates for the DNA polymerase and to be extended incorporating haptens that are detected subsequently by means of conjugated fluorescent antibodies. These authors analyzed the cuts generated by enzymes as *Hpa*II and *Msp*I (whose target sites are very frequent in the CpG islands). The patterns obtained clearly resembled that of R banding.

The prior evidences (*in situ* digestion with restriction enzymes, nick translation and modified G banding) of the existence of euchromatic banding (G, R) in some fish species, strongly supported the interest to apply the S-PRINS technique to the *Oncorhynchus mykiss* and *Anguilla anguilla* chromosomes. The results obtained showed a clear pattern of R banding in the chromosomes of these two species. These data seem to indicate a regional distribution of the genes in the genome of these fish species, where the locating of these CpG islands had been described previously (Viñas *et al.*, unpublished).

In order to tentatively locate the CpG islands in the human genome, Craig and Bickmore (1994) obtained material with a variable concentration of islands by means of DNA genomic digestion with restriction enzymes and then used the smallest fragments of DNA with greater density of CpG islands as probes for *in situ* hybridization. The technique produced a hybridization pattern similar to that obtained when the material from isochores H3 is hybridized, and that reflects a pattern of R banding.

The same strategy was applied to birds (McQueen *et al.*, 1996) and rodents (Cross *et al.*, 1997) genomes, showing a compartmentalization of the CpG islands location in these genomes, being more marked in birds than in rodents. Also, in these cases, the pattern of banding is of R type.

These results suggest developing these procedures to isolate CpG islands from fish genome in species where resolutive euchromatic banding (G and R) had been previously obtained by means of other techniques (*in situ* enzymatic digestion, *in situ* digestion/nick translation, processing with trypsin and S-PRINS).

The genomic DNA digestion of *Anguilla anguilla* and *Oncorhynchus mykiss* with various restriction enzymes indicated that the use of the enzyme *Hpa*II offers the clearest results, with band profile ranging between 500-1000 pb. The coverage of this DNA and its hybridization onto the chromosomes showed a regional locating of distribution of the CpG islands in these species (Viñas *et al.*, unpublished).

Keeping in mind the results obtained from euchromatic banding and the locating of the CpG islands in the chromosomes of fish species, as well as the publication of frequent examples of G banding in different species of this group (Luo, 1998; Maistro *et al.*, 1999; Ueda and Naoi, 1999), it was decided to re-examine the paradigm of Bernardi (1995, 2000) that the genome of the cold-blooded and warm-blooded vertebrates is clearly differentiated by its isochore structure.

Sequence Data and Isochore Structure

The studies of biased codon usage in the genes of vertebrates showed a positive correlation among the percentage of G + C in the third codon position (GC3) and the percentage G + C of the genome portion surrounding each gene (Ikemura, 1985; Aota and Ikemura, 1986; Ikemura and Aota, 1988). At around the same time Bernardi (1989), using CsCl centrifugation, had proposed the theory of the isochores and established that the distribution of the GC3 content of the coding sequences reproduces the distributions of the isochore classes obtained by centrifugation.

Until 2000, the sequence analysis of isochore structure was based on a few model organisms (human, chicken, mouse, *Xenopus*), in general, corroborating the findings of centrifugation studies (Bernardi and Bernardi, 1991; Mouchiroud and Bernardi, 1993). It was generally considered that GC-rich isochores were absent from cold-blooded vertebrates. Homothermous vertebrate genomes showed a large variation in GC content, a pattern apparently absent in poikilotherm vertebrates (Bernardi and Bernardi, 1990a, b; Bernardi, 1995). Fishes, amphibians, and reptiles would have highly homogeneous genomes lacking regions of exceptionally high GC content (Bernardi and Bernardi, 1990a, b).

In the last years, the increment of sequences delivered to databases allowed us to make the analysis of the compositional pattern in several species of the three cold-blooded vertebrates: fishes, amphibians and reptiles. We carried out the analysis of the compositional distribution of the third codon positions (GC3) in 1510 coding sequences (CDS) of fishes, 1414 CDS of amphibians and 320 CDS of reptiles. A relevant representation of the very GC rich coding sequences in cold-blooded vertebrates was found (Fig. 4.3.1).

However, an important diversity in compositional distribution was observed between different groups of cold-blooded vertebrates (Table 4.3.1).

The results reported here indicate the existence of a great heterogeneity between species of the same group. We found the most extreme differences in fish where the observed figures of CDS with GC3 being higher than 75%, ranged from 3.8% (Ostariophysi) to 80.5% (Cephalaspidomorphi). Despite this heterogeneity, the variation appeared mostly related to phylogenetic components (12 orders from 9 classes), more variation between rather than within orders being observed (Fig. 4.3.2). In amphibians and reptiles, this observation could be not so consistent because of the small representation of some taxonomic groups in databases.

Our results also revealed higher GC content in coding than in non-coding sequences for cold-blooded vertebrates (Fortes *et al.*, 2006). This phenomenon had been exhibited in warm-blooded vertebrates, where introns appeared to increase in GC in parallel with the corresponding exons, but showing systematically lower GC content rather than coding sequences (about 5-8% less; Vinogradov, 2001). We have compared the mean values of GC3 *vs* GCI (means GC content of introns) and total GC *vs* GCI, evidencing a high GC difference between exons and introns (10, 20% respectively, for both fishes and amphibians). The larger difference between GC content for introns and exons with respect to the homeotherm genomes could explain the discrepancies between sequence and gradient ultracentrifugation analysis of the DNA of these cold-blooded vertebrates. The relatively low GC content of non-coding sequences, as an essential part of isochores in genomes of fishes and amphibians, might mask some compositional variation such as the

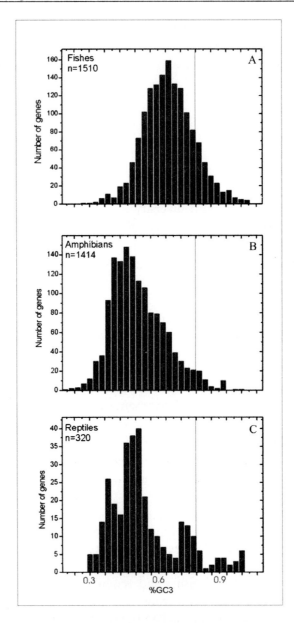

Fig. 4.3.1 Histograms of compositional distribution of third codon position (GC3) for genes of (A) fishes, (B) amphibians and (C) reptiles. Number of genes analyzed are represented against GC level of third codon position. A 2.5% GC window was used. The broken line at 75%GC is shown as a tentative reference of the very GC-rich compartments in vertebrate genomes (isochores H3; Mouchiroud *et al.*, 1991). n: total number of coding sequences (CDS) analyzed.

Table 4.3.1 GC content in third codon position of genes located in different isochore classes predicted according to its GC3 content.

Taxonomy groups	Number of genes	%GC3		
		<57%	57-75%	>75%
Fish				
Cephalaspidomorphi	66	3	16.5	80.5
Elasmobranchii	10	50	50	0
Sarcopterygii	10	90	10	0
Elopomorpha	39	12.8	46.2	41
Ostariophysi	713	37.7	58.5	3.8
Protacanthopterygii	260	11.1	77.0	11.9
Paracanthopterygii	17	11.8	17.6	70.6
Atherinomorpha	117	26.0	61.4	12.6
Percomorpha	278	15.3	65.0	19.7
Amphibians				
Ranoidea	105	71.4	26.7	1.9
Pipoidea	1230	76.3	21.4	2.3
Salamandroidea	79	35.5	43.0	21.5
Reptiles				
Lepidosauria	259	77.3	18.9	3.8
Crocodylidae	37	45.9	13.6	40.5
Testudinidae	24	56.4	35.3	8.3

genomic presence of very GC-rich coding sequences in these organisms. This could explain the difficulty to find GC-rich isochores in cold-blooded vertebrates by ultracentrifugation.

To conclude, the correlation of GC content between coding and non-coding sequences arguments are in favour of the existence of isochore structure in the genome of cold-blooded vertebrates in the presence of very GC-rich regions. It could indicate a common compositional organization for the vertebrate genomes where the most striking differences in genome organization between cold- and warm-blooded vertebrates could be more pronounced in non-coding rather than in coding sequences. The dichotomy between warm-blooded and cold-blooded vertebrates does not easily explain the evolution of vertebrate GC-rich isochores. However, it is likely that GC-rich isochores were present before the evolution of endothermy in a GC-rich ancestral vertebrate genome; from this ancestral condition different patterns of genome organization could have ocurred along vertebrate evolution maintaining the GC richness in coding sequences.

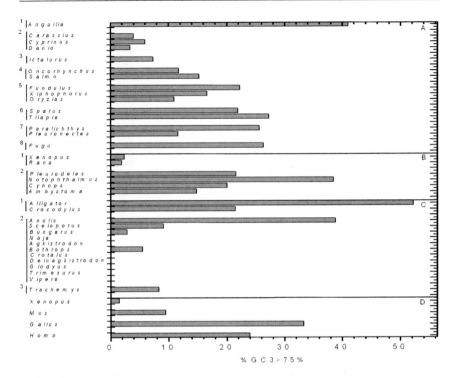

Fig. 4.3.2 Percentage of very GC3-rich sequences (>75%) in species from (A) eight orders of fishes: [1]Anguilliformes (*Anguilla*, n=41), [2]Cipriniformes (*Carassius*, n=101; *Cyprinus*, n=100; *Danio*, n=474), [3]Siluriformes (*Ictalurus*, n=27), [4]Salmoniformes (*Oncorhynchus*, n=188; *Salmo*, n=59), [5]Cyprinodontiformes (*Fundulus*, n=18; *Xiphophorus*, n=18; *Oryzias*, n=82), [6]Perciformes (*Sparus*, n=32; *Tilapia*, n=22), [7]Pleuronectiformes (*Paralichthys*, n=43; *Pleuronectes*, n=22), [8]Tetraodontiformes (*Fugu*, n=77), (B) two orders of amphibians: [1]Anura (*Xenopus*, n=1230; *Rana*, n=105) and [2]Caudata (*Pleurodeles*, n=14; *Nothopthalmus*, n=13; *Cynops*, n=25; *Ambystoma*, n=27), (C) three orders of reptiles: [1]Crocodilia (*Alligator*, n=23; *Crocodyllus*, n=14); [2]Squamata (*Anolis*, n=18; *Sceloporus*, n=11; *Bungarus*, n=36; *Naja*, n=62; *Agkistrodon*, n=19; *Bothrops*, n=18; *Crotalus*, n=20; *Deinagkistrodon*, n=13; *Glodyus*, n=18; *Trimesurus*, n=31; *Vipera*, n=13) and [3]Testudines (*Trachemys*, n=24). Data of (D) *Xenopus* (n= 547), *Mus* (n=2491), *Gallus* (n=677) and *Homo* (n=4270) from Bernardi, 1995, are shown for comparison.

In the near future, the knowledge of evolution of compositional patterns of vertebrate genomes will greatly benefit from studies of comparative genomics and from sequencing of complete genomes, avoiding the simplification of using a few model organisms for describing isochore structure in vertebrates.

References

Abuín, M., P. Martínez and L. Sánchez. 1995. Restriction endonuclease/nick translation procedure on fixed chromosomes of the Atlantic salmon fish cell line. *Chromosome Research* 3: 379–385.

Abuín, M., P. Martínez and L. Sánchez. 1996. G-like banding pattern in two salmonid species: *Oncorhynchus mykiss* and *Oncorhynchus kisutch*. *Chromosome Research* 4: 471–473.

Adolph, S. and H. Hameister. 1990. *In situ* nick translation of human metaphase chromosomes with the restriction enzymes *Msp*I and *Hpa*II reveals an R-band pattern. *Cytogenetics and Cell Genetics* 54: 132–136.

Aissani, B., G. D'onofrio, D. Mouchiroud, K. Gardiner, C. Gautier and G. Bernardi. 1991. The compositional properties of human genes. *Journal of Molecular Evolution* 32: 497–503.

Amaro, R. and L. Sánchez. 1997. Chromosomal analysis of two established salmonid cell lines: CHSE-214 (*Oncorhynchus tschawytscha*) and RTG-2 (*Oncorhynchus mykiss*). *In Vitro Cellular and Developmental Biology-Animal* 33: 662–664.

Amaro, R., M. Abuín and L. Sánchez. 1996. Chromosomal evolution in Salmonids: A comparison of Atlantic salmon, brown trout and rainbow trout R-band chromosomes. *Genetica* 98: 297–302.

Andersen, C.L., J. Koch and E. Kjeldsen. 1998. CpG islands detected by self-primed *in situ* labeling (SPRINS). *Chromosoma* 107: 260–266.

Aota, S. and T. Ikemura. 1986. Diversity in G+C content at the third position of codons in vertebrate genes and its cause. *Nucleic Acids Research* 14: 6345–6355.

Babu, A. and R.S. Verma. 1987. Chromosome structure: euchromatin and heterocrhomatin. *International Review of Cytology* 108: 1–60.

Babu, A., A.K. Agarwal and R.S. Verma. 1988. A new approach in recognition of heterochromatic regions of human chromosomes by means of restriction endonucleases. *American Journal of Human Genetics* 42: 60–65.

Barbin, A., C. Montpellier, N. Kokalj Vokac, A. Gibaud, A. Nivelau, B. Malfoy, B. Dutrillaux and C.A. Bourgeois. 1994. New sites of methylcytosine-rich DNA detected on metaphase chromosomes. *Human Genetics* 94: 684–692.

Bernardi, G. 1989. The isochore organization of the human genome. *Annual Review of Genetics* 23: 637–661.

Bernardi, G. 1995. The human genome: Organization and evolutionary history. *Annual Review of Genetics* 29: 445–476.

Bernardi, G. 2000. Isochores and the evolutionary genomics of vertebrates. *Gene* 241: 3–17.

Bernardi, G. and G. Bernardi. 1990a. Compositional patterns in the nuclear genome of cold-blooded vertebrates. *Journal of Molecular Evolution* 31: 256–281.

Bernardi, G. and G. Bernardi. 1990b. Compositional transitions in the nuclear genome of cold-blooded vertebrates. *Journal of Molecular Evolution* 31: 282–293.

Bernardi, G. and G. Bernardi. 1991. Compositional properties of nuclear genes from cold-blooded vertebrates. *Journal of Molecular Evolution* 33: 57–67.

Bernardi, G., B. Olofsson, J. Filipski, M. Zerial, J. Salinas, G. Cuny, M. Meunier-Rotival and F. Rodier. 1985. The mosaic genome of warm-blooded vertebrates. *Science* 228: 953–958.

Brum, M.J.I. and P.M. Galetti. 1997. Teleostei ground plant karyotype. *Journal of Comparative Biology* 2: 91–102.

Burkholder, G.D. 1989. Morphological and biochemical effects of endonucleases on isolated mammalian chromosomes *in vitro*. *Chromosoma* 97: 347–355.

Caspersson, T., S. Farber, G.E. Foley, J. Kudynowski, E.J. Modest, E. Simonsson, U. Wagh and L. Zech. 1968. Chemical differentiation along metaphase chromosomes. *Experimental Cell Research* 49: 219–222.

Clay, O., S. Caccaio, S. Zoubak, D. Mouchiroud and G. Bernardi. 1996. Human coding and non-coding DNA: Compositional correlations. *Molecular Phylogenetics and Evolution* 5: 2–12.

Commings, D.E. 1978. Mechanisms of chromosome banding and implications for chromosome structure. *Annual Review of Genetics* 12: 25–46.

Craig, J.M. and W.A. Bickmore. 1993. Chromosome bands—Flavours to savour. *BioEssays* 15: 349–354.

Craig, J.M. and W.A. Bickmore. 1994. The distribution of CpG islands in mammalian chromosomes. *Nature Genetics* 7: 376–382.

Cross, S.H. and A.P. Bird. 1995. CpG islands and genes. *Current Opinion in Genetics and Development* 5: 309–314.

Cross, S.H., P. Kovarik, J. Schmidtke and A.P. Bird. 1991. Non-methylated islands in fish genomes are GC-poor. *Nucleic Acids Research* 19: 1469–1474.

Cross, S.H., M. Lee, V.H. Clark, J.M. Craig, A.P. Bird and W.A. Bickmore. 1997. The chromosomal distribution of CpG islands in the mouse: Evidence for genome scrambling in the rodent lineage. *Genomics* 40: 454–461.

De La Torre, J., C. López-Fernández, P. Herrero and J. Gosálvez. 1993. *In situ* nick translation of meiotic chromosomes to demonstrate homologous heterochromatin heterogeneity. *Genome* 36: 268–270.

Dutrillaux, B., J. Couturier, C.L. Richer and E.Viegas-Péguinot. 1976. Sequence of DNA replication in 277 R- and Q-bands of human chromosomes using a BrdU treatment. *Chromosoma* 58: 51–61.

Fortes G., C. Bouza, P. Martinez and L. Sanchez. 2006. Diversity in isochore structure among cold-blooded vertebrates based on GC content of coding and non-coding sequences. *Genetica* (in press).

Galetti, P.M. 1998. Chromosome diversity in Neotropical fishes: NOR studies. *Italian Journal of Zoology* 65(Supplement S): 53–56.

Gold, J.R., Y.C. Li, N.S. Shirpley and P.K. Powers. 1990. Improved methods for working with fish chromosomes with a review of metaphase chromosome banding. *Journal of Fish Biology* 37: 563–575.

Holmquist, G.P. 1992. Chromosome bands, their chromatin flavors, and their functional features. *American Journal of Human Genetics* 51: 17–37.

Ikemura, T. 1985. Codon usage and tRNA content in unicellular and multicellular organisms. *Molecular Biology and Evolution* 2: 13–34.

Ikemura, T. and S. Aota. 1988. Global variation in G+C content along vertebrate genome DNA. *Journal of Molecular Biology* 203: 1–13.

Ikemura, T. and K.M. Wada. 1991. Evident diversity of codon usage patterns of human genes with respect to chromosome banding patterns and chromosome numbers; relation between nucleotide sequence data and cytogenetic data. *Nucleic Acids Research* 19: 4333–4339.

Ikemura, T., K.N. Wada and S. Aota. 1990. Giant G+C% mosaic structures of the human genome found by arrangement of genebank human DNA sequences according to genetic positions. *Genomics* 8: 207–216.

Kerem, B.S., R. Goitein, G. Diamond, H. Cedar and M. Marcus. 1984. Mapping of DnaseI sensitive regions on mitotic chromosomes. *Cell* 38: 493–499.

Levan, A. 1946. Heterochromaty in chromosomes during their contraction phase. *Hereditas* 32: 449–468.

Lloyd, M.A. and G.H. Thorgaard. 1988. Restriction endonuclease banding of rainbow trout chromosomes. *Chromosoma* 96: 171–177.

Lozano, R., C.R. Rejón and M.R. Rejón. 1991. An analysis of coho salmon chromatin by means of C-banding, Ag- and fluorochrome staining, and *in situ* digestion with endonucleases. *Heredity* 66: 403–409.

Luo, C. 1998. Multiple chromosomal banding in grass carp, *Ctenopharyngodon idellus*. *Heredity* 81: 481–485.

Maistro, E.L., F. Foresti and C. Oliveira. 1999. R- and G-band patterns in *Astyanax scabripinnis paranae* (Pisces, Characiformes, Characidae). *Genetics and Molecular Biology* 22: 201–204.

Martínez, P., A. Viñas, C. Bouza, J. Castro and L. Sánchez. 1993. Quantitative analysis of the variability of NOR region in *Salmo trutta*. *Genome* 36: 1119–1123.

Mcqueen, H.A., J. Fantes, S.H. Cross, V.h. Clark, A.l. Archibald and A.P. Bird. 1996. CpG islands of chicken are concentrated on microchromosomes. *Nature Genetics* 12: 321–324.

Medrano, L., G. Bernardi, J. Couturier, B. Dutrillaux and G. Bernardi. 1988. Chromosome banding and genome compartmentalization in fishes. *Chromosoma* 96: 178–183.

Metzner, R. 1894. Beitrage zur Granulalehre. I Kern und Kerntheilung. *Archiv fur Physiologie* 18: 309–348.

Mezzanotte, R. and L. Ferrucci. 1984. Alterations induced in mouse chromosomes by restriction endonucleases. *Genetica* 64: 123–128.

Miller, D.A., I. Choi and O.J. Miller. 1983. Chromosome localization of highly repetitive human DNAs and amplified ribosomal DNA with restriction enzymes. *Science* 219: 395–397.

Mouchiroud, D. and G. Bernardi. 1993. Compositional properties of coding sequences and mammalian phylogeny. *Journal of Molecular Evolution* 37: 109–116.

Mouchiroud, D., G. Fichant and G. Bernardi. 1987. Compositional compartmentalization and gene composition in the genome of vertebrates. *Journal of Molecular Evolution* 26: 198–204.

Mouchiroud, D., G. D'onofrio, B. Aissani, G. Macaya, C. Gautier and G. Bernardi. 1991. The distribution of genes in the human genome. *Gene* 100: 181–187.

Saccone, S., A. De Sario, G. Della Valle and G. Bernardi. 1992. The highest gene concentrations in the human genome are in T bands of metaphase chromosomes. *Proceedings of the National Academy of Sciences of the United States of America* 89: 4913–4917.

Saccone, S., A. De Sario, J. Wiegant, A.K. Raap, G. Della Salle and G. Bernardi. 1993. Correlations between isochores and chromosomal bands in the human genome. *Proceedings of the National Academy of Sciences of the United States of America* 90: 11929–11933.

Sánchez, L., P. Martínez, A. Viñas and C. Bouza. 1990. Analysis of the structure and variability of nucleolar organizer regions of *Salmo trutta* by C-Ag and restriction endonuclease banding. *Cytogenetics and Cell Genetics* 54: 6–9.

Sánchez, L., P. Martínez, C. Bouza and A. Viñas. 1991. Chromosomal heterochromatin differentiation in *Salmo trutta* with restriction enzymes. *Heredity* 66: 241–249.

Sola, L., B. Camerini and S. Cataudella. 1984. Cytogenetics of Atlantic eels: C- and G-banding, nucleolus organizer regions, and DNA content. *Cytogenetics and Cell Genetics* 38: 206–210.

Sumner, A.T. 1990. *Chromosome Banding.* Unwin Hyman Ltd., London.

Sumner, A.T., H.J. Evans and R.A. Buckland. 1971. New technique for distinguishing between human chromosomes. *Nature New Biology* 232: 31–32.

Ueda, T. and H. Naoi. 1999. BrdU-4Na-EDTA-Giemsa band karyotypes of 3 small freshwater fish, *Danio rerio, Oryzias latipes*, and *Rhodeus ocellatus*. *Genome* 42: 531–535.

Viñas, A., C. Gómez, P. Martínez and L. Sánchez. 1994. Induction of G-bands on *Anguilla anguilla* chromosomes by the restriction endonucleases HaeIII, HinfI and MseI. *Cytogenetics and Cell Genetics* 65: 79–81.

Viñas, A., C. Gómez, P. Martínez and L. Sánchez. 1996. Replication banding in the chromosomes of the european eel, *Anguilla anguilla*. *Genetica* 98: 107–110.

Viñas, A., C. Gómez, P. Martínez and L. Sánchez. 1998. Analysis of the European eel (*Anguilla anguilla*) chromosomes after treatment with TfiI and AvaI restriction endonucleases. *Journal of Applied Ichthyology* 14: 113–115.

Vinogradov, A. 2001. Within-intron correlation with base composition of adjacent exons in different genomes. *Gene* 276: 143–151.

Wiberg, U.H. 1983. Sex determination in the European eel (*Anguilla anguilla* L.). *Cytogenetics and Cell Genetics* 36: 589–598.

Chromosomes and Repetitive DNAs: A Contribution to the Knowledge of the Fish Genome

Cesar Martins

INTRODUCTION

Primary studies on whole sequenced genomes focused on single genes and little attention was directed to repetitive DNAs, and duplicated segments. Single- and few-copy sequences correspond to a small fraction of the genomes. For example, only 1.5% of the human genome is composed of coding sequences (Horvath *et al.*, 2001). On the other hand, the eukaryote genome contains several types of DNA sequences present in multiple copies that, in some instances, can represent large portions of the genome (Charlesworth *et al.*, 1994). Although extensively studied for the past three decades, the molecular forces that propagate and maintain repetitive DNAs in the genome are still being discussed.

Address for Correspondence: UNESP - Universidade Estadual Paulista, Instituto de Biociências, Departamento de Morfologia, CEP 18618-000, Botucatu, SP, Brazil.
E-mail: cmartins@ibb.unesp.br

Repeated DNA elements are often considered to be 'selfish' DNA (Doolittle and Sapienza, 1980; Orgel and Crick, 1980) or 'junk DNA' (Nowak, 1994) in the genome and they make no biological contributions to the carriers. Although the repetitive DNA sequences were thought to be functionless, their involvement in some diseases (Kazazian et al., 1988), gene regulation and repair (Messier et al., 1996), as well as in sex chromosome differentiation (Anleitner and Haymer, 1992; Kraemer and Schmidt, 1993) has been suggested. The repetitive sequence importance in gene regulation and chromosome physiology may be illustrated by mutations in some minisatellite sequences that may contribute to as many as 10% of all cases of breast, colorectal and bladder cancer, and acute leukemia (Devlin et al., 1993). The most significant function for the repetitive sequences is related to the head-to-tail tandem repeats of the centromere and telomere of eukaryote chromosomes. Centromeres, the primary constriction of each metaphase eukaryotic chromosome, are critical to the proper sorting of chromosomes during cell division; if it does not work correctly, it can result in cancer, defects in development, or similar misfortunes. Telomeres, the end of linear eukaryotic chromosomes, also play a critical role in maintaining chromosomal stability and function. Taken together, these data show that repeated DNAs seem to have an important role in the genome organization and evolution, leading to a significant impact on the speciation process.

In many species, repeated sequences comprise a large portion of the genomes. Ninety-five percent of the onion genome (Flavell et al., 1974) and 50% or much more of the human genome corresponds to repeated sequences (The Genome International Sequencing Consortium, 2001). The first question that arises is: what is the reason for the large number of repeated DNAs in the genome? During the course of eukaryotic evolution, genes and genome segments seem to have duplicated, leading to an increase of the DNA content in the cell nucleus. The variation in genome size of different eukaryotes is often reported as differences in the amount of repeated DNA sequences (Cavalier-Smith, 1985; Brenner et al., 1993). Recently, advances in studies concerning non-coding repetitive DNA sequences have shown that such sequences are extremely important in the structural and functional organization of the genome (Schueler et al., 2001).

Repetitive DNAs include the tandemly-arrayed satellite, minisatellite and microsatellite sequences, and dispersed repeats such as transposons and retrotransposons (Charlesworth et al., 1994). Some large satellite DNAs, often called macrosatellite DNA, have also been reported (Takahashi et al., 1994). There are also multigene families composed of hundreds to thousands of copies that code important molecules such as the ribosomal RNAs (rRNA), for example. The most known non-coding repetitive sequences are the head-to-tail tandem-arrayed repetitions found in the centromeres and telomeres. Most repetitive sequences are known to be highly unstable in the genomes (Charlesworth et al., 1994). The magnitude of genome instability due to head-to-tail tandem repeats is often much higher than that due to transposons (Gondo et al., 1998). These repetitive DNA sequences are thought to arise by many mechanisms, from direct sequence amplification by unequal recombination of homologous DNA regions to the reverse flow of genetic information using an intermediate RNA molecule. Due to the hypervariability of the tandem repeats, such genome segments are highly polymorphic and considered to be good molecular markers for genotyping of individuals and populations (Jeffreys et al., 1985).

A general overview of the human genome, for instance, shows that the majority of genes are located in an interstitial chromosome position far away from the centromere and telomere heterochromatins (Horvath et al., 2001). On the other hand, the repetitive rich chromosome regions are poorly understood. Among entire sequenced genomes, the repetitive areas remain as gaps because of the difficulty in determining their correct positioning and array in the genome. Even chromosomes reportedly 'sequenced to completion' such as chromosome 22 of humans have multiple gaps in the pericentromeric regions (Dunham et al., 1999). A complete understanding of the relationship between chromosome structure and function requires the repetitive segments to be fully resolved.

Studies of the repetitive fraction of the genome can contribute to the knowledge of the complex organization of DNA in the cell nucleus. Also, the integration of DNA sequences with chromosome mapping of repetitive DNAs can provide a better landscape of the genome, which is not yet clearly defined even in the wholly sequenced genomes. The repetitive DNA sequences can also provide chromosome markers useful in studies of species evolution, identification of specific chromosomes, homologous chromosomes, chromosome rearrangements, sex chromosomes, and applied genetics. In this chapter, I will focus on the understanding of the structure of the fish genome based on the repetitive sequences and their organization in the chromosomes.

Tandem Repeats and Dispersed Elements

Satellite DNAs

Although in the last two decades cytogenetic studies have been carried out in a large number of fish species, such analyses were mainly directed to the knowledge of basic karyotype structure and few research works have been conducted in the field of organization of DNA sequences in the chromosomes. Concerning chromosomal organization of repetitive DNA sequences, one of the most studied fractions of the fish genome is the satellite DNA. Satellite DNAs are tandemly arrayed, highly repetitive DNA sequences found in the eukaryotic genomes. These sequences can vary from 1,000 to over 100,000 copies of a basic motif or repeat unit, commonly 100-300 base pairs (bp), which occur at a few loci on the genome. These sequences are organized as large clusters mainly in the centromeric and telomeric regions of the chromosomes, and are the principal component of heterochromatins. Satellite DNA families can correspond to 0-66% of certain mammalian genomes and their composition and number of unrelated families can vary greatly (Beridze, 1986). Different species generally present a divergence among satellite DNA families as a result of concerted evolution mechanisms (Arnheim, 1983), leading to species-specific satellite DNA sequences. On the other hand, there are a few exceptions in which a group of species, or even a whole family or order, shares the same satellite DNA family. The most interesting example is the centromeric alpha satellite DNA that is preserved in the primate order, most probably because of its centromeric function (Schueler et al., 2001). Alpha satellite-like sequences were also detected in the chicken and zebrafish, showing an interesting sequence identity with human alpha sequences (Li and Kirby, 2003).

Satellite DNAs are useful for molecular cytogenetic analysis such as the identification of homologous chromosomes and chromosomal abnormalities by in situ hybridization. The molecular organization, chromosomal location, and possible functions of satellite DNAs have been studied in several groups of animals (Brutlag, 1980; Singer, 1982; Arnason et al., 1984; Hummel et al., 1984; Clabby et al., 1996). These studies have indicated that satellite-like sequences may play an important role at the chromosomal and nuclear level (Singer, 1982; Haaf and Schmid, 1991; Larin et al., 1994; Sart et al., 1997).

Although the chromosome distribution of the heterochromatin—where the satellite DNAs are supposed to be concentrated—has been extensively studied in fishes using cytological methods of staining or chromosome banding, molecular data on satellite DNAs are restricted to a few species (Table 4.4.1). The first descriptions of satellite DNA families in fishes correspond to the end of the 1980s (Datta et al., 1988; Moyer et al., 1988; Monaco et al., 1989). The data on satellite DNAs in fishes show that these sequences are mainly located in the centromeric region of chromosomes (Table 4.4.1). A HindIII satellite DNA family isolated from the sturgeon Acipenser naccarii genome was preserved in the pericentromeric regions of the chromosomes of six species of the genus Acipenser and one of the genus Huso (Lanfredi et al., 2001). Centromeric satellite DNA families were also isolated from the genome of the gobiid Gobius cobitis (Canapa et al., 2002) and the Nile tilapia Oreochromis niloticus (Franck and Wright, 1993). Particularly in the Nile tilapia, the satellite family was present in the centromeres of all chromosomes of the complement (Oliveira and Wright 1998) (Fig. 4.4.1a, b). The tilapia satellite was also preserved in the genome of other tilapiine and haplochromine species (Franck et al., 1994), thus leading to the possible hypothesis that this satellite sequence originated in an ancestor of the group, and subsequently, has been maintained in the centromeres of all chromosomes due to its functionality. One interesting feature of most centromeric satellite repeats is their basic length unit (155-180 bp) corresponding to the range of nucleosomal unit lengths (Henikoff et al., 2001). Larger repeat lengths may encompass two nucleosomes. An extensive analysis of centromeric satellite DNAs of vertebrates showed the presence of short A-rich motifs that are typical of centromere satellite (Vinãs et al., 2004). Short A-rich motifs have been identified in the centromeric satellite DNAs of various fish species (Wright, 1989; Denovan and Wright, 1990; Garrido-Ramos et al., 1994, 1995; Kato, 1999; Canapa et al., 2002; Viñas et al., 2004). These short sequences are quite similar and show considerable homology to other centromeric motifs found in humans (Vissel et al., 1992), mice (Wong and Rattner, 1988), and reptiles (Cremisi et al., 1988), suggesting that such sequences might also play some important role in the structure and function of the fish centromere.

Table 4.4.1 DNA tandem repeats and their chromosomal distribution in fishes.

Fish families and species	Repeat size (bp)	Chromosome position	References
Acipenseridae			
Acipenser naccarii	180	centromeric	Garrido-Ramos et al., 1997; Lanfredi et al., 2001
Acipenser gueldenstaedtii	180	centromeric	Lanfredi et al., 2001
Acipenser baerii	180	centromeric	Lanfredi et al., 2001
Acipenser transmontanus	180	centromeric	Lanfredi et al., 2001
Acipenser ruthenus	180	centromeric	Lanfredi et al., 2001
Huso huso	180	centromeric	Lanfredi et al., 2001
Adrianichthyidae			
Oryzias latipes	600	sex-linked	Matsuda et al., 1998
Anostomidae			
Leporinus elongatus	174, 729	chromosomes Z and W	Nakayama et al., 1994
Leporinus obtusidens	483	pericentromeric	Koehler et al., 1997
Channichthydae			
Chionodraco hamatus	1,000	centromeric and telomeric	Capriglione et al., 1994
Characidae			
Astyanax scabripinnis	51	heterochromatins	Mestriner et al., 2000; Mantovani et al., 2004
Cichlidae			
Oreochromis niloticus	237	centromeric	Franck et al., 1992; Oliveira and Wright, 1998
Oreochromis niloticus	1,900	short arm of chromosome four	Frank and Wright, 1993; Oliveira and Wright, 1998
Cyprinidae			
Carassius auratus langsdorfi	137		Murakami and Fujitani, 1997
Danio rerio	180, 191	centromeric	Ekker et al., 1992; He et al., 1992; Sola and Gornung, 2001
Danio rerio	200	AT-rich heterochromatins	Phillips and Reed, 2000
Danio rerio	92	GC-rich heterochromatins	Phillips and Reed, 2000
Erythrinidae			
Hoplias malabaricus	333-366	centromeric	Haaf et al., 1993
Hoplias malabaricus	356-360	centromeric	Martins et al. (unpublished)
Gobiidae			
Gobius cobitis	332	centromeric	Canapa et al., 2002
Gobius paganellus	332	centromeric	Canapa et al., 2002

(Table 4.4.1 contd.)

(Table 4.4.1 contd.)

Heptapteridae			
Imparfinis schubarti	2	telomeric	Vanzela *et al.*, 2002
Loricariidae			
Rineloricaria latirostris	2	near NOR	Vanzela *et al.*, 2002
Parodontidae			
Parodon hilarii	200	terminal heterochromatins, W chromosome	Vicente *et al.*, 2003
Pimelodidae			
Steindachneridion scripta	2	telomeric, dispersed	Vanzela *et al.*, 2002
Poecilidae			
Poecilia reticulata	4	Y chromosome	Nanda *et al.*, 1990
Prochilodontidae			
Prochilodus lineatus	441	pericentromeric	Jesus *et al.*, 2003
Prochilodus lineatus	900	pericentromeric and supernumeraries	Jesus *et al.*, 2003
Prochilodus lineatus	5	telomeric	Hatanaka *et al.*, 2002
Prochilodus marggravii	5	telomeric	Hatanaka *et al.*, 2002
Salmonidae			
Salvelinus alpinus	72, 127, 200, 400	centromeric	Hartley and Davidson, 1994
Salvelinus namaycush	140	centromeric	Reed and Phillips, 1995
Salvelinus namaycush	120	centromeric	Reed and Phillips, 1995
Salmo trutta	359	NOR	Abuín *et al.*, 1996
Salmo salar	380, 442, 923	NOR, rDNA	Goodier and Davidson, 1994; Abuín *et al.*, 1996
Salmo salar	260	pericentromeric	
Salmo salar	42	interstitial	Pérez *et al.*, 1999
Salmo salar	28	telomeric, pericentromeric, centromeric	Pérez *et al.*, 1999
Salmo salar	34	interstitial	Pérez *et al.*, 1999
Oncorhynchus tshawytscha	939	subtelomeric	Devlin *et al.* 1991, 1998
Sparidae			
Sparus aurata	186	centromeric	Garrido-Ramos *et al.*, 1994
Pagrus pagrus	186	subtelomeric	Garrido-Ramos *et al.*, 1998
Pagrus aurica	186	subtelomeric	Garrido-Ramos *et al.*, 1998
Pagellus erythrinus	186	subtelomeric and telomeric	Garrido-Ramos *et al.*, 1998
Tetraodontidae			
Tetraodon nigroviridis	118	(peri)centromeric	Crollius *et al.*, 2000
Tetraodon nigroviridis	10	heterochromatins	Crollius *et al.*, 2000

(Table 4.4.1 contd.)

(Table 4.4.1 contd.)

Tetraodon nigroviridis	100	heterochromatins	Fischer *et al.*, 2004
Tetraodon nigroviridis	104	heterochromatins	Fischer *et al.*, 2004
Fugu rubripes	118	centromeric	Brenner *et al.*, 1993; Elgar *et al.*, 1999
Cyclostomata			
Eptatretus okinoseanus	90	interstitial	Kubota *et al.*, 1993
Eptatretus burgeri	57 and 64	interstitial	Kubota *et al.*, 2001
Petromyzon marinus		centromeric and pericentromeric	Bóan *et al.*, 1996

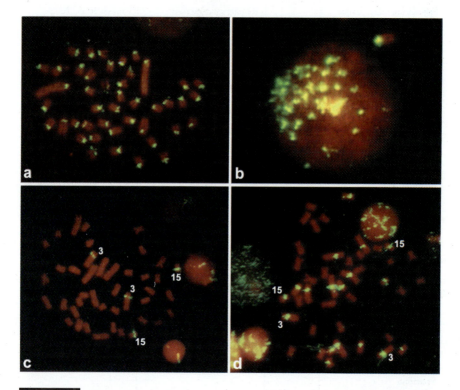

Fig. 4.4.1 Fluorescence *in situ* hybridization of SATA satellite DNA family in the chromosomes (a) and in the interphasic nucleous (b) of *Oreochromis niloticus*. The SATA maps into the centromeres of all chromosomes and has a non-random distribution in the interphasic nucleous. Fluorescence *in situ* hybridization of the true-5S rDNA (c) and variant-5S rDNA (d) repeats in the chromosomes of *H. malabaricus*, under highly stringent conditions. The chromosome pairs number 3 and 15 are indicated. (Courtesy of Irani A. Ferreira).

Particular attention has been directed to the identification of satellite DNAs related to sex and supernumerary chromosomes in fishes. Satellite DNAs have been isolated and mapped in sex chromosomes of several species, such as *Leporinus elongatus* (Nakayama *et al.*, 1994), *Chiondraco hamatus* (Capriglione *et al.*, 1994), *Poecilia reticulata* (Nanda *et al.*, 1990), and *Oncorhynchus tschawytscha* (Devlin *et al.*, 1991; Stein *et al.*, 2001), among others.

Morphologically differentiated sex chromosomes occur in several fish species. The South American fishes of the genus *Leporinus* (Anostomidae, Characiformes) represent a good model for studying sex chromosome differentiation. A clear ZZ/ZW sex chromosome system has been described for seven out of 40 studied species (*L. conirostris, L. trifasciatus, L. obtusidens, L. elongatus, L. macrocephalus, L. reinhardti* and *L.* aff. *elongatus*) (Galetti *et al.*, 1995). In these species, the subtelocentric W chromosome, only present in the female karyotype, is large and almost fully heterochromatic. In contrast, the Z chromosome, which is present in both sexes, is heterochromatic on the distal third part of the long arm. The shared morphological features of chromosomes Z and W among these seven ZW *Leporinus* suggest a common origin (Galetti *et al.*, 1995) and an initial heterochromatinization could have been the first step in the differentiation of these sex chromosomes (Galetti and Foresti, 1986). On the other hand, a novel ZW chromosome system, morphologically differentiated from the ZZ/ZW system previously detected, was described for *Leporinus* sp2 (Venere *et al.*, 2004).

Two satellite DNA families were isolated from *Leporinus elongatus* by subtractive hybridization and cloning (Nakayama *et al.*, 1994). One of them was located in both Z and W chromosomes, whereas the second family was specific to the W chromosome, which permits the sex identification of individuals of the species. Three atypical individuals, one ZW female, recognized as male with the W-specific probe sequence, and two ZW males, one of them showing satellite-pattern of males, were detected in the analysis. These three individuals seem to have originated from genetic exchange between regions of the chromosomes Z and W, yielding the atypical patterns observed. These data on sex chromosome in *Leporinus* allowed the construction of a model for the W chromosome differentiation that offers important contributions to the knowledge of sex-determination mechanisms in fishes.

Among salmonids, extensive studies have been conducted with the purpose of isolating DNA sequences to be used as markers for sex identification, since the heteromorphism of the X and Y chromosomes is not easily detected. An Y-specific repetitive DNA sequence with 300 copies organized in approximately six distinct clusters was isolated from the genome of the salmon *Oncorhynchus tshawytscha* (Devlin *et al.*, 1991, 1998; Stein *et al.*, 2001). Contrary to what was observed in several eukaryotes where long repetitive sequences originated from the duplication of short repeat units, the 8 kb repeat isolated from salmon does not contain relics of internal short repeats.

Another interesting issue is the presence of supernumerary chromosomes that have been described in case of several fish species. A satellite DNA correlated to a supernumerary chromosome was firstly isolated in *Astyanax scabripinnis*, a small headwater fish that has become a model species not only because of numeric and structural chromosome variations but also due to the occurrence of supernumerary chromosomes. In this species, the isolated repetitive DNA family, named As51, had repeats of 51 bp and was located in the non-centromeric heterochromatins, in the NORs and in the supernumerary chromosome. Interestingly, the As51 satellite family showed 58.8% similarity with a segment of the retrotransposon RT2 of *Anopheles gambiae* and a lower similarity with the transposase gene of the transposon TN4430 of *Bacillus thuringiensis*, suggesting that this sequence might have arisen from a mobile element. The presence of As51 satellite in the NOR region suggests that this repetitive family can be interspersed in the spacers of the 45S rDNA—being inserted in this region by transposition—as described before for other organisms. Although the chromosome hybridization detected fluorescent signals in almost the whole extension of the supernumerary chromosome, it was possible to verify that the satellite DNA As51 is organized in small clusters interspersed with other DNA types. The symmetric distribution in both arms of the supernumerary chromosome of *A. scabripinnins* and its meiotic behavior suggest that this chromosome is an isochromosome (Mestriner *et al.*, 2000).

Analyses of the nucleotide composition of supernumerary chromosome of other species showed that they are mainly composed of repetitive DNAs. Such data is related to the heterochromatic nature of the supernumeraries.

Satellite DNAs were also isolated from *Prochilodus lineatus,* which presents from zero to five small supernumerary chromosomes (Pauls and Bertollo, 1983). Two satellite DNA families, with monomeric units of 441 and 900 bp, were isolated from the genome of this species (Jesus *et al.*, 2003). Both satellite families were located in the pericentromeric region of several chromosomes of the A complement and the 900 bp satellite was also located in several supernumeraries. Double-chromosome hybridization using the 441- and the 900-bp satellites as probes showed that both families co-localize in the pericentromeric region of some chromosomes. Several sub repeats were detected in the 900-bp satellite, suggesting its origin from small repeat units. Such results demonstrate that the supernumerary chromosomes of this species have originated from A chromosomes that harbor the 900-bp satellite DNA family (Jesus *et al.*, 2003).

Studies on satellite DNAs have proved useful in clarifying a myriad of questions, including centromeric structure, and origin and evolution of sex and supernumerary chromosomes. Satellite DNAs could also find applications in the physical mapping of the genome, contributing to the development of genetic markers of significant importance to fundamental and applied biology of fish species.

Minisatellite DNAs

The definition of a minisatellite repeat is not particularly well standardized. Minisatellite includes all tandem repeats that are not sufficiently large to be included as a satellite repeat nor small enough to be considered a microsatellite sequence. Most scientists consider minisatellite to be tandem arrays (5-50 repeats) of moderately repetitive short DNA sequences (10-60 bases) found dispersed throughout the genome and clustered near telomeres. The intense evolutionary dynamics of minisatellites generate complex arrays of such sequences in the genome that can be used as specific markers for an individual. Microsatellites were first used in forensic applications and in paternity assessment in humans (Jeffreys *et al.*, 1985) and were later widely applied to population and identification of organisms from bacteria to mammals. Although minisatellites have been described for many fish species (Goodier and Davidon, 1998), the chromosomal mapping of minisatellites was only described for a few species (Table 4.4.1). In the Atlantic salmon (*Salmo salar*) (Pérez *et al.*, 1999), three minisatellites with core sequences of 42, 28 and 34 nucleotides were mapped onto the chromosomes. A single locus was detected for the 42-mer and 34-mer microsatellites and a multilocus pattern was detected for the 28-mer microsatellite. Analyses of the pufferfish (*Tetraodon nigroviridis*) genome allowed the identification of two major classes of tandem repeat elements: one corresponding to a 118-mer satellite that was mapped in the centromeric region of all chromosomes and the other with 10-mer minisatellite that was mapped in the complete length of the short arm of 10 subtelocentric chromosomes (Crollius *et al.*, 2000).

Microsatellite DNAs

Microsatellites consist of tandem repeats sequences of short DNA sequences with one to five or six bases interspersed in the eukaryotes genome (Tautz and Renz, 1984). The tandem array of the microsatellites are usually less than 100 bases long and have been used for linkage map analysis. A linkage map analyses of the zebrafish *Danio rerio* genome using 200 microsatellite markers of the CA and GT types identified that the $(CA)_n$ and $(GT)_n$ repeats are more clustered in the centromeric and telomeric regions (Shimoda *et al.*, 1999). Clustering of microsatellites in the centromeres and telomeres was also detected by *in situ* chromosome mapping. In three species of the genus *Prochilodus* (Prochilodontidae, Characiformes), the $(AATTT)_n$ repeats revealed 13 alleles while chromosome mapping showed signals predominantly in the telomeric regions of several chromosomes (Hatanaka *et al.*, 2002). $(GA)_n$ microsatellite repeats were highly clustered in the telomeric regions of *Imparfinis schubarti* (Heptapteridae, Siluriformes), dispersed along the chromosome arms with enhanced signals in some telomeric regions in *Steindachneridion scripta* (Pimelodidae, Siluriformes), and near to the NOR sites in *Rineloricaria latirostris* (Loricariidae, Siluriformes) (Vanzela *et al.*, 2002). *In situ* chromosome hybridization detected rich segments of the $(GACA)_n$ repetitive sequence in the heterochromatic portion of the W and Y chromosomes of *Poecilia reticulata* (Poecilidae, Cyprinodontiformes) (Nanda *et al.*, 1990). The microsatellites are essentially located in the heterochromatic regions (telomeres, centromeres and sex chromosomes) of fish genome, where a significant fraction of the repetitive DNAs is believed to be localized.

Transposons

Most repeated sequences of the genome are derived from transposable elements. In humans, 45% of the genome belongs to this class of repeats (The Genome International Sequencing Consortium, 2001). Among vertebrates, two classes of transposon types can be recognized. The first class (i) includes three types that transpose through RNA intermediates: long interspersed elements (LINEs), short interspersed elements (SINEs), and long terminal repeat (LTR) retrotransposons; and the second class (ii) includes those sequences that transpose directly as DNA:DNA transposons. Fish genomes contain all types of known transposable elements (Volff et al., 2003; Dettai et al. present volume, Chapter 4) and some of these elements were mapped onto the chromosomes (Table 4.4.2).

The small and compact genomes of the pufferfish Tetraodon nigroviridis and Takifugu rubripes are interesting for studying the repetitive sequences. The total content of transposons in the genome of T. nigroviridis is only 0.9%, a large fraction of which is constituted by LINE elements (0.4%) (Crollius et al., 2000). The genomes of the two pufferfish contain a low repeat content (Aparicio et al., 2002; Fischer et al., 2004). On the other hand, T. nigroviridis contains a high diversity of transposable elements not observed in larger genomes such as those of the human and mouse (Aparicio et al., 2002; Volff et al., 2003). The transposable elements are compartmentalized in heterochromatins and are not randomly distributed in the genome of this pufferfish (DaSilva et al., 2002; Bouneau et al., 2003; Fischer et al., 2004) (Table 4.4.2). Moreover, the distribution of repeats in the genome of T. nigroviridis is different from the one observed in humans, where repeat sequences comprise an important fraction of euchromatic DNA, and is more similar to the distribution observed in Drosophila melanogaster and Arabidopsis thaliana also having small genomes (Fischer et al., 2004. Repetitive sequences such as minisatellites and transposable elements are clearly located in AT-rich chromosome regions in T. nigroviridis (Fischer et al., 2004). Such compartmentalization of the pufferfish genome seems not to be a rule for fishes and might represent a characteristic of small and compact genomes.

Although cytogenetic studies of transposable elements in fishes are just starting, the preliminary results suggest that these elements can greatly contribute to the knowledge of fish genome evolution. For instance, a LINE element denominated CiLINE2, isolated from O. niloticus genome (Oliveira et al., 1999) was detected by hybridization in

Table 4.4.2 Dispersed repetitive elements and their chromosomal distribution in fish species.

Fish orders and species	Element type	Chromosome position	References
Aulopiformes			
Aulopus japonicus		W chromosome	Ota et al., 2003
Cypriniformes			
Alburnus alburnus	Gypse, Ty3	B chromosome	Ziegler et al., 2003
Cyprinodontiformes			
Xiphophorus maculatus	XIR LTR-like	Y chromosome	Nanda et al., 2000
Perciformes			
Artedidraco shackletoni	Rex1, Rex3	Dispersed	Ozouf-Costaz et al., 2004
Bovichtus angustifrons	Rex1, Rex3	Dispersed	
Chionodraco hamatus	Tc1-like	Pericentric, telomeric, interstitial	Capriglione et al., 2002
Chianodraco hamatus	Rex1, Rex3	Dispersed	Ozouf-Costaz et al., 2004
Dissostichus mawsoni	Rex1, Rex3	Dispersed	Ozouf-Costaz et al., 2004
Gobius niger	Mariner-like	Overlapping NORs	Mandrioli et al., 2001
Gymnodraco acuticeps	Rex1, Rex3	Dispersed	Ozouf-Costaz et al., 2004
Gymnodraco victori	Rex1, Rex3	Dispersed	Ozouf-Costaz et al., 2004
Neopagetopsis ionah	Rex1, Rex3	Dispersed	Ozouf-Costaz et al., 2004
Notothenia coriiceps	Rex1, Rex3	Dispersed	Ozouf-Costaz et al., 2004
Oreochromis niloticus	CiLINE2	Chromosome one and dispersed	Oliveira et al., 1999
Oreochromis niloticus	Ron1	Chromosome one and dispersed	Bryden et al., 1998; Oliveira et al., 2003
Oreochromis niloticus	Ron2	Dispersed	Oliveira et al., 2003
Oreochromis niloticus	On2318	Chromosome one and dispersed	Harvey et al., 2003
Oreochromis niloticus	On239, Tc1-like	Centromeric, telomeric and dispersed	Harvey et al., 2003
Patagonotothen tessellata	Rex1, Rex3	Dispersed	Ozouf-Costaz et al., 2004
Trematomus hansoni	Rex1, Rex3	Dispersed	Ozouf-Costaz et al., 2004
Trematomus newnesi	Rex1, Rex3	Dispersed	Ozouf-Costaz et al., 2004
Trematomus bernacchii	Rex1, Rex3	Dispersed	Ozouf-Costaz et al., 2004
Trematomus pennellii	Rex1, Rex3	Dispersed	Ozouf-Costaz et al., 2004
Tetraodontiformes			
Tetraodon fluviatilis	Mariner-like	NOR-associated heterochromatins	Mandrioli and Manicardi, 2001
Tetraodon nigroviridis	Dm-Line	Heterochromatins	DaSilva et al., 2002
Tetraodon nigroviridis	Tc1-like	Heterochromatins	DaSilva et al., 2002
Tetraodon nigroviridis	Zebulon	Heterochromatins	Bouneau et al., 2003
Tetraodon nigroviridis	Tol2	Heterochromatins	Fischer et al., 2004
Tetraodon nigroviridis	Buffy1	4-5 chromosomes	Fischer et al., 2004
Tetraodon nigroviridis	Rex3	Heterochromatins	Fischer et al., 2004
Tetraodon nigroviridis	Babar	Heterochromatins	Fischer et al., 2004

the genomic DNA of all Tilapiini species tested from the genera *Oreochromis*, *Tilapia*, and *Sarotherodon*. It is interesting to note that DNA from *Oreochromis* and *Sarotherodon* species produced a hybridization pattern different from that of *Tilapia* species, thus suggesting the possibility that the CiLINE2 probe could be used to distinguish fishes of the genera *Tilapia* from those of *Oreochromis* and *Sarotherodon*. Fluorescence *in situ* hybridization with the CiLINE2 element evidenced in *O. niloticus* very small signals distributed more or less randomly over the chromatids of all chromosomes, but strikingly enriched along the terminal two-thirds of the long arm of chromosome pair one (Oliveira *et al.*, 1999) that corresponds to the putative XY sex chromosomes.

The distribution of two SINE sequences denominated ROn-1 and the ROn-2, on chromosomes of the Nile tilapia, investigated by fluorescence *in situ* hybridization by Oliveira *et al.* (2003), showed that both SINE sequences are organized in small clusters and dispersed in all the chromosomes. Moreover, the ROn-1 element is almost exclusively distributed in interstitial regions of chromosomes and copies of ROn-2 are localized near the telomeric region of several chromosomes. A large cluster of ROn-1 is found in the middle of the long arm of chromosome pair one. No similarity was observed in the distribution of SINEs and LINEs between the Nile tilapia chromosomes and mammalian chromosomes.

Whole chromosome probes obtained through microdissection and DOP-PCR-amplification from chromosome one of *O. niloticus*, hybridized more intensely in the long arm of chromosome one, suggesting the presence of large numbers of repetitive elements in this chromosome region (Harvey *et al.*, 2002). Cloning and sequencing of the microdissected DNA from the XY sex chromosomes (chromosome pair one) of the same species proved that these chromosomes are enriched with repetitive sequences, most of them transposable elements (Harvey *et al.*, 2003). Moreover, the distribution of repetitive elements in the sex chromosomes of *O. niloticus* suggests that there are significant differences between the X and Y chromosomes. Considering that the main differences detected between the X and Y chromosomes reside in the long arm (Foresti *et al.*, 1993), the development of new genetic markers capable of distinguishing the X and Y chromosomes will be of considerable value for aquaculture purposes.

The genome dynamics and chromosomal localization of two non-long terminal repeat retrotransposons (*Rex1* and *Rex3*) were studied in 13 species of notothenioid Antarctic fishes (Ozouf-Costaz *et al.*, 2004) (Table 4.4.2). Both *Rex1* and *Rex3* transposon elements were spread all over the chromosomes with accumulation in some particular regions, such as in the Y sex chromosome of *Chionodraco hamatus*. The presence of another transposon-like element (*Tc1*) (Capriglione *et al.*, 2002) was also identified in the Y chromosome of *C. hamatus*. Particularly, the *Tc1*-like element hybridizes interstitially to the long arm of the Y sex chromosome, which is supposed to originate in tandem or by Robertsonian fusion (Morescalchi *et al.*, 1996). This suggests that transposon elements might have been involved in notothenioid sex-chromosome differentiation.

The presence of transposon-derived elements are also common in the B chromosomes of diverse organisms from plants such as *Brachycone dichronosonidica* (Franks *et al.*, 1996) and *Rye* (Langdon *et al.*, 2000), and animals such as the insect *Nasonia vitripennis* (McAllister, 1995) and the fish *Astyanax scabripinnis* (Mestriner *et al.*, 2000). A specific retrotransposon-like element isolated in *Alburnus alburnus*, by AFLP, has been found abundant in B chromosomes and absent in the normal A chromosomes (Ziegler *et al.*, 2003).

Taken altogether, the results of studies on repetitive DNAs seem to have considerable value in clarifying several issues concerning the origin and evolution of sex and supernumerary chromosomes among the organisms and the genome evolution. The constant presence of repeated sequences in the sex and supernumerary chromosomes in fish species indicates that the repetitive DNAs have played an important role in the evolution of their genomes.

5S rDNA and 5S rDNA Variant Repeats

Chromosomal organization of 5S rDNA

Studies on ribosomal RNA genes have gained prominence in a broad range of animals and plants, especially in relation to species or population characterization, evolutionary relationships and genome structuring. In higher eukaryotes, ribosomal RNA (rRNA) genes are organized as two distinct multigene families of tandemly arrayed repeats composed of hundreds to thousands of copies. One class is represented by the 45S rDNA, which consists of a transcriptional unit encoding for the 18S, 5.8S and 28S rRNAs, and an intergenic non-transcribed spacer (IGS). Multiple copies of this array correspond to the nucleolar organizer regions (NORs). The other class (5S rDNA) consists of a highly conserved sequence of 120 base pairs (bp) coding for the 5S rRNA which is separated from each transcriptional unit by a variable non-transcribed spacer (NTS) (reviewed in Long and Dawid, 1980). While the rRNA genes are conserved even among non-related taxa, the non-transcribed spacers show extensive length and sequence variation, which can give an accentuated dynamism to the rRNA genes.

The chromosomal location of the 5S rRNA genes has been described for more than 60 fish species, representing distinct groups such as Acipenseriformes, Anguilliformes, Cypriniformes, Characiformes, Salmoniformes, Perciformes, and Tetraodontiformes and has been shown to be of great importance in the comprehension of the structure and organization of repeated sequences in their chromosomes (for review, see Martins and Wasko, 2004).

In most eukaryotes, the 5S rRNA genes are generally detected in distinct areas of the genome, organized as one or more tandemly repeated clusters, and the number of 5S rRNA genes ranges from 100 to 300,000 copies, which is usually higher than the number of 45S rRNA genes (Hadjiolov, 1985). In many vertebrates, 5S rRNA genes are located on a single chromosome pair, while 45S rDNA is often present on multiple chromosomes (Suzuki et al., 1996; Makinem et al., 1997). In amphibian (Schmid et al., 1987; De Lucchini et al., 1993) and fish species (Martins and Wasko, 2004; Mazzei et al., 2004), the 5S rRNA genes can be found on several chromosomes. For most fish species, 5S rRNA genes have an interstitial position in the chromosomes, which suggests that such localization could represent some advantage related to the organization of these genes in the genome. Moreover, 45S and 5S rDNA loci may assume a syntenical organization in the chromosome (Pendás et al., 1994; Móran et al., 1996; Mazzei et al., 2004) or can be detected in different chromosome pairs (Martínez et al., 1996; Martins and Galetti, 1999). However, the divergent locations of NORs and 5S rDNA loci seem to be the most common situation observed in fish and is by far the most frequent distribution pattern observed in vertebrates (De Lucchini et al., 1993, Suzuki et al., 1996). It was suggested that the distinct chromosome locations of 5S and 45S rDNA could represent some advantage compared to the linked condition (Martins and Galetti, 1999). Syntenic location of

5S and 45S clusters could facilitate translocations between the 45S and 5S arrays, causing disruptive interference in the structure and function of such genes. This could explain why most vertebrates have these 5S rDNA clusters on distinct chromosomes.

In the characiform *Leporinus*, two classes of 5S rDNA, one consisting of monomeric repeat units around 200 bp and another one with monomers of 920 bp were identified (Martins and Galetti, 2001). Each of these different-sized 5S rDNA classes was characterized by distinct NTS sequences and clustered in distinct chromosome pairs. Several studies of 5S rDNA sequences among fish species have identified variant types of the 5S rDNA tandem repeats characterized by remarkable differences in the NTSs. The presence of two types of tandem repeats of this ribosomal DNA has been observed in Characiformes (Martins and Galetti, 2001), Perciformes (Martins *et al.*, 2002) and Salmoniformes (Pendás *et al.*, 1994). In the tilapiine cichlid fish *O. niloticus,* two distinct 5S rDNA units were identified, each characterized by distinct NTSs that varied in nucleotide sequence and length between the loci. The first class has monomers of 1.405 bp (denominated 5S rDNA type I) and the second one has monomers of 475 bp (denominated 5S rDNA type II). An inverted 5S rRNA putative pseudogene and two putative 5S rRNA bona fide genes (one of them inverted) were also detected in the 5S rDNA type I (Martins *et al.*, 2002). Both classes were clustered in distinct chromosomes. While the 5S rDNA type I was detected in an interstitial position in the long arm of a subtelo-acrocentric chromosome pair (chromosome 3), the 5S rDNA type II was identified interstitially in the long arm of a different subtelo-acrocentric pair and at the terminal region of the short arm of another subtelo-acrocentric chromosome pair (chromosomes 9 and 13). The exhaustive investigation of NTS segments of 5S rDNA type I and type II allowed detection of the presence of only one type of NTS in the 5S rDNA type I and two subtypes of NTS in the 5S rDNA type II (Alves-Costa, *et al.*, 2006). The subtypes detected in NTS of 5S rDNA type II are related to the presence of a 'TG' microsatellite expansion/deletion. Interestingly, the 5S rDNA type I is located in just one chromosome locus while the 5S rDNA type II is located in two different chromosome loci. Such data prove that the homogenization of the repeats in 5S rDNA can occur just within a specific locus, whereas different loci in the same genome can be highly differentiated in the nucleotide sequence and size of the repeat units. The assignment of different classes of 5S rDNA to distinct chromosome loci reinforces the idea that distinct 5S rDNA classes occupy different chromosome positions and seem to evolve independently in individual nuclear environments.

Dynamic of 5S rDNA and variant-5S rDNA in the genome of Hoplias malabaricus

An interesting model to demonstrate the intense dynamism of the 5S rDNA repeats in the genome is the fish *Hoplias malabaricus*. Two tandem repetitive families denominated true-5S rDNA and variant-5S rDNA were isolated and characterized in the genome of this species (Martins *et al.*, 2006). The true-5S rDNA repeats contain entire coding regions for 5S rRNA while the variant-5S rDNA repeats contain a truncated coding region for the 5S rRNA (Fig. 4.4.2). Similarities were also observed in the NTS of both classes.

Chromosome hybridization was carried out using the true-5S rDNA and the variant-5S rDNA sequences as probes. Under low-stringent conditions, both probes hybridized in the centromeric region of 18 chromosomes and near the centromeres in the short arm of chromosome pairs 3 and 15. Under high-stringent conditions, the true-5S rDNA probe hybridized to the short arm of chromosome pairs 3 and 15 and the variant-5S rDNA probe to the centromeric region of 18 chromosomes (Figs. 4.4.1c, d).

The true-5S rDNA repeats of *H. malabaricus* were nearly identical with a low value for the mean genetic distance (0.001) between the repeats, suggesting that such sequences are governed by strong selective pressure. On the other hand, the high (0.045) mean genetic distance between the repeats of the variant-5S rDNA suggests that these sequences are free of selection pressure. One evidence of the intense

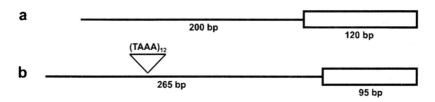

Fig. 4.4.2 Schematic representation of the true-5S rDNA (a) and the variant-5S rDNA (b) repeats isolated in *Hoplias malabaricus*. The boxes indicate segments with similarity to the coding region and the horizontal lines the NTS. The size of the NTS and the 5S rRNA coding region, and the TAAA microsatellite are indicated.

dynamism of the variant-5S rDNA sequences is the presence of the expanded TAAA microsatellite. To explain the difference between the genetic distance of the true-5S rDNA and the variant-5S rDNA satellite sequence, it is possible to hypothesize that a transfer of true-5S rDNA units to the centromeric position changed the status of the selective pressure under the 5S rRNA genes, making them free to multiply and spread over the centromeres of several chromosome pairs, as has been demonstrated for other centromeric satellite sequences.

The organization and evolution of tandem repetitive DNAs is governed by particular patterns of evolution such as unequal exchange, transposition, RNA-mediated transposition and gene conversion (Dover, 1986). Drouin and Moniz de Sá (1995) suggested the hypothesis that RNA-mediated transposition is the mechanism responsible for the unusual linkage of 5S rRNA genes to other tandemly repeated multigene families. According to the authors, the RNA-mediated transposition could be responsible for the dispersion of single copies of 5S rDNA repeats, whereas covalently closed circular DNA (cccDNA) molecules containing 5S rRNA genes would be expected to sometimes lead to the insertion of several 5S RNA gene copies within other sequences in the genome. Such cccDNA molecules have been found in many eukaryotes species, including mammals, chicken, *Drosophila*, and plants (reviewed in Renault et al., 1993). Several classes of cccDNAs have been found in *D. melanogaster* embryos, one of whom contains a variable number of sequences homologous to 5S rRNA genes (Pont et al., 1987). Therefore, in *H. malabaricus*, the first copies of the variant-5S rDNA could have transferred to centromeric position throughout cccDNAs. Alternatively, the variant-5S rDNA satellite sequence could have originated in the centromeric region of chromosomes 3 or 15 by duplication or chromosome inversion involving some adjacent 5S rDNA copies present in these chromosomes. The first variant-5S rDNA copies could have associated with other repetitive sequences in the centromeric heterochromatin that facilitated its dispersion to the other chromosomes due to concerted evolutionary mechanisms.

It cannot yet be addressed whether these repeats may confer some structural or functional advantage to the chromosomes as a component of the centromeric DNA in *H. malabaricus*. Centromeres have been recognized as evolutionary dynamic regions of the genome (Eichler and Sankoff, 2003); but although they have been well investigated in cases from animals to fungi, important tasks remain to be understood (Henikoff et al., 2001). The centromere is vital to the correct sorting of chromosomes during cell division, being essential for the appropriate maintenance and segregation of the genetic material. Although this role is conserved throughout evolution, the DNA sequences found in centromeric regions are often variable (Henikoff et al., 2001). Disturbances in the structural and functional organization of the centromeres are critical in terms of leading to problems such as developmental defects and cancer. The centromeric regions are rich in repetitive DNAs, which is a common trait in humans (Willard and Wayne, 1987), mice (Kipling et al., 1991; Narayanswami et al., 1992), maize (Kaszás and Birchler, 1996), Drosophila (Murphy and Karpen, 1995), Neurospora (Centola and Carbon, 1994), and yeast (Clark, 1990). One interesting finding is that the expanded TAAA motif in the variant-5S rDNA is similar to the short A-rich motifs identified in the centromeric satellite DNAs of different fish species, as previously reported in this chapter. These short sequences are quite similar, showing considerable homology to other centromeric motifs found in humans (Vissel et al., 1992), mice (Wong and Rattner, 1988), and reptiles (Cremisi et al., 1988), a fact which suggests that such sequences play an important role in the structure and function of the H. malabaricus centromere.

Several previous studies have found evidence of 45S rDNA-related sequence elements either dispersed or clustered throughout eukaryotic genomes. These elements have been characterized mainly as non-coding, small-unit tandem repeats of variable copy number and have been identified in various eukaryotic species, including yeast (Childs et al., 1981), animals (Arnheim et al., 1980; Kominami and Muramatsu, 1987; De Lucchini et al., 1988; Lohe and Roberts, 1990), and plants (Unfried et al., 1991; Falquet et al., 1997). The results presented here for the fish H. malabaricus show that similar elements also may have originated from 5S rDNA. Dispersed 5S rDNA variants and pseudogenes seem to be common in mammals (Emerson and Roeder, 1984; Doran et al., 1987; Leah et al., 1990). On the other hand, certain interesting features of the variant-5S rDNA repeats of H. malabaricus include high copy number, the tandem array, and their centromeric positioning.

Repetitive DNA sequences are subject to the action of several molecular mechanisms and are thought to be the most rapidly evolving components of eukaryotic genomes. The results discussed for

H. malabaricus also represent an apt example of the fluidity of repetitive sequences in providing novelties to the genomic organization of the centromeric region of vertebrates. The satellite variant-5S rDNA family has propagated within the centromeric region of several chromosomes and has been favored during the evolution due to a possible role in the centromere structure and function. Once again, it seems clear that studies on the repetitive sequences can provide interesting insights for the comprehension of the genome structuring and evolution.

Chromosomal Dynamics of Repetitive Sequences

Repetitive sequences, which correspond to a large fraction of the genomes, are governed by particular patterns of evolution such as unequal exchange, transposition, RNA mediated transposition and gene conversion that lead to a non-Mendelian segregation of repeats. It has been believed that multigene families evolve according to homogenization processes governed by molecular drive and concerted evolution (Dover, 1986; Elder and Turner, 1995), resulting in a sequence similarity of the repeat units that is greater within rather than between species. According to the results observed for the 5S rDNA repeat organization in the *Leporinus* (Martins and Galetti, 2001) and *Oreochromis niloticus* (Martins *et al.*, 2002; Alves-Costa *et al.*, 2006) genome, the homogenization in the repeat units is greater within a specific cluster. Repeats of different clusters, located at distinct chromosomes for example, can differ considerably. Although several mechanisms have been proposed as driving homogeneity among repetitive families, gene conversion and unequal exchange are the major drivers, since they occur in meiosis as well as mitosis (Crease and Lynch, 1991). Unequal exchange occurs when there is incomplete alignment between two chromosomes. One chromosome will gain extra genetic material while the other will lose DNA. When a mutation occurs in one member of a multigene family, the variant can be lost or maintained. If the variant is not lost, unequal exchange can increase the copy number of this variant in the multigene family. This new member of the repetitive family can spread across a population by several evolutionary ways such as natural selection, genetic drift, migration and bottleneck effect.

Ohno *et al.* (1968) postulated that gene duplication was the main driving force of vertebrate evolution. Once a gene was duplicated, one copy was no longer constrained by selection and any mutation that occurred in the duplicated copy could potentially lead to new expression patterns or altered functions, leaving the original copy to provide its required function. Over recent years genetic studies, including the sequencing of the human genome, have verified that duplications of the genome have led to the complexity of human genes when compared to flies and worms (Horvath *et al.*, 2001). With the knowledge of the entire sequence of several genomes from bacteria to humans, it has become clear that the increase in the genome size is correlated with the gain of duplicated non-coding sequences. In fact, whereas bacteria have small genomes with the presence or not of few duplicated DNA segments, the higher eukaryotes, like humans and other vertebrates have large genomes with the presence of large amounts of repetitive sequences. Thereby, the evolution and differentiation of the genomes seem to have occurred by acquisition of duplicated segments. These sequences can play a role in the chromosome structure, segregation, and evolution, nuclear architecture, species evolution, and repression/activation of gene transcription. The presence of large amounts of repetitive sequences seems to act as a buffer by acting in the compartmentalization of genes, linkage groups and chromosomes, contributing to the necessary conditions that make the genome a perfect functional unit.

Acknowledgments

The author is grateful to Dr A.P. Wasko for the critical review of this manuscript and to FAPESP (Fundação de Amparo à Pesquisa do Estado de São Paulo) for financial support.

References

Abuín, M., P. Martínez and L. Sánchez. 1996. Localization of the repetitive telomeric sequence (TTAGGG)$_n$ in four salmonid species. *Genome* 39: 1035–1038.

Alves-Costa, F., A.P. Vasko, C. Oliveira, F. Foresti and C. Martins. 2006. Genomic organization and evolution of the 5S ribosomal DNA in Tiladiini fishes. *Genetica* (in press).

Anleitner, J.E. and D.S. Haymer. 1992. Y enriched and Y specific DNA sequence from the genome of the Mediterranean fruit fly, *Ceratis capitata*. *Chromosoma* 101: 271–278.

Aparício, S., J. Chapman, E. Stupka, N. Putnam, J-M. Chia, P. Dehal, A. Christoffels, S. Rash, S. Hoon, A. Smit, M.D.S. Gelpke, J. Roach, T. Oh, I.Y. Ho, M. Wong, C. Detter, F. Verhoef, P. Predki, A. Tay, S. Lucas, P. Richardson, S.F. Smith, M.S. Clark, Y.J.K Edwards, N. Doggett, A. Zharkikh, S.V. Tavtigian, D. Pruss, M. Barnstead, C. Evans, H. Baden, J. Powell, G. Glusman, L. Rowen, L. Hood, Y.H. Tan, G. Elgar, T.

Hawkins, B. Venkatesh, D. Rokhsar and S. Brenner. 2002. Whole genome shotgun assembly and analysis of the genome of *Fugu rubripes*. *Science* 297: 1301–1310.

Arnason, U., M. Hoglund and B. Widegreen. 1984. Conservation of highly repetitive DNA in cetaceans. *Chromosoma* 89: 238–242.

Arnheim, N. 1983. Concerted evolution of multigene families. In: *Genes and Evolution of Genes and Proteins*, M. Nei and R.K. Koehn (eds.). Sinauer, Sunderland, pp. 36–61.

Arnheim, N., P. Seperack, J. Banerji, R.B. Lang, R. Miesfeld and K.B. Marcu. 1980. Mouse rDNA nontranscribed spacer sequences are found flanking immunoglobulin CH genes and elsewhere throughout the genome. *Cell* 22: 179–185.

Beridze, T.G. 1986. *Satellite DNA*. Springer-Verlag, Heidelberg.

Bóan, F., A. Viñas, J.M. Rodriguez, L. Sánchez and J. Gómez-Márquez. 1996. A new EcoRI family of satellite DNA in lampreys. *FEBS Letters* 394: 187–190.

Bouneau, L., C. Fischer, C. Ozouf-Costaz, A. Froschauer, O. Jaillon, J-P. Coutanceau, C. Körting, J. Weissenbach, A. Bernot and J-N. Volff. 2003. An active non-LTR and retrotransposon with tandem structure in the compact genome of the pufferfish *Tetraodon nigroviridis*. *Genome Research* 13: 1686–1695.

Brenner, S., G. Elgar, R. Sandford, A. Macrae, B. Venkatesh and S. Aparício. 1993. Characterization of the pufferfish (*Fugu*) genome as a compact model vertebrate genome. *Nature (London)* 366: 265–268.

Brutlag, D.L. 1980. Molecular arrangement and evolution of heterochromatic DNA. *Annual Review of Genetics* 14: 121–144.

Bryden, L., E.M. Denovan-Wright and J.M. Wright. 1998. ROn-1 SINEs: a tRNA-derived, short interspersed repetitive DNA element from *Oreochromis niloticus* and its species-specific distribution in Old World cichlid fishes. *Molecular Marine Biology and Biotechnology* 7: 48–54.

Canapa, A., P.N. Cerioni, M. Barucca, E. Olmo and V. Caputo. 2002. A centromeric satellite DNA may be involved in heterochromatin compactness in gobiid fishes. *Chromosome Research* 10: 297–304.

Capriglione, T., G. Odierna, V. Caputo, A. Canapa and E. Olmo. 2002. Characterization of a Tc1-like transposon in the Antarctic ice-fish, *Chionodraco hamatus*. *Gene* 295: 193–198.

Capriglione, T., A. Morescalchi, E. Olmo, L. Rocco, L. Stingo and S. Manzo. 1994. Satellite DNAs heterocromatin and sex chromosomes in *Chionodraco hamatus* (Channichthyidae, Perciformes). *Polar Biology* 14: 285–290.

Cavalier-Smith, T. 1985. *The Evolution of Genome Size*. John Wiley & Sons, New York.

Centola, M. and J. Carbon. 1994. Cloning and characterization of centromeric DNA from *Neurospora crassa*. *Molecular and Cellular Biology* 14: 1510–1519.

Charlesworth, B., P. Snlegowski and W. Stephan. 1994. The evolutionary dynamics of repetitive DNA in eukaryotes. *Nature (London)* 371: 215–220.

Childs, G., R. Maxson, R.H. Cohn and L. Kedes. 1981. Orphons: Dispersed genetic elements from tandem repetitive genes of eukaryotes. *Cell* 23: 651–663.

Clabby, C., U. Goswami, F. Flavin, N.P. Wilkins, J.A. Houghton and R. Powell. 1996. Cloning, characterization and chromosomal location of a satellite DNA from the Pacific oyster, *Crassostrea gigas*. *Gene* 168: 205–209.

Clark, L. 1990. Centromeres of budding and fission yeasts. *Trends in Genetics* 6: 150-154.

Crease, T.J. and M. Lynch. 1991. Ribosomal DNA variation in *Daphnia pulex. Molecular Biology and Evolution* 8: 620–640.

Cremisi, F., R. Vignali, R. Batistoni and G. Barsacchi. 1988. Heterochromatic DNA in *Triturus* (Amphibia, Urodela) II. A centromeric satellite DNA. *Chromosoma* 97: 204–211.

Crollius, H.R., O. Jaillon, C. DaSilva, C. Ozouf-Costaz, C. Fizames, C. Fischer, L. Bouneau, A. Billault, F. Quetier, W. Saurin, A. Bernot and J. Weissenbach. 2000. Characterization and repeat analysis of the compact genome of the freshwater pufferfish *Tetraodon nigroviridis. Genome Research* 10: 939–949.

Datta, U., P. Dutta and K. Mandal. 1988. Cloning and characterization of a highly repetitive fish nucleotide sequence. *Gene* 62: 331–336.

DaSilva, C., H. Hadji, C. Ozouf-Costaz, S. Nicaud, O. Jaillon, J. Weissenbach and H.R. Crollius. 2002. Remarkable compartmentalization of transposable elements and pseudogenes in the heterochromatin of the *Tetraodon nigroviridis* genome. *Proceedings of the National Academy of Sciences of the United States of America* 99: 13636–13641.

De Lucchini, S., F. Andronico, M. Andreazzoli, M. Giuliani, R. Savino and I. Nardi. 1988. Extra-chromosomal spacer sequences in *Triturus. Journal of Molecular Biology* 204: 805–813.

De Lucchini, S., I. Nardi, G. Barsacchi, R. Batistoni and F. Andronico. 1993. Molecular cytogenetics of the ribosomal (18S + 28S and 5S) DNA loci in primitive and advanced urodele amphibians. *Genome* 36: 762–773.

Denovam, E.M. and J.M. Wright. 1990. A satellite DNA family from pollock (*Pollachius virens*). *Gene* 87: 279–283.

Devlin, B., T. Krontiris and N. Risch. 1993. Population genetics of the HRAS1 minisatellite locus. *American Journal of Human Genetics* 53: 1298–1305.

Devlin, R.H., B.K. McNeil and E.M. Donaldson. 1991. Isolation of a Y-chromosomal DNA probe capable of determining sex in Chinook salmon. *Canadian Journal of Fisheries and Aquatic Science* 48: 1606–1612.

Devlin, R.H., G.W. Stone and D.E. Smailus. 1998. Extensive direct-tandem organization of a long repeat DNA sequence on the Y chromosome of chinook salmon (*Oncorhynchus tshawytscha*). *Journal of Molecular Evolution* 46: 277–287.

Doolittle, W.F. and C. Sapienza. 1980. Selfish genes, the phenotype paradigm and genome evolution. *Nature (London)* 284: 601–603.

Doran, J.L., W.H. Bingle and K.L. Roy. 1987. The nucleotide sequence of two human 5S rRNA pseudogenes. *Nucleic Acids Research* 15: 6297.

Dover, G.A. 1986. Linkage disequilibrium and molecular drive in the rDNA gene family. *Genetics* 122: 249–252.

Drouin, G. and M. Moniz de Sá. 1995. The concerted evolution of 5S ribosomal genes linked to the repeat units of other multigene families. *Molecular Biology and Evolution* 12: 481–493.

Dunham, I., N. Shimizu, B.A. Roe, S. Chissoe, A.R. Hunt, J.E. Collins, R. Bruskiewich, D.M. Beare, M. Clamp and L.J. Smink *et al.* 1999. The DNA sequence of the human chromosome 22. *Nature (London)* 402: 489–495.

Eichler, E.E. and D. Sankoff. 2003. Structural Dynamics of eukaryotes chromosome evolution. *Science* 301: 793–797.

Ekker, M., A. Fritz and M. Westerfield. 1992. Identification of two families of satellite-like repetitive DNA sequences from the zebrafish (*Brachydanio rerio*). *Genomics* 13: 1169–1173.

Elder Jr., J.F. and B.J. Turner. 1995. Concerted evolution of repetitive DNA sequences in eukaryotes. *Quarterly Review of Biology* 70: 277–320.

Elgar, G., M.S. Clark, S. Meek, S. Smith, S. Warner, Y.J.K. Edwards, N. Bouchireb, A. Cottage, G.S.H. Yeo, Y. Umrania, G. Williams and S. Brenner. 1999. Generation and analysis of 25 Mb of genomic DNA from the pufferfish *Fugu rubripes* by sequence scanning. *Genome Research* 9: 960–971.

Emerson, B.M. and R.G. Roeder. 1984. Isolation and genomic arrangement of active and inactive forms of mammalian 5S RNA genes. *Journal of Biological Chemistry* 259: 7916–7925.

Falquet, J., R. Creusot and M. Dron. 1997. Molecular analysis of DNA homologous to IGS subrepeats. *Plant Physiology and Biochemistry* 35: 611–622.

Fischer, C., L. Bouneau, J-P. Coutanceau, J. Weissenbach, J-N. Volff and C. Ozouf-Costaz. 2004. Global heterochromatic colocalization of transposable elements with minisatellites in the compact genome of the pufferfish *Tetraodon nigroviridis*. *Gene* 336: 175–184.

Flavell, R.B., M.D. Bennett, J.B. Smith and D.B. Smith. 1974. Genome size and the proportion of repeated nucleotide sequence DNA in plants. *Biochemical Genetics* 12: 257–269.

Foresti, F., C. Oliveira, P.M. Galetti Jr. and L.F. Almeida-Toledo. 1993. Synaptonemal complex analysis in spermatocytes of tilapia, *Oreochromis niloticus* (Pisces, Cichlidae). *Genome* 36: 1124–1128.

Franck, J.P.C. and J.M. Wright. 1993. Conservation of a satellite DNA sequence (SATB) in the tilapiine and haplochromine genome (Pisces: Cichlidae). *Genome* 36: 187–194.

Franck, J.P.C., J.M. Wright and B.J. McAndrew. 1992. Genetic variability in a family of satellite DNAs from tilapia (Pisces: Cichlidae). *Genome* 35: 719–725.

Franck, J.P.C., I. Kornfield and J.M. Wright. 1994. The utility of SATA satellite DNA sequences for inferring phylogenetic relationships among the three major genera of tilapiine cichlid fishes. *Molecular Phylogenetics and Evolution* 3: 10–16.

Francks, A., A. Houben, C.R. Leach and J.N. Timmis. 1996. The molecular organization of a B chromosome tandem repeat sequence from *Brachycome dichromosomatica*. *Chromosoma* 105: 223–230.

Galetti Jr., P.M. and F. Foresti. 1986. Evolution of the ZZ/ZW system in *Leporinus* (Pisces, Anostomidae). *Cytogenetics and Cell Genetics* 43: 43–46.

Galetti Jr., P.M., N.R.W. Lima and P.C. Venere. 1995. A monophyletic ZW chromosome system in *Leporinus* (Anostomidae, Characiformes). *Cytologia* 60: 375–382.

Garrido-Ramos, M.A., M. Jamilena, R. Lozano, C. Ruiz Rejón and M. Ruiz Réjon. 1994. Cloning and characterization of a fish centromeric satellite DNA. *Cytogenetics and Cell Genetics* 65: 233–237.

Garrido-Ramos, M.A., M. Jamilena, R. Lozano, C. Ruiz Rejón and M. Ruiz Rejón. 1995. The *EcoRI* centromeric satellite DNA of the Sparidae family (Pisces, Perciformes) contains a sequence motive common to other vertebrates centromeric satellite DNAs. *Cytogenetics and Cell Genetics* 71: 345–351.

Garrido-Ramos, M.A., M.C. Soriguer, R.M. Hérran, C. Jamilena, A. Ruiz Rejón, J. Domezain, A. Hernando and M. Ruiz Rejón. 1997. Morphometric and genetic analysis as proof of the existence of two sturgeon species in the Guadalquivir river. *Marine Biology* 129: 33–39.

Garrido-Ramos, M.A., R. Herrán, C. Ruiz Rejón and M. Ruiz Rejón. 1998. A satellite DNA of the Sparidae family (Pisces, Perciformes) associated with telomeric sequences. *Cytogenetics and Cell Genetics* 83: 3–9.

Gondo, Y., T. Okada, N. Matsuyama, Y. Saitoh, Y. Yanagisawa and J.E. Ideda. 1998. Human megasatellite DNA RS447: Copy-number polymorphisms and interspecies conservation. *Genomics* 54: 39–49.

Goodier, J.L. and W.S. Davidson. 1994. Characterization of a repetitive element detected by *NheI* in the genomes of *Salmo* species. *Genome* 37: 639–645.

Goodier, J.L. and W.S. Davidson. 1998. Characterization of novel minisatellite repeat loci in Atlantic salmon (*Salmo salar*) and their phylogenetic distribution. *Journal of Molecular Evolution* 46: 245–255.

Haaf, T. and M. Schmid. 1991. Chromosome topology in mammalian interphase nuclei. *Experimental Cell Research* 192: 325–332.

Haaf, T., M. Schmid, C. Steinlein, P.M. Galetti Jr. and H.F. Willard. 1993. Organization and molecular cytogenetics of a satellite DNA family from *Hoplias malabaricus* (Pisces, Erythrinidae). *Chromosome Research* 1: 77–86.

Hadjiolov, A.A. 1985. The nucleolus and ribosome biogenesis. In: Alferd, M., Benmann, W., Goldstein, L., Porter, K.R. and Site, P. (eds.) *Cell Biology Monographs*. Springer-Verlag, New York, Vol. 12, pp. 1–263.

Hartley, S.E. and W.S. Davidson. 1994. Characterization and distribution of genomic repeat sequences from Arctic charr (*Salvelinus alpinus*). In: *Genetics and Evolution of Aquatic Organisms*. Chapman and Hall, London, pp. 271–280.

Harvey, S.C., J. Masabanda, L.A.P. Carrasco, N.R. Bromage, D.J. Penman and D.K. Griffin. 2002. Molecular cytogenetic analysis reveals sequence differences between the sex chromosomes of *Oreochromis niloticus*: evidence for an early stage of sex-chromosome differentiation. *Cytogenetic and Genome Research* 97: 76–80.

Harvey, S.C., C. Boonphakdee, R. Campos-Ramos, M.T. Ezaz, D.K. Griffin, N.R. Bromage and D.J. Penman. 2003. Analysis of repetitive DNA sequences in the sex chromosomes of *Oreochromis niloticus*. *Cytogenetic and Genome Reseach* 101: 314–319.

Hatanaka, T., F. Henrique-Silva and P.M. Galetti Jr. 2002. A polymorphic, telomeric-like sequence microsatellite in the Neotropical fish *Prochilodus*. *Cytogenetic and Genome Research* 98: 308–310.

He, L., A. Ahu, A.J. Faras, K.S. Guise, P.B. Hackett and A.R. Kapusciniski. 1992. Characterization of *AluI* repeats of zebrafish (*Brachidanio rerio*). *Molecular Marine Biology and Biotechnology* 1: 125–135.

Henikoff, S., K.K. Ahmad and H.S. Malik. 2001. The centromere paradox: stable inheritance with rapidly evolving DNA. *Science* 293: 1098–1102.

Horvath, J.E., J.A. Bailey, D.P. Locke and E.E. Eichler. 2001. Lessons from the human genome: transitions between euchromatin and heterochromatin. *Human Molecular Genetics* 10: 2215–2223.

Hummel, S., W. Meyerhhof, E. Korge and W. Knochel. 1984. Characterization of highly and moderately repetitive 500 bp EcoRI fragments from *Xenopus laevis* DNA. *Nucleic Acids Research* 12: 4921–4937.

Jeffreys, A.J., V. Wilson and S.L. Thein. 1985. Individual-specific 'fingerprints' of human DNA. *Nature (London)* 316: 76–79.

Jesus, C.M., P.M. Galetti Jr., S.R. Valentini and O. Moreira-Filho. 2003. Molecular characterization and chromosomal location of two families of satellite DNA in *Prochilodus lineatus* (Pisces, Prochilodontidae), a species with B chromosomes. *Genetica* 118: 25–32.

Kaszás, E. and J.A. Birchler. 1996. Misdivision analysis of centromere structure in maize. *EMBO Journal* 15: 5246–5255.

Kato, M. 1999. Structural bistability of repetitive DNA elements featuring CA/TG dinucleotide steps and mode of evolution of satellite DNA. *European Journal of Biochemistry* 265: 204–209.

Kazazian, H.H., C. Wong, H. Youssoufian, A.F. Scott, D.G. Phillips and S.E. Antonarakis. 1988. Haemofilia: A resulting from the novo insertion of L1 sequences represents a novel mechanism of mutation in man. *Nature (London)* 332: 164–166.

Kipling, D., H.E. Ackford, B.A. Taylor and H.J. Cooke. 1991. Mouse minor satellite DNA genetically maps to the centromere and is physically linked to the proximal telomere. *Genomics* 11: 235–241.

Koehler, M.R., D. Dehm, M. Guttenbach, I. Nanda, T. Haaf, W.F. Molina, P.M. Galetti Jr. and M. Schmid. 1997. Cytogenetics of the genus *Leporinus* (Pisces, Anostomidae). 1. Karyotype analysis, heterochromatin distribution and sex chromosomes. *Chromosome Research* 5: 12–22.

Kominami, R. and M. Muramatsu. 1987. Amplified ribosomal spacer sequence: structure and evolutionary origin. *Journal of Molecular Biology* 193: 217–222.

Kraemer, C. and E.R. Schmidt. 1993. The sex determination region of *Chironomus thummi* is associated with highly repetitive DNA and transposable elements. *Chromosoma* 102: 553–562.

Kubota, S., M. Kuro-O, S. Mizuno and S. Kohno. 1993. Germ line-restricted highly repeated DNA sequences and their chromosome localization in a japanese hagfish (*Eptatretus okinoseanus*). *Chromosoma* 102: 163–173.

Kubota, S., J. Takano, R. Tsuneishi, S. Kobayakawa, N. Fujikawa, M. Nabeyama and S. Kohno. 2001. Highly repetitive DNA families restricted to germ cells in a Japanese hagfish (*Eptatretus burgeri*): A hierarchical and mosaic structure in eliminated chromosomes. *Genetica* 111: 319–328.

Lanfredi, M., L. Congiu, M.A. Garrido-Ramos, R. Herrrán, M. Leis, M. Chicca, R. Rossi, J. Tagliavini, M. Ruiz Rejón and F. Fontana. 2001. Chromosomal location and evolution of a satellite DNA family in seven sturgeon species. *Chromosome Research* 9: 47–52.

Langdon, T., C. Seago, R.N. Jones, H. Ougham, H. Thomas, J.W. Forster and G. Jenkins. 2000. De novo evolution of satellite DNA on the rye B chromosome. *Genetics* 154: 869–884.

Larin, Z., M.D. Fricker and C. Tyler-Smith. 1994. *De novo* formation of several features of a centromere following introduction of an Y alphoid YAC into mammalian cells. *Human Molecular Genetics* 3: 689–695.

Leah, R., S. Frederiksen, J. Engberg and P.D. Sorensen. 1990. Nucleotide sequence of a mouse 5S rRNA variant gene. *Nucleic Acids Research* 18: 7441–7441.

Li, Y.X. and M.L. Kirby. 2003. Coordinated and conserved expression of alphoid repeat and alphoid repeat-tagged coding sequence. *Developmental Dynamics* 228: 72–81.

Lohe, A.R. and P.A. Roberts. 1990. An unusual Y chromosome of *Drosophila simulans* carrying amplified rDNA spacer without RNA genes. *Genetics* 125: 399–406.

Long, E.O. and I.D. Dawid. 1980. Repeated genes in eukaryotes. *Annual Review of Biochemistry* 49: 727–764.

Mäkinem, A., C. Zijlstra, N.A. De Haan, C.H.M. Mellink and A.A. Bosma. 1997. Localization of 18S plus 28S and 5S ribosomal RNA genes in the dog by fluorescence *in situ* hybridization. *Cytogenetics and Cell Genetics* 78: 231–235.

Mandrioli, M. and G.C. Manicardi. 2001. Cytogenetics and molecular analysis of the pufferfish *Tetraodon fluviatilis* (Osteichthyes). *Genetica* 111: 433–438.

Mandrioli, M., G.C. Manicardi, N. Machella and V. Caputo. 2001. Molecular and cytogenetic analysis of the goby *Gobius niger* (Teleostei, Gobiidae). *Genetica* 110: 73–78.

Mantovani, M., L.D.S. Abel, C.A. Mestriner and O. Moreira-Filho. 2004. Evidence of the differentiated structural arrangement of constitutive heterochromatin between two populations of *Astyanax scabripinnis* (Pisces, Characidae). *Genetics and Molecular Biology* 27: 536–542.

Martínez, J.L., P. Móran, E. Garcia-Vásquez and A.M. Pendás. 1996. Chromosomal localization of the major and minor 5S rRNA genes in the European eel (*Anguilla anguilla*). *Cytogenetics and Cell Genetics* 73: 149–152.

Martins, C. and P.M. Galetti Jr. 1999. Chromosomal localization of 5S rRNA genes in *Leporinus* fish (Anostomidae, Characiformes). *Chromosome Research* 7: 363–367.

Martins, C. and P.M. Galetti Jr. 2001. Organization of 5S rDNA in species of the fish *Leporinus*: Two different genomic locations are characterized by distinct nontranscribed spacers. *Genome* 44: 903–910.

Martins, C. and A.P. Wasko. 2004. Organization and evolution of 5S ribosomal DNA in the fish genome. In: *Focus on Genome Research*, C. R. Williams (ed.). Nova Science Publishers, Hauppauge, NY., pp. 289–318.

Martins, C., A.P. Wasko, C. Oliveira, F. Porto-Foresti, P.P. Parise-Maltempi, J.M. Wright and F. Foresti. 2002. Dynamics of 5S rDNA in the tilapia (*Oreochromis niloticus*) genome: Repeat units, inverted sequences, pseudogenes and chromosome loci. *Cytogenetic and Genome Research* 98: 76–85.

Martins, C., I.A. Ferreira, C. Oliveira, F. Foresti and P.M. Galetti Jr. 2006. A tandemly repetitive centromeric DNA sequence of the fish *Hoplias malabaricus* is derived from 5S rDNA. *Genetica*. (in press).

Matsuda, M., C. Matsuda, S. Hamaguchi and M. Sakaizumi. 1998. Identification of the sex chromosomes of the medaka, *Oryzias latipes*, by fluorescence *in situ* hybridization. *Cytogenetics and Cell Genetics* 82: 257–262.

Mazzei, F., L. Ghigliotti, C. Bonillo, J-P. Coutanceau, C. Ozouf-Coustaz and E. Pisano. 2004. Chromosomal patterns of major and 5S ribosomal DNA in six icefish species (Perciformes, Notothenioidei, Channichthyidae). *Polar Biology* 28: 47–55.

McAllister, B.F. 1995. Isolation and characterization of a retroelement from B chromosome (PSR) in the parasitic wasp *Nasonia vitripennis*. *Insect Molecular Biology* 4: 253–262.

Messier, W., S.H. Li and C.B. Stewart. 1996. The birth of microsatellites. *Nature (London)* 381: 483.

Mestriner, C.A., P.M. Galetti Jr., S.R. Valentini, I.R.G. Ruiz, L.D.S. Abel, O. Moreira-Filho and J.P.M. Camacho. 2000. Structural and functional evidence that a B chromosome in the characid fish *Astyanax scabripinnis* is an isochromosome. *Heredity* 85: 1–9.

Mônaco, P.J., K.F. Swan, E.M. Rasch and P.R. Musich. 1989. Characterization of a repetitive DNA in the unisexual fish *Poecilia formosa*. I. Isolation and cloning of the MboI family. *Evolution and Ecology of Unisexual Vertebrates* 466: 123–131.

Móran, P., J.L. Martínez, E. Garcia-Vásquez and A.M. Pendás. 1996. Sex linkage of 5S rDNA in rainbow trout (*Oncorhynchus mykiss*). *Cytogenetics and Cell Genetics* 75: 145–150.

Morescalchi, A., T. Capriglione, R. Lanna, M.A. Morescalchi, G. Odierna and E. Olmo. 1996. Genome structure in notothenioid fish from the Ross Sea. *Proceedings of the Third Meeting on Antarctic Biology*, Santa Margherita Ligure, 13–15 December, 1996. pp. 365–379.

Moyer, S.P., D.P. Ma, T.L. Thomas and J.R. Gold. 1988. Characterization of a highly repeated satellite DNA from the cyprinidae fish *Notropis lutrensis*. *Comparative Biochemistry and Physiology* 91B: 639–646.

Murakami, M. and H. Fujitani. 1997. Polyploid-specific repetitive DNA sequences triploid ginbuna (Japanese silver crucian carp, *Carassius auratus langsdorfi*). *Genes Genetics and Systematics* 72: 107–113.

Murphy, T.D. and G.H. Karpen. 1995. Interactions between the nod[+] kinesin-like gene and extracentromeric sequences are required for transmission of a *Drosophila* minichromosome. *Cell* 81: 139–148.

Nakayama, I., F. Foresti, R. Tewari, M. Schartl and D. Chourrout. 1994. Sex chromosome polymorphism and heterogametic males revealed by two cloned DNA probes in the ZW/ZZ fish *Leporinus elongatus*. *Chromosoma* 103: 31–39.

Nanda, I., W. Feichtinger, M. Schmid, J.H. Schroder, H. Zischler and J.T. Epplen. 1990. Simple repetitive sequences are associated with the differentiation of the sex chromosomes in the guppy fish. *Journal of Molecular Evolution* 30: 456–462.

Nanda, I., J-N. Volff, S. Weis, C. Körting, A. Froschauer, M. Schmid and M. Schartl. 2000. Amplification of a long terminal-like element on the Y chromosome of the platyfish, *Xiphophorus maculatus*. *Chromosoma* 109: 173–180.

Narayanswami, S., N.A. Doggett, L.M. Clark, C.E. Hildebrand, H.U. Weier and B.A. Hamkalo. 1992. Cytological and molecular characterization of centromeres in *Mus domesticus* and *Mus spretus*. *Mammalian Genome* 2: 186–194.

Nowak, R. 1994. Mining treasures from 'junk DNA'. *Science* 263: 608–610.

Ohno, S., U. Wolf and N.B. Atkin. 1968. Evolution from fish to mammals by gene duplication. *Hereditas* 59: 169-187.

Oliveira, C. and J.M. Wright. 1998. Molecular cytogenetic analysis of heterochromatin in the chromosomes of tilapia, *Oreochromis niloticus* (Teleostei: Cichlidae). *Chromosome Research* 6: 205–211.

Oliveira, C., J.S.K. Chew, F. Porto-Foresti, M.J. Dobson and J.M. Wright. 1999. A LINE2 repetitive element from the cichlid fish, *Oreochromis niloticus*: Sequence analysis and chromosomal distribution. *Chromosoma* 108: 457–468.

Oliveira, C., Y. Wang, L.J. Bryden and J.M. Wright. 2003. Short interspersed repetitive elements (SINEs) from the cichlid fish, *Oreochromis niloticus*, and their chromosomal localization by fluorescent *in situ* hybridization. *Caryologia* 56: 177–185.

Orgel, L.E. and F.H.C. Crick. 1980. Selfish DNA: The ultimate parasite. *Nature (London)* 284: 604–607.

Ota, K., Y. Tateno and T. Gojobori. 2003. Highly differentiated and conserved sex chromosome in fish species (*Aulopus japonicus*: Teleostei, Aulopidae). *Gene* 317: 187-193.

Ozouf-Costaz, C., J. Brandt, C. Körting, E. Pisano, C. Bonillo, J-P. Coutanceau, J-N. Volff. 2004. Genome dynamics and chromosomal localization of the non-LTR retrotransposons *Rex1* and *Rex3* in Antarctic fish. *Antarctic Science* 16: 51–57.

Pauls, E. and L.A.C. Bertollo. 1983. Evidence for a system of supernumerary chromosome in *Prochilodus scrofa* (Pisces, Prochilodontidae). *Caryologia* 36: 307–314.

Pendás, A.M., P. Móran, J.P. Freije and E. Garcia-Vásquez. 1994. Chromosomal location and nucleotide sequence of two tanden repeats of the Atlantic salmon 5S rDNA. *Cytogenetics and Cell Genetics* 67: 31–36.

Pérez, J., P. Móran and E. García-Vásquez. 1999. Physical mapping of three minisatellite sequences in the Atlantic salmon (*Salmo salar*) genome. *Animal Genetics* 30: 371–374.

Phillips, R.B. and K.M. Reed. 2000. Localization of repetitive DNAs to zebrafish (*Danio rerio*) chromosomes by fluorescence *in situ* hybridization (FISH). *Chromosome Research* 8: 27–35.

Pont, G., F. DeGroote and G. Picard. 1987. Some extrachromosomal circular DNAs from *Drosophila* embryos are homologous to tandemly repeated genes. *Journal of Molecular Biology* 195: 447–451.

Reed, K.M. and. R.B. Phillips. 1995. Molecular characterization and cytogenetic analysis of highly repeated DNAs of lake trout, *Salvelinus namycush*. *Chromosoma* 104: 242–251.

Renault, S., F. DeGroote and G. Picard. 1993. Identification of short tandemly repeated sequences in extrachromosomal circular DNAs from *Drosophila melanogaster* embryos. *Genome* 36: 244–254.

Sart, D., M.R. Cancilla, E. Earle, J.I. Mao, R. Saffery, K.M. Tainton, P. Kalitsis, J. Martyn, A.E. Barry and K.H. Choo. 1997. A functional neo-centromere formed through activation of a latent human centromere and consisting of non-alpha-satellite DNA. *Nature Genetics* 16: 144–153.

Schmid, M., L. Vitelli and R. Batistoni. 1987. Chromosome banding in Amphibia. IV. Constitutive heterochromatin, nucleolus organizers, 18S+28S and 5S ribosomal RNA genes in Ascaphidae, Pipidae, Discoglossidae and Pelobatidae. *Chromosoma* 95: 271–284.

Schueler, M.G., A.W. Higgins, M.K. Rudd, K. Gustashaw, H. Willard. 2001. Genomic and genetic definition of a functional human centromere. *Science* 294: 109–115.

Shimoda, N., E.W. Knapik, J. Ziniti, C. Sim, E. Yamada, S. Kaplan, D. Jackson, F. Sauvage, H. Jacob and M.C. Fishman. 1999. Zebrafish genetic map with 200 microsatellite markers. *Genomics* 58: 219–232.

Singer, M.F. 1982. Highly repetitive sequences in mammalian genomes. *International Review of Cytology* 76: 67–112.

Sola, L. and E. Gornung. 2001. Classical and molecular cytogenetics of the zebrafish, *Danio rerio* (Cyprinidae, Cypriniformes): An overview. *Genetica* 111: 397–412.

Stein, J., R.B. Phillips and R.H. Devlin. 2001. Identification of the Y chromosome in chinook salmon (*Oncorhynchus tshawytscha*). *Cytogenetics and Cell Genetics* 92: 108–110.

Suzuki, H., S. Sakurai and Y. Matsuda. 1996. Rat rDNA spacer sequences and chromosomal assignment of the genes to the extreme terminal region of chromosome 19. *Cytogenetics and Cell Genetics* 72: 1–4.

Takahashi, Y., K. Mitani, K. Kuwabara, T. Hayashi, M. Niwa, N. Miyashita, K. Moriwaki and R. Kominami. 1994. Methylation imprinting was observed of mouse mo-2 macrosatellite on the pseudoautosomal region but not on chromosome 9. *Chromosoma* 103: 450–458.

Tautz, D. and M. Renz. 1984. Simple sequences are ubiquitous repetitive components of eukaryotes genomes. *Nucleic Acids Research* 12: 4127–4138.

The Genome International Sequencing Consortium. 2001. Initial sequencing and analysis of the human genome. *Nature (London)* 409: 860–921.

Unfried, K., K. Schiebel and V. Hemleben. 1991. Subrepeats of rDNA intergenic spacer present as prominent independent satellite DNA in *Vigna radiata* but not *Vigna angularis*. *Gene* 99: 63–68.

Vanzela, A.L.L., A.C. Swarça, A.L. Dias, R. Stolf, P.M. Ruas, C.F. Ruas, I.J. Sbalqueiro and L. Giuliano-Caetano. 2002. Differential distribution of $(GA)_9+C$ microsatellite on chromosomes of some animal and plant species. *Cytologia* 67: 9–13.

Venere, P.C., I.A. Ferreira, C. Martins and P.M. Galetti Jr. 2004. A novel ZZ/ZW sex chromosome system for the genus *Leporinus* (Pisces, Anostomidae, Characiformes). *Genetica* 121: 75–80.

Vicente, V.E., L.A.C. Bertollo, S.R. Valentini and O. Moreira-Filho. 2003. Origin and differentiation of a sex chromosome system in *Parodon hilarii* (Pisces, Parodontidae). Satellite DNA, G- and C-banding. *Genetica* 119: 115–120.

Viñas, A., M. Abuín, B.G. Pardo, P. Martínez and L. Sánchez. 2004. Characterization of a new *Hpa*I centromeric satellite DNA in *Salmo salar*. *Genetica* 121: 81–87.

Vissel, B., A. Nagy and K.H.A. Choo. 1992. A satellite III sequence shared by human chromosomes 13, 14 and 21 that is contiguous with alpha satellite DNA. *Cytogenetics and Cell Genetics* 61: 81–86.

Volff, J-N., L. Bouneau, C. Ozouf-Costaz and C. Fischer. 2003. Diversity of retrotransposable elements in compact pufferfish genomes. *Trends in Genetics* 19: 674–678.

Willard, H.F. and J.S. Wayne. 1987. Hierarchical order in chromosomal-specific human alpha satellite DNA. *Trends in Genetics* 3: 192–198.

Wong, A.K.C. and J.B. Rattner. 1988. Sequence organization and cytological localization of the minor satellite of mouse. *Nucleic Acids Research* 16: 11645–11661.

Wright, J.M. 1989. Nucleotide sequence, genomic organization and the evolution of a major repetitive DNA family in tilapia (*Oreochromis mossambicus/hornorum*). *Nucleic Acids Research* 17: 5071–5079.

Ziegler, C.G., D.K. Lamatsch, C. Steinlein, W. Engel, M. Schartl and M. Schmid. 2003. The giant B chromosome of the cyprinid fish *Alburnus alburnus* harbours a retrotransposon-derived repetitive DNA sequence. *Chromosome Research* 11: 23–35.

4.5

Application of Fluorescence *In Situ* Hybridization (FISH) to Genome Mapping in Fishes

Ruth B. Phillips

INTRODUCTION

Molecular cytogenetics has made important contributions to genomic studies of fishes. Perhaps one of the most useful applications is the ability to document the existence of duplicate loci and to determine if duplicates are found in tandem or on different chromosomes. Genetic linkage maps are being connected to specific chromosomes in several species using linkage group specific probes, which will be very useful for tracing chromosome evolution in related species. In many fish species, recombination is suppressed near centromeres; so physical mapping of clones gives a more accurate picture of where genes are located compared with genetic mapping.

This article begins by summarizing the genome mapping projects that are completed or in progress for fish species. These projects are generating large insert clones that can be used for Fluorescence *In Situ* hybridization

---

Address for Correspondence: Department of Biological Sciences, Washington State University of Vancouver, Vancouver, Washington 98686, USA. E-mail: phllipsr@vancouver.wsu.edu

(FISH), so we can expect that the number of fish genomic studies will increase in the near future. This chapter summarizes current work on mapping of single copy genes, identification and mapping of sex chromosomes, production of chromosome specific probes and application of these probes to chromosome evolution.

Genome mapping projects are either completed or in progress for a number of fish species including *Danio rerio* (zebrafish), *Fugu rubripes* (Japanese pufferfish), *Tetraodon nigrovidis* (green spotted pufferfish), *Xiphophorus* (platyfish), *Oryzias lapites* (medaka), *Oreochromis niloticus* (tilapia), *Gasterostres aculeatus* (stickleback), *Ictalurus punctatus* (catfish) and various salmonids including *Oncorhynchus mykiss* (rainbow trout), *O. tshawytscha* (chinook salmon), *O. kisutch* (coho salmon), *Salmo salar* (Atlantic salmon), *Salmo trutta* (brown trout), and *Salvelinus alpinus* (Arctic char). These projects involve mapping a large number of AFLP (anonymous fragment length polymorphisms) or microsatellite markers to produce detailed genetic maps and usually include production of large insert bacterial chromosome (BAC) libraries and physical mapping. The two pufferfish genomes have been sequenced (Aparicio *et al.*, 2002; Jaillon *et al.*, 2004), and zebrafish, medaka, and stickleback are in progress (Clark, 2003).

Chromosomal location of protein coding genes using FISH has been done in several fish species including pufferfish (Crollius *et al.*, 2000), zebrafish (Phillips, 2001 and unpublished), stickleback (Cresko *et al.*, unpublished), several species of Antarctic fish from the family Nototheniidae (Pisano *et al.*, 2003; Mazzei *et al.*, 2004), rainbow trout (Phillips *et al.*, 2005b and unpublished), Atlantic salmon (Mitchell *et al.*, unpublished), chinook salmon and coho salmon (Phillips, unpublished). Mapping large insert clones (lambda, cosmid, PAC, and BAC) containing genes or genetic markers that are on the genetic map allows assignment of linkage groups to specific chromosomes and this has been completed for zebrafish and rainbow trout.

Chromosome evolution in closely related species can be examined using FISH with large insert probes linked to the genetic map, because application of this technique allows interspecific homology of chromosome arms to be determined. This is currently being done for salmonid fishes and will probably be applied to other fish groups, as these large insert libraries become more widely available. Another approach to determining chromosomal arm homologies uses paint probes produced by microdissection or flow sorting of chromosomes. Many studies tracing chromosome evolution in mammals have used the human paint probes produced by flow sorting of human chromosomes (Weinberg, 2005). For successful flow sorting, chromosomes must vary substantially in size and AT/GC content which is uncommon in fishes, so most fish paint probes to date have been prepared using microdissection (reviewed in Phillips, 2001). This technique requires identification of specific chromosomes in Giemsa preparations, so it has been difficult to produce a set of probes for each chromosome. However, future technical improvements in flow sorting may allow production of paint probes from single chromosomes (Gribble *et al.*, 2004) and make this technique more widely applicable to fishes. In some cases, entire genomes have been used to prepare paint probes in analysis of fish hybrids.

The sex chromosomes have been analyzed using genomic approaches in a number of fish species including medaka (Matsuda *et al.*, 1998), stickleback (Peichel *et al.*, 2004), platyfish (Froschauer *et al.*, 2002), tilapia (Ezaz *et al.*, 2004) as well as several salmonids (Artieri *et al.*, 2005; Phillips *et al.*, 2005a).

Localization of Single Copy Genes

Single copy genes have been cytogenetically mapped to chromosomes of zebrafish, Japanese pufferfish, green spotted pufferfish, stickleback, several notothenioid fishes, and salmonids including rainbow trout, Atlantic salmon, chinook and coho salmon. To localize single copy genes to chromosomes, clones from large insert libraries such as cosmids, PACs, BACs or YACs are usually required. With the exception of zebrafish (Koch *et al.*, 2004) and pufferfish, these libraries have only recently been prepared (Clark, 2003). For example, there are now four BAC libraries for cichlid fishes including Nile tilapia (*Oreochromis niloticus*) (Katagiri *et al.*, 2005), *Haplochromis chilotes* from Lake Victoria(Watanabe *et al.*, 2003), *Metriaclima zebra* from Lake Malawi (T. Kocher, pers. com.), and *Astatotilapia burtoni* (C. Amemiya, pers. com.). Other fishes for which BAC libraries are available include catfish (Quiniou *et al.*, 2003), carp (Katagiri *et al.*, 2001), rainbow trout (Katagiri *et al.*, 2001; Phillips *et al.*, 2003; Palti *et al.*, 2004), Atlantic salmon (Thorsen *et al.*, 2005), chinook salmon (Devlin, pers. com.), medaka (Matsuda *et al.*, 2001), stickleback (Reusch *et al.*, 2004), platyfish (Froschauer *et al.*, 2002), hagfish (Suzuki *et al.*, 2004), and red sea bream (Katagiri *et al.*, 2002). Cosmid libraries are available for a number of other fishes including coho salmon (Devlin, pers. com.), Arctic char (Davidson, pers. com.), and lamprey (Burgtorf et al., 1998). Before BAC libraries were available, cosmid clones containing sex-linked single copy genes were used as probes in FISH experiments to identify the sex chromosomes in platyfish (Nanda et al., 2000) and medaka (Matsuda et al., 1998). Cosmid clones containing the mesoderm specific transcript (MEST) were localized to single chromosome pairs in *Fugu rubripes* and *Tetraodon nigroviridis* (Brunner et al., 2000) and a cosmid containing the tRNA^{met} genes was mapped to a single chromosomal location in Atlantic salmon and brown trout (Perez et al., 2000).

Protein-coding genes present in clusters of multiple copies such as hemoglobins or histones can be mapped using lambda clones. The alpha and beta globin genes were mapped to a cluster on a single chromosome pair in four species of nototheniidae fishes with red blood, but no signals were obtained when these same probes were used on the white blooded Antarctic icefishes (Pisano et al., 2003) (Fig. 4.5.1b). Localization of the IgH locus in various species of the same family revealed that although many species had a single IgH locus, some of the species including *Trematomus scotti* had signals on two chromosome pairs, suggesting a duplication of the IgH locus (Fig. 4.5.1a). Similarly, the histone genes were mapped in rainbow trout, brown trout and Atlantic salmon using lambda clones (Pendas et al., 1994).

Several single copy genes from specific linkage groups have been localized in stickleback using *in situ* hybridization with BAC clones (Cresko et al., unpublished). For zebrafish, two PAC or BAC clones containing single copy genes from each of the 25 genetic linkage groups have been mapped using FISH (Phillips et al., 2006a). Figure 4.5.2a shows the localization of a PAC containing smo (LG4) to the short arm of chromosome 3 in zebrafish. The size of the long arm of chromosome 3 is variable because it contains a cluster of 5S rDNA. Chromosome identification was done using relative chromosome size and chromosome arm ratios (Gornung et al., 1997), although the presence of specific repetitive DNAs at centromeres (Phillips and Reed, 2000) can also be used. Zebrafish large insert clones containing genes near telomeres often hybridized to telomeres of multiple chromosome pairs, suggesting the presence of shared subtelomeric repetitive DNAs near telomeres. There is evidence from comparative gene mapping in medaka, zebrafish and humans that duplicate proto-chromosomes are present in zebrafish and

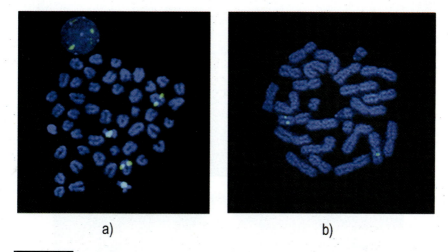

a)   b)

Fig. 4.5.1 *In situ* hybridization on chromosomes of Antarctic fish species using various probes: 1a—double IgH locus in *Trematomus scotti;* 1b—alpha beta globin genes in *Notothenia angustata.*

medaka (Naruse *et al.*, 2004). However, these duplicate linkage groups are not associated with chromosomes of similar size or morphology. This suggests that considerable chromosome restructuring occurred subsequent to the genome duplication in teleosts. The only vertebrate proto-linkage group that corresponds to one zebrafish linkage group is LG 1. This linkage group corresponds to chromosome 6, the only large metacentric chromosome pair in zebrafish. This chromosome was probably created by a chromosome fusion, so may explain the reason why only one zebrafish LG corresponds to this vertebrate proto-linkage group.

In case of rainbow trout, BAC clones have been isolated containing single copy genes or microsatellite loci from each of the 30 genetic linkage groups and at least one per linkage groups have been mapped using FISH (Phillips *et al.*, 2006b). Most of the linkage groups have at least two BAC clones mapped, so the orientation of the genetic map on the chromosome map is known. Chromosome identification was done using a combination of relative lengths and chromosome arm ratios and centromere probes (Reed *et al.*, 1995; Phillips *et al.*, 2006a). Figure 4.5.2b shows a dual hybridization with two different centromere probes in rainbow trout: 66L6 in green and 10h19 in red. Centromeres of some chromosome pairs have only 66L6 (4, 8 and 14); some have only 10h19 (7, 20), and others have

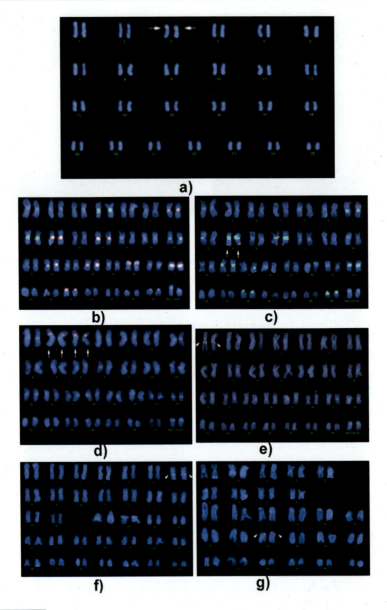

Fig. 4.5.2 *In situ* hybridization on fish chromosomes with various probes: (a) zebrafish with PAC clone containing smo (mapped to LG4 in red); (b) rainbow trout with two centromere probes, 10h19 in red and 66L6 in green; (c) rainbow trout with 10h19 in green and a BAC clone containing msSsa197 (mapped to LG 21) in red; (d) rainbow trout with BAC clone containing TAPBPR (in red) that hybridizes to both duplicate loci; (e) rainbow trout with BAC containing FGF6 in red; (f) chinook salmon with BAC containing FGF6; and (g) New Brunswick Atlantic salmon with BAC containing FGF6.

both (9, 11, 16, 18, 22, 25 and 29). Figure 4.5.2c illustrates a dual hybridization of 10h19 in green and a BAC in red that contains microsatellite locus Ssa197 (mapped to LG 21). In salmonids, the recombination rate is much higher in females compared to males, especially in regions near the centromeres (Sakomoto *et al.*, 2000). It has been observed that the ratio of female: male recombination rate is much higher in certain linkage groups than others (Danzmann *et al.*, 2005) and these linkage groups correspond to the larger metacentric pairs. An entire genome duplication occurred in the early ancestor of salmonid fishes 50-100 million years ago, so many genes are present in duplicates. The most common diploid karyotype in teleosts is 48 to 50 acrocentric (single armed) chromosomes, and rainbow trout karyotypes vary from 58-64, with primarily metacentric (bi-armed) chromosomes, suggesting that many centric fusions occurred between the duplicated chromosomes following the tetraploid event. The duplicated chromosome pairs in salmonid fishes are referred to as homeologues or homeologous chromosomes in accordance with the terminology used for sister chromosomes resulting from polyploidy in plants (Allendorf and Danzmann, 1997 and Wolfe, 2003). BAC clones containing duplicate genes usually mapped to a single chromosomal location, suggesting that most of the chromosomes have diploidized (Phillips *et al.*, 2003). Duplicate genes mapping closer to the telomeres were more likely to hybridize to two chromosome pairs (Fig. 4.5.2d). In both cases, duplicate genes were usually found at a similar location on the chromosome arm. This data suggests that most of the chromosome fusions following tetraploidization were of the centric type and did not involve the duplicated homeologous chromosomes (see also Figs. 4.5.2e-g and 4.5.3a-d).

Genome-specific Chromosome Paint Probes

These probes are produced by labeling genomic DNA by nick translation or by amplification by PCR of sheared genomic DNA. Species-specific chromosome paint probes have been used to study chromosome elimination in hagfish *Eaptatretus okinosearius* (Kubota *et al.*, 1993) and salmonid hybrids (Fujiwara *et al.*, 1997; Phillips, unpublished). In the case of masu/rainbow hybrids, the paternal chromosomes are preferentially eliminated. These hybrids could be used for localizing genes to specific chromosomes.

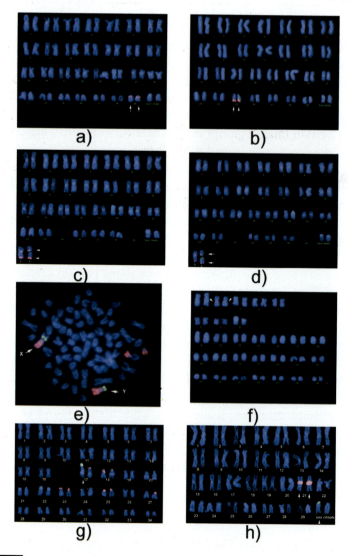

Fig. 4.5.3 *In situ* hybridization to salmonid fish chromosomes with various probes. Figures a-d illustrate the identification using chromosome-specific BAC probes of the chromosomes pairs involved in intraspecific chromosome fusions in rainbow trout and Figures e-h illustrate probes to salmonid sex chromosomes: (a) rainbow trout of OSU strain with BAC probe to G9 (red) which hybridizes to chromosome 29; (b) rainbow trout of OSU strain with BAC probe to TCRbeta (red) which hybridizes to chromosome 25; (c) hybrid between OSU and Clearwater with probes to G9 and TCRbeta; (d) hybrid between OSU and Swanson with probes to G9 and TCRbeta; (e) lake trout male with paint probes to Yp (green) and Yq (red) (f) Norwegian Atlantic salmon with probe to Oneu202 (closely linked to SEX) in red; (g) chinook salmon with 5S rDNA (red) and male-specific GHY (green); (h) coho salmon with 5S rDNA (red) and GHY (green).

Autosome-specific Chromosome Paint Probes

Several chromosomes or chromosome-arm specific probes have been made for rainbow trout and one made to the largest acrocentric chromosome pair also hybridizes to the telomere of the homologous metacentric chromosome pair (Phillips, 2001). The paint probes made from rainbow trout can be applied to all species of salmonids examined to date, so if a complete set were made, they could be used to trace chromosome evolution in salmonids. Although chromosome numbers vary from 2N=52-84 in salmonids, most of them have 100 chromosome arms (NF=100), further suggesting that chromosome rearrangements have predominately involved Robertsonian centric fusions/fissions. Atlantic salmon (2N=54, NF=74) is an exception and the large acrocentrics in the karyotype have been assumed to be the result of tandem fusions. This finding is supported by results with a couple of the chromosome arm-specific paint probes prepared from rainbow trout which paint only half an arm in the large acrocentrics in Atlantic salmon (Phillips, 2001).

An additional use of paint probes for salmonids is to help identify the homologous chromosome regions which occasionally exchange genes at the telomeres in male meiosis, thus preserving the duplicated gene regions (Allendorf and Danzmann, 1997; Sakomoto et al., 2000). One of the chromosome paint probes prepared from rainbow trout also highlights the telomeric region of a second chromosome pair, which presumably corresponds to its homeologous chromosome pair. Again, specific BAC probes are probably more useful for this purpose, since they divulge information on the specific genes involved in the duplicate regions.

Analysis of Sex Chromosomes with Sex Chromosome-specific Chromosome Paint Probes and BACs

In tilapia, paint probes were made to the terminal portion of the long arm from both XX and YY individuals and they paint that particular region in both X and Y chromosome. Dual hybridization with both probes (one labeled in red and the other in green) results in yellow color due to the superimposition of the two signals mainly in the middle of the chromosome arm, suggesting that the terminal region may contain some sex-specific sequences (Harvey et al., 2003). O. niloticus bacterial artificial chromosome clones, containing sex-linked AFLP markers, hybridized to the long arm of chromosome 1 (Ezaz *et al.*, 2004). This confirmed previous evidence, based on meiotic chromosome pairing (Campos-Ramos *et al.*, 2001) and fluorescence *in situ* hybridization probes obtained through chromosome microdissection (Harvey *et al.*, 2002), that chromosome pair 1 is the sex chromosome pair.

Sex chromosome-specific paint probes have been prepared for lake trout (Reed *et al.*, 1995; Phillips *et al.*, 2001) and tilapia (Harvey *et al.*, 2002). In the case of lake trout, the Yp probe was prepared from the short arm of the Y chromosome by microdissection and amplification, and the Yq probe that was prepared from the largest acrocentric chromosome arm in rainbow trout happened to correspond to the long arm of the Y chromosome (Yq) in lake trout. The probes go to both X and Y chromosomes, suggesting that most of the chromosome is pseudoautosomal (Fig. 4.5.3e). The lake trout Yp probe highlights the sex chromosome in only lake trout and brook trout (Phillips *et al.*, 2001, 2002), but not in any other salmonid fish species. This is consistent with genetic mapping studies (Woram *et al.*, 2003) that have shown that the sex chromosome linkage group corresponds to a different rainbow trout autosomal linkage group in almost every species.

The sex chromosome pair in Atlantic salmon was identified using a BAC probe containing a sex-linked microsatellite locus, Oneu202 (Artieri *et al.*, 2005). This locus maps to the middle of the long arm of chromosome 2, adjacent to a large distal block of heterochromatin (Fig. 4.5.3f). All of the Pacific salmon except for sockeye salmon have a male-specific pseudogene (GH-Y) which has been used to identify the sex chromosome pair (Figs 4.5.3g and h). Recently, we used *in situ* hybridization with a BAC clone containing Omy7INRA, a marker in the pseudoautosomal region of chinook salmon, to show that the Y chromosomes of chinook and coho are not homologous even though they share male-specific markers on the small short arms (Phillips *et al.*, 2005a). This BAC clone hybridized close to the centromere of the sex chromosome pair in chinook salmon, but on an autosome in coho salmon.

In medaka, the sex chromosomes were identified using BAC clones containing markers closely linked to sex (Matsuda *et al.*, 2002). Dual hybridization with two BAC clones, one containing the sex-linked marker caspase 6 and the other the male-determining gene dmrtY showed that these BACs co-localized on a morphologically similar chromosome in the closely related O. *curvinotus*. However, they went to different chromosome pairs in O. *luzonensis*, which has only the autosomal Dmrt1 gene. Thus, it appears that the dmrtY gene arose less than ten million years ago. The DMRTY locus appears to have arisen by duplication in the last 4-10MY because it is only present in close relatives of the medaka (Kondo *et al.*, 2004). DMRT 1 is a conserved gene in the sex determination cascade from flies and worms to vertebrates (Zarkower, 2001). Apparently, the duplication of this gene led to an increase in expression and elevated it to the status of the master sex-determining gene in medaka (Schartl, 2004). A recent origin for the sex-determining gene has also been shown for tilapia. In most fishes with sex chromosomes, the male-specific region is small and of recent origin. The production of new sex chromosomes can occur due to the emergence of new sex-determining genes as has been shown in medaka or by translocation or transposition of the male determining region which appears to be occurring in salmonid fishes (Woram *et al.*, 2003; Phillips *et al.*, 2005a). The frequent origin of new sex chromosomes would counter the degeneration usually found on the Y chromosome and obviates the need for evolving mechanisms of dosage compensation.

Focusing on the Chromosome Evolution in Salmonid Fishes

BAC probes obtained from rainbow trout have been hybridized to chromosomes of chinook salmon, coho salmon and Atlantic salmon. In addition, BAC probes have been used to identify the intraspecific translocations in both rainbow trout and Atlantic salmon. Results have shown that BACs usually hybridize to the same relative location on chromosome arms of different species, suggesting that intrachromosomal rearrangements such as inversions have been minimal. For example, if the BAC maps just proximal to the centromere in one species, it will usually be located there in the other species (Fig. 4.5.2e-g). Rainbow trout probes often localize to a chromosome of a similar size and morphology in coho salmon, but not in the other two species. This is not surprising, because these two species have similar chromosome numbers: $2N=58-64$ in rainbow trout, and $2N=60$ in coho salmon. In the case of chinook salmon $(2N=68)$, probes going to metacentrics in rainbow went to acrocentrics in chinook and vice versa, suggesting that a large number of whole chromosome arm fusions have occurred between these two species. The karyotype of Atlantic salmon has undergone a large number of tandem fusions compared to the other species (reviewed in Phillips and Rab, 2001), but still a significant number of probes map to the same location on the chromosome arm, suggesting relatively few inversions have occurred.

In rainbow trout, a survey of karyotypes from throughout the range (Thorgaard, 1983; Ostberg and Thorgaard, 1999) showed that most of the interior trout have 2N=58, and coastal Washington and Oregon had 2N=60, while 2N=64 was more common in coastal California. We have examined the specific chromosomes involved in the difference between two interior 2N=58 strains: Clearwater from Idaho and Swanson from Alaska using BAC probes containing genes from specific linkage groups Figs 4.5.3a,b,c,d) (Phillips *et al.*, 2005b). Results showed that both 2N=58 strains have the same chromosomal fusion involving chromosomes 25 and 29 from the OSU 2N=60 strain. This further suggests that the interior trout may all have the same karyotype. Currently, we are testing two 2N=64 strains (one from southern Oregon, and the other from northern California) to see if they will have the same rearrangement. Similar experiments are planned with European and North American Atlantic salmon. The karyotypes of North American Atlantic salmon differ by both centric fusions and tandem fusions (Phillips and Rab, 2001).

To summarize, genome projects have generated large clones that can be readily localized on chromosomes using FISH. Single copy genes have been localized in a number of species using these probes. Production of paint probes by microdissection has also been useful for identification of sex chromosomes and production of chromosome paints. However, this technique has been limited to chromosomes that are easily identified in unstained preparations. Development of new techniques for flow sorting of chromosomes (Gribble *et al.*, 2004) should facilitate production of chromosome paint probes for fishes. Such probes would be very useful for fish chromosome evolution studies. Accurate chromosome identification that can be used in connection with localization of single copy genes is a major deficiency that needs to be addressed. Preliminary work with salmonid fishes suggests that it will be possible to develop chromosome-specific centromere probes that could fill this need and allow new clones to be assigned quickly to linkage groups using FISH.

Acknowledgments

This work was supported by grants from NIH and the NRI Competitive Grants Program of USDA.

References

Allendorf, F.W. and R.G. Danzmann. 1997. Secondary tetrasomic segregation of MDH-B and preferential pairing of homologues in rainbow trout. *Genetics* 145: 1083–1092.

Aparicio, S., J. Chapman, E. Stupka, N. Putnam, J.M. Chia, P. Dehal, A. Christoffels, S. Rash, S. Hoon, A. Smit, M.D. Gelpke, J. Roach, T. Oh, I.Y. Ho, M. Wong, C. Detter, F. Verhoef, P. Predki, A. Tay, S. Lucas, P. Richardson, S.F. Smith, M.S. Clark, Y.J. Edwards, N. Doggett, A. Zharkikh, S.V. Tavtigian, D. Pruss, M. Barnstead, C. Evans, H. Baden, J. Powell, G. Glusman, L. Rowen, L. Hood, Y.H. Tan, G. Elgar, T. Hawkins, B. Venkatesh, D. Rokhsar and S. Brenner. 2002. Whole-genome shotgun assembly and analysis of the genome of *Fugu rubripes*. *Science* 297: 1301–1310.

Artieri, C.G., L.A. Mitchell, S.H.S. Ng, S.E. Parisotto, R.G. Danzmann, B. Hoyheim, R.B. Phillips, M. Morasch, B. F. Koop, and W.S. Davidson. 2006. Identification of the sex-determining locus of Atlantic salmon (*Salmo salar*) on chromosome 2. *Cytogenetic and Genome Research*. (In Press).

Brunner, B., F. Grutzner, M.-L. Yaspo, H.-H. Ropers, T. Haaf and V.M. Kalscheuer. 2000. Molecular cloning and characterization of the *Fugu rubripes* MEST/COPG2 imprinting cluster and chromosomal localization in *Fugu* and *Tetraodon nigroviridis*. *Chromosome Research* 8: 465–476.

Burgtorf, C., K. Welzel, R. Hasenbank, G. Zehetner, S. Weis and H. Lehrach. 1998. Gridded genomic libraries of different chordate species: A reference library system for basic and comparative genetic studies of chordate genomes. *Genomics* 52: 230–232.

Campos-Ramos, R., S.C. Harvey, J.S. Masabanda, L.A. Carrasco, D.K. Griffin, B.J. McAndrew, N.R. Bromage and D.J. Penman. 2001. Identification of putative sex chromosomes in the blue tilapia, *Oreochromis aureus*, through synaptonemal complex and FISH analysis. *Genetica* 111: 143–153.

Clark, M. 2003. Genomics and mapping of Teleostei (bony fish). *Comparative and Functional Genomics* 4: 182–193.

Crollius, H.R., O. Jaillon, C. DaSilva, C. Ozouf-Costaz, C. Fizames, C. Fischer, L. Bouneau, A. Billault, F. Quetier, W. Saurin, A. Bernot and J. Weissenbach. 2000. Characterization and repeat analysis of the compact genome of the freshwater pufferfish *Tetraodon nigroviridis*. *Genome Research* 10: 939–949.

Danzmann, R.G., M. Cairney, W.S. Davidson, M.M. Ferguson, K. Gharbi, R. Guyomard, L.E. Holm, E. Leder, N. Okamoto, A. Ozaki, C.E. Rexroad 3rd, T. Sakamoto, J.B. Taggart, and R.A. Woram. 2005. A comparative analysis of the rainbow trout genome with 2 other species of fish (Arctic charr and Atlantic salmon) within the tetraploid derivative Salmonidae family (subfamily: Salmoninae). *Genome* 48: 1037–51.

Ezaz, M.T., S.C. Harvey, C. Boonphakdee, A.J. Teale, B.J. McAndrew and D.J. Penman. 2004. Isolation and physical mapping of sex-linked AFLP markers in Nile tilapia (*Oreochromis niloticus* L.). *Marine Biotechnology* 6: 435–445.

Froschauer, A., C. Korting, T. Katagiri, T. Aoki, S. Asakawa, N. Shimizu, M. Schartl and J.N. Volff. Construction and initial analysis of bacterial artificial chromosome (BAC) contigs from the sex-determining region of the platyfish *Xiphophorus maculatus*. *Gene* 295: 247–254.

Fujiwara, A., S. Abe, E. Yamaha, F. Yamazaki and C. YM. 1997. Uniparental chromosome elmination in the early embryogenesis of the inviable salmonid hybrids between masu salmon female and rainbow trout male. *Chromosoma* 106: 44–52.

Gornung, E., I. Gabrielli, S. Cataudella and L. Sola. 1997. CMA₃-banding pattern and fluorescence *in situ* hybridization with 18S rRNA genes in zebrafish chromosomes. *Chromosome Research* 5: 40–46.

Gribble, S., B.L. Ng, E. Prigmore, D.C. Burford and N.P. Carter. 2004. Chromosome paints from single copies of chromosomes. *Chromosome Research* 12: 143–153.

Harvey, S.C., C. Boonphakdee, R. Campos-Ramos, M.T. Ezaz, D.K. Griffin, N.R. Bromage and D.J. Penman. 2003. Analysis of repetitive DNA sequences in the sex chromosomes of *Oreochromis niloticus*. *Cytogenetic and Genome Research* 101: 314–319.

Harvey, S.C., J. Masabanda, L.A. Carrasco, N.R. Bromage, D.J. Penman and D.K. Griffin. 2002. Molecular-cytogenetic analysis reveals sequence differences between the sex chromosomes of *Oreochromis niloticus*: Evidence for an early stage of sex-chromosome differentiation. *Cytogenetic and Genome Research* 97: 76–80.

Jaillon, O., J.-M. Aury, J.-L. Petit, N. Stange-Thomann, E. Mauceli, L. Bouneau, F. Brunet, C. Fischer, C. Ozouf-Costaz, A. Bernot, S. Nicaud, D. Jaffe, S. Fisher, G. Lutfalla, C. Dossat, B. Segurens, C. Dasilva, M. Salanoubat, M. Levy, N. Boudet, S. Castellano, V. Anthouard, C. Jubin, V. Castelli, M. Katinka, B. Vacherie, C. Biémont, Z. Skalli, L. Catolico, J. Poulain, S. Duprat, P. Brottier, J.-P. Coutanceau, J. Gouzy, G. Parra, G. Lardier, C. Chapple, K.J. McKernan, P. McEwan, S. Bosak, J.-N. Volff, R. Guigó, M. Zody, J. Mesirov, K. Lindblad-Toh, B. Birren, C. Nusbaum, D. Kahn, M. Robinson-Rechavi, V. Laudet, V. Schaechter, F. Quetier, W. Saurin, C. Scarpelli, P. Wincker, E.S. Lander, J. Weissenbach and H. Roest Crollius. 2004. Genome duplication in the teleost fish *Tetraodon nigroviridis* reveals the early vertebrate proto-karyotype. *Nature (London)* 431: 946–957.

Katagiri, T., S. Asakawa, S. Minagawa, N. Shimizu, I. Hirono and T. Aoki. 2001. Construction and characterization of BAC libraries for three fish species; rainbow trout, carp and tilapia. *Animal Genetics.* 32: 200–204.

Katagiri, T., C. Kidd, E. Tomasino, J.T. Davis, C. Wishon, J.E. Stern, K.L. Carleton, A.E. Howe, T.D. K. 2005. A BAC-based physical map of the Nile tilapia genome. BMC *Genomics* 6: 89.

Katagiri, T., S. Minagawa and I. Hirono. 2002. Construction of a BAC library for the Red Sea bream *Pagras major*. *Fisheries Science* 68: 942–944.

Koch, R., G.J. Rauch, S. Humphray, T.R. Geisler and R. Plasterk. 2004. Bacterial artificial chromosome (BAC) clones and the current clone map of the zebrafish genome. *Methods in Cell Biology* 77: 295–304.

Kondo, M., I. Nanda, U. Hornung, M. Schmid and M. Schartl. 2004. Evolutionary origin of the medaka Y chromosome. *Current Biology* 14: 1664–1669.

Kubota, S., O. M. Kuro, S. Mizuno and S. Kohno. 1993. Germ line restricted highly repeated DNA sequences and their chromosome localization in a Japanese hagfish (*Eeptatretus okinoseanus*). *Chromosoma* 102: 163–173.

Matsuda, M., N. Kawato, S. Asakawa, N. Shimizu, Y. Nagahama, S. Hamaguchi, M. Sakaizumi and H. Hori. 2001. Construction of a BAC library derived from the inbred Hd-rR strain of the teleost fish, *Oryzias latipes*. *Genes and Genetic Systems* 76: 61–63.

Matsuda, M., C.S.H. Matsuda and M. Sakaizumi. 1998. Identification of the sex chromosomes of the medaka, *Oryzias latipes*, by fluorescence *in situ* hybridization. *Cytogenetics and Cell Genetics* 82: 257–262.

Mazzei, F., O. Umberto, V. Alfieri, M.R. Coscia, C. Ozouf-Costaz, L. Ghigliotti and E. Pisano. 2004. Comparative cytogenetic mapping of IgH locus in Antarctic fish species (Suborder Notothenioidei, Family Nototheniidae). *Cytogenetic and Genome Research* 106: 20.

Nanda, I., J.-N. Volff, S. Weis, C. Koring, A. Froschauer, M. Schmid and M. Schartl. 2000. Amplification of a long terminal repeat-like element on the Y chromosome of the platyfish, *Xiphophorus maculatus*. *Chromosoma* 109: 173–180.

Naruse, K., M. Tanaka, K. Mita, A. Shima, J. Postlethwait and H. Mitani. 2004. A medaka gene map: The trace of ancestral vertebrate proto-chromosomes revealed by comparative gene mapping. *Genome Research* 14: 820–824.

Ostberg, C.O. and G.H. Thorgaard. 1999. Geographic distribution of chromosome and microsatellite DNA polymorphisms in *Oncorhynchus mykiss* native to western Washington. *Copeia* 1999: 287–298.

Palti, Y., S.A. Gahr, J.D. Hansen and C.E. Rexroad. 2004. Characterization of a new BAC library for rainbow trout: Evidence for multi-locus duplication. *Animal Genetics* 35: 130–133.

Peichel, C.L., J.A. Ross, C.K. Matson, M. Dickson, J. Grimwood and J. Schmutz. 2004. The master sex-determination locus in three-spine sticklebacks is on a nascent Y chromosome. *Current Biology* 14: 1416–1424.

Pendas, A.M., P. Moran and E. Garcia-Vazquez. 1994. Organization and chromosomal location of the major histone cluster in brown trout, Atlantic salmon and rainbow trout. *Chromosoma* 103: 147–152.

Perez, J., P. Moran and E. Garcia-Vazquez. 2000. Isolation, characterization, and chromosomal location of the tRNA(Met) genes in Atlantic salmon (*Salmo salar*) and brown trout (*Salmo trutta*). *Genome* 43: 185–190.

Phillips, R.B. and K.M. Reed. 2000. Localization of repetitive DNAs to zebrafish chromosomes using multi-color fluorescence *in situ* hybridization. *Chromosome Research* 8: 27–35.

Phillips, R.B. 2001. Application of fluorescence *in situ* hybridization (FISH) to fish genetics and genome mapping. *Marine Biotechnology* 3: 5145–5152.

Phillips, R.B. and P. Rab. 2001. Chromosome evolution in the Salmonidae (Pisces): An update. *Biological Reviews* 76: 1–25.

Phillips, R.B., N.R. Konkol, K.M. Reed and J.D. Stein. 2001. Chromosome painting supports lack of homology among sex chromosomes in *Oncorhynchus*, *Salmo* and *Salvelinus* (Salmonidae). *Genetica* 111: 119–123.

Phillips, R.B., M.P. Matsuokaeg and K.M. Reed. 2002. Characterization of charr chromosomes using fluorescence *in situ* hybridization. *Environmental Biology of Fishes* 64: 223–228.

Phillips, R.B., A. Zimmerman, M. Noakes, Y. Palti, M. Morasch, L. Eiben, S. Ristow, G.H. Thorgaard and J.D. Hansen. 2003. Physical and genetic mapping of the rainbow trout major histocompatibility regions: Evidence for duplication of the class I region. *Immunogenetics* 91: 561–569.

Phillips, R.B., M.R. Morasch, L.K. Park, K.A. Naish and R.H. Devlin. 2005a. Identification of the sex chromosome pair in coho salmon (*Oncorhynchus kisutch*): lack of conservation of the sex linkage group with chinook salmon (*Oncorhynchus tschawytscha*). *Cytogenetic and Genome Research.* (In press).

Phillips, R.B., M.R. Morasch, P.A. Wheeler and G.H. Thorgaard. 2005b. Rainbow trout (*Oncorhynchus mykiss*) of Idaho and Alaskan origin (2n=58) share a chromosome fusion relative to trout of California origin (2n=60). *Copeia* 2005: 660–663.

Phillips, R.B., A. Amores, M.R. Morasch, C. Wilson, and J.H. Postlethwait. 2006a. Assignment of zebrafish genetic linkage groups to chromosomes. *Cytogenetic and Genome Research* 114: 155–162.

Phillips, R.B. K.M. Nichols, J.J. DeKoning, M.R. Morasch, K.A. Keatley, C. Rexoad, S. Gahr, R.G. Danzmann, R.E. Drew and G.H. Thorgaard. 2006b. Assignment of rainbow trout linkage groups to specific chromosomes. *Genetics* (in press).

Pisano, E., E. Cocca, F. Mazzei, L. Ghigliotti, G. di Prisco, H.W.I. Detrich and C. Ozouf-Costaz. 2003. Mapping of alpha and beta-globin genes on Antarctic fish chromosomes by fluorescence in-situ hybridization. *Chromosome Research* 11: 663–640.

Quiniou, S.M., T. Katagiri, N.W. Miller, M. Wilson, W.R. Wolters and G.C. Waldbieser. 2003. Construction and characterization of a BAC library from a gynogenetic channel catfish *Ictalurus punctatus*. *Genetics, Selection, Evolution* 35: 673–683.

Reed, K.M., S.K. Bohlander and R.B. Phillips. 1995. Microdissection of the Y chromosome and FISH analysis of the sex chromosomes of lake trout (*Salvelinus namaycush*). *Chromosome Research* 3: 221–226.

Reusch, T.B.H., H. Schaschl and K.M. Wegner. 2004. Recent duplication and inter-locus gene conversion in major histocompatibility class II genes in a teleost, the three-spined stickleback. *Immunogenetics* 56: 427–437.

Sakomoto, T., R.G. Danzmann, K. Gharbi, P. Howard, A.Ozaki, S.K. Khoo, R.A. Woram, N. Okamoto, M.M. Ferguson, L-E. Holm, R. Guyomard and B. Hoyheim. 2000. A microsatellite linkage map of rainbow trout (*Oncorhynchus mykiss*) characterized by large sex-specific differences in recombination rates. *Genetics* 155: 1331–1345.

Schartl, M. 2004. Sex chromosome evolution in non-mammalian vertebrates. *Current Opinion in Genetics and Development* 14: 634–641.

Suzuki, T., T. Ota, A. Fujiyama and M. Kasahara. 2004. Construction of a bacterial artificial chromosome library from the inshore hagfish, *Eptatretus burgeri*: A resource for the analysis of the agnathan genome. *Genes and Genetic Systems* 79: 251–253.

Thorgaard, G.H. 1983.Chromosomal differences among rainbow trout populations. *Copeia* 1983: 650–662.

Thorsen, J., B. Zhu, E. Frengen, K. Osegawa, P.J. de Jong, B.F. Koop, W.S. Davidson and B. Hoyheim. 2005. A highly redundant BAC library of Atlantic salmon (*Salmo salar*): An important tool for salmon projects. *BMC Genomics* 6: 50.

Watanabe, M., N. Kobayashi, A. Fujiyama and N. Okada. 2003. Construction of a BAC library for *Haplochromis chilotes*, a cichlid fish from Lake Victoria. *Genes and Genetic Systems* 78: 103–105.

Weinberg, J. 2005. Fluorescence in situ hybridization to chromosomes as a tool to understand human and primate genome evolution. *Cytogenetic and Genome Research* 108: 139–160.

Woram, R.A., K. Gharbi, R.G. Danzmann, T. Sakamoto and B. Hoyheim. 2003. Comparative genome analysis of the primary sex determining locus in salmonid fishes. *Genome Research* 13: 272–280.

Zarkower, D. 2001. Establishing sexual dimorphism: Conservation amidst diversity? *Nature Reviews Genetics* 2: 175–185.

Molecular Markers in Cartilaginous Fish Cytogenetics

Lucia Rocco

INTRODUCTION

Elasmobranch fishes (sharks and rays) are considered to be ancient vertebrates, having diverged from the stem lineage leading to crown vertebrates early in the diversification of craniates. In fact, the first fossil sharks that we know date back to about 450 million years ago (Zangerl, 1981).

The higher interrelationships of living selachian fishes have been intensively studied, but even now remain unresolved. Investigations of elasmobranch morphological features have led to the compilation of several controversial classifications. Early hypotheses based on morphology suggested a relationship between laterally gilled shark-like and ventrally gilled ray-like elasmobranchs. Some hypotheses propose an unresolved radiation of several lineages and leave many doubts concerning the evolution of Chondrichthyes and the systematic

Address for Correspondence: Dipartimento di Scienze della Vita, Seconda Università di Napoli, Via Vivaldi 43 – 81100 Caserta, Italy. E-mail: lucia.rocco@unina2.it

relationships among living elasmobranchs, mostly at order and superorder levels (Compagno, 1973, 1999; Maisey, 1982; de Carvalho, 1996; Shirai, 1996).

Based on certain osteological features shared by most forms, Compagno (1973) placed the elasmobranchs in a more inclusive group, the 'Euselachii'. This class is composed of all recent and hybodont fossil sharks that were abundant during the Paleozoic and Mesozoic strata. He further suggested that living sharks and rays have evolved from a common ancestor (such as Schaeffer's (1967) 'hybodont level') within these fossil taxa. Compagno divided the recent taxa into four superorders, Squalomorphii, Batoidea, Galeomorphii and Squatinomorphii, suggesting a monophyletic origin of the living elasmobranchs.

Shirai (1996) divided the living elasmobranchs into two lineages, the Galea, (Compagno's Galeomorphii) and Squalea, that included Compagno's Squalomorphii, Batoidea and Squatinomorphii.

Cytogenetic data in cartilaginous fishes are very poor, if we consider the fact that the karyotype morphology of only about 6% of 1,100 living species is presently known (Stingo and Rocco, 2001). Even if the classical cytogenetic investigation does not allow any conclusive indications to be drawn on the mechanisms of chromosomal evolution in cartilaginous fishes, the available cytogenetic information supports some general considerations.

The most widespread and primitive species in the different chondrichthyan taxa, as found in the other lower vertebrates (cyclostomes and primitive actynopterygians) have a high diploid number (up to 106), with many telocentric elements, and sometimes microchromosomes. As the species evolved a progressive reduction in the diploid number, an increase in the number of biarmed chromosomes and the disappearance of microchromosomes can be observed, probably as a consequence of polyploidy followed by subsequent diploidization events and Robertsonian rearrangements (reviewed in Stingo and Rocco, 2001).

Only in the last few years, different molecular approaches—to begin with physical mapping on metaphase chromosomes—have been used to investigate the cytotaxonomic relationship existing among the living elasmobranchs (Rocco et al., 2001, 2002, 2005).

Physical mapping involves *in situ* hybridization of specific segments of genomic DNA to their physical location on chromosomes. It is extremely useful in terms of gaining an insight into structural arrangements within the genome, such as distribution of repetitive DNA families, i.e., specific short interspersed nuclear elements (SINE), long interspersed nuclear elements (LINE) and satellite DNA among chromosomes and identification of multi-gene family locations in the genome (i.e., ribosomal DNA clusters). The use of these molecular cytogenetic techniques has often either suggested new taxonomical implications or confirmed the existing phylogenetic relationships among the different fish species analyzed (Perez et al., 1999; Viñas et al., 2004).

Such molecular cytogenetic techniques were particularly useful in elasmobranch karyology because, in these fishes, genome compartmentalization is not present, as has also been demonstrated in the majority of teleostean species (Medrano et al., 1988; Bernardi, 1993).

Due to their basal position in the vertebrate phylogenetic tree, the study on elasmobranch genetics and cytogenetics can provide interesting information on the mechanisms underlying the evolution of all vertebrates.

The present chapter reviews the data collected using the very recent molecular approach based on both physical chromosome mapping and genomic hybridization of some multicopy genes and repetitive sequences in Chondrichthyes.

The molecular markers provide information on the karyotype evolution in this class of vertebrates and identify some chromosome regions (i.e., centromeres or telomeres) as sites in which particular repeated DNA sequences accumulate.

MAJOR AND MINOR RIBOSOMAL GENES (rDNA)

In Chondrichthyes, chromosomal localization of major ribosomal genes by using fluorescence in situ hybridization (FISH) has often been associated with fluorochromes and silver-staining techniques. Chromomycin A_3 (CMA_3) is a GC-rich DNA specific fluorochrome dye and detects the nucleolar organizer regions (NORs) independently of their activity (Schmid and Guttenbach, 1988) in cartilaginous fishes. Silver (Ag-NOR) staining detects only NORs that are functionally active during the previous interphase (Howell, 1977).

The rDNA FISH probes are able to reveal all rRNA loci on the chromosomes. Table 4.6.1 reports the number and the chromosome location of Ag-NORs and CMA_3 sites as well as of 18S and 5S rRNA

Table 4.6.1 Number of sites and chromosomal location of major and minor ribosomal genes in Elasmobranchs so far studied.

SUPERORDER ORDER SPECIES	Ag-NOR		CMA$_3$		rDNA 18S		rDNA 5S		References
	Number of sites	Location	Number of sites	Location	Number of sites	Location	Number of sites	Location	
Batoidea Torpediniformes *Torpedo ocellata*	3	Telomeres of uniarmed chromosomes and subtelomeres of a biarmed one	–	–	–	–	–	–	Stingo et al., 1995
Torpedo marmorata	3	Telomeres of uniarmed chromosomes	–	–	–	–	–	–	Stingo et al., 1995
Rajiformes *Raja asterias*	6 + 2	Telomeres and centromeres of uniarmed chromosomes	–	–	–	–	–	–	Rocco et al., 2002
Raja montagui	10 + 2	Telomeres and centromeres of uniarmed chromosomes	12	Telomeres and centromeres of uniarmed chromosomes	12	Telomeres and centromeres of uniarmed chromosomes	4	Telomeres of uniarmed chromosomes	Rocco et al., 2005
Myliobatiformes *Taeniura lymma*	3 + 1	Telomeres of two biarmed chromosomes and pericentromeres of one/two biarmed ones	4 + 2	Telomeres of two biarmed chromosomes and pericentromeres of two biarmed ones + telomeres of an uniarmed pair	6	Telomeres of two biarmed chromosomes, pericentromeres of two biarmed ones and telomeres of an uniarmed pair	4	Telomeres of two biarmed chromosomes and pericentromeres of two biarmed ones	Rocco et al., 2005

(Table 4.6.1 contd.)

(Table 4.6.1 contd.)

Galeomorphii Carcharhiniformes *Scyliorhinus canicula*	2	Telomeres of biarmed chromosomes	–	–	–	–	–	–	Rocco et al., 2002; Schmid et al., 1982
Scyliorhinus stellaris	2	Telomeres of biarmed chromosomes	2	Telomeres of biarmed chromosomes	2	Telomeres of biarmed chromosomes	–	–	Rocco et al., 2002
Orectolobiformes *Ginglymostoma cirratum*	4 + 2	Telomeres of biarmed chromosomes + telomeres of uniarmed ones	6	Telomeres of biarmed chromosomes and telomeres of uniarmed ones	4 + 2	Telomeres of biarmed chromosomes + telomeres of uniarmed ones	4	Telomeres of biarmed chromosomes	Rocco et al., unpublished
Chiloscyllium punctatum	4	Telomeres of biarmed chromosomes	8	Telomeres of bi- and uni-armed chromosomes	4	Telomeres of biarmed chromosomes	4	Telomeres of biarmed chromosomes	Rocco et al., unpublished

genes in the elasmobranch species studied so far. Figures 4.6.1 and 4.6.2 show the ideograms of the involved chromosomes in the representative species analyzed.

Major rDNA has been studied in five batoids and four galeoids. In the Batoidea, some multiple loci have been observed. It is particularly remarkable in the genus *Raja* (Rajiformes, a primitive superorder), where this number is very high. The number of chromosome pairs bearing major rDNA notably decreases in the more advanced species, *Torpedo ocellata* and *T. marmorata* (Torpediniformes) and *Taeniura lymma* (Myliobatiformes).

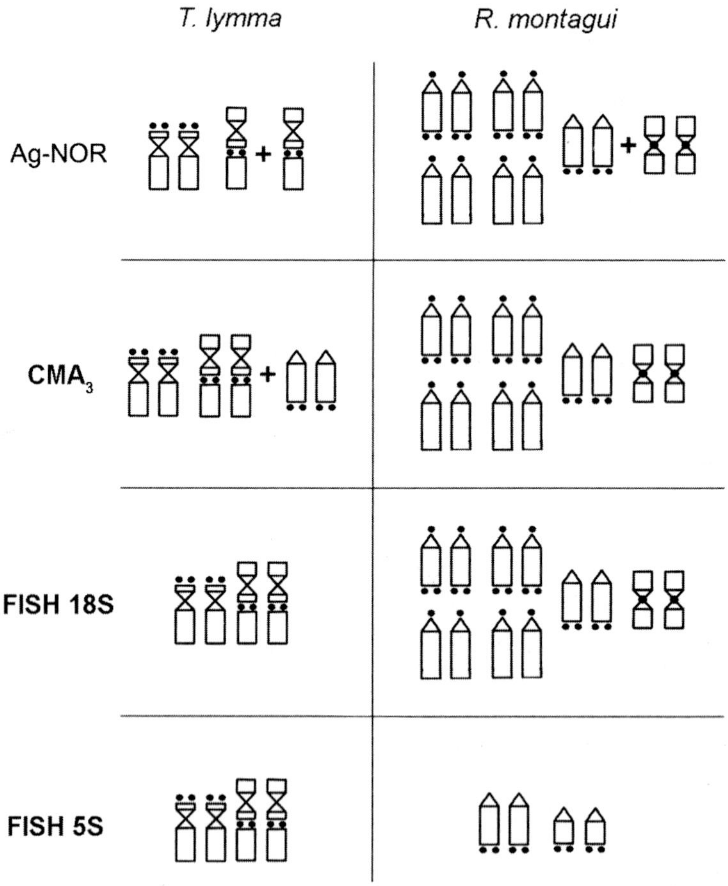

Fig. 4.6.1 Ideograms showing the chromosome pairs bearing the ribosomal clusters (black circles) in some representative species of Batoidea.

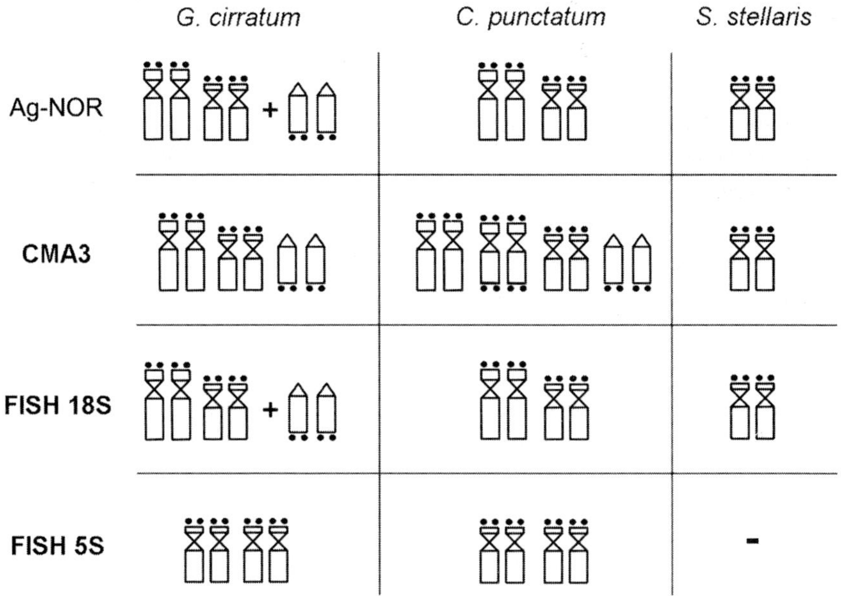

Fig. 4.6.2 Ideograms showing the chromosome pairs bearing the ribosomal clusters (black circles) in some representative species of Galeomorphii.

In Galeomorphii, the two species of Carcharhiniformes (*Scyliorhinus stellaris* and *S. canicula*) showed a single pair of chromosomes bearing major rRNA genes in their chromosome sets. This condition is common to several vertebrate groups such as mammals and amphibians (Hsu *et al.*, 1975) and to numerous species of bony fish (Gold and Amemiya, 1986; Amemiya and Gold, 1988). In the two species of Orectolobiformes (*Ginglymostoma cirratum* and *Chiloscyllium punctatum*), some additional sites have been observed. The presence of multiple chromosomal sites of major rRNA genes would appear to be a frequent characteristic in elasmobranch fishes.

Some hypotheses on the mechanism responsible for new rDNA sites in cartilaginous fish have been formulated (Rocco *et al.*, 2002, 2005). The most probable of these theories concern the transposition events brought about by mobile elements of the genome, followed by the subsequent amplification of the transposed sequences that might have given rise to the formation of these new sites.

The chromosomal location of 5S rDNA clusters has been studied in four species. The hybridization signal is present on two chromosome pairs in each species. In *T. lymma*, one of the two 5S sites is co-located with the major ribosomal cluster, at the interstitial level along the chromosome. Such an interstitial chromosomal location of 5S rDNA sequences seems frequent in fishes and it has also been observed in mammals and amphibians (De Lucchini *et al.*, 1993; Mäkinem *et al.*, 1997), suggesting that this type of pattern is not accidental, but could possibly provide some advantage to the organization of these genes in the genome of vertebrates (Martins and Galetti, 2001; Martins, chapter 4.4, present volume).

In fish species, as in amphibians, the genes for 5S rRNA can be localized on more than one chromosome pair (Schmid *et al.*, 1987; Martins and Galetti, 1999; Martins, present volume). One explanation for the presence of multiple chromosome sites bearing the gene for 5S rRNA in Elasmobranchs might be the extensive chromosomal rearrangements that occurred during the evolution in cartilaginous fishes.

The pattern of chromosomal localization of 5S rDNA strengthens the hypothesis that the karyotype of the Chondrichthyes evolved through fusion between acrocentric elements, forming bi-armed ones. The chromosomes bearing the minor ribosomal clusters might, in fact, have been involved in the Robertsonian events that took place in the karyological evolution of these fishes, stemming from a progenitor with more primitive karyotype characteristics such as those of *R. montagui* and bearing 5S rDNA sequences at the telomeric position (Fig. 4.6.3).

As far as 5S rDNA molecular organization is concerned, in the two batoid species studied, the results of the amplification products by PCR are similar to those reported in literature for other vertebrates. The length of the bands obtained in the different species varies due to the different size and base composition of the non-transcribed spacer (NTS) region. Moreover, the NTS region contains some TATA-like trinucleotides and other short sequences, such as $(TGC)_n$ trinucleotides, $(CA)_n$ dinucleotides and $(GTGA)_n$ tetranucleotides, which could be involved in the regulation of the gene itself (Rocco *et al.*, 2005), as it has been described for some species of teleosteans (Martins and Galetti, 2001).

Telomeric Sequences

The human telomeric sequence repeats $(TTAGGG)_n$, detected by FISH, were assayed on chromosomes of three species belonging to Batoidea superorder and of one galeoid species (Table 4.6.2).

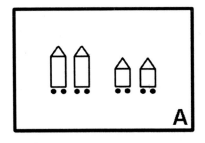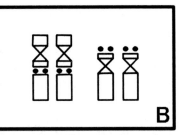

Fig. 4.6.3 Schematic representation of the chromosome pairs bearing clusters of the minor ribosomal genes (black circles) in two batoid species with more primitive (A, *Raja montagui*) and more advanced (B, *Taeniura lymma*) karyological characteristics.

Table 4.6.2 *In situ* localization of telomeric sequence $(TTAGGG)_n$ in Elasmobranchs so far studied.

Superorder Order Species	$(TTAGGG)_n$ sequence localization		References
	Telomeric	Interstitial	
Batoidea Rajiformes R. asterias	+	–	Rocco *et al.*, 2001
Torpediniformes T. ocellata	+	+	Rocco *et al.*, 2001
Myliobatiformes T. Lymma	+	+	Rocco *et al.*, 2002
Galeomorphii Carcharhiniformes S. stellaris	+	–	Rocco *et al.*, 2002

FISH allowed the detection of fluorescent signals on the telomeres of both uni- and bi-armed elements in the four species. In two of them (*T. ocellata* and *T. lymma*), an interstitial and/or paracentromeric labeling on four bi-armed chromosomes was also evident. This interstitial FISH pattern might represent further evidence supporting the hypothesis that karyotype evolution in Chondrichthyes occurred by a progressive reduction of chromosome number due to centric fusions (Stingo and Rocco, 1991; Stingo *et al.*, 1995). In fact, species that underwent karyotype rearrangements as a result of Robertsonian fusions also exhibit non-telomeric sites of the $(TTAGGG)_n$ sequences in addition to the telomeric ones (Meyne *et al.*, 1990; Nanda *et al.*, 1995).

Furthermore, the presence of additional interstitial sites of the $(TTAGGG)_n$ sequences in *T. ocellata* and *T. lymma* could indicate that, from a karyological point of view, these two species are in a phase of active evolutionary change. In fact, the interstitial sites containing $(TTAGGG)_n$ sequences are often involved in recombination events (Meyne *et al.*, 1990).

SINE AND LINE SEQUENCES

Retroposons represent a significant portion of repetitive DNA in Eukaryotes. They are mobile genetic elements that are amplified by a reverse transcription of an intermediate RNA (Boeke and Devine, 1998). For a review on trasposable elements see Dettai *et al.*, Chapter 4.1, present volume.

Two large families of retroposons—first identified as interspersed repeated sequences—belong to this genomic component. The LINE (Long Interspersed Nuclear Elements) family comprises long sequences, while the SINE (Short Interspersed Nuclear Elements) family has short sequences, both irreversibly inserted in the genome (Okada, 1991).

The SINE family was amplified in the Elasmobranch genome by PCR (Rocco *et al.*, unpublished) using the primers taken from both salmonids (*Hpa* I family) (Kido *et al.*, 1991) and humans (*Alu* sequences) (Kariya *et al.*, 1987).

FISH performed with the *Hpa* I-like SINE probe on *T. ocellata* chromosomes have revealed hybridization signals at the centromeric and/ or the paracentromeric region on several bi-armed chromosomes, while conspicuous fluorescent signals at the centromeric and/or telomeric level on the acrocentric ones were produced (Fig. 4.6.4A), providing a hybridization pattern that is not coincident with that evidenced by C-banding (Stingo *et al.*, 1995). Such a result is similar to that evidenced by Perez *et al.* (1999) in studies conducted on Atlantic salmon and rainbow trout karyotypes where the members of this SINE family are located only in the euchromatic regions of the chromosomes.

The location of *Alu*-like sequences on the chromosomes of *T. ocellata* is prevalently at the centromeric level, as in humans, and sometimes at the telomeric level, probably as a result of intraspecific polymorphism (Fig. 4.6.4B).

Marçais *et al.* (1991) and Prades *et al.* (1996) observed that the block of alphoid DNA at the centromeres of human chromosomes is susceptible

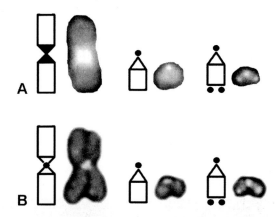

Fig. 4.6.4 Representative chromosomes of *T. ocellata* and corresponding ideograms on which *Hpa* I-like SINE (A) and *Alu*-like (B) sequences have been localized by fluorescence *in situ* hybridization.

to variations in length created by jumping amplification, with an unequal exchange of large alphoid domains between homologous chromosomes, and deletions of large DNA segments. Probably these *Alu*-like sequences retrotransposed within the *T. ocellata* genome to specific sites such as telomeric regions particularly exposed to a new insertion of transposable elements (Korenberg and Rykowsky, 1988).

The *Hpa* I-like SINE and *Alu*-like sequences were also hybridized on genomic DNA of four species of Batoidea (*Torpedo marmorata, T. ocellata, Raja asterias* and *R. montagui*) and two species of Galeomorphii (*Mustelus asterias* and *Scyliorhinus stellaris*) which provided other phylogenetic and systematic-evolutionary implications. In fact, they confirmed the taxonomic position that each species occupies within the class and the relationships present among the species attributed to the different superorders examined (Compagno, 1999).

Ogiwara *et al.* (1999) identified and characterized a LINE (HER I) family in 15 different elasmobranch species and used these sequences for a phylogenetic analysis, comparing them with those found in other vertebrates, especially birds and reptiles.

Recently, FISH performed with a probe consisting of a 800 bp fragment, taken from the final portion of this LINE element, and not including the SINE sequence, showed that these interesting sequences are located prevalently at the telomeric and centromeric level of several chromosome pairs. In particular, they are present on uni-armed elements in species belonging to primitive superorders such as *R. asterias,* and on bi-armed chromosomes in more evolved species, such as *Taeniura lymma* and *Scyliorhinus stellaris* (Rocco *et al.,* unpublished) (Fig. 4.6.5).

SATELLITE DNA

This particular repeated DNA is generally organized in tandem and preferentially localized on centromeric regions of chromosomes or on other heterochromatic regions, such as telomeres (Brutlag, 1980). As the satellite DNA families diverged rapidly during evolution (Wichman *et al.,*

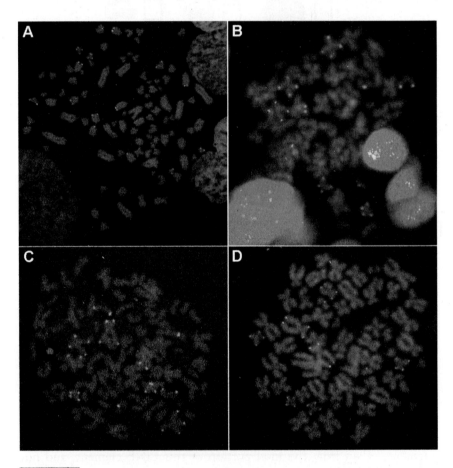

Fig. 4.6.5 *In situ* localization of the terminal portion of the LINE HER 1 element, not including the SINE sequence, in four Elasmobranch species: *Raja montagui* (A), *Taeniura lymma* (B), *Ginglymostoma cirratum* (C) and *Scyliorhinus stellaris* (D).

1991), they showed extreme diversity in type and chromosomal distribution even among closely related species (Miklos, 1985; Charlesworth et al., 1994) thus providing a valuable tool in disclosing taxonomical and phylogenetic relationships among related taxa (Garrido-Ramos et al., 1995; de la Herran et al., 2001; Lanfredi et al., 2001).

The genomic DNA of R. montagui showed a highly repeated component after digestion with the restriction endonuclease Hind III (Rocco et al., 1996). The monomer consisted of 311 bp and was rich in AT (61%). The conservation of the satellite monomeric unit was demonstrated by using both hybridization on genomic DNA and FISH on metaphase plates in some selachian species (Rocco et al., 1996, and unpublished).

The Southern Blot analysis showed an extremely similar repetitive pattern in the two species of Raja genus examined (R. asterias and R. montagui). In the two torpedoes (T. marmorata and T. ocellata), only the multimeric units with the highest molecular weight were present. In the two galeoid species (M. asterias and S. stellaris), the hybridization, under the same stringent experimental conditions did not provide any signals. For this reason, it was hypothesized that the origin of the satellite lays in a common progenitor from which the Batoidea and the Galeomorphii diverged, probably dating back to the upper Jurassic (Rocco et al., 1996).

FISH, with the homologous monomeric unit as a probe, showed a brilliant labeling at the telomeres of several chromosome pairs in the two batoid species R. montagui and T. lymma and in the two galeoid ones S. stellaris and Chiloscyllium punctatum (Rocco et al., unpublished).

The monomeric unit sequence of the Hind III satellite revealed a 76% degree of identity with a non coding portion of the acetylcholinesterase gene of Torpedo californica and with the same gene of T. marmorata (Sikorav et al., 1988; Maulet et al., 1990) and a 90% degree of identity with the 3'-end segment of the HER 1 LINE family—without the HE 1 element—identified and characterized by Ogiwara et al. (1999) in several cartilaginous fishes. The high degree of identity with the acetylcholinesterase gene led to the hypothesis that this satellite family originated from a pseudogene, which might have duplicated and then re-incorporated into another site of the genome of an elasmobranch progenitor. Afterwards, it might have undergone amplification becoming highly repetitive DNA.

CONCLUDING REMARKS

The molecular cytogenetic evidence collected so far confirms the particular genomic organization of selachians. The characteristics of this genomic organization are very particular regarding both the karyotypes and the size and composition of selachian DNAs. In fact, the data for this group exhibit more differences than similarities with teleosts. The few studies on genome composition have revealed marked differences between cartilaginous fishes and teleosteans in the ratio of the amount of GC-rich DNA to the total increase in genome. High values of compositional heterogeneity are present in Chondrichthyes (Stingo and Rocco, 1991), while Teleostei exhibit a low degree of base pair heterogeneity and substantially symmetric density gradient profiles (Hudson *et al.*, 1980; Bernardi and Bernardi, 1990).

The physical distribution of the different types of molecular markers used in elasmobranch cytogenetics provide valuable information about genome organization and the evolutionary status of a given species. Their use as probes for FISH makes it possible to compare different genomes and further contributes to standardizing the karyotype of the species studied. In fact, *in situ* hybridization of these sequences provides a clearly recognizable pattern allowing the identification of several chromosomes, thus helping in assigning unambiguous homologies and homeologies.

Moreover, they single out some preferential chromosome regions such as centromeres or telomeres. These last regions are particularly interesting because in almost all the selachian species also, the constitutive heterochromatin shown by C-banding, is prevalently concentrated at this level (Stingo *et al.*, 1995; Rocco *et al.*, 2002).

The results of physical mapping of almost all the sequences examined to date corroborate the primary hypothesis that karyotype evolution in Chondrichthyes occurred by way of a progressive reduction of chromosome number due to centric fusions (Stingo and Rocco, 2001).

The study of organization and chromosomal localization of ribosomal genes and repeated sequences reviewed in this chapter undoubtedly makes it possible to obtain better insight into molecular bases of chromosome structure and function, thus providing valuable information on the arrangements that are possibly involved in processes of speciation.

Finally, the molecular techniques reveal specific linkage groups on definite chromosome pairs in the Elasmobranch genome. Thus, some chromosome pairs are to be considered as chromosome markers themselves representing new tools for comparative cytogenetic and evolutionary studies.

REFERENCES

Amemiya, C.T. and J.R. Gold. 1988. Chromosomal NORs phenotypes as taxonomic and systematic characters in North American cyprinid fish. *Genetica* 76: 81–90.

Bernardi, G. 1993. The Vertebrate Genome: Isochores and Chromosomal Bands. In: *Chromosomes Today*, A.T. Sumner and A.C. Chandle (eds.). Chapman and Hall, London. Vol. 11, pp. 49–60.

Bernardi, G. and G. Bernardi. 1990. Compositional patterns in the nuclear genomes of cold-blooded vertebrates. *Journal of Molecular Evolution* 31: 265–281.

Boeke, J.D. and S.E. Devine. 1998. Yeast retrotransposons: Finding a nice quiet neighborhood. *Cell* 93: 1087–1089.

Brutlag, D.L. 1980. Molecular arrangement and evolution of heterochromatic DNA. *Annual Review of Genetics* 14: 121–44.

Charlesworth, B., P. Sniegowski and W. Stephan. 1994. The evolutionary dynamics of repetitive DNA in eukaryotes. *Nature (London)* 371: 215–220.

Compagno, L.J.V. 1973. Interrelationships of living elasmobranchs. In: *Interrelationships of Fishes*, P.H. Greenwood, R.S. Miles and C. Patterson (eds.). Academic Press, New York, pp. 15–61.

Compagno, L.J.V. 1999. Systematics and body form. In: *Sharks, Skates and Rays*, W.C. Hamlett (ed.). The John Hopkins University Press, Baltimore, pp. 1–42.

de Carvalho, M.R. 1996. Higher-level elasmobranch phylogeny, basal squalians, and paraphyly. In: *Interrelationships of Fishes*, M.L.J. Stiassny, L.R. Parenti and G.D. Johnson (eds.). Academic Press, San Diego, pp. 35–62.

de la Herran, R., C.R. Rejon, M.R. Rejon and M.A. Garrido-Ramos. 2001. The molecular phylogeny of the Sparidae (Pisces, Perciformes) based on two satellite DNA families. *Heredity* 87: 691–697.

De Lucchini, S., I. Nardi, G. Barsacchi, R. Batistoni and F. Andronico. 1993. Molecular cytogenetics of the ribosomal (18S + 28S and 5S) DNA loci in primitive and advanced urodele amphibians. *Genome* 36: 762–773.

Garrido-Ramos, M.A., M. Jamilena, R. Lozano, C. Ruiz Rejon and M. Ruiz Rejon. 1995. The Eco RI centromeric satellite DNA of the Sparidae family (Pisces, Perciformes) contains a sequence motive common to other vertebrate centromeric satellite DNAs. *Cytogenetics and Cell Genetics* 71: 345–351.

Gold, J.R. and C.T. Amemiya. 1986. Cytogenetic studies in North American minnows (Cyprinidae). XII. Patterns of chromosomal nucleolus organizer region variation among 14 species. *Canadian Journal of Zoology* 64: 1869–1877.

Howell, W.M. 1977. Visualization of ribosomal gene activity: Silver stains proteins associated with rRNA transcribed from oocyte chromosomes. *Chromosoma* 62: 361–367.

Hsu, T.C., S.E. Spirito and M.L. Pardue. 1975. Distribution of 18S 1 28S ribosomal genes in mammalian genomes. *Chromosoma* 53: 25–36.

Hudson, A.P., G. Cuny, J. Cortadas, A.E.V. Haschmeyer and G. Bernardi. 1980. An analysis of fish genomes by density gradient centrifugation. *European Journal of Biochemistry* 112: 203–210.

Kariya, Y., K. Kato, Y. Hayashizaki, S. Himeno, S. Tarui and K. Matsubara. 1987. Revision of consensus sequence of Human *Alu* repeats—A review. *Gene* 53: 1–10.

Kido, Y., M. Aono, T. Yamaki, K. Matsumoto, S. Murata, M. Saneyoshi and N. Okada. 1991. Shaping and reshaping of salmonid genomes by amplification of tRNA-derived retroposons during evolution. *Proceedings of the National Academy of Sciences of the United States of America* 88: 2326–2330.

Korenberg, J.R. and M.C. Rykowski. 1988. Human genome organization: *Alu*, LINEs, and the molecular structure of metaphase chromosome bands. *Cell* 53: 391–400.

Lanfredi, M., L. Congiu, M.A. Garrido-Ramos, R. de la Herran, M. Leis, M. Chicca, R. Rossi, J. Tagliavini, C. Ruiz Rejon, M. Ruiz Rejon and F. Fontana. 2001. Chromosomal location and evolution of satellite DNA family in seven sturgeon species. *Chromosome Research* 9: 47–52.

Maisey, J.G. 1982. The anatomy and interrelationships of Mesozoic Hybodont sharks. *American Museum Novitates* 2724: 1–48.

Mäkinen, A., C. Zijlstra, N.A. de Haan, C.H. Mellink and A.A. Bosma. 1997. Localization of 18S plus 28S and 5S ribosomal genes in the dog by fluorescence *in situ* hybridization. *Cytogenetics and Cell Genetics* 78: 231–235.

Marçais, B., J.P. Charlieu, B. Allain, E. Brun, M. Bellis and G. Roizes. 1991. On the mode of evolution of alpha satellite DNA in human populations. *Journal of Molecular Evolution* 33: 42–48.

Martins, C. and P.M. Galetti. 1999. Chromosomal localization of 5S rDNA genes in *Leporinus* fish (Anastomidae, Chraciformes). *Chromosome Research* 7: 363–367.

Martins, C. and P.M. Galetti. 2001. Two 5S rDNA array in Neotropical fish species: Is it a general rule for fishes? *Genetica* 111: 439–446.

Maulet, Y., S. Camp, G. Gibney, T.L. Rachinsky, T.J. Ekstrom and P. Taylor. 1990. Single gene encodes glycophospholipid-anchored and asymmetric acetylcholinesterase forms: Alternative coding exons contain inverted repeat sequences. *Neuron* 4: 289–301.

Medrano, L., G. Bernardi, J. Couturier, B. Dutrillaux and G. Bernardi. 1988. Chromosome banding and genome compartmentalization in fishes. *Chromosoma* 96: 178–183.

Meyne, J., R.J. Baker, H.H. Hobart, T.C. Hsu, O.A. Ryder, O.G. Ward, J.E. Wiley, D.H. Wurster-Hill, T.L. Yates and R.K. Moyzis. 1990. Distribution of non-telomeric sites of the (TTAGGG)$_n$ telomeric sequence in vertebrate chromosomes. *Chromosoma* 99: 3–10.

Miklos, G.L.G. 1985. Localized highly repetitive DNA sequences in vertebrate and invertebrate genomes. In: *Molecular Evolutionary Genetics*, RJ MacIntyre (ed.). Plenum Press, New York, pp. 241–321.

Nanda, I., S. Schneider-Rasp, H. Winking and M. Schmid. 1995. Loss of telomeric sites in the chromosomes of Mus musculus domesticus (Rodentia, Muridae) during Robertsonian rearrangements. Chromosome Research 3: 399–409.

Ogiwara, I.M., K. Ohshima and N. Okada. 1999. Retropositional parasitism of SINEs on LINEs: Identification of SINEs and LINEs in elasmobranchs. Molecular Biology and Evolution 16: 1238–1250.

Okada, N. 1991. SINEs. Current Opinion in Genetics and Development 1: 498–504.

Perez, J., E. Garcia-Vásquez and P. Móran. 1999. Physical distribution of SINE elements in the chromosomes of Atlantic salmon and rainbow trout. Heredity 83: 575–579.

Prades, C., A.M. Laurent, J. Puechberty, Y. Yurov and G. Roizes. 1996. SINE and LINE within human centromeres. Journal of Molecular Evolution 42: 37–43.

Rocco, L., V. Stingo and M. Belletti. 1996. Cloning and characterization of a repetitive DNA detected by Hind III in the genome of Raja montagui (Batoidea, Chondrichthyes). Gene 176: 185–189.

Rocco, L., D. Costagliola and V. Stingo. 2001. (TTAGGG)$_n$ telomeric sequence in selachian chromosomes. Heredity 87: 583–588.

Rocco, L., M.A. Morescalchi, D. Costagliola and V. Stingo. 2002. Karyotype and genome characterization in four cartilaginous fishes. Gene 295: 289–298.

Rocco L., D. Costagliola, M. Fiorillo, F. Tinti and V. Stingo 2005. Molecular and chromosomal analysis of ribosomal cistrons in two cartilaginous fish, Taeniura lymma and Raja montagui (Chondrichthyes, Batoidea). Genetica 123: 245–253.

Schaeffer, B. 1967. Comments on elasmobranchs evolution. In: Sharks, Skates and Rays, P.W. Gilbert, R.F. Mathewson and D.P. Rall (eds.). The Johns Hopkins University Press, Baltimore, pp. 3–35.

Schmid, M. and M. Guttenbach. 1988. Evolutionary diversity of reverse fluorescent chromosome bands in vertebrates. Chromosoma 97: 101–114.

Schmid, M., C. Loser, J. Schmidtke and W. Engel. 1982. Evolutionary conservation of a common pattern of activity of nucleolus organizers during spermatogenesis in vertebrates. Chromosoma 86: 149–179.

Schmid, M., L. Vitelli and R. Batistoni. 1987. Chromosome banding in Amphibia. IV. Constitutive heterochromatin, nucleolus organizers, 18S + 28S and 5S ribosomal RNA genes in Ascaphidae, Pipidae, Discoglossidae and Pelobatidae. Chromosoma 95: 271–284.

Shirai, S., 1996. Phylogenetic interrelationships of neoselachians (Chondrichthyes, Euselachii). In: Interrelationships of Fishes, M.L.J. Stiassny, L.R. Parenti and G.D. Johnson (eds.). Academic Press, San Diego. pp. 9–34.

Sikorav, J., L. Duval, A. Anselmet, S. Bon, E. Krejci, C. Legay, M. Osterlund, B. Reimund and J. Massoulie. 1988. Complex alternative splicing of acetylcholinesterase transcripts in Torpedo electric organ; primary structure of the precursor of the glycolipid-anchored dimeric form. EMBO Journal 7: 2983–2993.

Stingo, V. and L. Rocco. 1991. Chondrichthyan cytogenetics: A comparison with teleosteans. Journal of Molecular Evolution 33: 76–82.

Stingo, V. and L. Rocco. 2001. Selachian cytogenetics: A review. Genetica 111: 329–347.

Stingo, V., L. Rocco, G. Odierna and M. Bellitti. 1995. NOR and heterochromatin analysis in two cartilaginous fishes by C-, Ag- and RE (restriction endonuclease)-banding. *Cytogenetics and Cell Genetics* 71: 228–234.

Viñas, A., M. Abuín, B.G. Pardo, P. Martínez and L. Sánchez. 2004. Characterization of a new *Hpa I* centromeric satellite DNA in *Salmo salar*. *Genetica* 121: 81–87.

Wichman, H.A., C.T. Payne, O.A. Ryder, M.J. Hamilton, M. Maltbie and R.J. Baker. 1991. Genomic distribution of heterochromatic sequences in equids: Implications to rapid chromosomal evolution. *Journal of Heredity* 82: 369–377.

Zangerl, R. 1981. Chondrichthyes. I. Paleozoic Elasmobranchii. In: *Handbook of Paleoichthyology*, H.P. Schultze (ed.). Gustav Fischer Verlag, Stuttgart, Vol. 3A, pp. 115.

Index

This index is not including all taxa cited in the book. Only those for which a full chapter or a complete paragraph have been developed are listed.